Proceedings in Technology Transfer

The book series "Proceedings in Technology Transfer" provides a valuable platform for researchers, practitioners, and policymakers to share their experiences and insights related to technology transfer. Our aim is to facilitate knowledge exchange between academic researchers and industry professionals, accelerating the introduction of new technologies and driving economic growth.

Our series publishes books and conference proceedings that focus on original research findings, best practice recommendations, emerging trends, and technology evaluations. The series covers all topics which explore the transfer of technology in various industry sectors that require innovations and economic growth.

Join us in advancing the technology transfer and its impact on innovation. Publish your proceedings with us and offer insights and expertise to professionals in your field.

Djihed Berkouk · Uday Chatterjee ·
Tallal Abdelkarim Bouzir · Imed Ben Dhaou ·
Samar Altarteer
Editors

Proceedings of the 2nd International Conference on Creativity, Technology, and Sustainability

CCTS 2025, 30 April – 1 May, Jeddah, Saudi Arabia

Springer

Editors
Djihed Berkouk
Department of Architecture
Dar Al-Hekma University
Jeddah, Saudi Arabia

Tallal Abdelkarim Bouzir
Department of Architecture
Biskra University
Biskra, Algeria

Samar Altarteer
Department of Visual Communication
Dar Al-Hekma University
Jeddah, Saudi Arabia

Uday Chatterjee
Department of Geography
Bhatter College Dantan
Medinipur, West Bengal, India

Imed Ben Dhaou
Department of Computer Science
Dar Al-Hekma University
Jeddah, Saudi Arabia

ISSN 2948-2321 ISSN 2948-233X (electronic)
Proceedings in Technology Transfer
ISBN 978-981-95-7606-7 ISBN 978-981-95-7607-4 (eBook)
https://doi.org/10.1007/978-981-95-7607-4

This Springer imprint is published by the registered company Springer Nature Singapore Pte Ltd.
The registered company address is: 152 Beach Road, #21-01/04 Gateway East, Singapore 189721, Singapore

If disposing of this product, please recycle the paper.

Preface

This book presents the selected proceedings of the Second International Conference on Creativity, Technology, and Sustainability (CCTS 2025), held at Dar Al-Hekma University in Jeddah, Saudi Arabia, from April 30 to May 1, 2025. As a multidisciplinary academic forum, the CCTS 2025 conference explores the synergies between creative innovation, cutting-edge technology and sustainability strategies. This second edition builds on the success of the first and highlights the growing importance of collaborative research, deeply connected to today's sustainable development challenges, as defined by the United Nations' Sustainable Development Goals (SDGs) and the Kingdom of Saudi Arabia's Vision 2030.

The conference received 130 full paper submissions from researchers representing 25 countries and 76 institutions worldwide. Each submission underwent a rigorous peer-review process, evaluated by at least two independent experts. Following this meticulous review, 53 papers were accepted for inclusion in these proceedings, ensuring both academic rigor and a diverse exploration of the conference themes.

The various chapters contained in this book offer case studies, theoretical frameworks and scientific contributions that provide a comprehensive overview of how technology, stimulated by creativity, can be effectively mobilized to address multidimensional sustainability challenges. These contributions span a broad spectrum of relevant topics, carefully organized into three distinct parts:

- Harnessing Technology for a Sustainable World: This section highlights how technological advancements can serve as powerful tools for sustainability. It features contributions on AI-driven design and planning, sustainable housing and urban development, data science for environmental monitoring, advancements in sustainable materials and construction, as well as broader reflections on the role of technology in addressing global sustainability challenges.
- Diverse Applications of Technology and Creative Solutions: The second section showcases the wide-ranging impact of technology across various sectors. This section features applications of AI and machine learning, the role of technology in education and well-being, the intersection of human factors and technology, advancements in environmental and industrial technology, and considerations for information systems and future trends.
- Shaping a Sustainable Future: The Role of Creativity in Societal and Economic Transformation: The final section focuses on the crucial role of innovation and strategic thinking in achieving sustainability. This part addresses economic policy and its impact on sustainability, the power of creativity and innovation for sustainable solutions, the application of technology for sector-specific sustainability, the integration of sustainability into business strategy, the importance of security and automation for sustainability, and an analysis of emerging trends.

Throughout this book, the contributions show how important it is to use creativity and technology together to deal with the challenges of sustainable development. The research and ideas shared aim to increase awareness, support the use of sustainable practices, and encourage cooperation between researchers, professionals, and decision-makers. Each part of the book connects to key global sustainability priorities. Part I addresses SDG 9 and SDG 11, focusing on sustainable infrastructure, innovation, and urban development. Part II connects to SDG 4, SDG 3, and SDG 8, emphasizing quality education, public health, and inclusive economic growth enabled by technology. Part III reflects SDG 12, SDG 13, and SDG 17, highlighting responsible innovation, climate action, and the importance of global cooperation. Together, these chapters provide useful insights and practical steps, offering a clear roadmap for progress toward a more sustainable, equitable, and technology-driven future.

We extend our deepest gratitude to the leadership of Dar Al-Hekma University, particularly the President, Dr. Abeer Aldoghaither, and the Vice President of Graduate Studies, Research and Business Division Dr. Norah Farooqi, for their unwavering support and visionary guidance. We are equally grateful to the authors for their valuable contributions and dedication to advancing knowledge in these critical fields. We are profoundly indebted to the international panel of reviewers and Program Committee members for their rigorous, constructive, and timely evaluations. Our sincere appreciation goes to the dedicated conference organizers whose meticulous efforts ensured the event's success. We are especially grateful to Dar Al-Hekma University for providing an exceptional, collaborative environment as host institution, fostering the rich exchange of ideas central to CCTS 2025.

It is our firm hope that the innovative ideas, research findings, and practical solutions within this volume will inspire further interdisciplinary collaboration, spark creative partnerships, and catalyze impactful actions towards building a more sustainable, equitable, and technologically empowered future.

Djihed Berkouk
Samar Altarteer
Uday Chatterjee
Tallal Abdelkarim Bouzir
Imed Ben Dhaou

Contents

Harnessing Technology for a Sustainable World

Strategic Site Selection for Constructing a Space Center and Spacecraft Launch Station in Saudi Arabia: A Multi-criteria Evaluation ... 3
Abdulaleem Aboud, Omar Dakhil, and Hassan A. Abas

Transformation of Affordable Housing in Saudi Arabia Toward Sustainable Development Goals ... 12
Waleed S. Alzamil

Sustainable Urban Greening Using Innovative Water Solutions in Madina Al Munawara ... 21
Hassan A. Abas, Wed Makki, Yussra Aljohani, Braa Alhrbi, and Hussein Chekfehand

Smart Cities in Egypt: Advancing Digital Transformation and Urban Sustainability ... 29
Anne A. Soliman, Amr Salem, Aya Ezzat, Mostafa Khaled Ahmed, Rana Elshafei, Samah El Khateeb, and Eman Farouk

Smart Sustainable Cities: Innovative AI Technologies to Improve Urban Development and Quality of Life ... 37
Muna Ahmed

Developing a Custom AI Model for Vegetation Health Monitoring and Survival Prediction in Arid Regions ... 46
Ahmed Ahmed and Elshaimaa Nada

Smart Waste Management System for Makkah City Using Artificial Intelligence and Internet of Things ... 56
Rawabi Saeed Al Qurashi, Maram Meshal Almanjumi, Teef Lahg Alghamdi, Amjad Hasan Almalki, Shahad Saed Alharthi, Shahad Mashal Althobuti, Alanoud Saad Alharthi, and Maha A. Thafar

A Climate-Aided Methodological Approach for Regenerating Cultural Heritage in Fragile Landscape ... 66
Angelo Figliola and Khaled Mohamed

An Analysis of Different Types of Gum Used in Concrete Applications: A Review ... 75
Alaa M.E.A. Saleem, Tareg Abdalla Abdalla, and Abdullah Albogami

Thermal Evaluation of Wax-Expanded Polystyrene Composite in Concrete Blocks ... 82
Aisha Hassan, Ethar Alawad, Shorouq Alsirhani, Haitham Rashed, Hassan A. Abas, and Hussain Alsadiq

Sustainable Concrete Solutions: Harnessing CornCob as a Renewable Resource ... 91
Arwa Fallatah and Shifana Fatima Kaafil

Digital Transformation Innovations for Sustainability: Advancing Workforce Training in the Digitalization Era ... 98
Azza AlGhamdi and Alia Aljindi

Optimizing Solar Energy Integration in Educational Buildings ... 107
Ahmed Ahmed, Elshaimaa Nada, and Eman Ahmed

The Role of Digital and Green Marketing in Promoting Sustainable Consumer Attitudes ... 116
Rasha Abdulkhalek

Towards Sustainable AI Development: A Focus on Transparency and Explainability ... 125
Shahina Begum, Mobyen Uddin Ahmed, and Mosarrat Farhana

When War Threatens Sustainability: The Legal Fight to Protect Our Environment ... 134
Adnan Mahmutovic

Sustainable Precision Agriculture: IoT-Based Soil Monitoring and ML-Driven Crop Recommendations ... 145
Basnt Melibari, Jomana Alhothly, Joud Alshareef, Lenah Alsaeedi, Arwa Alsubhi, and Sarah Al-Shareef

Diverse Applications of Technology and Creative Solutions

CNN-Based Deep Learning for Automated Kidney Stone Classification in Medical Imaging (CT Scan) ... 157
Vishesh Tanwar, Bhisham Sharma, and Imed Ben Dhaou

Generative Artificial Intelligence Tools in Higher Education: Systematic Literature Review 168
Maha M. Hassan and Sanaa S. Askool

Role of Artificial Intelligence in Redefining Modern Speech Therapy Practices in Saudi Arabia 176
Amjad Alsaadi and Areej Alasseri

A Review on the Use of Artificial Intelligence for Anomaly Detection in SCADA Systems 190
Lujain Ali Alqahtani, Sara khaed Alnatah, Maram Abdullah Alzahrani, Dalal Khalid Alkhaldi, Fatimah Saeed Aqaqah, Maryam M. Aldossary, and Mustafa Youldash

Hybrid Data Augmentation model for Sequential Recommendation 197
Jian Gang Gao, Shun Zheng, Syed Bilal Hussain Shah, and Arshad Ahmed Mir

Hybrid Machine Learning Models for DDoS Detection: A Performance Evaluation on the CICIDS2017 Dataset 206
Varkha K. Jewani, Pragati V. Thawani, Prafulla E. Ajmire, Mohammad Atique, and Imed Ben Dhaou

Innovation-Driven Education Reforming Engineering Education in the AI Era 215
Jouni Isoaho, Hannu Tenhunen, Imed Ben Dhaou, Xing Guo, and Zhuo Zou

The Impact of Intensive Training on General Education Teachers for Including Students with Autism Spectrum Disorder (ASD) in Regular Classrooms in Saudi Arabia 223
Mona Al-Haddad

VitalTwin: A Predictive and Management Prototype Digital Twin-Based System for Health and Wellbeing of Hajj and Umrah Pilgrims 231
Fatema Sabeen Shaikh, Rayhanah Jassim Aldolaim, Abrar Saleh AlGhamdi, Aisha Saleh AlKatheer, Reem Zaki AlMahfoud, Solaf Mohammed Alhotailah, Rawan Hamed Alghamdi, and Abdullah Ali M. Alqahtani

Workplace Toxicity and Its Shadow: Investigating Inner Resignation Through Employee Well-Being and Creativity in Saudi Academia 238
Komal Khalid

Exploring the Impact of Instant Gratification Factors on Long-Term Sustainable Financial Planning Among Saudi Youth 247
Fatima Balfaqih and Samar Altarteer

The Use of Zeta Technology to Modify Vermiculite Clay for the Adsorptive Removal of Anionic Detergent Residues from Wastewater 256
Saad Mohamed Elsaid and Ali ben Saleh Albarak

Assessing the Combined Effects of Climate Change and Urbanization on Water Resources in Arid Environments: A Comprehensive Machine Learning and Data Analysis Approach—Case Study Saudi Arabia 264
Heba Fathi

A Review of Capture the Flag (CTF) Platforms for OT/ICS Security Training 274
Maryam M. Aldossary, Sara F. Aldossary, Zahra A. Albeesh, Rahaf A. Al-Ghamdi, Shujon A. Alshehri, Dalal M. Al-Thawadi, Khalid A. Alissa, and Abdulrahman M. Alharby

SWOT Analysis of MIS in Higher Education: A Case Study of KKU 282
Nouf S. Aldahwan

A Study on the Global Hype of Metaverse for Sustainable Growth in the Future 295
Saima Abdullah, Umara Zahid, and Zahid Rashid

Sustainable Development of Rural and Remote Areas: A Case Study from Saudi Arabia 304
Salem Ahmed Alabdali, Salvatore F. Pileggi, and Dilek Cetindamar

Shaping a Sustainable Future: The Role of Creativity in Societal and Economic Transformation

Effects of Public Spending on Economic Growth: Evidence from the GCC Countries 315
Salwa Abdulraquib and Rozina Shaheen

The Impact of ESG on Financial Performance and Market Value in Saudi Arabia 324
Roba Al Sharif and Karima Saci

Renewable Energy Investments and their Impact on Stabilizing Inflation 333
Karima Saci and Saida Khalifa

Mawhiba Creative Programs for Young Talented Minds and their Sustainable Impact ... 341
Fahad Alfaiz, Saad Bakry, and Nada Alqahtani

Enhancing Innovation Through Interdisciplinary Collaboration: A Study on Interior Design and Construction Management Students in Augmented Reality Learning Environments ... 349
Amani Khan, Nathaniel B. Walker IV, and Henry Wanakuta

Fog Computing-Powered Driving Assistant System Under Fog Weather to Support Sustainable Tourism in Rural Area ... 357
Tami Alwajeeh

Adapting New Technologies Within Cultural Heritage in Islamic Studio Ceramics ... 366
Zelal Basodan

Identifying and Reducing Waste in Parkinson's Disease Diagnosis: A Lean Approach to Sustainable Healthcare ... 375
Dannah AlSafran, Mostafa Fawzy, Raneem AlQahtani, Sarah AlSafran, and Maha AlOtaibi

A Comprehensive Assessment of Sustainable Construction Materials in Saudi Arabia's Construction Industry ... 384
Maram Ghaleb and Mohammed M. Gomaa

Blending Tradition with Security: Branding and Cybersecurity Strategies for Success in Saudi Arabia ... 397
Ahmed Nasseraldin and Homam El-Taj

Exploring Collaborative Marketing Experience Between Small and Medium Enterprises and Startups in Saudi Arabia ... 403
Amira Masood and Samar Altarteer

Corporate Sustainability Reporting and Blockchain a Legal Analysis of Mandatory ESG Reporting in Saudi Arabia ... 413
Mohammed Alsudais

Impact of Knowledge Sharing, Information and Communication Technology Usage, and Green Market Orientation on Sustainable Performance ... 424
Jalal Rajeh Hanaysha

The Influence of Latin Typography on Arabic-Speaking Female Consumers' Perception of Brand Identity ... 435
Abdelqader Kamal Alnobani, Tahar Lazhar Ayed, Ahmed Khaled Nasseraldin, Lama Alkhail, Afnan Bahri, Farah Alfityani, and Joud Alhussan

Sustainable AI Adoption: Legal Considerations for Cybersecurity in Saudi Arabia's Context ... 443
Homam El-Taj and Firas Najjar

AI and Sustaining Legal Education for Creative Originality in the Gulf States: Saudi Arabia, Qatar, and United Arab Emirates ... 452
Samuel Samiái Andrews

Towards Automation in OT Risk Assessment: A Review of Methods, Machine Learning Techniques, Challenges, and Gaps ... 459
Amal A. Alahmadi, Aryam A. Alnemari, Deema A. Alfaleh, Fatema A. Alqahtani, Nouf M. Alabdallah, Ghada M. Alrashedi, and Abdulrahman A. Alharby

Cybersecurity for Sustainable Supply Chains in the Fourth Industrial Revolution ... 468
Fatima Al-ammari and Imed Ben Dhaou

The Effect of Vaccination and Investor Sentiment on the Crypto-Currency Market ... 477
Sonia Sayari and Faten Sameer Hanbali

Author Index ... 487

About the Editors

Dr. Djihed Berkouk
is Associate Professor in the Architecture Department within the School of Engineering, Computing, and Design at Dar Al-Hekma University, Saudi Arabia. As Holder of the Habilitation à Diriger des Recherches (HDR) from Biskra University in Algeria, he is qualified to supervise doctoral-level research. His research investigates the relationship between the physical environment and space occupancy, human comfort, soundscape, multisensory perception, noise pollution, and urban morphology. Dr. Berkouk has published extensively in high-impact international journals indexed in Scopus and Web of Science, including those by Elsevier, MDPI, Springer, SAGE, and AIP Publishing. He is actively engaged in the global academic community, serving as Reviewer for several Scopus-indexed journals and as Guest Editor for two Scopus journals and the CCTS Proceedings. He is supervising several PhD theses and has supervised Master's research in architecture and environmental design, with a focus on multisensory perception, thermal comfort, heat-health risk assessment, and multi-objective optimization of building performance.

Dr. Samar Altarteer
holds BA, MA, MPhil, PhD. She is Director of the Scientific Research Center and Assistant Professor at the School of Engineering, Computing, and Design at Dar Al-Hekma University, Jeddah, Saudi Arabia. Her expertise lies in Human-Computer Interaction (HCI), User Experience (UX), and immersive technologies such as virtual and augmented reality. She holds a PhD in Human-Computer Interaction from Glasgow Caledonian University (2015), an MPhil in HCI for E-Commerce (2011), an MA in 3D Design for Virtual Environments (2010), and a BA in Art from King Abdulaziz University (2006). Dr. Altarteer has held multiple leadership positions in academia, including Department Chair of Visual Communication, Program Director of Visual Communication, and Program Coordinator for the Master of Design in Game Design. She previously served as Designer at Awad Altarteer Engineering Consultations and as Founding Board Member and

VP of the Saudi Digital Game Association. Her research contributions span UX, HCI, VR/AR, 3D visualization, and digital education. She regularly publishes in high-impact journals and presents at international conferences. She has supervised a substantial number of capstone projects and student research papers, many of which have been published in reputable journals and conference proceedings. Dr. Altarteer is Certified Academic Quality Assurance Practitioner and Autodesk Certified Professional. She has organized scientific conferences and senior exhibitions and serves as Reviewer for indexed journals, including IEEE Xplore and Multimedia Tools and Applications. Her recognitions include the Dar Al-Hekma University Outstanding Performance Award (AY 2023–2024), the University's Community Service Award, and multiple awards of academic excellence from the UK Saudi Cultural Bureau.

Dr. Uday Chatterjee

is Assistant Professor at the Department of Geography, Bhatter College, Dantan, Paschim Medinipur, West Bengal, India and Applied Geographer with a Doctoral Degree in Applied Geography at Ravenshaw University, Cuttack, Odisha, India. His areas of research interest cover Urban Planning, Social and Human geography, Applied Geomorphology, Hazards & Disasters, Environmental Issues, disaster governance, community-based disaster risk management, climate change adaptation, urban risk management, and disaster. He has delivered 10 invited lectures in University Grants Commissions (UGC) sponsored national seminars and various academic departments of different colleges in India. In addition, he presented 20 papers in national and international seminars/conferences held in India as well as chaired and co-chaired more than 5 technical sessions. He has successfully guided project dissertations to undergraduate students. He has also conducted (Convener) one Faculty Development Programme on 'Modern methods of teaching and advanced research methods' sponsored by Indian Council of Social Science Research (ICSSR), Govt. of India. Currently, Dr. Uday Chatterjee has completed the Special Issue (S.I) of Urbanism, Smart Cities and Modelling, Geojournal, Springer as Lead Editor, book series editor Development in Environmental Science, Elsevier (https://www.elsevier.com/books-and-journals/book-series/developments-in-environmental-science). His research work has been funded by the West Bengal Pollution Control Board (WBPCB) Govt. of West Bengal, India. He has served as

Reviewer for many International journals. Currently, Dr. Chatterjee is doing the international project, in collaboration with Indonesia, Malaysia, and Japan funded by APN (Asia Pacific-Global Change Research), and he is Guest Editor of the special issue Social Ecology, Human-Well-being and Sustainability, Global Social Welfare journal, Springer. He has published 30 research papers, 10 edited books, and 2 author books, 15 book chapters, and 2 conference proceedings.

Dr. Tallal Abdelkarim Bouzir
is Associate Professor at the Department of Architecture of the University of Biskra in Algeria, and his area of expertise is the analysis of urban and architectural atmospheres. Specifically, he concentrates on examining how the sound, visual, and thermal environment interacts with human perception, comfort, and public health. His research aims to deepen our understanding of how the urban and architectural environment impacts the quality of life of individuals.

Prof. Imed Ben Dhaou
holds a Ph.D. from the Royal Institute of Technology, obtained in 2002. He is currently Full Professor of Embedded Systems at Dar Al-Hekma University, where he has served as Secretary of the Scientific Council since 2024. Additionally, he is Adjunct Professor at the University of Turku. In 2024, he was listed among the world's top 2% of scientists by Stanford University and Elsevier. With over 120 published papers, Dhaou has received several accolades, including the Best Paper Award at the 1997 Finnish Symposium on Signal Processing. He is Editor of the textbook IoT Enabled-DC Microgrids: Architecture, Algorithms, Applications, and Technologies (CRC Press). He has supervised and co-supervised several Ph.D., Master's, and capstone students and has also served as External Evaluator for Ph.D. theses from institutions in Italy, India, and Saudi Arabia. Since 2014, he has been Editor for the Microelectronics Journal and has actively contributed as TPC Chair or Member for numerous conferences. His managerial experience includes serving as Director of a private university and providing consultancy to the Deanship for Graduate Studies at Qassim University.

List of Contributors

Hassan A. Abas Prince Mugrin University, Madinah, Saudi Arabia;
University of Prince Mugrin, Madinah, Kingdom of Saudi Arabia;
Prince Mugrin University, Madinah, Kingdom of Saudi Arabia

Tareg Abdalla Abdalla Department of Civil Engineering, Faculty of Engineering Sciences, Omdurman Islamic University, Omdurman, Sudan

Rasha Abdulkhalek Université La Sagesse, Beirut, Lebanon

Saima Abdullah Department of Computer Science, International Islamic University, Islamabad, Pakistan

Salwa Abdulraquib Dar Al-Hekma University, Jeddah, Saudi Arabia

Abdulaleem Aboud University of Prince Mugrin, Madinah, Kingdom of Saudi Arabia

Ahmed Ahmed Taibah University, Almadinah, Saudi Arabia

Eman Ahmed Taibah University, Almadinah, Saudi Arabia

Mobyen Uddin Ahmed School of Innovation, Design and Engineering, Mälardalen University, Västerås, Sweden

Mostafa Khaled Ahmed Department of Urban Design and Planning, Faculty of Engineering, Ain Shams University, Cairo, Egypt

Muna Ahmed Jazan University, Jazan, Saudi Arabia

Prafulla E. Ajmire Computer Science Department, SGBAU Amravati University, Mumbai, Maharashtra, India

Salem Ahmed Alabdali Department of Management Information Systems, Jazan University, Jazan, Saudi Arabia;
School of Computer Science, University of Technology Sydney, Ultimo, NSW, Australia

Nouf M. Alabdallah Imam Abdulrahman bin Faisal University, Dammam, Saudi Arabia

Amal A. Alahmadi Imam Abdulrahman bin Faisal University, Dammam, Saudi Arabia

Fatima Al-ammari Cybersecurity Department, Dar Al-Hekma University, Jeddah, Saudi Arabia

Areej Alasseri School of Education, Health, and Behavioral Sciences, Department of Health and Behavioral Sciences, Dar Al-Hekma University, Jeddah, Saudi Arabia

Ethar Alawad Prince Mugrin University, Madinah, Saudi Arabia

Ali ben Saleh Albarak Medical Applied College, Qassim university, Buraidah, Saudi Arabia

Zahra A. Albeesh Imam Abdulrahman bin Faisal University, Dammam, Saudi Arabia

Abdullah Albogami Department of Civil Engineering, Faculty of Engineering, Al-Baha University, Al-Baha, Saudi Arabia

Nouf S. Aldahwan Department of Information System, King Khalid University, Abha, Saudi Arabia

Rayhanah Jassim Aldolaim Department of Computer Information Systems, Imam Abdulrahman Bin Faisal University, Dammam, Saudi Arabia

Maryam M. Aldossary Imam Abdulrahman bin Faisal University, Dammam, Saudi Arabia

Sara F. Aldossary Imam Abdulrahman bin Faisal University, Dammam, Saudi Arabia

Fahad Alfaiz King Saud University, Riyadh, Saudi Arabia

Deema A. Alfaleh Imam Abdulrahman bin Faisal University, Dammam, Saudi Arabia

Farah Alfityani Dar Al-Hekma University, Jeddah, Kingdom of Saudi Arabia

Rahaf A. Al-Ghamdi Imam Abdulrahman bin Faisal University, Dammam, Saudi Arabia

Abrar Saleh AlGhamdi Department of Computer Information Systems, Imam Abdulrahman Bin Faisal University, Dammam, Saudi Arabia

Azza AlGhamdi Faculty of Education, King Abdulaziz University, Jeddah, Saudi Arabia

Rawan Hamed Alghamdi Department of Computer Information Systems, Imam Abdulrahman Bin Faisal University, Dammam, Saudi Arabia

Teef Lahg Alghamdi College of Computers and Information Technology, Computer Science Department, Taif University, Taif, Kingdom of Saudi Arabia

Mona Al-Haddad Dar Al-Hekma University, Jeddah, Saudi Arabia

Abdulrahman A. Alharby Imam Abdulrahman bin Faisal University, Dammam, Saudi Arabia

Abdulrahman M. Alharby Imam Abdulrahman bin Faisal University, Dammam, Saudi Arabia

Alanoud Saad Alharthi College of Computers and Information Technology, Computer Science Department, Taif University, Taif, Kingdom of Saudi Arabia

Shahad Saed Alharthi College of Computers and Information Technology, Computer Science Department, Taif University, Taif, Kingdom of Saudi Arabia

Solaf Mohammed Alhotailah Department of Computer Information Systems, Imam Abdulrahman Bin Faisal University, Dammam, Saudi Arabia

Jomana Alhothly Computer Science & Artificial Intelligence Department, Umm Al-Qura University, Makkah, Saudi Arabia

Braa Alhrbi Prince Mugrin University, Madinah, Kingdom of Saudi Arabia

Joud Alhussan Dar Al-Hekma University, Jeddah, Kingdom of Saudi Arabia

Khalid A. Alissa Imam Abdulrahman bin Faisal University, Dammam, Saudi Arabia

Alia Aljindi Faculty of Education, King Abdulaziz University, Jeddah, Saudi Arabia

Yussra Aljohani Prince Mugrin University, Madinah, Kingdom of Saudi Arabia

Aisha Saleh AlKatheer Department of Computer Information Systems, Imam Abdulrahman Bin Faisal University, Dammam, Saudi Arabia

Lama Alkhail Dar Al-Hekma University, Jeddah, Kingdom of Saudi Arabia

Dalal Khalid Alkhaldi Imam Abdulrahman bin Faisal University, Dammam, Saudi Arabia

Reem Zaki AlMahfoud Department of Computer Information Systems, Imam Abdulrahman Bin Faisal University, Dammam, Saudi Arabia

Amjad Hasan Almalki College of Computers and Information Technology, Computer Science Department, Taif University, Taif, Kingdom of Saudi Arabia

Maram Meshal Almanjumi College of Computers and Information Technology, Computer Science Department, Taif University, Taif, Kingdom of Saudi Arabia

Sara khaed Alnatah Imam Abdulrahman bin Faisal University, Dammam, Saudi Arabia

Aryam A. Alnemari Imam Abdulrahman bin Faisal University, Dammam, Saudi Arabia

Abdelqader Kamal Alnobani Dar Al-Hekma University, Jeddah, Kingdom of Saudi Arabia

Maha AlOtaibi Najran Education Administration, Ministry of Education, Najran, Kingdom of Saudi Arabia

Raneem AlQahtani King Fahd University of Petroleum and Minerals, Dhahran, Kingdom of Saudi Arabia

Abdullah Ali M. Alqahtani Department of Computer Information Systems, Imam Abdulrahman Bin Faisal University, Dammam, Saudi Arabia

Fatema A. Alqahtani Imam Abdulrahman bin Faisal University, Dammam, Saudi Arabia

Lujain Ali Alqahtani Imam Abdulrahman bin Faisal University, Dammam, Saudi Arabia

Nada Alqahtani Mawhiba, Riyadh, Saudi Arabia

Rawabi Saeed Al Qurashi College of Computers and Information Technology, Computer Science Department, Taif University, Taif, Kingdom of Saudi Arabia

Ghada M. Alrashedi Imam Abdulrahman bin Faisal University, Dammam, Saudi Arabia

Amjad Alsaadi School of Education, Health, and Behavioral Sciences, Department of Health and Behavioral Sciences, Dar Al-Hekma University, Jeddah, Saudi Arabia

Hussain Alsadiq Prince Mugrin University, Madinah, Saudi Arabia

Lenah Alsaeedi Computer Science & Artificial Intelligence Department, Umm Al-Qura University, Makkah, Saudi Arabia

Dannah AlSafran Dar Al-Hekma University, Jeddah, Kingdom of Saudi Arabia

Sarah AlSafran Imam Abdulrahman Bin Faisal University, Dammam, Kingdom of Saudi Arabia

Sarah Al-Shareef Computer Science & Artificial Intelligence Department, Umm Al-Qura University, Makkah, Saudi Arabia

Joud Alshareef Computer Science & Artificial Intelligence Department, Umm Al-Qura University, Makkah, Saudi Arabia

Roba Al Sharif Dar Al-Hekma University, Jeddah, Kingdom of Saudi Arabia

Shujon A. Alshehri Imam Abdulrahman bin Faisal University, Dammam, Saudi Arabia

Shorouq Alsirhani Prince Mugrin University, Madinah, Saudi Arabia

Arwa Alsubhi Computer Science & Artificial Intelligence Department, Umm Al-Qura University, Makkah, Saudi Arabia

Mohammed Alsudais Al Yamamah University, Riyadh, Saudi Arabia

Samar Altarteer Dar Al-Hekma University, Jeddah, Kingdom of Saudi Arabia

Dalal M. Al-Thawadi Imam Abdulrahman bin Faisal University, Dammam, Saudi Arabia

Shahad Mashal Althobuti College of Computers and Information Technology, Computer Science Department, Taif University, Taif, Kingdom of Saudi Arabia

Tami Alwajeeh Department of Computer Science, College of Computing and Information, Al-Baha University, Alaqiq, Saudi Arabia

Maram Abdullah Alzahrani Imam Abdulrahman bin Faisal University, Dammam, Saudi Arabia

Waleed S. Alzamil Department of Urban Planning, King Saud University, Riyadh, Saudi Arabia

Samuel Samiái Andrews Albany State University College of Professional Studies, Albany, Georgia, USA

Fatimah Saeed Aqaqah Imam Abdulrahman bin Faisal University, Dammam, Saudi Arabia

Sanaa S. Askool Dr. Hussein AlSayyed Research & Innovation Center, Dept. of SRGS, University of Prince Mugrin, Madinah, Saudi Arabia;
College of Computer and Cyber Sciences, University of Prince Mugrin, Madinah, Saudi Arabia

Mohammad Atique Computer Science Department, SGBAU Amravati University, Mumbai, Maharashtra, India

Tahar Lazhar Ayed Dar Al-Hekma University, Jeddah, Kingdom of Saudi Arabia

Afnan Bahri Dar Al-Hekma University, Jeddah, Kingdom of Saudi Arabia

Saad Bakry King Saud University, Riyadh, Saudi Arabia

Fatima Balfaqih Dar Al-Hekma University, Jeddah, Kingdom of Saudi Arabia

Zelal Basodan University of Jeddah, Jeddah, Saudi Arabia

Shahina Begum School of Innovation, Design and Engineering, Mälardalen University, Västerås, Sweden

Dilek Cetindamar School of Computer Science, University of Technology Sydney, Ultimo, NSW, Australia

Hussein Chekfehand Prince Mugrin University, Madinah, Kingdom of Saudi Arabia

Omar Dakhil University of Prince Mugrin, Madinah, Kingdom of Saudi Arabia

Imed Ben Dhaou Computer Science Department, Dar Al-Hekma University, Jeddah, Saudi Arabia;
Department of Computer Science, Hekma School of Engineering Computing and Design, Jeddah, Saudi Arabia;
Computer Science Department, School of Engineering, Computing, and Design, Dar Al-Hekma University, Jeddah, Saudi Arabia;
Dar Al-Hekma University, Jeddah, Saudi Arabia

Samah El Khateeb Smart and Future Cities Laboratory for Sustainable Solutions, Faculty of Engineering, Ain Shams University, Cario, Egypt;
Smart and Sustainable Cities Center, University of Business and Technology, Jeddah, Saudi Arabia

Saad Mohamed Elsaid Private Colleges for Engineering and Information Technique, Onaizah College, Onaizah, Saudi Arabia

Rana Elshafei Smart and Future Cities Laboratory for Sustainable Solutions, Faculty of Engineering, Ain Shams University, Cario, Egypt

Homam El-Taj Department of Cybersecurity, Dar Al-Hekma University, Jeddah, Saudi Arabia;
Dar Al Hekma University, Jeddah, Kingdom of Saudi Arabia

Aya Ezzat Architecture Department, Modern University for Technology and Information (MTI), Cairo, Egypt

Arwa Fallatah Hekma School of Engineering, Computing and Design, Dar Al-Hekma University, Jeddah, Saudi Arabia

Mosarrat Farhana School of Business Society and Engineering, Mälardalen University, Västerås, Sweden

Eman Farouk Architecture and Urban Design Program, German University in Cairo, Cario, Egypt

Heba Fathi College of Arts and Humanities, Jazan University, Jazan, Saudi Arabia

Mostafa Fawzy Dar Al-Hekma University, Jeddah, Kingdom of Saudi Arabia

Angelo Figliola PDTA Department Planning Design Technology of Architecture, Sapienza University of Rome, Rome, RM, Italy

Jian Gang Gao Tercelo Tire Group CO., LTD, Weifang, Shandong, China

Maram Ghaleb Department of Architecture, Dar Al-Hekma University, Jeddah, Saudi Arabia

Mohammed M. Gomaa Department of Architecture, Dar Al-Hekma University, Jeddah, Saudi Arabia;
Department of Architecture, Faculty of Engineering, Aswan University, Aswan, Egypt

Xing Guo Fudan University, Shanghai, China

Jalal Rajeh Hanaysha School of Business, Horizon University College, Ajman, United Arab Emirates

Faten Sameer Hanbali Department of Business and Administration, College of Administration and Finance, Saudi Electronic University, Jeddah, Saudi Arabia

Aisha Hassan Prince Mugrin University, Madinah, Saudi Arabia

Maha M. Hassan Civil Engineering Department, University of Prince Mugrin, Madinah, Saudi Arabia;
Dr. Hussein AlSayyed Research & Innovation Center, Dept. of SRGS, University of Prince Mugrin, Madinah, Saudi Arabia

Jouni Isoaho University of Turku, Turku, Finland

Varkha K. Jewani Computer Science Department, SGBAU Amravati University, Mumbai, Maharashtra, India

Shifana Fatima Kaafil Department of Architecture, Dar Al-Hekma University, Jeddah, Saudi Arabia

Komal Khalid King Abdulaziz University, Jeddah, Saudi Arabia

Saida Khalifa Dar Al-Hekma University, Jeddah, Kingdom of Saudi Arabia

Amani Khan Dar Al-Hekma University, Jeddah, Kingdom of Saudi Arabia

Adnan Mahmutovic Department of Law, Al-Yamamah University, Riyadh, Saudi Arabia

Wed Makki Prince Mugrin University, Madinah, Kingdom of Saudi Arabia

Amira Masood Dar Al-Hekma University, Jeddah, Kingdom of Saudi Arabia

Basnt Melibari Computer Science & Artificial Intelligence Department, Umm Al-Qura University, Makkah, Saudi Arabia

Arshad Ahmed Mir University Centre, Leeds, United Kingdom

Khaled Mohamed Via Teano 331, Rome, RM, Italy

Elshaimaa Nada Zagazig University, Zagazig, Egypt

Firas Najjar Department of Information Security and Cybersecurity Department, Philadelphia University, Amman, Jordan

Ahmed Nasseraldin Dar Al Hekma University, Jeddah, Kingdom of Saudi Arabia

Ahmed Khaled Nasseraldin Dar Al-Hekma University, Jeddah, Kingdom of Saudi Arabia

Salvatore F. Pileggi School of Computer Science, University of Technology Sydney, Ultimo, NSW, Australia

Haitham Rashed Prince Mugrin University, Madinah, Saudi Arabia

Zahid Rashid Technology Management Economics and Policy Program, College of Engineering, Seoul National University, Seoul, South Korea

Karima Saci Dar Al-Hekma University, Jeddah, Kingdom of Saudi Arabia

Alaa Saleem Department of Civil Engineering, Faculty of Engineering, University of Khartoum, Khartoum, Sudan

Amr Salem Department of Urban Design and Planning, Faculty of Engineering, Ain Shams University, Cairo, Egypt

Sonia Sayari Department of Business and Administration, College of Administration and Finance, Saudi Electronic University, Jeddah, Saudi Arabia;
Higher Institute of Accounting and Business Administration, University of Manouba, Tunis, Tunisia

Syed Bilal Hussain Shah University Centre, Leeds, United Kingdom

Rozina Shaheen Effat University, Jeddah, Saudi Arabia

Fatema Sabeen Shaikh Department of Computer Information Systems, Imam Abdulrahman Bin Faisal University, Dammam, Saudi Arabia

Bhisham Sharma Centre of Research Impact and Outcome, Chitkara University, Rajpura, Punjab, India

Anne A. Soliman Department of Architecture, Faculty of Engineering, Misr International University, Cairo, Egypt

Vishesh Tanwar Chitkara University Institute of Engineering and Technology, Chitkara University, Rajpura, Punjab, India

Hannu Tenhunen KTH-Royal Institute of Technology, Stockholm, Sweden;
Dar Al-Hekma University, Jeddah, Saudi Arabia

Maha A. Thafar College of Computers and Information Technology, Computer Science Department, Taif University, Taif, Kingdom of Saudi Arabia

Pragati V. Thawani Computer Science Department, SGBAU Amravati University, Mumbai, Maharashtra, India

Nathaniel B. Walker IV Stephen F. Austin State University, Nacogdoches, USA;
Dar Al-Hekma University, Jeddah, Kingdom of Saudi Arabia

Henry Wanakuta Stephen F. Austin State University, Nacogdoches, USA

Mustafa Youldash Imam Abdulrahman bin Faisal University, Dammam, Saudi Arabia

Umara Zahid Department of Computer Science, International Islamic University, Islamabad, Pakistan

Shun Zheng Tercelo Tire Group CO., LTD, Weifang, Shandong, China

Zhuo Zou Fudan University, Shanghai, China

Harnessing Technology for a Sustainable World

Strategic Site Selection for Constructing a Space Center and Spacecraft Launch Station in Saudi Arabia: A Multi-criteria Evaluation

Abdulaleem Aboud(✉), Omar Dakhil, and Hassan A. Abas

University of Prince Mugrin, Madinah, Kingdom of Saudi Arabia
{4020025,o.dakhil,h.abbas}@upm.edu.sa

Abstract. The selection of a strategic site for constructing a space center and spacecraft launch station is essential for advancing Saudi Arabia's space exploration capabilities. A well-planned location ensures operational efficiency, safety, and sustainability. This study aims to identify an optimal site that meets international engineering practices by evaluating environmental conditions, construction feasibility, transportation accessibility, security measures, and sustainability. Three sites were shortlisted for comparison: Jabal Hassan Island, Abo Ali Island, and Mastura Village. A multi-criteria decision analysis (MCDA) framework was applied to ensure a structured and objective assessment. The methodology involved geospatial data analysis, environmental assessments, soil investigations, and transportation analysis. Satellite imagery examined land stability, while climate reports provided insight into weather conditions. Geotechnical studies and transportation networks were evaluated for accessibility. Security measures were analyzed based on site remoteness, natural barriers, and risk factors. The evaluation applied weighted criteria, assigning 30% each to environmental conditions and construction feasibility and 20% each to accessibility, transportation, and security measures. The results showed that Mastura Village achieved the highest score of 90.7, followed by Jabal Hassan Island at 68.3 and Abo Ali Island at 61.3. The findings align with international benchmarks and provide a solid foundation for future aerospace development in Saudi Arabia.

Keywords: Spacecraft Station · Space Center · Sustainability · Aerospace Development

1 Introduction

Space exploration contributes to scientific discoveries, technological advancements, and innovations, particularly in tracking the climate and managing resources. Modern space centers that launch spacecraft, conduct research, and position satellites in orbit fuel economic growth, generate high-skilled jobs, advance aerospace industries, facilitate global cooperation, and improve global strategic positioning. Successful space center development involves accurate site selection based on stringent engineering requirements and sustainability objectives to ensure safety, and durability. Strategic site selection is critical to a space center's overall performance and efficiency [1, 2].

D. Berkouk et al. (Eds.): CCTS 2024, PTT, pp. 3–11, 2026.
https://10.1007/978-981-95-7607-4_1

Environmental conditions, soil stability, and seismic conditions are some of the factors that have a critical role to play in ensuring that the structure of the facility remains stable and long-term viable. Additionally, proximity to transportation networks, such as ports, highways, and airports, is essential for logistical efficiency and cost-effective construction and maintenance. Security is also a critical issue with locations that minimize risks while providing controlled access. Sustainability is now a priority with space centers that incorporate renewable energy systems and minimize their ecological impact. In the long term, a carefully chosen site assures a space center's operating success and resilience.

Saudi Arabia's strategic strengths, economic and geographic make it an excellent candidate for hosting and operate space center. Its open space, low population densities in remote areas, and favorable weather conditions. These factors make it ideal to launch spacecraft, while proximity to the equator maximizes cost-effectiveness and payload capacity [3]. Economically, Vision 2030 facilitates technological development, infrastructure building and aerospace investment. Strategically, being located centrally between Asia, Africa, and Europe maximizes global cooperation and logistics efficiency. Adding a stable political situation and a focus on sustainability to that equation positions Saudi Arabia to become a leader in space exploration globally.

Construction of a space center requires adherence to stringent standards that consider construction aspects, environmental footprint, and operational considerations. For safe takeoff and landing of rockets, sites require stable climatic conditions, controlled wind patterns, and low seismic activity [4, 5]. Ning et al. 2019 reported that the noise reduction and contamination control are needed to reduce environmental degradation [6]. In addition, operational benchmarks address the importance of proximity to transportation networks. Commitment on Sustainability standards promote renewable energy, green materials, and water efficiency [7, 8]. Environmental impact studies ensure ecological conservation during construction and operation project's life cycle.

The aim of this study is to identify a strategic site in Saudi Arabia to construct a space center and launching station. Three sites are examined based on environmental conditions, construction feasibility, transportation connectivity, security, and sustainability. The study considers climatic conditions, logistics facilities, and renewable energy opportunities in addition to comparisons with global benchmarks to check against standards. The research aims to provide a comprehensive evaluation that will serve as a basis to construct a sustainable space center in Saudi Arabia.

2 Methodology

This study employs a multi-criteria evaluation framework to select candidate location for a space center launch in Saudi Arabia consistent with Vision 2030's vision. A comparative analysis of NASA and SpaceX spaceport locations was conducted to determine their site selection criteria. The research's methodology particularly includes semi-structured expert interviews and tailored weighting of criteria that is especially pertinent to Saudi Arabia's Vision 2030 and diverges from NASA and SpaceX's generic international methodologies.

Second, selection criteria were environmental conditions, construction feasibility, transportation and access, and security factors. The factors have weighted percentage values assigned to each of them based on their contribution to project success. Third, information collection is via satellite imagery, geospatial data, and environmental studies to analyze climatic trends, wind stability, and seismic activity. Transportation infrastructure like distance to ports, highways, and airports is analyzed to consider logistics efficiency. Security factors take into account remoteness of location, restricted access, and natural barriers. Finally, the research concludes with a recommendation report listing the best site and expected development challenges.

2.1 Expert Insights on Site Selection

Seven professionals from diverse fields participated in a semi-structured interview to evaluate key factors in site selection. The interviewees included experts in astronomy, astrophysics researchers from King Abdulaziz University, and professionals from the aviation and space sectors. Each participant provided insights based on their area of expertise. The interview served as the foundation for the initial selection and shortlisting of potential sites for the proposed space center. Figure 1. illustrates the results of expert input on site selection

The results showed differing views on the feasibility of construction of a space center. About 43% of participants rated feasibility as 3 out of 5, while 29% gave it a rating of 4 or 5 out of 5. These responses were based on factors related to environmental, logistical, and operational challenges.

In terms of location preference, 57% of respondents favored the construction of the space center on an island rather than in a city. Their reasoning was based on factors such as safety, minimal environmental impact, and operational efficiency. Another key consideration was whether the site should be along the Arabian Gulf or the Red Sea. The findings showed that 57% preferred a Red Sea location due to its geological stability, and political advantages.

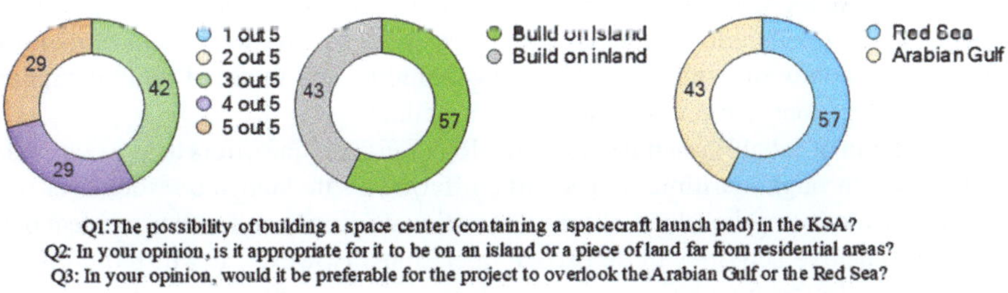

Fig. 1. Expert Insights on Space Center Site Selection in Saudi Arabia.

3 Potential Sites

In this study, three sites were shortlisted to be compared: Jabal Hassan Island, Abu Ali Island, and Mastura Village. Figure 2 shows the three candidate locations. They were compared against key factors such as environmental conditions, construction feasibility, accessibility, and security factors. The next section provides a comprehensive evaluation of each site with its unique characteristics and their suitability to host the space center development.

Fig. 2. Location of the three potential sites Jabal Hassan Island, Abu Ali Island, and Mastura Village

3.1 Jabal Hassan Island

Off northwest Saudi Arabia near Umluj city, Jabal Hassan Island offers favorable conditions to accommodate massive infrastructure projects like space launch facilities. Its flat and expansive topography minimizes construction complexity and lowers site preparation costs. Low population in the area enhances safety with low risks to human life. A dry weather condition with clear skies, low humidity, and minimal rainfall supports continuous operations and predictable launch schedules.

The location of Jabal Hassan Island on the Red Sea coastline offers heavy equipment transportation through maritime routes while offering a safe launch corridor over open waters. The isolation of the island offers extra national security and is thus an ideal place to conduct sensitive activities of aerospace operations. However, it will require heavy investment in infrastructure because the land does not have roads, utilities, or facilities to function. Also protecting local wildlife and water resources is essential for sustainable development.

3.2 Abo Ali Island

Abo Ali Island in the eastern region, close to industrial centers such as Dammam and Jubail, has high potential for mega-size infrastructure projects. The topography of flat terrain allows low-cost construction with minimal site preparation. The transportation infrastructure of highways, railways, and seaports supports efficient logistics and access to resources. The weather conditions of heat and dryness allow year-round operations, with heat management necessary to preserve equipment efficiency. The population density in the eastern region requires planning to maintain safety in construction stage and launch operations. Urbanization requires tight security and limited access, while compliance with global airspace rules is essential to control overflights over the Gulf.

3.3 Mastura Village

Mastura Village, located in a lightly populated desert area near the Red Sea, offers a great site for large-scale infrastructure projects such as spaceports. Its flat and open landscape minimizes construction complexity with little land preparation while maximizing safety due to low population density. The region's dry climate, clear skies, and stable weather conditions allow uninterrupted operations, making it an ideal location for rocket launches. Additionally, its proximity to the Red Sea provides logistical benefits, including marine transportation for heavy equipment and secure overwater routes that reduce risks to inhabited areas and ensure safe launch operations.

Mastura Village's remoteness offers security through limited access and low threat levels. However, the absence of existing infrastructure means significant investment is needed for highways and operational facilities. This challenge also presents an opportunity to design customized infrastructure tailored to specific technical requirements for aerospace centers. With its combination of space, security, and logistical advantages, Mastura Village stands as a strategically viable option for projects that prioritize operational efficiency and long-term sustainability.

4 Results and Discussion

This section presents a comprehensive and structured list of comparison items to ensure a thorough evaluation of the three potential sites for construction aerospace centers in Saudi Arabia: Jabal Hassan Island, Abo Ali Island, and Mastura Village. Each main criterion includes subitems to guide data collection and analysis, with all subitems rated on a scale of 1 to 5.

4.1 Environmental Conditions

The assessment of environmental conditions includes temperature ranges, precipitation, and humidity, all which impact construction feasibility and operational efficiency. Stable temperatures help ensure material durability, while moderate humidity minimizes the risk of equipment corrosion. Wind speeds and seasonal variations are evaluated for their potential effects on launches and infrastructure stability. Extreme weather events, such as storms or flooding, are analyzed for their impact on safety and scheduling. Additionally, proximity to the sea is considered in terms of storm surge risks and water resource availability. Figure 3 presents a summary of these factors across the three sites.

4.2 Construction Feasibility

Construction feasibility is assessed by evaluating the flatness of the site and its accessibility to minimize grading efforts, as well as the impact of natural barriers like mountains or valleys on infrastructure development. Being close to local construction materials reduces costs and logistical challenges, improving overall efficiency. Soil load-bearing capacity and bedrock depth are analyzed to ensure stable foundations, while susceptibility to erosion is examined for long-term stability. The site's proximity to fault lines, seismic zones, and historical earthquake data are also reviewed to ensure safety. Figure 4 summarizes these criteria for the three sites.

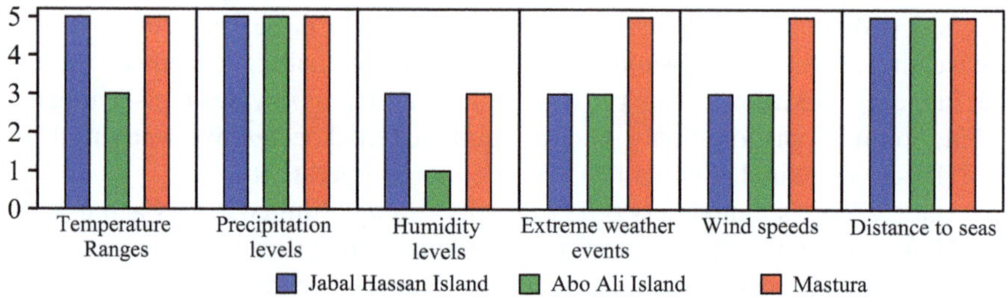

Fig. 3. Comparative Evaluation of Environmental Conditions Across the Three Sites.

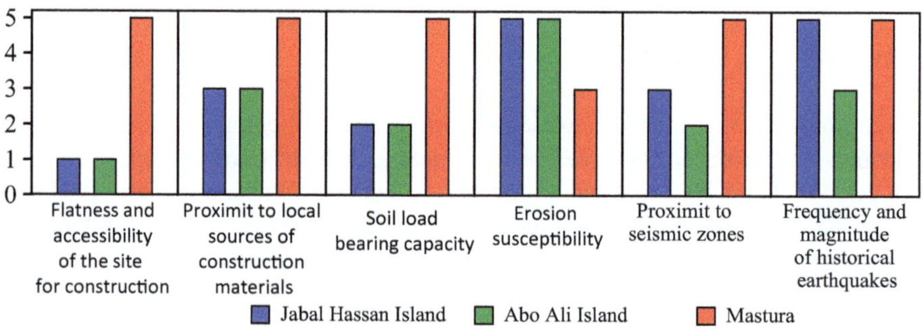

Fig. 4. Comparative Evaluation of Construction Feasibility Across the Three Sites.

4.3 Accessibility and Transportation

Accessibility and transportation are fundamental considerations in site selection. The presence of ports nearby ensures efficient transportation of heavy equipment, while good highway connectivity allows convenient transportation of supplies and personnel. The presence of airports with cargo facilities near the potential site allows speedy delivery of resources. Figure 5 considers the three locations based on these factors.

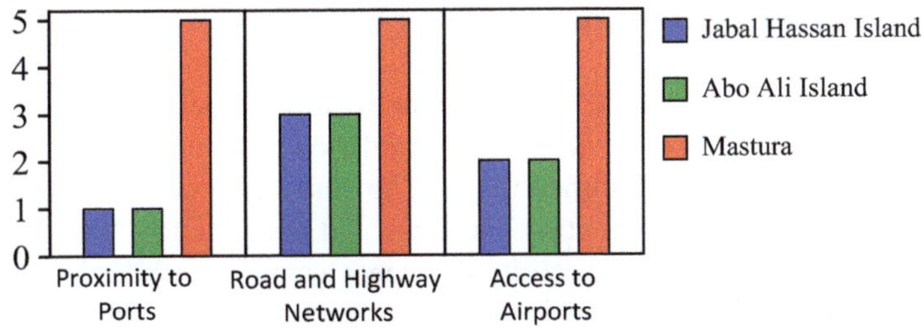

Fig. 5. Comparative Assessment of Accessibility and Transportation Across the Three Sites

4.4 Security Measures

Security is essential to operations. Urban proximity is examined to balance against threats and reduce risks, with areas farther from cities being more desirable. Natural barriers like mountains or rivers enhance security by restricting unauthorized entry. Risk assessments identify threats like natural disasters and human intervention to ensure safety and resilience at the location. Figure 6 provides a comparative analysis of the three locations against such considerations.

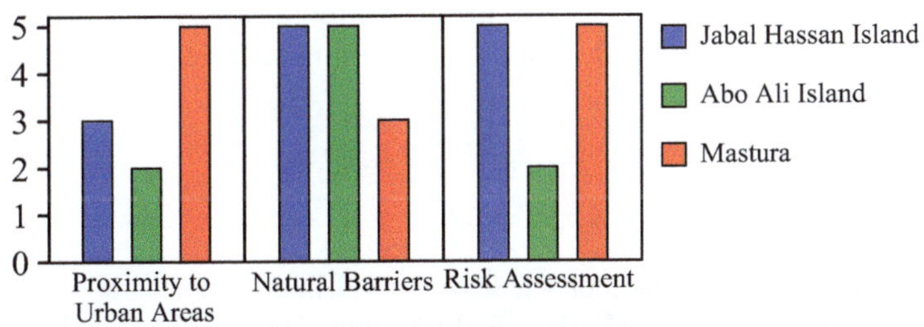

Fig. 6. Comparative Evaluation of Security Measures Across the Three Sites

The evaluation between three considered potential sites. The study used weighted factors with environmental conditions and construction feasibility at 30%, and transportation, security measures, and accessibility at 20% each. The result indicates that Mastura Village scored a high of 90.7, Jabal Hassan Island 68.3, and Abo Ali Island 61.3. Mastura Village shows stable environmental conditions, good construction feasibility, and well-balanced security measures make it the best location.

The research methodology and results of this study provide pragmatic lessons that can be generalized and applied to other areas or countries with similar space center projects. In particular, areas with similar environmental conditions, geographical features, and infrastructural settings can apply this multi-criteria decision framework to make effective and sustainable site selection.

5 Conclusion and Recommendations

Selecting a suitable location for a space center requires a detailed study of a number of factors. In this study, three sites were shortlisted to be compared: Jabal Hassan Island, Abu Ali Island, and Mastura Village. The study adopted evaluation based on four weighted factors. Environmental conditions and construction feasibility account for 30%, while transportation and security measures account for 20%. Mastura Village had a maximum of 90.7%, followed by Jabal Hassan Island with 68.3% and Abo Ali Island with 61.3%. Several factors make Mastura Village a suitable location to host the proposed space center.

Mastura Village provides stable weather conditions with low rainfall, clear skies, and low humidity. The weather conditions allow best launch operations throughout the year. Its open and flat land minimizes construction complexity and reduces site preparation costs. The location offers ample space for construction infrastructure while keeping it at a safe distance from cities. This minimizes threats to human life and surrounding areas. The proximity to the Red Sea offers convenient maritime transportation of heavy equipment and allows safe overwater paths to enhance logistics and safety.

The research methodology and results of this study provide pragmatic lessons that can be generalized and applied to other areas or countries planning similar space center projects. Future studies must explore more advanced construction approaches suitable for desert environments. Studies should also investigate innovative renewable energy solutions and emerging technologies to improve operational efficiency and environmental sustainability.

References

1. Dachyar, M., Purnomo, H.: Spaceport site selection with analytical hierarchy process decision making. Indian J. Sci. Technol. **11**(10), 1–8 (2018) https://doi.org/10.17485/ijst/2018/v11i10/96506
2. Öz, İ.: Decision making in evaluation and selection of launch site with the Best and Worst method. Gazi Univ. J. Sci. **35**(4), 1521–1533 (2022) https://doi.org/10.35378/gujs.721823
3. Finger, G., McClesky, C.: Spaceport performance measures. In: SpaceOps 2010 Conference. American Institute of Aeronautics and Astronautics, Reston (2010). https://doi.org/10.2514/6.2010-2148
4. Johnson, D.L., Rawlins, M.A., Batts, G.W.: A characterization of the terrestrial environment of Kodiak Island, Alaska for the design, development and operation of launch vehicles. Unpublished Report (2000)
5. Vaughan, W.W., Brown, S.C.: Natural environment considerations for Space Shuttle system development support. J. Spacecr. Rocket. **22**(3), 355–360 (1985). https://doi.org/10.2514/3.25756
6. Ning, X., Qi, J., Wu, C., Wang, W.: Reducing noise pollution by planning construction site layout via a multi-objective optimization model. J. Clean. Prod. **222**, 218–230 (2019). https://doi.org/10.1016/j.jclepro.2019.03.018
7. Castiglioni, A.G., et al.: Spaceship Earth. Space-driven technologies and systems for sustainability on ground. Acta Astronaut. **115**, 195–205 (2015). https://doi.org/10.1016/j.actaastro.2015.05.029
8. Williams, A.: Kennedy Space Center Five Year Sustainability Plan. NASA, Florida (2016)

Transformation of Affordable Housing in Saudi Arabia Toward Sustainable Development Goals

Waleed S. Alzamil(✉)

Department of Urban Planning, King Saud University, Riyadh, Saudi Arabia
waalzamil@ksu.edu.sa

Abstract. This paper deals with the analysis of the theoretical concepts of affordable housing in the context of the sustainable development goals SDGs. Locally, there are great efforts to provide housing programs that are consistent with the economic capacity of low-income people. However, there is a deficiency in understanding the holistic perspective of affordable housing, as most housing programs focus on the economic aspect. In light of the sustainable development goals, the paper attempts to propose transformation principles to achieve the duality of sustainable affordable housing. This paper relied on the theoretical approach and critical analysis of a wide range of studies that addressed the concept of affordable housing within the framework of sustainability. The study concluded with four strategic principles for achieving affordable housing in line with the sustainable development goals which are: sustainable regional planning, local urban legislation, sustainable housing production, and sustainable subsidies programs.

Keywords: Affordable housing · Sustainability · Sustainable development

1 Introduction

UN statistics estimate that 1.1 billion urban dwellers live in slums or inadequate housing conditions [1]. Statistics predict that an additional 2 billion people are expected to live in slums in developing countries. Access to affordable housing has become one of the main challenges facing developing countries in recent decades [2]. Housing affordability is subject to several factors including housing supply, income levels, legislative framework, and subsidies and financing programs [3]. Low affordability negatively impacts quality of life and household spending, including transportation, utilities, health, and education [4]. In other words, affordable housing provides spatial and economic stability and social convenience, leading to improved living standards [5]. Locally, the construction sector is still unable to develop innovative solutions that meet the economic and social needs of families. Producing affordable housing does not only mean providing housing environments that are compatible with the economics of the population; but also, with the extent to which they are compatible with economic, social and environmental perspective. Therefore, integrating sustainable urban development goals into city strategic plans requires developing the urban planning infrastructure including the legislative, procedural and executive framework [6]. These structural transformations will lead to the

D. Berkouk et al. (Eds.): CCTS 2024, PTT, pp. 12–20, 2026.
https://10.1007/978-981-95-7607-4_2

sustainability of affordable housing production, which will enhance cities' commitment to a sustainable urban future. This paper seeks to stimulate affordable housing production within a framework consistent with the Sustainable Development Goals SDGs by developing strategic principles for reforming the local housing sector infrastructure including legislative, procedural and implementation aspects.

Saudi Arabia has witnessed urban growth over the past four decades, with an urbanization rate of 82% [7]. Urban growth has been concentrated in some major cities such as Riyadh, Jeddah, and Al Khobar due to the growth of economic investments. The volume of housing demand has increased in major cities, and land prices. Despite the United Nations commitment to providing everyone with equal access to fair housing, the housing sector still fails to reach the balanced production in line with the quality necessary to meet the demand for affordable homes [8]. The rise in land prices and rents is an inevitable result of the limited supply of housing that is compatible with the economic capacity of low-income people and the weakness of housing support programs. These urban problems can increase in light of the rapid urban transformations that Saudi cities are witnessing [9]. Therefore, there is a need to activate the role of affordable housing within a framework that serves the goals of sustainable development.

2 Theoretical Framework

2.1 The Concept of Affordable Housing

Many cities around the world struggle to provide adequate housing stock for middle- and low-income people. Affordable housing is described as housing units that do not burden low- or middle -income families and do not affect the cost of living [10]. The concept of affordable housing refers to a "housing market" that maintains a housing supply that is adequate for middle- and low-income families [11]. The measure of affordability varies between countries, with the proportion of housing costs varying between 30–35% of the capital invested in housing, whether mortgage or rent [12].

2.2 Affordable Housing and SDGs

The concept of affordable housing remains unclear as some studies focus on the economic side of households [3]. Therefore, many literatures suggest that affordable housing should be consistent with a sustainability framework to ensure its success including achieving functional quality, economic diversity, and social suitability [4, 13]. Affordable housing means more than just physical housing but also providing privacy, accessibility, adequate space, security, stability, infrastructure, lighting and heating, environmental quality, and a suitable location with affordable costs [14]. Sustainable cities provide equitable urban policies that promote affordable housing options, thus preserving city demographics and preventing the displacement of vulnerable populations due to rising costs. Sustainable cities develop housing programs that adapt to all economic groups and monitor the housing stock in the city. Sustainable cities establish regulations to prevent real estate speculation to ensure a fair housing market. In addition, sustainable cities encourage investors to provide economical housing alternatives and exploit vacant lands within the city through fair urban legislation [1]. Target 11.1 states "ensuring access for all to adequate, safe and affordable housing and basic services and improving slums." [15]. To achieve SDG 11, sustainable urban development policies emphasize allowing communities to access basic services, affordable housing, efficient transportation and green spaces in a way that serves current and future generations without depleting or disrupting resources [15] (see Fig. 1).

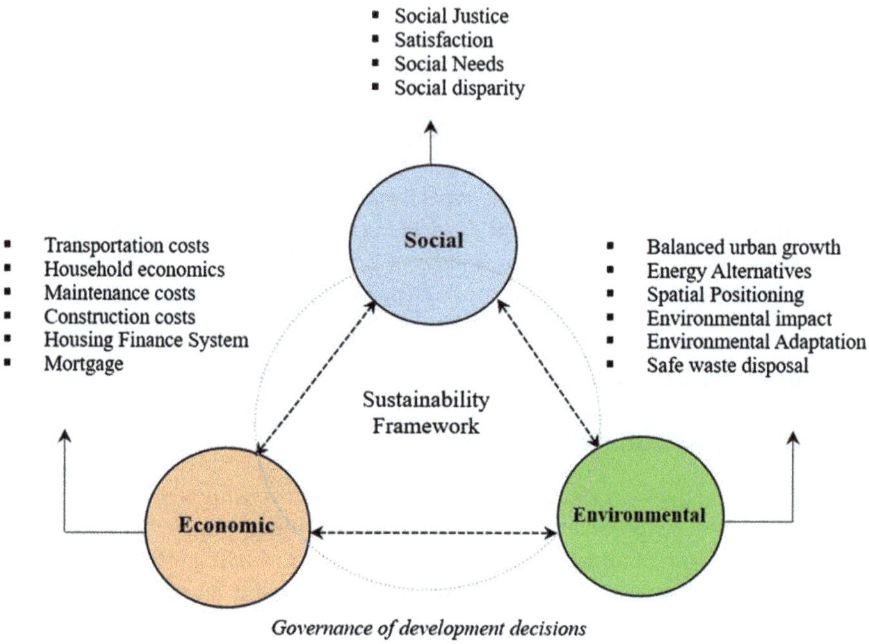

Fig. 1. Affordable Housing in the Context of Sustainability (Author based on [1, 3, 4, 11])

Based on the above, cities that achieve the Sustainable Development Goals SDGs are those cities that respond to the needs of society, economic variables and environmental conditions through effective urban policies, development programs and flexible plans [16]. sustainable policies work to develop an effective governance framework for development decision-making that serves all segments of society and achieves equal opportunities for low-income economic groups, including providing housing support programs and a sufficient stock of affordable housing to meet current and future demand. Sustainable Development Goals SDGs intersect with affordable housing concepts to achieve a balance between environmental, social and economic dimensions:

- Environmental: Developing housing policies that encourage rational consumption, use of alternative energy systems, waste disposal systems, and adapt to the natural environment. In addition, protect agricultural areas from urban sprawl and random extensions [17].
- Economic: Developing a balanced housing market that takes into account spatial and economic disparities between cities without negative impacts on vulnerable groups,

including providing housing programs for low-income people and financing in line with economic capacity limits [18, 15].

- Social: Enhance community participation in development decision-making, including measuring community satisfaction and household needs in affordable housing projects [17].

3 Materials and Methods

This study was based on the theoretical approach through an in-depth analysis of the United Nations reports on affordable housing and a review of the Sustainable Development Goals (Target 11). The study compared local efforts in the context of providing affordable housing and their alignment with the Sustainable Development Goals. Local data was collected based on reports from the General Authority for Statistics and the Ministry of Municipalities and Housing, then applied to the case of Riyadh, Jeddah, and Al Khobar. The study concluded with a set of strategic principles for transformation of affordable housing within the framework of sustainability that can be applied at the local level. (See Fig. 2).

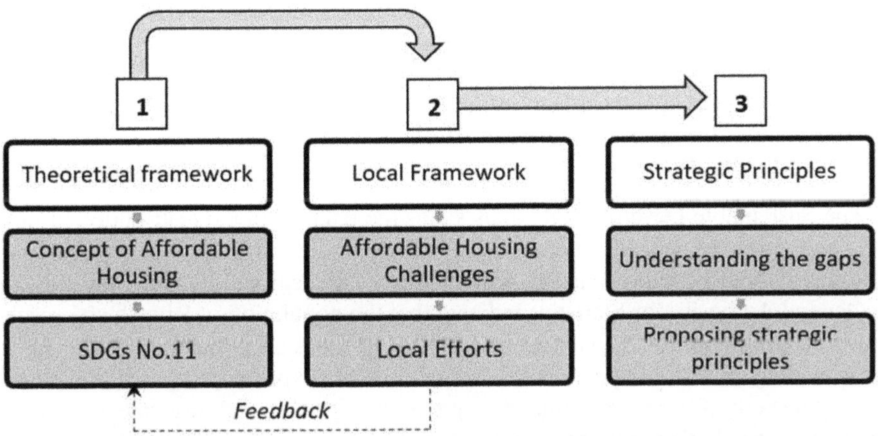

Fig. 2. Scheme of the research methodology (Author based on [19])

4 Local Context

There is a clear disparity in the housing market in Saudi Arabia, with land and housing unit prices rising in Riyadh, Jeddah, and Al Khobar. The estimated population of these cities is 11,419,392 which is equivalent to 40% of the total population in Saudi Arabia [20]. In Riyadh, apartment prices have been on the rise since Q3 2020, with sales values around 11.7% annually over the past four years. The average price reached SAR 5,000 per square meter at the end of Q2 2024. On the other hand, residential villas have been on the rise since 2019, with the average value of villas reaching SAR 5,824 per square meter at the end of 2024 [21]. In contrast, villa prices in Jeddah contracted by 0.8% at the end of Q2 2024, with the average villa price reaching SAR 5,707 per square meter and average apartment SAR 3,945 [21, 22]. Moreover, average villa prices Al Khobar increased by 1.8% and 3.2% compared to 2022 with SAR 3,612 per square meter at the end of 2023. In addition, the average apartment prices saw a 2.9% increase to SAR 3,424 per square meter at the end of 2023 [21] (see Fig. 3).

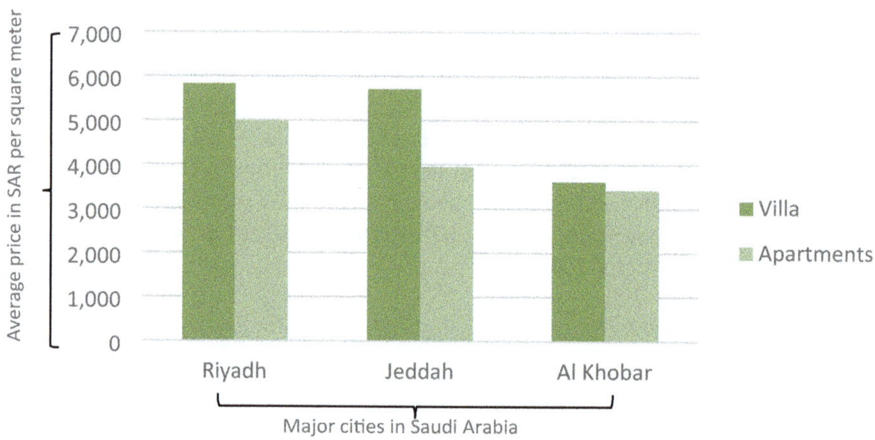

Fig. 3. Average prices of villas and apartments in Riyadh, Jeddah, and Al Khobar (Author based on [21, 22]).

The Sustainable Development Goals states the balance in the distribution of population and resources. Locally, there is an imbalance in the distribution of population and resources without a comprehensive understanding of the regional context. Statistics from the General Authority for Statistics indicate that the population of Riyadh city reached 7,009,120 people in 2022, that constitutes 82% of the total population of the Riyadh region, which is 8,591,748 people [23]. Major cities are expected to see a rise in housing prices as a result of increased demand, population migration and investment concentration. Large cities have expanded in response to housing demand, and this expansion has contributed to urban problems that are not in line with sustainable development goals, such as accessibility, traffic congestion, and rising prices. In addition, some affordable housing projects have been located on the outskirts of major cities due to lower prices or in small cities without a comprehensive program for urban renewal of city centers or maintenance of dilapidated housing [24].

The Sustainable Development Goals indicate that housing production should achieve social justice, social inclusion, and partnership among stakeholders especially for low-income social groups [25]. Therefore, the production of affordable housing is subject to understanding the economic, social and environmental variables in cities. Locally, many housing plans have been developed with a real estate objective to accommodate demand without taking into account the economic diversity of the population, the level of quality, the type of services, environmental considerations, and social patterns [4]. This ultimately generated urban patterns that are not in line with the social, economic, and environmental diversity within the framework of the SDGs. The Ministry of Municipalities and Housing launched the "Sakani" program to support middle-income families and a number of initiatives to support low-income people such as cooperative housing and development housing [26]. Efforts in Saudi Arabia have tried to increase the rate of family home ownership in line with the goals of the National Vision 2030. Statistics from the General Authority for Statistics indicate that the housing stock in Saudi Arabia amounts to 8,174,674 units, of which 20,901 units are classified as public housing, which represents 0.2% of the total housing stock [23]. Moreover, the percentage of Saudi families owning homes has reached (around 60% to 63.74%), 36.4% of whom own homes without a mortgage or loan. More than 24% of families still need to pay off mortgage loans [23, 27]. Yet, there are many challenges related to providing affordable housing due to high production costs, increasing demand, and weak support programs for low-income people [9]. Many middle-income households pay more than 35% of their income for housing [4, 24, 28]. Low-income households do not have the income to obtain a mortgage and rely entirely on developmental housing or cooperative programs, which are still limited [9].

5 Principles of Transformation Towards SDGs

There are significant challenges facing affordable housing in Saudi Arabia in the context of the Sustainable Development Goals SDGs, including regional disparities in the housing market, the effectiveness of housing finance and subsidies programs, urban development policies and the housing production process. Therefore, the study proposes the application of the following transformation principles: First, the housing sector should be part of comprehensive regional development in line with sustainability concepts. This includes restructuring the population in cities to ensure balanced development, enhance connectivity between small and large cities, and encourage reverse migration to small cities. The regional imbalance contributes to increased migration rates to major cities and the growth of informal settlements [29]. Sustainable regional planning requires the development of municipal structures to make decisions that support resilience, social justice, and a balance between economic investment and environmental protection.

Second, strategic plans in cities should include sustainability principles, as the horizontal urban growth of large cities increases the demand for land and housing. Instead, smart growth can be adopted as a sustainable approach that emphasizes the optimal use of resources and reduces unbalanced urban sprawl. Developing urban legislation in cities in line with sustainable development goals, including the Saudi Building Code, which emphasizes rationalization of economic spending, energy, and waste disposal systems. In addition, update urban legislation in a way that is consistent with the social, economic, and environmental diversity of each city.

Third, the process of producing housing locally should be in line with the principle of social and economic diversity. Yet, the real estate development system still deals with physical construction (quantitative production) without developing a model for improving living standards in residential neighborhoods (qualitative production). The development of RED procedures entails requirements for real estate companies to limit the traditional approach based on economic investment [24]. These requirements may include allocating areas for housing projects that increase the housing stock in the market and maintain the balance in prices and requiring the development of a percentage of the plan completely, starting from the infrastructure to the urban structure, to reduce real estate speculation on vacant lands. Fourth, improving housing subsidy programs involves building a sustainable "business model" that emphasizes partnership and community participation within the framework of governance [29]. Housing subsidy programs should target eligible groups based on a housing stock allocated to these programs, such as cooperative housing, urban renewal, and public housing projects. The shortage of available land for trading affects the housing stock in cities and increases prices as many landowners invest in land. Although vacant land is available in cities, it is owned by real estate companies [30]. Therefore, government land must be allocated for affordable housing and developed through public-private partnerships.

6 Conclusion

The paper addressed the principles of transformation of affordable housing in Saudi Arabia towards the sustainable development goals. The study was based on an in-depth review of SDG 11 and comparing it in the context of the local housing situation. The study concluded with four strategic principles for transformation, which are: sustainable regional planning, local urban legislation,sustainable housing production, and sustainable subsidies programs. Implementing these policies faces challenges in aligning the housing sector's structural framework with sustainability indicators. Therefore, the study recommends developing a procedural model to verify the implementation of SDGs in national, regional and local plans.

References

1. UN: Sustainable Development Goals. Goal 11: Make cities inclusive, safe, resilient and sustainable. https://www.un.org/sustainabledevelopment/cities/ (2024)
2. Anacker, K.B.: Introduction: Housing affordability and affordable housing. Int. J. Hous. Policy. **19**(1), 1–16 (2019)
3. Mazáček, D.: Concepts of housing affordability measurements. FFA Working Paper. **8**, 1–15 (2023)
4. Alzamil, W.: Evaluation of Affordable Housing and Subsidy Programs in Saudi Arabia. LAP LAMBERT Academic Publishing, Saarbrücken (2016)
5. Rubin, R.: Affordable housing and resident health. J. Afford. Hous. Community Dev. Law. **27**(2), 263–317 (2018)
6. Reid, A.: Closing the affordable housing gap: identifying the barriers hindering the sustainable design and construction of affordable homes. Sustainability **15**(11) (2023)
7. UN-Habitat: Future Saudi Cities Program: Saudi Arabia. United Nations Human Settlements Programme, Riyadh (2016)
8. Alzamil, W.: A conceptual framework for the activation of sustainable cooperative housing production in Saudi Arabia. Sustainability. **16**(19), 1–20 (2024)
9. Abelson, P.: Affordable housing: Concepts and policies. J. Appl. Econ. Policy. **28**(1), 1–12 (2009)
10. Lin, J.: The Development of Affordable Housing: A Case Study in Guangzhou City, China. Master's Thesis, Stockholm University, Stockholm (2011)

11. Fernandez, R.: Understanding affordable housing in the context of sustainable development. In: No Poverty, pp. 1087–1097. Springer, Cham (2021)
12. Luffman, J.: Measuring housing affordability. Statistics Canada (2006)
13. Habitat for Humanity: Children, Cities and Housing Rights and Priorities. Habitat for Humanity, Atlanta (2022)
14. The Global Goals: Make cities and human settlements inclusive, safe, resilient and sustainable. https://www.globalgoals.org/goals/11-sustainable-cities-and-communities/ (2024). Accessed 29 Dec 2024
15. United Nations: The Sustainable Development Goals Report 2023. United Nations, New York (2023)
16. Cohen, S.: Defining the sustainable city. In: The Sustainable City, pp. 3–14. Columbia University Press, New York (2018)
17. Rogers, P., Jalal, K., Boyd, J.: An Introduction to Sustainable Development. Earthscan, London (2007)
18. Yin, R.K.: Case Study Research: Design and Methods, 5th edn. SAGE Publications, New York (2013)
19. General Authority for Statistics: Social statistics. https://www.stats.gov.sa/statistics?index=119025 (2025). Accessed 01 Jan 2025/
20. Green, M., Alsharif, H.: Saudi Arabia residential market notes. CBRE Research (2024)
21. CBRE: Saudi Arabia residential market review Q3 2023. CBRE Research (2023)
22. General Authority for Statistics: Housing and residences. https://portal.saudicensus.sa/portal/public/8/24?type=TABLE (2022). Accessed 26 March 2024
23. Alhajri, M.F.: Transformation of the Saudi housing sector through an enabling approach to affordable housing. Land. **13**, 1–18 (2024)
24. Habitat for Humanity: Housing and the Sustainable Development Goals–The transformational impact of housing. Habitat for Humanity, Atlanta (2024)
25. Ministry of Municipalities and Housing: Cooperative Housing Initiative. https://momrah.gov.sa/en/node/6745 (2024). Accessed 13 May 2024
26. Ministry of Municipalities and Housing: The Housing Program. https://momah.gov.sa/en/news/annual-report-housing-program (2024). Accessed 03 Feb 2025
27. Alhajri, M.F.: Housing challenges and programs to enhance access to affordable housing in the Kingdom of Saudi Arabia. Ain Shams Eng. J. **13**(6), 1–15 (2022)
28. Alzamil, W.: Characteristics of squatter settlements near Grand Mosque of Mecca: Goz AlNakasa case study. J. Archit. Plan., 247–273 (2018)
29. Rohe, W.M.: Converting public housing to cooperatives: The experience of three developments. Hous. Policy Debate. **6**(2), 439–479 (1995)
30. UN-Habitat: Comprehensive Urban Vision for the City of Riyadh. Saudi Cities Future Program, Riyadh (2020)

Sustainable Urban Greening Using Innovative Water Solutions in Madina Al Munawara

Hassan A. Abas(✉), Wed Makki, Yussra Aljohani, Braa Alhrbi, and Hussein Chekfehand

Prince Mugrin University, Madinah, Kingdom of Saudi Arabia
h.abbas@upm.edu.sa

Abstract. Madina Al Munawara faces severe water scarcity due to its low rainfall, and high evaporation rates. The city depends heavily on energy-intensive desalinated water to meet its growing demand. This study aims to develop innovative water management solutions by reusing ablution water from the Prophet's Mosque. It also seeks to integrate these solutions into the design of multifunctional public green spaces to enhance urban sustainability. The methodology involves a multi-phase approach that includes data collection, system design, and material selection. The system design incorporates reservoirs made from volcanic stones to optimize water storage and minimize evaporation. A graded layering system from volcanic stones was introduced to increase retention capacity. The findings demonstrate that reusing ablution water can reduce Madina Al Munawara's dependence on desalinated water. The integration of these green spaces helps improve air quality and alleviate the urban heat island effect. Additionally, the project enhances social well-being by establishing public recreational spaces that assist community involvement and social interaction. These multifunctional green areas provide a healthier and more connected urban environment. The study presents a replicable model tailored to the needs of Madina Al Munawara, supporting both sustainable water management and urban resilience in arid environments.

Keywords: Water scarcity · Ablution Water · Volcanic Stones

1 Introduction

Madina Al Munawara experiences persistent water scarcity due to its arid climate, low rainfall, and high evaporation rates. The city depends heavily on desalinated water transported from coastal plants. This reliance on desalination water imposes high energy costs and environmental impact. In last decades, Rapid population growth, and rising visitor numbers have significantly increased water demand. The number of visitors expanded from 6.3 million in 2014 to over 14 million in 2023, representing growth of more than 220%. Notwithstanding the significant demand, alternate water sources, such as rainfall and greywater reclamation, are still inadequately exploited. To achieve sustainable infrastructure and long-term water security in Madina, these difficulties must be addressed using innovative water management systems.

D. Berkouk et al. (Eds.): CCTS 2024, PTT, pp. 21–28, 2026.
https://10.1007/978-981-95-7607-4_3

Sustainable water management is critical for urban resilience and environmental sustainability. Effective strategies incorporate advanced technologies, policy frameworks, and community engagement to optimize water use [1]. Approaches such as rainwater harvesting, and wastewater reclamation help mitigate water scarcity. These approaches also reduce reliance on desalination and support ecosystem sustainability in arid countries like Kingdom of Saudi Arabia (KSA). Fito and Van Hulle (2020) highlight that using reclaimed wastewater for irrigation conserves freshwater and decreases dependence on desalination [2]. The integration of technological innovations with regulatory frameworks is essential to support a sustainable balance between urban development and the natural water cycle.

Ablution water generated daily from the Prophet's Mosque in Madina Al Munawara offers a significant opportunity for water reuse. This opportunity is especially relevant given the city's ongoing water scarcity. Approximately 600,000 cubic meters of water are pumped daily to the mosque and its surrounding area. A large portion of this water is used for ablution. The relatively clean water from these activities can be reused for non-potable applications such as irrigation, landscaping, and groundwater recharge. Implementing a system to collect, treat, and redistribute this water can reduce the overall demand for desalinated water.

Urban greening improves livability, air quality, mitigates the urban heat island effect, and supports environmental health. It is difficult to sustain green spaces in semi-arid and arid regions due to high water demand. Gourfi et al. (2022) quoted vegetation as a promising solution to mitigate worsening urban heat island (UHI) effect. Nevertheless, vegetation coverage continues to decline due to augmented urbanization and deteriorating water scarcity [3]. According to Jamali et al. (2021), limited water resources constrain green infrastructure development and maintenance in semi-arid cities [4]. Water solutions that address such challenges must be both sustainable and compatible with such regions' climatic conditions.

This study will design innovative water management solutions through ablution water collection at the Prophet's Mosque in Madina Al Munawara. It will also incorporate such solutions into multifunctional public green space design. The incorporation of treated ablution water into green space infrastructure can reduce the reliance of Madina Al Munawara on desalinated water. The integrated system offers a replicable model that can increase environmental sustainability and urban resilience in water-stressed regions like Madina Al Munawara.

2 Methodology

The study adopts a structured, multi-phase approach to design and evaluate a sustainable water reuse and urban greening system for Madina Al Munawara. Data on water consumption and existing treatment practices were collected from local authorities. The system design focuses on integrating ablution water reuse, storage, and landscape development. A schematic layout was developed to illustrate the collection, treatment, and redistribution of ablution water. The design includes reservoirs constructed with porous volcanic stones to enhance water retention. A graded layering system was incorporated to improve storage efficiency. Native drought-tolerant plants propose for urban greening to minimize irrigation demands and support local biodiversity.

3 Design of Reservoirs and Water Reuse System

This research proposes a sustainable approach to improve water management and enhance urban greening in Madina. The strategy includes three key components. The first part involves reusing ablution water from the Prophet's Mosque. The second part proposes the design of environmentally friendly artificial reservoirs to store this water. The third part emphasizes using stored water to support urban green spaces. Figure 1 illustrates the overall concept of this approach. This strategy aligns with Madina's efforts to expand green areas and Saudi Arabia's goal to achieve sustainable water management. The following sections provide detailed descriptions of each part.

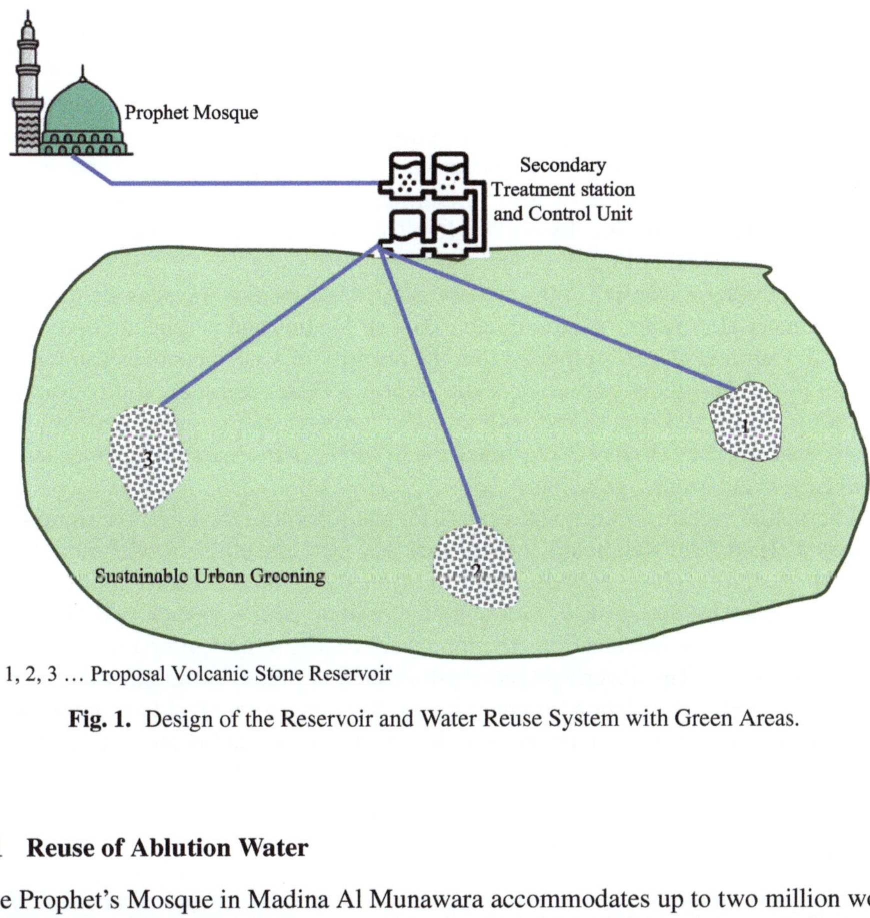

Fig. 1. Design of the Reservoir and Water Reuse System with Green Areas.

3.1 Reuse of Ablution Water

The Prophet's Mosque in Madina Al Munawara accommodates up to two million worshippers and hosts millions of visitors annually. This high visitor volume contributes to significant water consumption. The General Authority for Water reports that approximately 600,000 cubic meters of water are pumped daily to the mosque and its surrounding areas. A large portion of this water is allocated for ablution activities. Currently, ablution water is mixed with black water and transported to centralized treatment plants. This process limits the potential for direct reuse of this relatively clean water.

Having a mechanism to separate and reuse ablution water presents a valuable opportunity to enhance sustainable water management in Madina. Ablution water is relatively free of pollutants and requires minimal processing to meet non-potable water standards. The reduced need for complex treatment systems lowers both operational costs and infrastructure demands. Once treated, this water can be reused for various non-potable purposes, including landscaping, and groundwater recharge. These applications help reduce dependence on desalinated water and relieve pressure on the city's water supply network. Reusing ablution water improves resource efficiency and promotes environmental sustainability. Through minimizing water waste and maximizing the reuse of local water sources, this approach contributes to long-term urban water security. It also aligns with Saudi Arabia's national water strategy and broader sustainability goals.

3.2 Design of Innovative Artificial Reservoirs

The reservoirs are designed to store ablution water and provide a stable water supply for urban green spaces. The concept draws inspiration from sand dams, which are used in arid regions to store rainwater. Ritchie et al. (2021) reported that sand dams retain water in the voids between sand particles [5]. During rainy seasons, water fills these voids and can be withdrawn during dry periods [6]. This mechanism reduces evaporation and helps keep water quality [7].

This study suggests using volcanic stones instead of sand or gravel as the reservoir's primary material. Volcanic stones are abundant in Medina and nearby areas, as shown in Fig. 2. Fauria et al. 2017 reported that the porosity of volcanic stones can reach up to 64% [8]. Their highly porous structure improves water retention ability compared to sand or gravel. Water is stored both in the voids between the stones and within the porous material itself. Figure 3 illustrates the differences in water storage mechanisms between volcanic stones, gravel and sand.

The use of volcanic stones also allows for the incorporation of larger stone sizes, see Fig. 3. Their light weight and local availability facilitate efficient construction. The reservoir design employs a graded layering system to enhance water infiltration and retention. Coarser stones form the base and smaller stones are layered above to slow water percolation and improve retention. The top layer consists of agricultural soil to support vegetation growth. This layering strategy, illustrated in Fig. 4, optimizes water storage and creates a fertile foundation for green spaces. Piezometers are fitted to monitor water levels and quality. Water pumps are installed in an appropriate location for extracting water.

3.3 Urban Greening and Landscape Design

The reservoir surfaces will be transformed into gardens and recreational areas, as illustrated in Fig. 4. The design focuses on sustainability, ecological balance, and aesthetic enhancement. Drought-tolerant native plants will be selected to reduce irrigation needs and ensure long-term viability in the arid climate. These plants, adapted to low water availability, will minimize water consumption and support local biodiversity. The public green spaces will serve both recreational and social functions, fostering community engagement and improving urban livability. Recreational features, such as walking paths

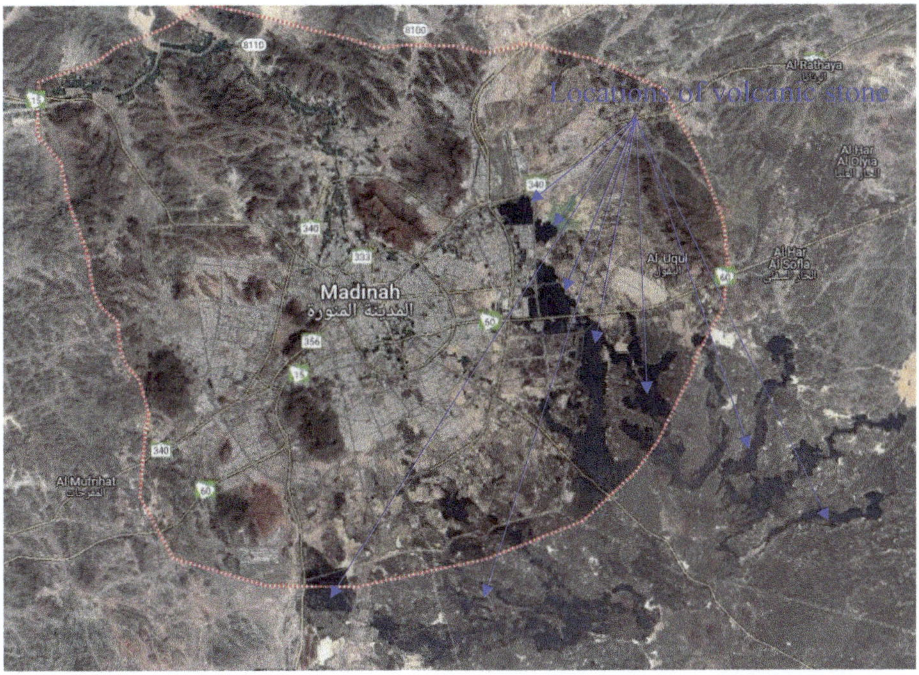

Fig. 2. Distribution of Volcanic Stones in and Around Al Madina Al Munawara.

1. Volcanic stones are characterized by large voids and irregular shape.
2. Gravel with smooth and solid surface.
3. Ability to use large volcanic stones.

Fig. 3. The differences in water storage mechanisms between volcanic stones, gravel and sand.

and seating areas, will be integrated to promote physical activity and relaxation. The paths will wind through various sections of the gardens, enabling visitors to interact with different plant species and scenic views.

The green spaces will be designed to support community events and gatherings, offering multifunctional landscapes for group activities and informal meetings. The design combines natural elements with functional facilities, using native and drought-tolerant plants to reduce water consumption and maintenance. Aesthetic features like flowering plants and art installations will enrich the sensory experience. This approach promotes environmental conservation, social interaction, and urban life quality.

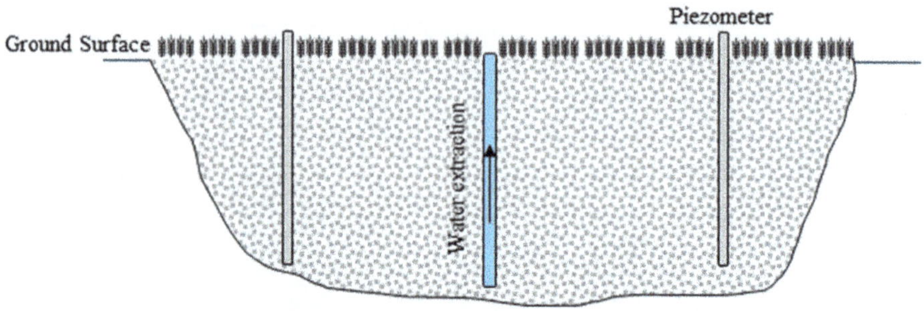

Fig. 4. Vertical Cross-Section of the Artificial Reservoir with Volcanic Gravel.

4 Discussion

4.1 Environmental and Social Benefits

The project improves urban ecosystems by expanding green cover and enhancing biodiversity. Vegetation improves air quality by filtering pollutants and releasing oxygen. Strategically positioned trees and plants reduce the urban heat island effect in Al Madina Al Munawara by creating cooler microclimates. The development of public green spaces encourages social interaction and strengthens community engagement. Recreational features, including walking paths and seating areas, promote physical activity and outdoor leisure. These facilities provide accessible areas for rest, exercise, and informal gatherings, supporting a more active and healthier lifestyle.

The project improves urban ecosystems by expanding green cover and enhancing biodiversity. Vegetation improves air quality by filtering pollutants and releasing oxygen. Strategically positioned trees and plants reduce the urban heat island effect in Al Madina Al Munawara by creating cooler microclimates. The development of public green spaces encourages social interaction and strengthens community engagement. Recreational features, including walking paths and seating areas, promote physical activity and outdoor leisure. These facilities provide accessible areas for rest, exercise, and informal gatherings, supporting a more active and healthier lifestyle.

4.2 Water Storage Efficiency

The project improves water storage through the use of volcanic stones in artificial reservoirs. These stones have high porosity, which increases their capacity to retain water. Water is stored both in the voids between the stones and within their porous structure.

This design enhances water infiltration and reduces evaporation, minimizing water loss over time.

A grading and layering strategy further improve water retention. Coarser stones form the base layer to facilitate rapid infiltration. Smaller stones are layered above to enhance retention and filtration. This arrangement reduces surface runoff by allowing water to gradually move through the layers. A top layer of soil supports vegetation and slows water percolation, ensuring a stable supply for plant growth.

The system reduces dependence on desalinated water sources, improving resource efficiency. It supports a closed-loop approach by recycling and reusing ablution water within the urban environment. This method conserves water and aligns with sustainable water management practices. The reliable water supply provided by the reservoirs increases the resilience of green spaces in water-scarce environments.

5 Conclusion

This study demonstrates that integrating ablution water into innovative water management strategies, particularly in support of urban greening, can help address water scarcity in Madina Al Munawara. Key findings highlight the potential of harvesting ablution water from the Prophet's Mosque in conjunction with volcanic stone reservoirs to sustain water efficiency. This approach reduces reliance on desalinated water and ensures a more sustainable supply for the city's expanding green spaces.

The study highlights the importance of utilizing drought-tolerant native plant species to support urban greening efforts in Madina's arid climate. These plants require minimal irrigation and promote biodiversity. The increased vegetation improves air quality, mitigates the urban heat island effect, and enhances thermal comfort across the city. In addition to environmental benefits, the project strengthens social cohesion through the creation of multifunctional public green spaces.

The integration of landscape design and water management provides a replicable model for sustainable urban development in regions facing similar climate challenges. The findings show that innovative solutions can enhance Madina Al Munawara's resilience to water scarcity while promoting environmental conservation and social well-being. These initiatives align with the Kingdom of Saudi Arabia's vision for sustainable development and resource management.

The local government should adopt the proposed ablution water reuse and reservoir system as part of urban planning initiatives in Madina Al Munawara. This system can enhance water sustainability by reducing reliance on desalinated water and supporting the city's expanding green infrastructure. Further research should focus on optimizing water reuse systems, including advanced treatment methods for ablution water. Studies should also explore additional applications of reclaimed water, such as its use in groundwater recharge and urban cooling strategies.

References

1. Bouramdane, A.-A.: Optimal water management strategies: Paving the way for sustainability in smart cities. Smart Cities. **6**(5), 2849–2882 (2023). https://doi.org/10.3390/smartcities6050128

2. Fito, J., Van Hulle, S.W.H.: Wastewater reclamation and reuse potentials in agriculture: Towards environmental sustainability. Environ. Dev. Sustain. **23**(3), 2949–2972 (2021). https://doi.org/10.1007/s10668-020-00732-y
3. Gourfi, A., Taïbi, A.N., Salhi, S., El Hannani, M., Boujrouf, S.: The surface urban heat island and key mitigation factors in arid climate cities, case of Marrakesh. Morocco. Remote Sens. **14**(16), 3935 (2022). https://doi.org/10.3390/rs14163935
4. Jamali, F.S., Khaledi, S., Razavian, M.T.: Priority areas for developing green infrastructure in semi-arid cities: A case study of Tehran. Environ. Urban. ASIA. **12**(1), 118–135 (2021). https://doi.org/10.1177/0975425321990326
5. Ritchie, H., Eisma, J.A., Parker, A.: Sand dams as a potential solution to rural water security in drylands: Existing research and future opportunities. Front. Water. **3** (2021). https://doi.org/10.3389/frwa.2021.651954
6. Yifru, B.A., Kim, M.-G., Lee, J.-W., Kim, I.-H., Chang, S.-W., Chung, I.-M.: Water storage in dry riverbeds of arid and semi-arid regions: Overview, challenges, and prospects of sand dam technology. Sustainability. **13**(11), 5905 (2021). https://doi.org/10.3390/su13115905
7. Castelli, G., et al.: Sand dams for sustainable water management: Challenges and future opportunities. SSRN Electron. J. (2022). https://doi.org/10.2139/ssrn.3998987
8. Fauria, K.E., Manga, M., Wei, Z.: Trapped bubbles keep pumice afloat and gas diffusion makes pumice sink. Earth Planet. Sci. Lett. **460**, 50–59 (2017). https://doi.org/10.1016/j.epsl.2016.11.055

Smart Cities in Egypt: Advancing Digital Transformation and Urban Sustainability

Anne A. Soliman[1], Amr Salem[2], Aya Ezzat[3], Mostafa Khaled Ahmed[2], Rana Elshafei[4], Samah El Khateeb[4,5(✉)], and Eman Farouk[6]

[1] Department of Architecture, Faculty of Engineering, Misr International University, Cairo, Egypt

[2] Department of Urban Design and Planning, Faculty of Engineering, Ain Shams University, Cairo, Egypt

[3] Architecture Department, Modern University for Technology and Information (MTI), Cairo, Egypt

[4] Architectural Engineering Department, University of Business and Technology, Jeddah, Saudi Arabia

samahelsayed@gmail.com

[5] Smart and Future Cities Laboratory for Sustainable urban Solutions, Faculty of Engineering, Ain Shams University, Cario, Egypt

[6] Architecture and Urban Design Program, German University in Cairo, Cario, Egypt

Abstract. Smart cities are transforming urban development by utilizing AI, IoT, and big data in resolving urbanization challenges, enhancing sustainability, and improving quality of life. Egypt, under Vision 2030, is initiating smart city projects to digitalize infrastructure, propel digital transformation, and attain inclusive, resilient cities. This study employs a qualitative mixed-methods approach, integrating policy analysis, case studies, and empirical surveys to examine Egypt's smart cities through the lenses of governance, education, healthcare, and mobility. A two-stage approach is employed to examine both macro-level regulatory regimes and micro-level smart city implementations and end-user experiences. The study examines progress through Digital Egypt and leading projects like the New Administrative Capital and New Alamein City. Despite progress, challenges persist including regulatory gaps together with scalability problems and digital accessibility barriers as well as socio-economic inequalities. The study establishes the need for stronger governance and private sector participation together with affordable and digital literacy-focused policies to enable equitable smart city development. The final recommendations emphasize regulatory clarity and stakeholder collaboration to overcome socio-economic barriers as a pathway to sustainable and inclusive urban transformation in Egypt.

Keywords: Urban Innovation · Smart Infrastructure · Technology-driven Urbanism · Sustainable Urban Transformation

D. Berkouk et al. (Eds.): CCTS 2024, PTT, pp. 29–36, 2026.
https://10.1007/978-981-95-7607-4_4

1 Introduction

The concept of a smart city has emerged as a revolutionary urban development strategy, combating urbanization, resource scarcity, and climate change. Smart cities, through the integration of AI, IoT, and big data, optimize sustainability, efficiency, and quality of life to the maximum. Since its origin in Silicon Valley in 1993, the concept has evolved to focus on inclusiveness, resilience, and citizen-driven governance [1].

Egypt, the world's 14th most populous country, faces increasing demand for efficient infrastructure and services as over 47 million people reside in urban areas. In response, Vision 2030 prioritizes smart city development, aligning with SDG 11 for inclusive, safe, and resilient cities [2]. Flagship projects like the New Administrative Capital (NAC) and New Alamein City integrate smart infrastructure, energy efficiency, and digital governance to create sustainable urban environments [3].

Despite these developments, successful implementation is held back by several challenges, such as budget constraints, poor infrastructure, and lack of skilled experts [4]. Further, socioeconomic hurdles, such as cultural barriers, poverty, and inequality, also limit the scope to take advantage of the benefits provided by smart cities [5].

This paper critically analyzes the reality of smart city construction in Egypt with a particular emphasis on the digital solutions for the transformation, the policies of the state, and the efforts on the ground.

2 Methodology

This study combines policy analysis, case studies, and user experience evaluations in a qualitative mixed-methods approach to evaluate Egypt's smart city projects. The processes of digital transformation, strategic planning, regulations, sectoral applications, and challenges are analyzed. Information is collected through an analysis of government documents, published academic work, and fieldwork on the state of digitalization in governance, education, healthcare, and mobility.

A two-stage approach involves examining the broader context of the institutional frameworks and policies being implemented to a more focused view of the smart city technologies and trends in their adoption. The study identifies technological gaps, infrastructure deficits, and economic barriers, alongside enablers like private-sector investment and youth-driven digital initiatives, infrastructural, and economic hurdles. Key projects, including the New Administrative Capital (smart governance, mobility), New Alamein City and Noor Smart City (renewable energy, smart housing), are evaluated. Conclusions are drawn through cross-validated data collection processes which include stakeholder participation and technology adoption metrics, acknowledging limitations due to Egypt's growing smart city ecosystem and lack of long-term performance data.

3 Results

3.1 Digital Transformation

Fueled by Egypt's Vision 2030 and Digital Egypt initiative, the nation's digital transformation aims to achieve economic improvement, enhancement in public services, and address issues spurred by urbanization through innovation [6]. A significant aspect is

ICT infrastructure improvement with the Decent Life Initiative upgrading fiber optic network connections to 33,000 governmental edifices and 4,500 villages, which will serve 3.5 million households [7]. This progress led to a 16% growth in the ICT sector (2020/2021) and earned Egypt the Speed Test accolade for the fastest fixed internet service in Africa [7]. The country's internet penetration is around 70% whilst affordability and the rural urban split remains problems. Egypt placed 77th in the Network Readiness Index, was 17th in active mobile broadband subscriptions, and came third in the Digital Inclusion Index 2020. To narrow such gaps, Egypt launched a new portal for e-government services and set up 311 technology centers, which are projected to render 250 different services by 2022 [8].

Some sectors grew with the development of e-learning through the Egyptian Knowledge Bank and smart classrooms, the implementation of telemedicine and electronic health records in the healthcare sector, and the expansion of e-commerce through Fawry and Souq.com, all enabled by mobile banking [9]. Governance has improved through e-government services promoting transparency and progress toward Sustainable Development Goals (SDGs). There is a drive among The National Telecommunications Regulatory Authority (NTRA) and MCIT to strengthen cybersecurity and data protection and tech startup support, while AI and IoT drive entrepreneurship [10]. These efforts aid Egypt's position as a regional hub for digital infrastructure development, yet challenges in digital literacy and infrastructure disparities remain.

3.2 Strategic Agendas

President Abdel-Fattah El-Sisi, in cooperation with the Ministry of Housing, NUCA, and the World Bank, launched "National Smart Cities" and "National Green Urbanism" projects under Egypt's first national smart cities strategy. The strategy aims to develop technologically advanced, sustainable, and inclusive cities aligned with international standards and SDGs [11, 12].

Egypt's National AI Strategy, launched in 2021, aims to use artificial intelligence to improve governance, promote development, build skills, and foster international partnerships balancing economic growth with ethical considerations. During its first phase, which ended in 2022, the focus was on training people in AI technologies and initiating pilot projects to test these new ideas. Now, in its second phase, the strategy aims to broaden AI applications across various sectors in the country [13]. In addition, Egypt Vision 2030, started in 2016, introduced a comprehensive digital transformation strategy emphasizing innovation, cybersecurity, and digital literacy to enhance governance, economic growth, and residents' quality of life [14].

3.3 Policies and Regulations

Egypt is aiming to improve urban services with the development of smart cities, economically and socially. The National Smart Cities Strategy integrates digital intelligence and sustainability standards, while the National Telecom Regulatory Authority introduced a comprehensive IoT service regulatory framework. The National Initiative for Green Smart Projects promotes environmentally friendly solutions. Egypt's Vision

2030 emphasizes digital transformation and green urbanization, but challenges like data accessibility and the need for comprehensive frameworks remain.

3.4 Case Studies of "Smart" in Egypt

The National Initiative for Smart Green Projects was launched during COP27. This initiative is targeting green investments in all the 27 governorates, finding ways to solve climate issues and support the green economy through digital technologies [15, 16]. It engages citizens and organizations in smart green projects, in a bottom-up approach. The Smart Cities Program was established in 2016 to make urban areas sustainable and to enhance the quality of life. It also seeks to make Egypt as a smart city leader [17]. The first phase of the project was implemented from 2016 to 2021, which targeted unsafe areas, and the project cost 39 billion EGP. It is important to note that we are currently in the second phase of the program, which began in 2021 and will continue until 2030.

City Scale: Smart cities foster efficiency by using digital solutions to enhance resource utilization, emissions, and energy efficiency. Smart cities consist of smarter transport grids, water supply, waste disposal, and lighting, in addition to supporting interactive governance and public security [18, 19].

Egypt plans to introduce fourth-generation cities to manage suffocating urban agglomerations, with a vision to upgrade existing urban development, construct infrastructure services, and preclude slum growth using sustainable urban development approaches [20].

The New Administrative Capital east of Cairo is Egypt's premier project in its urbanization plans with experiments to embrace smart technologies in all facets. The city is linked to all the neighboring urban cities using highly integrated public transport and smart government via a huge network of underground tunnels [20].

New Alamein City, Egypt's Northwest Coast, is the first privately owned "smart" city. The 48 km-long city comprises residential, cultural, tourism, industrial, and research areas. It features smart home facilities, security, automatic lighting, climate control, and a remote-control mobile application. Mobility, irrigation through renewable energy, and smart waste management are also part of the city, establishing a new standard for private investors [21]. New Alamein city is in an advanced construction phase, with completed beach towers, ongoing high-rise developments, a landmark tower, and a planned data center to enhance its smart city infrastructure.

Noor smart City, developed by Talaat Mostafa Group, will be Egypt's first private "smart" city under the ownership of investors. It will be constructed in the area of the New Administrative Capital and aims to be an international prototype for smart cities through the incorporation of advanced technology and green concepts [22].

Initiative Scale: Smart Mobility. Egypt is moving in the direction of smart transport, education, and government for efficiency and sustainability. International organizations are collaborating with Mwasalat Misr to offer high-quality bus services [23]. Cairo Bike is promoting sustainable urban mobility, while Rabbit app offers environmentally friendly short-distance travel [24].

Smart Education. Led by the Ministry of Communications and Information Technology, it seeks to equip the nation's young people with digital capabilities. The "Egypt Future Work is Digital (FWD)" initiative equips 200,000 Egyptian young people with the capacity to handle the fast-evolving global digital economy and address the requirements of the job market [25].

Smart Governance. Smart governance is one of the main pillars of Egypt's smart cities with a view to keeping pace with the evolution of telecommunication services and the shift towards a digital society. The National Institute for Governance and Sustainable Development (NIGSD) has initiated a training course in Portugal on capacity building towards digital transformation and smart governance [26].

Smart health. Vezeeta, an Egyptian online healthcare platform, is revolutionizing healthcare access and management by connecting patients with qualified doctors, offering convenient consultations, and facilitating appointment scheduling [27].

Smart Food Market. The Food Lab, a cloud kitchen operator, offers innovative solutions for expanding Egyptian food markets without opening new restaurants or delivery chains [28]. Talabat, which absorbed the first Egyptian online food delivery initiative Otlob.com, focuses on food delivery and has expanded to cover four GCC countries by 2013 [29].

Smart Solid Waste Management. Egyptian government aims to increase municipal solid waste collection coverage and efficiency to 80% and 90% by 2030 through e-waste recycling initiative, smart waste collection monitoring, and Eco Green Waste Management [30].

Fig. 1. Timeline of Digital Transformation in Egypt.

3.5 Case Studies Reflection

Egypt's 2050 urban development strategies place smart cities at the lead of development drivers. The government vision is to attract investment, promote competitiveness, and improve urban living by means of technology-driven, people-oriented solutions. (Refer to Fig. 1: Timeline of Digital Transformation in Egypt).

4 Discussion: Challenges and Opportunities

Egypt has been developing digital transformation since 2004 but fully developed smart cities remain a challenge [31]. Although it has 99.38 million mobile phone subscriptions as of December 2022, only 69% of them have access to the internet, revealing digital gaps among population segments [32, 33]. Shortage of skilled labor would also cause outsourcing and job displacement in specialized domains [34].

Despite e-government efforts and infrastructure upgrades, no Egyptian city has fully achieved smart city status, with several projects still under construction [35]. NAC and New Alamein City are in line with international trends but lag Singapore, Barcelona, and Dubai in AI uptake, IoT integration, and blockchain-based governance. Egypt's command-and-control model prioritizes cultural heritage, affordability, and security, and can act as a likely model for Africa and MENA but requires policy reforms and investment.

The development of smart cities in Egypt has been studied concerning various parameters including internet penetration, IoT, energy efficiency, transport, and governance. The study revealed that New Cairo was at the most developed level of ICT at 78%, yet economically developed Alexandria came at second position at 59% and required most governance improvement; the other cities were not far behind: New Assiut, at 51%, which is at the edge of becoming a high-growth potential, [36]. It was reported that in the first stage of maturity level one, NAC is still engaged with the introduction of smart transport projects which are not completed to a certain degree [37]. Therefore, only strong governance with public-private partnerships and could lead to the development of Egypt's smart cities.

5 Conclusion

Smart city initiatives in Egypt clearly showcase strong government commitment as well as robust infrastructure, particularly with flagship projects like the New Administrative Capital and New Alamein City. On the other hand, the smart city ecosystem is quite fragmented, as all the private-sector initiatives, although viable to stimulate other developments, are small and not fully integrated into a wider system. While private investment drives innovation, its limited scalability and accessibility hinder nationwide transformation. Hence to mitigate this, Egypt must boost its public-private partnerships, enhance its policy frameworks for regulation, and develop coordinated urban policies. Expanding new funding mechanisms for projects, improving coordination among these institutions, and prioritizing digital inclusion- and affordability-based investments will all be critical in constructing an integrated, scalable, and citizen-centered smart city model attuned to Vision 2030 and global sustainability goals.

References

1. Omayer, H.M., Mobarak, M.A., Selim, O.: Digital transformation: Towards sustainable and smart new cities in Upper Egypt (New Qena case study). IOP Conf. Ser. Earth Environ. Sci. **1113**, 012027 (2022)
2. El Khafif, M., Salem, N.: Smart cities–Policy and regulatory frameworks. In: Smart Cities and Digital Transformation, pp. 325–340. Springer, Cham (2019). https://doi.org/10.1007/978-3-030-64217-4_19
3. El Menchawy, A.S., Moustafa, H.H., Abdel-Hamid, N.I.: Smart capital cities: Towards a smart new administrative capital (NAC). F1000Research. **12**, 265 (2023). https://doi.org/10.12688/f1000research.130322.1

4. Ghobrial, S.B., Mandour, M.A., Sobhy, T.M.: KPIs for maintaining smart clean sustainable cities. Int. J. Eng. Mod. Technol. **8**, 109–125 (2023) https://doi.org/10.56201/ijemt.v8.no5.2022.pg109.125
5. Emam, W.: Evaluation of the possibility of applying artificial intelligence technology in Egyptian smart city planning. MSA Eng. J. **2**, 331–349 (2023) https://doi.org/10.21608/msaeng.2023.291876
6. Ministry of Communications and Information Technology: Digital Egypt. https://mcit.gov.eg/en/Digital_Egypt
7. Aboul-Dahab, K.: Assessing Egypt's progress toward its digital transformation and collaborative regulation. SSRN Electron. J. (2025). https://doi.org/10.2139/ssrn.5027167
8. ITU: Measuring digital development – ICT Development Index 2023. https://www.itu.int/itu-d/reports/statistics/IDI2023/
9. Mahmoud, S.S., Labib, G.M.: Digital transformation and the SDGs: A case study on Egypt's e-government and SDG-16. Upravlenie i Politika. **3**, 8–20 (2025) https://doi.org/10.24833/2782-7062-2024-3-4-8-20
10. Kamel, S., Rizk, N.: The role of innovative and digital technologies in transforming Egypt into a knowledge-based economy. In: Digital Transformation in the Global South. IGI Global, Hershey (2019). https://doi.org/10.4018/978-1-5225-7214-5.CH017
11. Ibrahim, A.: Building Egypt's future–Launching the first national smart cities strategy. https://wuf.unhabitat.org/event/wuf12/building-egypts-future-launching-first-national-smart-cities-strategy (2025). Accessed 28 Jan 2025
12. Nassar, M.: Sisi announces launch of "National Smart Cities Strategy," "National Green Urbanization Strategy" at WUF. CSR Egypt. https://www.csregypt.com/en/
13. Technical Committee of the National Council for Artificial Intelligence: Egypt National AI Strategy (2021)
14. Rashwan, D., Abdel Moneim, I., El-Saadany, S., Sabber Saleh, R., Mahmoud, N.: Foundation for Digital Egypt: A Soaring Progress (2023)
15. Government of Egypt: The National Initiative for Smart Green Projects in Egypt's Governorates–Round 2 Background (2023)
16. Egyptian Government: NISGP Concept Note. https://mped.gov.eg/Files/NISGP_Concept_note.pdf (2022)
17. Allam, H.: Egypt's smart cities: A vision for a more sustainable and livable future. https://www.developmentaid.org/news-stream/post/169491/egypts-smart-cities (2024). Accessed 11 May 2024
18. European Commission: Smart cities. https://commission.europa.eu/eu-regional-and-urban-development/topics/cities-and-urban-development/city-initiatives/smart-cities_en
19. Tourism Smart Cities–Turning Point Towards Sustainable Development in Egypt: Concepts, characteristics and applications. Pharos Int. J. Tour. Hosp. **1**, 21–30 (2022). https://doi.org/10.21608/pijth.2022.262065
20. ElMenshawy: 4th generation cities – The promising future of Egypt real estate. https://www.cityscape-events.com/en/insights/egypt/residential/4th-generation-cities-promising-future-egypt-real-estate.html (2025). Accessed 28 Jan 2025
21. Arab Contractors: New El Alamein City. https://www.arabcont.com/english/project-553 (2025). Accessed 19 March 2025
22. Noor Smart City–Capital Gardens. https://talaatmoustafa.com/communities/noor-city-capital-gardens/
23. Mwasalat Misr: Our Story. https://mwasalatmisr.com/our-story/ (2025). Accessed 28 Jan 2025
24. Ahram Online: Egypt launches its first-ever bike-sharing service "Cairo Bike" ahead of COP27. https://english.ahram.org.eg/NewsContentMulti/478186/Multimedia.aspx (2025). Accessed 19 March 2025

25. Ministry of Communications and Information Technology: Egypt Future Work is Digital (FWD). https://mcit.gov.eg/
26. Ministry of Planning: Egypt's NIGSD training program in Portugal. https://mped.gov.eg/singlenews?id=851&lang=en (2025). Accessed 19 March 2025
27. Daigavhane, M.K., Gundewar, S., Gadkari, A., Anpan, S.: Web-based smart healthcare system. Proc. IEEE Int. Conf. Adv. Eng. Inf. Technol., 1–6 (2024). https://doi.org/10.1109/idicaiei61867.2024.10842858
28. Hamed, A.: The Food Lab raises $4.5 million from global investors. https://thefoodlab.com/the-food-lab-egypts-leading-cloud-kitchen-operator-raises-pre-seed-round-of-4-5-million-from-global-investors/ (2025). Accessed 28 Jan 2025
29. Talabat Blog: 'Otlob' becomes "talabat" as Egyptian success story expands. https://blog.talabat.com/otlob-becomes-talabat-as-egyptian-success-story-continues-with-creation-of-50000-new-jobs-and-expansion-into-8-cities/ (2025). Accessed 28 Jan 2025
30. Alsobky, A., Ahmed, M., Al Agroudy, S., El Araby, K.: A smart framework for municipal solid waste management: A case study in Greater Cairo. ASU Eng. J., 102183 (2023)
31. Abdelfattah, S., Hosny El Assar, H., Ibrahim, J., Elsayeh, Y.: Tourism smart cities–Turning point towards sustainable development in Egypt. Pharos Int. J. Tour. **1** (2022)
32. CAPMAS: Egypt in Figures. https://www.capmas.gov.eg/Pages/Publications.aspx?page_id=5104&YearID=23410&Year=23410 (2024)
33. Elshamy, A.: Determinants of the digital divide and social effects in Egypt. Arab J. Media Commun. Res., 48–79 (2022)
34. Moussa, A.M., Tarek, S.: Digital transformation and its impact in Egypt: A comprehensive literature review. Int. J. Prof. Bus. Rev. **8**, e02755 (2023)
35. Moubarak, L.M., Bakeer, L.M., Rashed, A.Y.: Smart urban design in Egypt: Potentials and challenges. In: Proc. 2nd Int. Conf. on Sustainable Construction and Project Management, (2018)
36. Hassebo, A., Tealab, M., Hamouda, M.: From a traditional city to a smart city: Measuring cities' readiness for transition – Egypt as a case study. Urban Sci. **8**, 212 (2024). https://doi.org/10.3390/urbansci8040212
37. Elkhateeb, S., El-Shahat, M.: A roadmap for smart city in the Arab region—A paradigm shift in post-pandemic era. In: Proc. 9th ASCAAD Int. Conf, pp. 2–4. Springer, Cham (2021)

Smart Sustainable Cities: Innovative AI Technologies to Improve Urban Development and Quality of Life

Muna Ahmed(✉)

Jazan University, Jazan 45142, Saudi Arabia
mgahmed@jazanu.edu.sa

Abstract. Smart Cities are urban centers characterized by integrating advanced technologies, digital infrastructures convergence, and data-driven decision-making to enhance the quality of life for urban residents. They utilize innovative solutions to optimize urban operations, improve resource management, and foster sustainability. Innovation and Artificial Intelligence (AI) technologies are central to Smart Cities, they provide powerful tools that enable cities to optimize resource utilization, streamline operations, and respond to changing needs. Smart cities provide a platform for testing and implementing emerging technologies, such as 5G networks, Internet of Things (IoT) sensors, and (AI) solutions. This paper is an attempt to explore the intersection of AI and smart sustainable cities, focusing on the increasing role of AI-driven technologies in creating more sustainable urban environments, it presents initiatives from Arabic cities using (AI) applications to support sustainable urban development and improve citizens' quality of life. The method applied for this research included a literature review about definitions and key elements of a smart city, and selected case studies of Arabic cities ranked among the (SCI) Smart City Index 2024 of the International Institute for Management Development Switzerland (IMD), these cities serve as valuable examples, offering insights into the diverse applications of (AI), providing real-world illustrations of the integration of (AI) technologies with a specific focus on building sustainable and efficient urban settings. The research conclusion asserts the futuristic potential of (AI) to promote sustainable urban environments, address urban difficulties, boost efficiency, and improve quality of life within smart cities.

Keywords: Smart Sustainable City · Quality of Life · AI Technologies

1 Introduction

Cities worldwide are experiencing unprecedented urbanization, more than half of the world's population (57%) lives in cities in 2023 [1], that figure is expected to increase to (68,4%) by 2050 [2]. The rapid pace of urbanization creates complex urban challenges related to land management, urban planning, and the delivery of efficient public services. It necessitates proactive measures to prepare cities for future challenges in economy, environment, governance, and society [3]. Future cities need to be sustainable, economically vibrant, well governed and driven by innovations, such innovations are powerful

D. Berkouk et al. (Eds.): CCTS 2024, PTT, pp. 37–45, 2026.
https://10.1007/978-981-95-7607-4_5

tools to propel cities towards positive urban transformation. Smart cities are urban areas that integrate advanced technologies and data analytics to improve the quality of life for their citizens, enhance sustainability, and optimize the use of resources [4]. The smart city is a revolutionary concept based on the integration of several electronic modules to enhance the information for the easiness of users [5]. Smart sustainable cities are emerging as a key solution to address growing urban challenges. Governments and public services worldwide are increasingly turning to Artificial Intelligence (AI) technologies, especially at the local level, to manage city operations and data, to enhance efficiency and productivity in service delivery, automating processes, aiding in decision-making, streamlining routines, and tackling complex tasks [6].

2 Research Methodology

Digital technologies have become increasingly integral to the daily life of people, the research explores the intersection of Artificial Intelligence (AI) and sustainable cities, focusing on the increasing role of AI-driven technologies in creating efficient smart sustainable urban environments, the research main question is how does AI technologies contribute to sustainability in smart cities in Arab region?

The research methodology for this paper includes terms and key elements definitions of smart sustainable city. In the second part it presents case studies of selected Arabic cities ranked among the (SCI) Smart City Index 2024 of the International Institute for Management Development Switzerland (IMD), these cities serve as valuable examples, offering insights into the diverse applications of AI, providing real-world illustrations of the integration of AI technologies with a specific focus on building sustainable and efficient urban settings. The information gathered is synthesized and summarized to present the role of AI in smart sustainable Arabic cities.

3 Definition of the Smart Sustainable City

Smart sustainable city is defined as a city that utilizes ICT to provide present inhabitants needs, while sustaining future generations abilities to meet their needs [6, 7]. It is also described as innovative city in reference to the increasing role of AI-driven technologies to create efficient smart sustainable urban environments and enhance quality of life of inhabitants [8].

A set of key performance indicators (KPIs) for Smart Sustainable Cities (SSC) have been developed as part of the initiative ‘United for Smart Sustainable Cities’ (U4SSC) in 2016 by the UN Habitat in collaboration with the International Telecommunication Union (ITU) and the United Nations Economic Commission for Europe (UNECE) [9]. The (KPIs) are used to evaluate the contribution of ICT in making cities smarter, sustainable, and people-centered. Adopted by over 200 cities globally, the U4SSC KPIs are empowering cities and providing self-assessments of achieving United Nations sustainable development goals (SDGs).

3.1 Key Elements of a Smart Sustainable City

The key performance indicators (KPI's) for smart sustainable cities have been developed according to 'United for Smart Sustainable Cities' initiative (U4SSC) [9] to provide cities with a consistent and standardized method to collect data and measure the performance and progress of achieving the Sustainable Development Goals (SDGs). The (U4SSC - KPI's) are a comprehensive, globally recognized framework supported by 16 UN agencies, designed to provide cities with a standardized approach to collect data, evaluate performance, and drive people-centered digital transformation.

The structure of the key performance indicators is divided into three dimensions:

- Economy: ICT, Productivity, Infrastructure.
- Environment: Sustainability, Energy.
- Society and culture: Education, Health, Safety (Table 1).

Table 1. U4SSC - Key performance indicators (KPI's) Dimensions

Economy		Environment		Society and Culture	
Sub-dimension	KPI's	Sub-dimension	KPI's	Sub-dimension	KPI's
ICT Infrastructure	17	Environment	13	Education, health &Culture	12
Productivity	7	Energy	4	Safety, Housing & Social Inclusion	17
Infrastructure	21				

Source: compiled from Telecommunication Standardization Sector Of (ITU-T): Key performance indicators for smart sustainable cities to assess the achievement of sustainable development goals (Y.4903) (2022) [9]

3.2 Global Indexes for Assessment of Smart Sustainable Cities

A number of different global indexes are used to assess smart sustainable cities performance:

- IMD Smart City Index: It offers a balanced focus on economic and technological provisions of the city across five key areas: health and safety, mobility, activities, opportunities, and governance [10].
- Arcadis Sustainability Index: ranks 100 global cities across three pillars environmental, social, economic sustainability [11].
- World Air Quality Rankings: provides live air quality city ranking [12].
- Energy Transition Index: provide statistics about energy transition towards renewables [13].
- Mercer Quality of Living City Ranking: Quality of Life [14].
- Waste Atlas: Provides data about waste management, in addition to Kilograms of waste per population [15].

- Sustainable Cities Water Index: water Sustainability [16].
- Innovation Cities Index: Provides city ranking in human Infrastructure, network markets, communication of innovation [17].
- Robeco SAM Country Sustainability Ranking: gives score for countries based on 40 indicators, grouped in 15 criteria, with a 20% weighting given to environmental factors, 30% to social issues, and 50% to governance [18].
- The Top 50 Smart City Government Rankings: Smart City Ecosystem [19].

3.3 Smart Sustainable Cities in the Arab Region

Cities around the globe are transforming towards being smart and sustainable. The transformation process is often based on two different development paradigms [20]:

- The brownfield model; the transformation process takes place in existing cities (B).
- The greenfield model; cities are established as smart sustainable cities (G).
- Arab countries adopted both transformation models. The following Table 2. presents the smart sustainable cities status in the Arab region.

Table 2. Smart sustainable cities status in the Arab region

Country	Smart Sustainable Cities (SSC) project	SSC project Used Approach
Algeria	Algiers Smart City	Brownfield
	Cyberpark City of Sidi Abdellah	Green Field
Bahrain	Manama	Brownfield
Egypt	New Capital, Smart Village	Green Field
Jordan	Amman Smart City, Ecocity Feta	Brownfield
Kuwait	Kuwait City South Saad Al Abdullah City	Brownfield Green Field
Lebanon	Beit Misk Village	Green Field
Morocco	Rabat City, Casablanca Tangier Tech City Ben G.	Brownfield Green Field
Oman	Muscat	Brownfield
Qatar	Doha City Lusail City	Brownfield Green Field
Saudi Arabia	Riyadh City, Jeddah City, Al-Ahsa City King Abdullah Economic City, Knowledge Economic City, Prince Abdul Aziz Bin M. Economic City, Jizan Economic City, Neom Smart City	Brownfield Green Field Green Fiel Green Field
United Arab Emirates	Dubai City, Abu Dhabi City Masdar City, Dubai Silicon Oasis Desert Rose City	Brownfield Green Field

Source: Based on United Nations Economic and Social Commission for Western Asia: Smart sustainable cities and smart digital solutions for urban resilience in the Arab region Lessons from the pandemic, E/ESCWA/CL2.GPID/2021/TP.3(2021) [20]

4 Top 10 Arabic Cities Ranked Among the (SCI) Smart City Index 2024

The IMD Smart City Index [21] evaluates technological provision of the city across five key areas: health, safety, mobility, activities, opportunities and governance. It assesses the perceptions of residents on issues related to technology applications in their city. The method of the assessment categorizes world cities into four groups (quartile) based on the Global Data Lab's Human Development Index (HDI) score, within each HDI group, cities are assigned a 'rating scale' ranges from (AAA to D) based on the perceptions-score of a given city compared to the scores of all other cities within the same group (quartile). The rankings present the overall ranking (1 to 142), and also the detailed rating in each component.

In this section the top 10 Arabic cities ranked among the (SCI) Smart City Index 2024 of the International Institute for Management Development Switzerland (IMD) are presented in Table 3.

Table 3. Top 10 Smart sustainable cities in the Arab region in 2024

City	S C Rank 2024	S C Rating 2024	Structure 2024	Technology 2024	S C Rank 2023	Change	country
Abu Dhabi	10	BB	BB	BB	13	+3▲	United Arab Emirates
Dubai	12	BB	BB	BB	17	+5▲	
Riyadh	25	B	B	B	30	+5▲	Saudi Arabia
Doha	48	B	BB	B	59	+11▲	Qatar
Mecca	52	B	B	B	52	▬	Saudi Arabia
Jeddah	55	B	B	B	56	+1▲	
Medina	74	CCC	B	CCC	85	+11▲	
Muscat	88	B	B	CCC	96	+8▲	Oman
Al-Khobar	99	CCC	CCC	CC		NEW	Saudi Arabia
Cairo	114	C	C	CC	108	−6▼	Egypt

Source: compiled from (SCI) Smart City Index 2024 of the International Institute for Management Development Switzerland (IMD), https://www.imd.org/smart-city-observatory/home/ [21]

5 Best Practices and Initiatives From Arab Region

The next section presents selected case studies of Arabic cities ranked among the (SCI) Smart City Index 2024 of the (IMD), these cities serve as valuable examples, offering insights into the diverse applications of (AI), providing real-world illustrations of the integration of (AI)technologies with a specific focus on building sustainable and efficient urban settings.

5.1 Technological Innovations of Smart Cities in Saudi Arabia

Saudi Arabia has 5 cities ranked among the top 10 Arabic cities in the (SCI) Smart City Index 2024 of the International Institute for Management Development Switzerland (IMD). The remarkable achievement is a result of collaborative efforts between the Saudi Data & AI Authority (SDAIA) with a number of governmental entities [22] to leverage (AI) solutions in support of sustainable urban development, the following are related initiatives:

Environment, Water and Agriculture (AI) Center The center focuses on developing AI-powered sustainable solutions for the environment, water, and agriculture sectors. The initial phase prioritized monitoring environmental changes, providing open access to data and AI tools to maximize their potential, building capacity, and transferring knowledge. AI-powered solutions are used for predicting environmental phenomena, monitoring emissions, and optimizing energy use through integrated (AI) systems.

(AI) Center for Energy . The center aims to utilize (AI) technologies to drive energy sector initiatives and build a robust foundation in data and (AI), and also contributes to the goals of the Saudi Green Initiative to reduce carbon emissions, maximize renewable energy output, and improve operational efficiency at renewable energy production sites. Additionally it encourages cooperation with leading national companies such as TAQA, and international companies in the fields of energy and (AI) to localize technology and stimulate investment in the fields of renewable energy production, enhance grid stability, and optimize energy management.

The National Smart Cities Platform (Smart C) . The National Smart Cities Platform is concerned with developing smart solutions for various government entities in the Kingdom to enhance sustainability, efficiency, and improve the quality of life for city residents, through the following tools:

- An innovative tool for detecting visual distortions in the city by using AI-powered cameras supported by advanced analytics, to collect data on the urban landscape and visual distortions, and assist local authorities to identify and monitor the built environment components (roads, open spaces, buildings) conditions in order to enhance the quality of life in these cities.
- Employ congestion solutions using data and (AI) to reduce congestion and facilitate vehicle traffic, and support data-based decision-making to reduce level of congestion.

Global Smart Cities Forums . The Forum is a global platform for smart cities held in Riyadh, bringing together the world's leading experts in smart cities and (AI). The

Forum aims to shape an ambitious vision for the future of cities using smart solutions and establish fundamental principles that support development plans for achieving sustainable urban growth, ultimately enhancing the quality of life.

5.2 Technological Innovations of Smart Cities in the UAE

Abu Dhabi and Dubai ranked first, and second smart sustainable cities in the Arab region in (SCI) Smart City Index 2024. Smart cities in the UAE integrated innovative (AI) technologies in transportation, energy, health care, security, and governance. The following are related initiatives [23]:

The Digital Dubai Initiatives . Over 130 initiatives include the Dubai Data Initiative, the Dubai Blockchain Strategy, the Happiness Agenda, the Dubai AI Roadmap, and the Dubai Paperless Strategy. The strategies included partnership between government and private sector entities [26].

(AI) Geo Platform . The platform aims to provide an advanced AI-based tool for analyzing and managing spatial geographic data, benefiting local municipalities, architects, urban development planners, researchers, and academics.

6 Conclusions

It can be concluded that the world continues to change rapidly, cities are places where new solutions and ways to be future ready are emerging to face the increased uncertainties. The findings demonstrate the futuristic potential of AI to promote sustainable urban environments, address urban difficulties, boost efficiency, and improve quality of life within smart cities. The case studies initiatives present a range of valuable applications that include formation of national strategies, establishing institutional arrangements, and creating an enabling environment for a trustworthy use of AI in the public and private sectors. Further research is necessary to address the challenges of integrating innovative (AI) technologies to improve urban development and quality of life.

References

1. World Bank Group: Urban population (% of total population). https://data.worldbank.org/indicator/SP.URB.TOTL.IN.ZS (2025). Accessed 15 March 2025
2. United Nations: Department of Economic and Social Affairs, Population Division: World Urbanization Prospects: The 2018 Revision (ST/ESA/SER.A/420), p. 23, United Nations, New York (2019)
3. World Economic Forum: Inspiring Future Cities & Urban Services–Shaping the Future of Urban Development & Services Initiative, p. 10 (2016)
4. United Nations, Department of Economic and Social Affairs, Population Division: World Population Prospects 2024: Summary of Results (UNDESA/POP/2024/TR/NO. 9). United Nations, New York (2024)
5. Digital Government Authority: Smart Cities in Digital Government–Digital Brief (2023)

6. Sarker, M.N.I., Kamruzzaman, M.M., Huq, M.E., Zaman, R., Hossain, B., Khurshid, S.: Smart city governance through big data: Transformation towards sustainability. In: Proc. Int. Conf. of Women in Data Science (WiDSTaif), pp. 1–6. IEEE (2021)
7. Höjer, M., Wangel, J.: Smart sustainable cities: Definition and challenges. In: Hilty, L., Aebischer, B. (eds.) ICT Innovations for Sustainability. Advances in Intelligent Systems and Computing, vol. 310, pp. 333–349. Springer, Cham (2015). https://doi.org/10.1007/978-3-319-09228-7_20
8. Ben Dhaou, S., Isagah, T., Distor, C., Ruas, I.C.: Global Assessment of Responsible Artificial Intelligence in Cities: Research and Recommendations to Leverage AI for People-Centered Smart Cities. UN-Habitat, Nairobi (2024)
9. International Telecommunication Union (ITU): Key Performance Indicators for Smart Sustainable Cities to Assess the Achievement of Sustainable Development Goals (Y.4903). ITU, Geneva (2022)
10. IMD: Smart City Observatory 2024. https://www.imd.org/smart-city-observatory (2025). Accessed 19 March 2025
11. Arcadis: Sustainable Cities Index 2024: Key Insights. https://www.arcadis.com/en (2025). Accessed 19 March 2025
12. IQAir: World Air Quality Index (AQI) Ranking. https://www.iqair.com/world-air-quality-ranking (2025). Accessed 19 March 2025
13. Statista: Energy Transition Index (ETI) by Country 2024. https://www.statista.com (2025). Accessed 19 March 2025
14. Mercer: Quality of Living City Ranking 2024. https://www.mercer.com (2025). Accessed 19 March 2025
15. Waste Atlas: Waste management interactive data platform. https://www.atlas.d-waste.com (2025). Accessed 19 March 2025
16. City Water Index.: https://citywaterindex.com (2025). Accessed 19 March 2025
17. Innovation Cities: Innovation Cities™ Index–World 500 Rankings. https://www.innovation-cities.com (2025). Accessed 19 March 2025
18. Robeco: Country Sustainability Ranking. https://www.robeco.com (2025). Accessed 19 March 2025
19. Smart City Governments: Top 50 Smart City Governments. https://smartcitygovernments.com (2025). Accessed 19 March 2025
20. United Nations ESCWA: Smart Sustainable Cities and Smart Digital Solutions for Urban Resilience in the Arab Region–Lessons from the Pandemic. E/ESCWA/CL2.GPID/2021/TP.3 (2021)
21. IMD: Smart City Index 2024. https://www.imd.org/smart-city-observatory (2025). Accessed 15 Feb 2025
22. Saudi Data & AI Authority (SDAIA): Initiatives. https://sdaia.gov.sa/en/MediaCenter/Initiatives/Pages/default.aspx (2025). Accessed 15 Feb 2025
23. Dubai Digital Authority: Digital Dubai Initiatives. https://www.digitaldubai.ae (2025). Accessed 19 March 2025

Developing a Custom AI Model for Vegetation Health Monitoring and Survival Prediction in Arid Regions

Ahmed Ahmed[1](✉) and Elshaimaa Nada[2]

[1] Taibah University, Almadinah, Saudi Arabia
dr.aaahmed@hotmail.com
[2] Zagazig University, Zagazig, Egypt

Abstract. Reforestation in arid regions is critical for combating desertification, restoring ecosystems, and mitigating climate change. Leveraging AI models tailored for such environments offers a promising solution to monitor vegetation health and predict survival rates effectively. This study focuses on designing and implementing a Convolutional Neural Network (CNN)-based model using Sentinel-2 satellite data to support vegetation health monitoring and survival prediction. The developed model achieves state-of-the-art performance by utilizing vegetation indices such as NDVI and EVI and incorporating transfer learning techniques. Validation against field data and benchmarking with traditional models demonstrate its scalability and applicability for reforestation initiatives such as the Saudi Green Initiative (SGI). This is critical due to increasing desertification and urgent global climate targets.

Keywords: SGI · GEE · ArcMap · CNN · Remote Sensing · Environmental Sustainability

1 Introduction

The Saudi Green Initiative (SGI), introduced in 2021 as part of Saudi Arabia's Vision 2030, is a major environmental initiative to address climate change, desertification, and biodiversity loss while fostering sustainable development. Key objectives of the SGI include planting 10 billion trees, cutting carbon emissions by 278 million tons annually by 2030, and preserving marine and coastal ecosystems in the Red Sea and Arabian Gulf. The initiative harnesses cutting-edge technologies such as AI and big data to enhance reforestation, improve energy efficiency, and promote sustainable urban development. Additionally, SGI prioritizes international collaboration, contributing to regional projects through the Middle East Green Initiative (MGI) and supporting global climate targets. With advancements already seen in tree planting, renewable energy, and ecosystem conservation, SGI offers a replicable framework for tackling environmental issues [1–4]. Table 1 summarizes the main objectives and methodologies of the Saudi Green Initiative [1, 2].

D. Berkouk et al. (Eds.): CCTS 2024, PTT, pp. 46–55, 2026.
https://10.1007/978-981-95-7607-4_6

Integrating AI and big data in environmental initiatives like SGI represents a paradigm shift in how we approach ecological restoration. By leveraging these technologies, we can move from reactive to proactive strategies, enabling real-time monitoring, predictive analytics, and optimized resource allocation. This is particularly crucial in arid regions, where traditional methods often fall short due to the harsh climate and limited resources. These technologies empower resource-efficient ecological restoration in challenging environments, providing scalable solutions.

Artificial intelligence (AI) and big data technologies have emerged as powerful tools to address these challenges in recent years. AI leverages machine learning (ML) algorithms and predictive models to analyze large datasets, while big data provides the necessary volume, velocity, and variety of information. Together, they enable efficient decision-making, resource optimization, and monitoring of reforestation projects in arid environments [1, 2]. Table 2 presents AI Models in reforestation challenges in Arid Regions. Traditional methods, often labor-intensive and limited in scope, have been augmented by AI-driven technologies capable of analyzing extensive datasets to inform and optimize ecological restoration strategies. Applications of AI in Vegetation Monitoring may include: Automated Plant Classification [14], Remote Sensing and Image Analysis [5], Predictive Modeling [15], Integration AI with Internet of Things (IoT) [16], AI-Driven Ecological Restoration [17]. Arid and semi-arid regions, known for their challenging climates and scarce water resources, present significant obstacles to reforestation and vegetation monitoring. However, advancements in AI technologies have proven invaluable in overcoming these challenges in countries like Saudi Arabia.

This paper focuses on how AI and Big Data can be leveraged for monitoring and enhancing vegetation and reforestation under the Saudi Green Initiative (SGI). This section presents an introduction to the objectives and methodologies of the SGI and AI models for addressing reforestation challenges in arid regions. The next section presents the methodology of model design and its implementation. The last section presents the conclusion, challenges, and future work.

2 Methodology of Model Design and Implementation

2.1 Data Preparation

- Data Acquisition: Sentinel-2 multi-spectral data (10 m resolution) is acquired for selected arid regions (Riyadh, Saudi Arabia is chosen as a case study).
- Vegetation Indices: includes calculating NDVI and EVI as in Eqs. 1,2
- NDVI (Normalized Difference Vegetation Index):

$$NDVI = \frac{NIR - Red}{NIR + Red} \tag{1}$$

- EVI (Enhanced Vegetation Index):

$$EVI = 2.5 * \frac{NIR - Red}{NIR - 6 * Red - 7.5 * Blue + 1} \tag{2}$$

Table 1. Objectives and Methodologies of the Saudi Green Initiative (SGI)

Category	Details
Reforestation and Vegetation Enhancement	**Target:** -Plant 10 billion trees across the Kingdom over the coming decades.
	Purpose: -Combat desertification by reversing the effects of land degradation. -Increase vegetation cover to improve biodiversity and ecological resilience. -Enhance carbon sequestration to mitigate greenhouse gas emissions.
	Methodologies: -Leveraging AI and big data for site selection and monitoring. -Utilizing drought-resistant and native plant species.
Reducing Carbon Emissions	**Target:** Reduce Saudi Arabia's carbon emissions by more than 278 million tons annually by 2030.
	Purpose: -Transition to a circular carbon economy (CCE) by focusing on carbon capture, utilization, and storage (CCUS). -Promote renewable energy to reduce dependency on fossil fuels. -Enhance energy efficiency in industrial and residential sectors.
	Initiatives: Deployment of clean hydrogen and renewable energy projects. Expanding solar and wind energy capacity in the Kingdom.
Protecting Marine and Coastal Ecosystems	**Focus:** Restoring and conserving the biodiversity of the Red Sea and Arabian Gulf.
	Purpose: -Combat coral reef degradation and overfishing. -Protect endangered marine species and their habitats. -Establish marine protected areas to enhance sustainability.
Enhancing Livability and Urban Sustainability	**Goal:** Develop sustainable cities that prioritize human well-being and environmental conservation.

(*continued*)

Table 1. *(continued)*

Category	Details
	Examples: -Expanding green spaces in urban areas to improve air quality. -Promoting smart city technologies to optimize energy and resource consumption. -Adopting sustainable water management practices, including desalination and wastewater recycling.
Regional and Global Collaboration	**Goal:** -Position Saudi Arabia as a leader in global environmental initiatives.
	Collaborations: -Cooperation with Middle Eastern countries to plant an additional 40 billion trees regionally under MGI. -Active participation in international dialogues on climate change and sustainability.

- Preprocessing: Data is preprocessed using Google Earth Engine (GEE) to remove clouds, normalize bands, and generate time-series vegetation indices. We choose Riyadh as Saudi Arabia's study region as in Fig. 1. Since field data is huge, predictions will focus on satellite-derived vegetation indices (NDVI and EVI values in GeoTIFF formats). Time-series chart for NDVI and EVI values is presented in Fig. 2.
- Using ArcMap 10.8.1 combine NDVI and EVI into a single raster for more comprehensive analysis as shown in Fig. 3. The dataset of the composite layer is obtained in.csv format.

2.2 Model Development

Design and train a Convolutional Neural Network (CNN) AI model to detect vegetation health and predict survival rates. Experiment with transfer learning techniques to enhance model performance with limited training data. Fig. 4 presented the CNN architecture diagram while Fig. 5 shows the scatter plot for comparing model predictions with ground truth.

2.3 Model Performance

The model demonstrates robust scalability, with deployment potential in other arid and semi-arid regions globally as shown in Table 3. The custom CNN model, while slightly less accurate than the pretrained ResNet50, offers the advantage of being specifically tailored to the unique environmental conditions of arid regions, making it more adaptable for future applications.

Table 2. AI Models in reforestation challenges in Arid Regions

Application	AI Model	Description	Advantages	Limitations	References
Site Selection and Planning	GIS with AI-based Spatial Analysis	Integrates geospatial data with machine learning algorithms (e.g., Random Forest, SVM) to identify optimal reforestation sites.	High precision in site identification.	Requires high-quality geospatial data.	[5, 6]
			Processes large datasets, including satellite imagery.	Computationally intensive for large-scale projects.	
Species Selection and Survival Prediction	Decision Trees and Neural Networks	Evaluates environmental compatibility of species and predicts survival rates using environmental parameters.	Provides tailored species recommendations.	Neural Networks require extensive training data.	[3, 7]
			Handles complex interactions among variables.	Decision Trees may overfit with noisy data.	
Monitoring and Assessment	Convolutional Neural Networks (CNNs)	Processes satellite and drone-captured imagery for vegetation monitoring, deforestation detection, and stress identification.	High accuracy in image analysis.	Limited by image data availability and quality.	[8, 9]
			Automates change detection over time.	Requires substantial computational power.	

(*continued*)

Table 2. (*continued*)

Application	AI Model	Description	Advantages	Limitations	References
Resource Optimization	Reinforcement Learning (RL)	Optimizes irrigation schedules using weather patterns, soil moisture, and plant water needs.	Continuously improves through learning.	Data-intensive during initial implementation.	[10, 11]
			Adapts to environmental changes, minimizing water usage.	Requires frequent recalibration for variable environments.	
Tree Counting and Density Estimation	Deep Learning Models for Object Detection	Utilizes deep learning algorithms to analyze high-resolution satellite imagery, enabling the identification and counting of individual trees in arid and semi-arid regions.	Provides accurate tree counts over large areas.	Dependent on the resolution and quality of satellite imagery.	[12]
			Assists in assessing tree density and biomass estimation.	Requires significant computational resources for processing extensive datasets.	
Predictive Modeling for Reforestation Success	Ensemble Learning Models (e.g., Random Forest, Gradient Boosting)	Combines multiple machine learning algorithms to predict the success rates of reforestation efforts based on various environmental and anthropogenic factors.	Enhances predictive accuracy by aggregating multiple models.	Complexity in model interpretation and increased computational requirements.	[13]

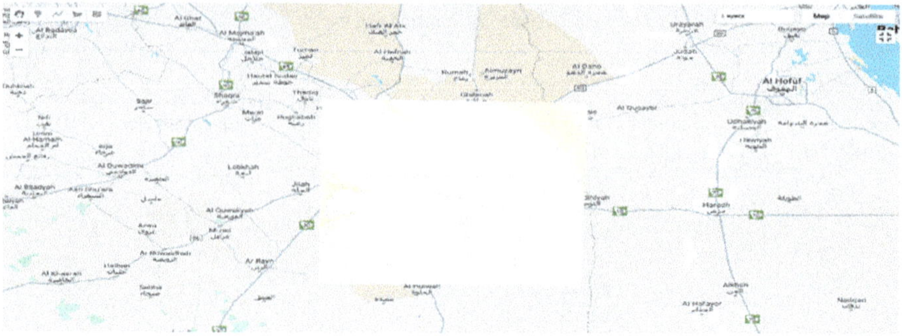

Fig. 1. Riyadh Region on GEE.

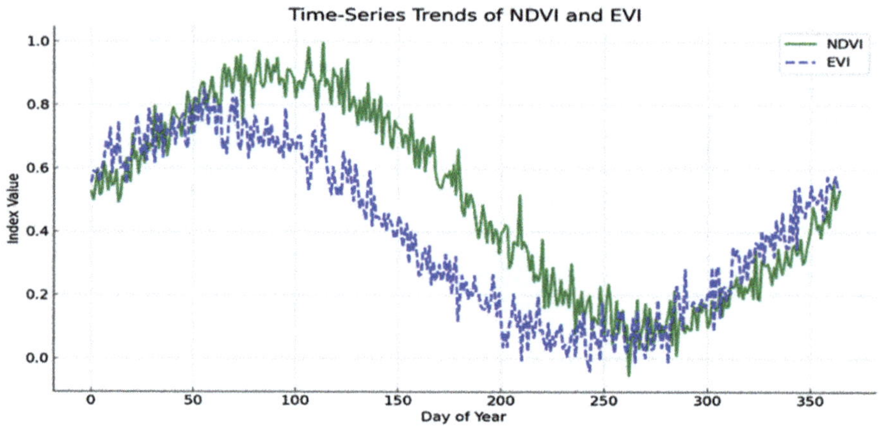

Fig. 2. Time-series chart for NDVI and EVI values

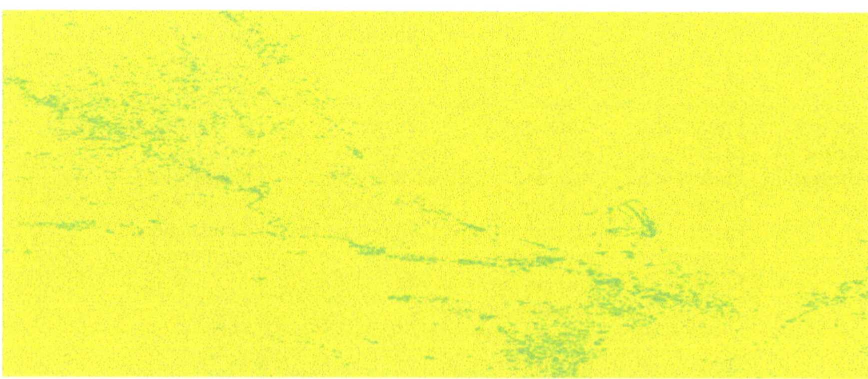

Fig. 3. Composite layer of NDVI and EVI using ArcMap

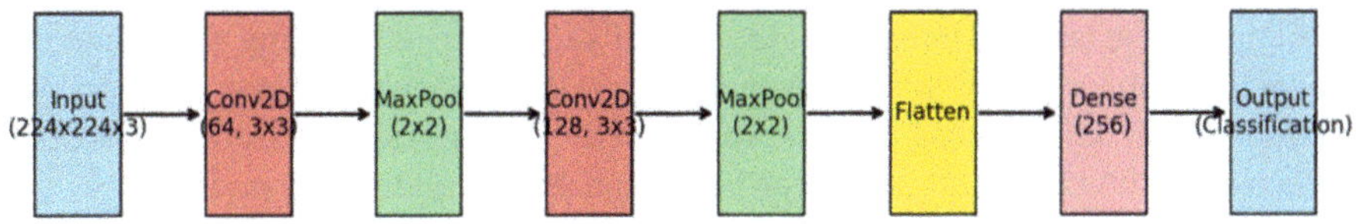

Fig. 4. CNN Architecture Diagram

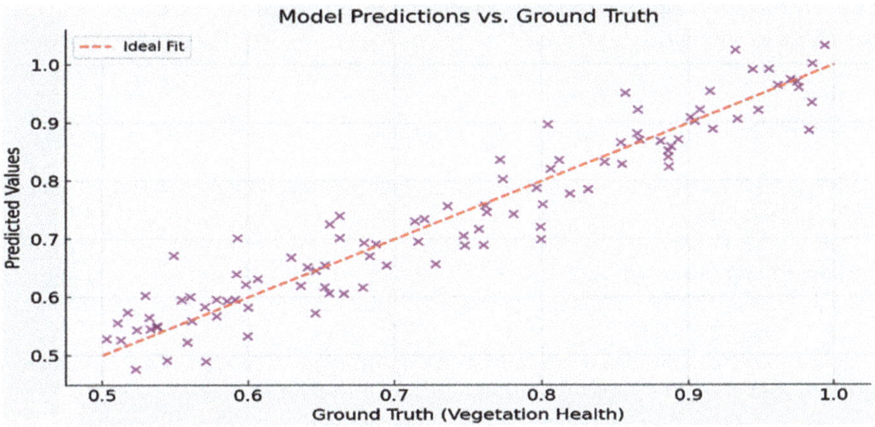

Fig. 5. Scatter plot for comparing model predictions with ground truth.

Table 3. Model Performance

Metric	Custom CNN	Pretrained ResNet50	Random Forest
Accuracy (Health)	92%	94%	85%
R^2 (Survival Prediction)	0.87	0.89	0.78
F1-Score (Health)	0.91	0.93	0.83

3 Conclusion

This study presents a scalable AI framework for monitoring vegetation health and predicting survival rates in arid regions using Sentinel-2 data where high-resolution Sentinel-2 data combined with tailored AI models provides a powerful approach for vegetation monitoring in arid regions. The developed CNN model outperforms traditional approaches and provides actionable insights for large-scale reforestation projects such as the Saudi Green Initiative. The developed framework supports real-time decision-making for environmental agencies.

The challenges of this work include the limited availability of ground-truth data and variability in vegetation patterns across different arid regions. Future work will focus on expanding the dataset and enhancing the model's usability, integrating additional environmental factors (e.g., soil moisture, temperature), and developing user-friendly interfaces for non-technical stakeholders. Additionally, incorporating AI-driven geospatial analytics and climate modeling can further enhance the predictive framework.

Furthermore, the integration of AI with IoT devices and the use of real-time data streams could further enhance the model's predictive capabilities, enabling more dynamic and adaptive reforestation strategies. Collaboration with international environmental organizations and the sharing of data and insights will also be crucial for scaling these solutions globally.

References

1. Vision 2030: Saudi Arabia's vision for a sustainable future. https://www.vision2030.gov.sa (2021). Accessed 2021
2. International Energy Agency (IEA): Achieving net-zero emissions by 2050. https://www.iea.org (2022). Accessed 2022
3. Saudi Green Initiative: Progress reports and updates. https://www.greeninitiatives.gov.sa (2023). Accessed 2023
4. United Nations Convention to Combat Desertification (UNCCD): Global models for desertification control. https://www.unccd.int (2022). Accessed 2022
5. Smith, J., Green, R.: Leveraging AI for sustainable reforestation. J. Environ. Sci. **45**(3), 234–245 (2022)
6. Lee, M.: Big data in environmental management. Environ. Data Sci. **12**(4), 567–578 (2021)
7. Saudi Green Initiative: Saudi Arabia's commitment to environmental sustainability. https://www.vision2030.gov.sa/en/explore/projects/saudi-green-initiative (2023). Accessed 2023
8. Brown, R.: AI for plant growth optimization. AI Agric. **6**, 145–157 (2023)
9. Kim, L., Zhao, H.: Predictive modeling in reforestation. Comput. Ecol. **19**(2), 88–102 (2022)
10. NEOM: The world's largest green hydrogen plant and reforestation projects. https://www.neom.com (2023). Accessed 2023
11. White, A.: Smart irrigation systems in arid regions. Water Resour. Manag. **28**(5), 345–358 (2023)
12. Davis, P.: AI-driven water management. J. Hydrol. **50**, 122–134 (2021)
13. United Nations Convention to Combat Desertification (UNCCD): The Great Green Wall Initiative. https://www.unccd.int (2022). Accessed 2022
14. Amri, E., et al.: Advancing automatic plant classification system in Saudi Arabia: Introducing a novel dataset and ensemble deep learning approach. Model. Earth Syst. Environ. **10**, 2693–2709 (2024). https://doi.org/10.1007/s40808-023-01918-9
15. Thompson, S., Garcia, E.: Ensemble modeling for predicting reforestation outcomes in semi-arid regions. Environ. Model. Softw. **131** (2020)
16. Tanmeya: Smart farming in Saudi Arabia–Integrating IoT and AI for better yields. https://www.tanmeya.com.sa/article/smart-farming-in-saudi-arabia-integrating-iot-and-ai-for-better-yields (2023). Accessed 2023
17. Experience AlUla: Green future is now – AlUla harnessing AI for comprehensive ecological restoration. https://www.experiencealula.com/en/about/features/green-future (2023). Accessed 2023
18. NEOM: NEOM launches regreening initiative for planting 100 million trees. https://www.neom.com/en-us/newsroom/neom-regreening-initiative (2022). Accessed 2022

Smart Waste Management System for Makkah City Using Artificial Intelligence and Internet of Things

Rawabi Saeed Al Qurashi, Maram Meshal Almanjumi, Teef Lahg Alghamdi, Amjad Hasan Almalki, Shahad Saed Alharthi, Shahad Mashal Althobuti, Alanoud Saad Alharthi, and Maha A. Thafar(✉)

College of Computers and Information Technology, Computer Science Department, Taif University, Taif, Kingdom of Saudi Arabia
m.thafar@tu.edu.sa

Abstract. Waste management is a critical global issue with significant environmental and public health implications. It has become more destructive during large-scale events such as the annual pilgrimage to Makkah, Saudi Arabia, one of the world's largest religious gatherings. This event's popularity has attracted millions worldwide, leading to significant and unpredictable accumulation of waste. Such a tremendous number of visitors leads to increased waste management issues at the Grand Mosque and other holy sites, highlighting the need for an effective solution other than traditional methods based on rigid collection schedules. To address this challenge, this research proposed an innovative solution that is context-specific and tailored to the unique requirements of pilgrimage season: a Smart Waste Management System, called TUHR, that utilizes the Internet of Things and Artificial Intelligence. This system encompasses ultrasonic sensors that monitor waste levels in each container at the performance sites. Once the container reaches full capacity, the sensor communicates with the microcontroller, which alerts the relevant authorities. Moreover, our system can detect harmful substances such as gas from the gas detector sensor. Such a proactive and dynamic approach promises to mitigate the environmental and health risks associated with waste accumulation and enhance the cleanliness of these sites. It also delivers economic benefits by reducing unnecessary gasoline consumption and optimizing waste management resources. Importantly, this research aligns with the principles of smart cities and exemplifies the innovative, sustainable, and health-conscious approach that Saudi Arabia is implementing as part of its Vision 2030 initiative.

Keywords: Smart Waste management · AI · Internet of Things · Smart City · Public health · Sustainability

1 Introduction

Waste management is crucial for urban sustainability, impacting public health, the environment, and economic efficiency [1–3]. It involves the collection, transportation, and processing of waste to maintain clean and healthy living spaces [4]. Waste management

D. Berkouk et al. (Eds.): CCTS 2024, PTT, pp. 56–65, 2026.
https://10.1007/978-981-95-7607-4_7

becomes especially challenging during large-scale events like the Hajj pilgrimage in Makkah, where millions of people contribute to substantial waste production. Traditional waste management systems, which rely on preset collection schedules, are ineffective in handling the high volume and unpredictability of waste, often resulting in premature pickups or bin overflows [5].

The rapid development of technologies such as internet of things (IoT) and artificial intelligence (AI) including machine learning (ML) and deep learning (DL), has opened up new opportunities in different sectors, including healthcare and medicine [6], transportation [7], energy [8], and smart cities [9]. These innovations hold the potential to enhance the effectiveness of systems across these domains. In the context of waste management, these technologies promise to enhance the efficiency of waste collection and transportation services, which are vital components of municipal cleaning programs [10, 11]. IoT-enabled smart sensors and communication systems allow real-time monitoring of bin fill levels, optimizing collection schedules and preventing inefficiencies [12]. This research focuses on designing and implementing an innovative waste detection and management system for Makkah during the Hajj season. By applying state-of-the-art technologies, we aim to enhance cleanliness, health, and the overall experience of one of the largest religious gatherings in the world. This work aligns with Saudi Arabia's Vision 2030 [13], which focuses on environmental sustainability, a vibrant society, and an innovative economy. While the immediate goal is to improve the Hajj experience, the broader aim is to contribute to realizing Vision 2030. This study contributes to waste management in several keyways:

- Makkah's Context: The system is designed to handle the unpredictable waste during the Hajj season, addressing Makkah's unique needs and supporting Vision 2030's goal of enhancing pilgrimage services.
- Efficient Waste Collection: The system ensures timely collection, improving urban cleanliness and promoting a healthier environment.
- Healthier Environment: By optimizing waste management, the system helps maintain a healthier environment, especially during crowded seasons.
- Harmful Substance Detection: The gas detector sensor improves the system's ability to monitor hazardous gases, enabling quick response to potential risks.

2 Background and Related Works

Recently, there has been a growing focus on research aimed at developing smart waste management systems. Several studies have been published to address this issue by leveraging AI and IoT technologies [11, 12, 14, 15]. These works highlight integrating smart technologies and IoT solutions to enhance waste management efficiency and sustainability in Makkah [15–17] and other cities worldwide [1, 18–20].

One of the studies was done by Kamm and coauthors [21], where they designed a case study to investigate the efficacy of smart waste management systems. They installed sensors in glass bins to measure the fill level of each bin. Their system included an ultrasonic sensor with a real-time clock and a battery capable of scheduling deep-sleep modes. They also used a LoRa chip for long-range data transmission. The software comprised a sensor application, the things network service, and a web application. The study presented the following constraints: city topology, network coverage and penetration, and energy consumption for data transmission. Another study [22] discussed waste collection in smart cities based on IoT. The implementation included hardware such as HC-SR04, an ultrasonic sensor to measure bin fill levels, and a GPS module to track each bin's location. The data from the sensors and GPS modules were communicated via a Raspberry Pi microcontroller to a cloud platform for storage and analysis. The cloud would then signal an Android application. An interesting feature is that after emptying each waste bin, a water pump was activated to clean the bin.

A different study [23] proposed a smart waste management system and demonstrated its effectiveness through a real-life case study. The IoT module's middleware, which included a decision unit, database, and management unit, served as the connecting hub for the My Waste Bin module (which included sensors like the HC-SR04 and devices such as the DHT11 temperature and humidity sensor and GPS), and the My Waste App module. The system was powered by an external rechargeable battery attached to a photovoltaic solar panel. However, the system lacked the ability to calculate the most efficient route for waste collection trucks. Another study in 2022 [17] focused on organic waste management in Saudi Arabia proposed an integrated system involving community education, public participation, policy development, and advanced technology. The system used IoT, ultrasonic sensors, GPS, Arduino, GPRS, and Google Maps. Smart bins sent fill-level data to the management center (MC) via GPRS, which then dispatched the appropriate waste collection truck based on bin location and weight capacity. A smart routing algorithm, the Genetic Algorithm, optimized the collection route. The study also emphasized organic waste as a renewable energy source and advocated for community education and supportive policies.

The last study was conducted by [15], who proposed a smart waste management solution for Makkah to address the rapid accumulation of waste contributing to air and water pollution. The system operates at three levels: city, municipalities, and district management centers. Each smart bin and truck are equipped with ultrasonic sensors, Arduino, GPRS, and GPS, all connected via a cloud server. The IoT-enabled system includes a monitoring app that allows bins to report their status and request collection once 90% full, identifying waste type. Trucks are notified based on location, free space, and waste type. Additionally, the system includes manual collection where the truck driver updates data via the app.

Although significant research has been conducted on smart waste management systems, their full integration into operational settings, particularly in Makkah, is still in progress. Many existing systems lack scalability, real-time responsiveness, and large-scale integration with city management. This highlights the need to move from theoretical solutions to practical deployment. Collaboration with local authorities is vital to adapt these technologies to Makkah's specific challenges, ensuring efficient and sustainable waste collection, and improving environmental and public health outcomes.

3 Methodology

The methodology presents the feasibility study. Then it outlines the approaches for designing and implementing the smart waste management system, including system architecture, IoT and AI techniques, and an administrative portal. The goal is to ensure the system is resilient, efficient, and user-friendly, effectively managing waste during high-demand periods.

4 Feasibility Study

The feasibility study highlights the environmental and sociological importance of the TUHR Smart Waste Management System based on survey data. The survey included over 300 participants, with 57.2% from the population and 42.8% from visitors. Among the respondents, 95.2% agreed that improved waste management during the Hajj season is crucial, and 95.5% supported the implementation of a smart waste management system in Makkah. Survey results also revealed that 87.8% of respondents noticed frequent waste accumulation during the Hajj season, emphasizing the need for timely waste collection. Additionally, 93.2% believed that smart containers could reduce pollution from waste transport truck exhausts. Moreover, 41% of respondents indicated that waste collection should occur twice a day during the Hajj period, with 38.5% supporting three times a day collection.

Among officials at the Grand Mosque and holy sites, 98.7% identified crowding as a major waste management challenge, and 94.9% pointed out the insufficiency of bins and ineffective waste collection methods. These findings support the need for a smart waste management system that can address these challenges.

4.1 Problem Formulation

In this study, the problem is formulated as follows:

- Input: The status of each waste bin, collected via IoT-enabled bins with sensors, and the geographic location of each bin, obtained using GPS.
- Output: A time-specific signal indicating when waste collection is needed and an optimized route for sanitation workers.
- Constraints: Ensure that each full bin is assigned to one worker, minimizing travel distance to improve efficiency and reduce fuel consumption, achieved through AI-powered route optimization with Google Maps.
- Goal: Ensure timely waste collection and optimize workers' routes.

4.2 System Overview

Figure 1 illustrates the workflow of our proposed smart waste management system's architecture, called TUHR. The Flutter mobile application serves as the interface between the users (i.e., workers) and the system server.

The TUHR System workflow can be summarized in the following steps:

1. Data Collection: Each smart bin, equipped with an ESP32 microcontroller and ultrasonic sensor, monitors the fill level and transmits data to a cloud server in real time.
2. Data Processing and Storage: Cloud servers process and store the data received from the smart bins.

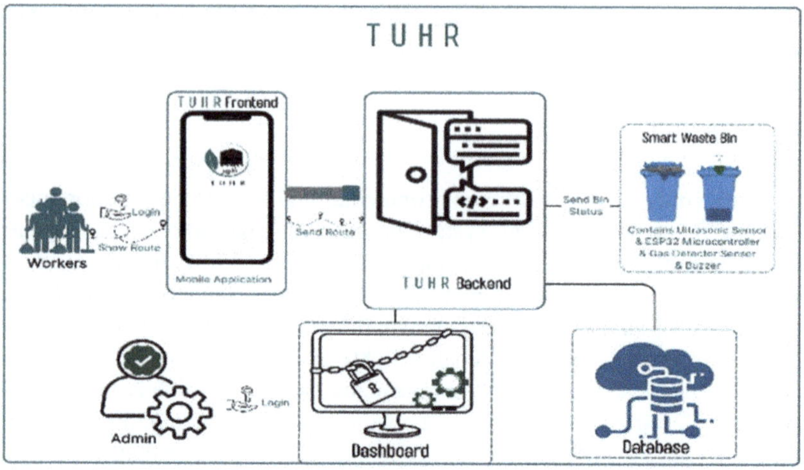

Fig. 1. Workflow of the proposed smart waste system.

3. Full Bin Detection and Alerting: The system continuously checks the bin's fill status and triggers an alert when a bin reaches capacity, initiating waste collection and route planning.
4. Route Optimization: The AI algorithm determines the optimal path for the worker to the bin using Google Maps.
5. Notification Dispatch: Workers receive notifications about the bin's status and location.
6. Bin Status Update: Once collected, the bin's status is updated to empty in real time, ensuring all users have up-to-date information.

5 Experiments and Results

This section discusses the simulation of the TUHR system, explaining the hardware setup and configuration, discusses implementing of TUHR application, building the admin dashboard, and integrating those system components.

5.1 TUHR Software Implementation

The TUHR system consists of two main software components: the first one is TUHR application for workers. The application is designed to cater to the needs of Worker employees by allowing them to modify their personal information and efficiently perform the required tasks. Workers can log into the application using their username and password, which were previously registered in the system database. The second component is administrative dashboard, which allows system administrators to manage the entire system through a centralized web interface. The administrator gains access to the dashboard by entering the designated username and password in the administrator login interface. Once logged in, the initial page that appears is the Dashboard, which contains several sections (see Fig. 2). These sections include My Profile, Manage Users,

Manage Zones, Manage Sensors, View Reads, and Contacts. The administrator has full access to these pages and can navigate them as needed. After completing their tasks, the administrator can log.

5.2 Integration of Hardware and Software Components

Initially, the hardware and software were designed separately based on the specific requirements of the smart waste management system, TUHR. Next, these components were integrated to ensure the smooth and accurate function of the system. This integration process involves several key factors: physically connecting the hardware components within the smart waste containers, enabling efficient and continuous communication, synchronizing timing, and establishing a unified database. The database serves as a central hub, ensuring that both the dashboard and the mobile application receive the same data from the sensors simultaneously. The hardware and software components are integrated to ensure seamless communication and compatibility across devices. Figure 2 illustrates the integration between the system components and how the dashboard and mobile application function simultaneously.

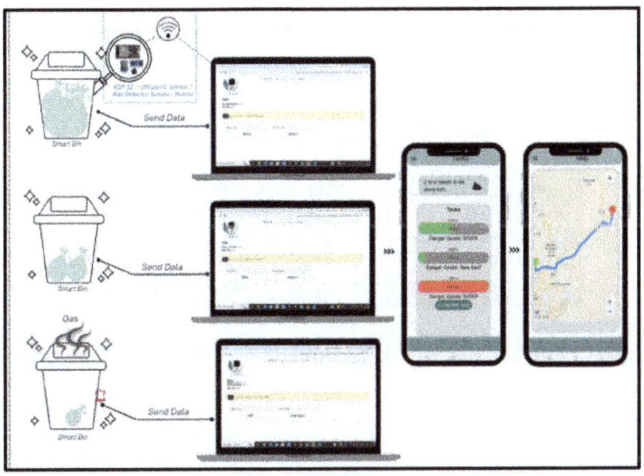

Fig. 2. Devices Integration with Dashboard and TUHR Application.

5.3 Experiment Setup and Procedure

This section outlines a series of experiments conducted to benchmark the system's performance in waste management. It describes the system setup and discusses the results to validate its usability in realistic scenarios.

Initial Step and Configuration: Figure 3 shows the integration of three key components: ultrasonic sensor, gas detector, and buzzer, into the microcontroller on the circuit board. Each component was programmed and tested. The setup involved three medium-sized containers, each equipped with the integrated components.

Fig. 3. Electrical Circuit

Experiments Scenarios and Evaluation We conducted a series of small-scale experiments to perform real-world operations with the TUHR System. First Experiment Scenario (Fig. 4).: To simulate different waste levels (e.g., 0% empty, 50% almost full, and 95% full), each container was filled with varying amounts of waste. This setup allowed us to demonstrate the system's capability to accurately estimate and display waste levels on the application's dashboard. The Second Experiment Scenario (Fig. 5): we conducted a fire experiment to test the functionality of the gas sensor and the buzzer alarm, ensuring that both components activated appropriately in response to simulated hazardous conditions.

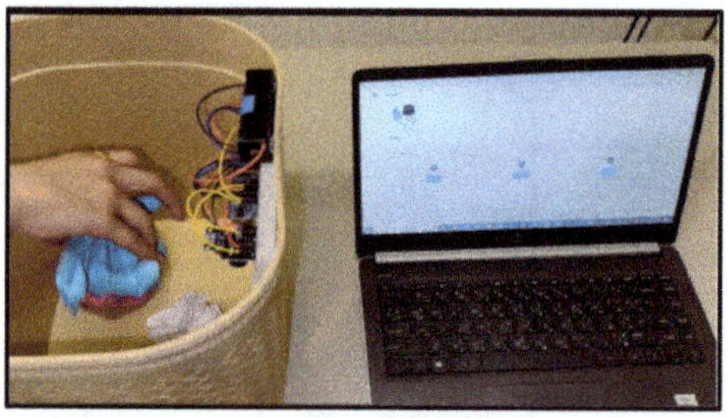

Fig. 4. First Experiment Scenario with administration dashboard showing in the screen.

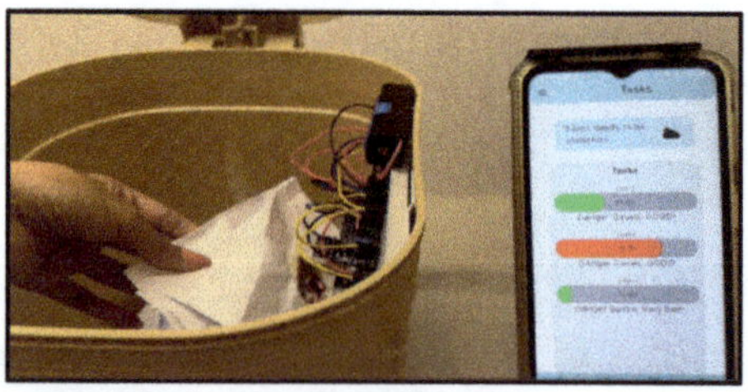

Fig. 5. Second Experiment Scenario of detecting fire, with mobile application notification.

6 Discussion and Conclusion

The TUHR Smart Waste Management System has shown significant promise in enhancing waste management efficiency during high-demand events like the Hajj pilgrimage. The system's use of real-time data transmission and smart sensors allows for better coordination of waste collection activities, which addresses common issues faced by traditional systems, such as premature or delayed pickups. Previous research on smart waste management, also emphasizes the role of IoT in improving waste management efficiency, especially in large-scale urban settings. However, unlike many existing systems that focus on specific applications or geographic areas, TUHR is tailored to the unique conditions of Makkah during the Hajj, a context that has not been widely studied in the existing literature.

One key strength of the TUHR system is its potential to prevent bin overflows, which have been a persistent issue in large-scale events. By leveraging IoT sensors, the system can accurately monitor fill levels and schedule collections in real time. This approach has the potential to significantly reduce inefficiencies, which are common in traditional systems, as highlighted in earlier studies. The integration of AI ML models further contributes to the system's effectiveness by allowing for data-driven decision-making that improves with time and usage.

A key limitation was the difficulty in obtaining real-world data for accurate testing, which impacted the system's evaluation under actual conditions and not fully reflect real-world complexities. The integration of multiple hardware components and ensuring smooth communication among them, especially in dynamic, large-scale environments, posed significant technical challenges. Additionally, the system's reliance on network connectivity, which can fluctuate in crowded settings, limited its overall reliability. Despite these limitations, our system demonstrates potential for improving waste management practices. This involves future work such as managing the hardware integration, refining sensor calibration, and testing within the real operation scenario. Further enhancements in data processing and analytics performance to address large datasets lead to improved and scalable performances. This advancement will go the extra mile toward cleanliness and optimize waste management practices at holy sites.

References

1. Sosunova, I., Porras, J.: IoT-enabled smart waste management systems for smart cities: A systematic review. IEEE Access. **10**, 73326–73363 (2022)
2. Addas, A., Khan, M.N., Naseer, F.: Waste management 2.0 leveraging internet of things for an efficient and eco-friendly smart city solution. PLoS One. **19**(7), e0307608 (2024)
3. Abuga, D., Raghava, N.S.: Real-time smart garbage bin mechanism for solid waste management in smart cities. Sustain. Cities Soc. **75**(103347), 103347 (2021)
4. Shukla, S., Hait, S.: Smart waste management practices in smart cities: Current trends and future perspectives. Adv. Organic Waste Manag. (2022)
5. Acharya, B., Dey, S., Zidan, M.: IoT-Based Smart Waste Management for Environmental Sustainability. CRC Press, London, England (2024)
6. Alshahrani, M., Thafar, M.A., Essack, M.: Application and evaluation of knowledge graph embeddings in biomedical data. PeerJ Comput. Sci. **7**, e341 (2021)

7. Van Hoang, T.: Impact of integrated artificial intelligence and Internet of Things technologies on smart city transformation. J. Tech. Educ. Sci. (2024)
8. Zhu, S., Ota, K., Dong, M.: Green AI for IIoT: Energy efficient intelligent edge computing for industrial internet of things. IEEE Trans. On Green Commun. Netw. **6**(1), 79–88 (2022)
9. Jagatheesaperumal, S.K., Bibri, S.E., Huang, J., Rajapandian, J., Parthiban, B.: Artificial intelligence of things for smart cities: advanced solutions for enhancing transportation safety. Comput. Urban Sci. **4**(1) (2024)
10. Hasan, M.K., Khan, M.A., Issa, G.F., Atta, A., Akram, A.S., Hassan, M.: Smart waste management and classification system for smart cities using deep learning. In: 2022 International Conference on Business Analytics for Technology and Security (ICBATS), pp. 1–7, Dubai, United Arab Emirates (2022)
11. Sallang, N.C.A., Islam, M.T., Islam, M.S., Arshad, H.: A CNN-based smart waste management system using TensorFlow Lite and LoRa-GPS shield in internet of things environment. IEEE Access. **9**, 153560–153574 (2021)
12. Wang, C., Qin, J., Qu, C., Ran, X., Liu, C., Chen, B.: A smart municipal waste management system based on deep-learning and Internet of Things. Waste Manag. **135**, 20–29 (2021)
13. Khoirunnisa, K., Nurhaliza, S.A.: Saudi vision 2030: Economic reforms and sustainable development in the kingdom. **Jurnal Public Policy** (2024)
14. Ghahramani, M., Zhou, M., Molter, A., Pilla, F.: IoT-based route recommendation for an intelligent waste management system. IEEE Internet Things J. **9**(14), 11883–11892 (2022)
15. Abdullah, N., et al.: IoT-based waste management system in formal and informal public areas in mecca. Int. J. Environ. Res. Public Health. **19**(20), 13066 (2022)
16. Elhassan, R.M.: Organic waste sorting proposed system in mecca city. In: 2021 International Conference on Sustainable Energy and Future Electric Transportation (SEFET), pp. 1–6, Hyderabad, India (2021)
17. Abdullah, N., et al.: Integrated approach to achieve a sustainable organic waste management system in Saudi Arabia. Foods. **11** (2022)
18. Wang, B., Farooque, M., Zhong, R.Y., Zhang, A., Liu, Y.: Internet of Things (IoT)-Enabled accountability in source separation of household waste for a circular economy in China. J. Clean. Prod. **300**(126773), 126773 (2021)
19. Cheela, V.R.S., Ranjan, V.P., Goel, S., John, M., Dubey, B.: Pathways to sustainable waste management in Indian Smart Cities. J. Urban Manag. **10**(4), 419–429 (2021)
20. Vishnu, S., et al.: IoT-enabled solid waste management in smart cities. Smart Cities. (2021)
21. Kamm, M., Gau, M., Schneider, J., Vom Brocke, J.: Smart waste collection processes - A case study about smart device implementation. In: Proceedings of the Annual Hawaii International Conference on System Sciences (2020)
22. Ravi, R.V., Dutta, P.K., Lai, W.C.: Internet of Things-based garbage collection for smart cities. IET Conf. Proc. **2022**(26), 412–416 (2023)
23. Pardini, K., Rodrigues, J.J.P.C., Diallo, O., Das, A.K., de Albuquerque, V.H.C., Kozlov, S.A.: A Smart Waste Management Solution Geared towards Citizens. Sensors (Basel). **20**(8) (2020)

A Climate-Aided Methodological Approach for Regenerating Cultural Heritage in Fragile Landscape

Angelo Figliola[1](✉) and Khaled Mohamed[2]

[1] PDTA Department Planning Design Technology of Architecture, Sapienza University of Rome, Rome, RM 00185, Italy
angelo.figliola@uniroma1.it
[2] Via Teano 331, Rome, RM 00185, Italy

Abstract. The notion of building construction, survival, and destruction undeniably intersects with intricate climatic, political, social, and economic agendas. From ancient Mesopotamian cities until the end of time, buildings have not only been generators of meanings and stories but they have also been encapsulating these stories. Therefore, the destruction of architecture, along with being a very unsustainable practice, fuels the loss of the collective spatial memory of an urban settlement and empties it of meanings related to identity and belonging. The paper focuses on the importance of architectural preservation and rehabilitation practice of modernist buildings that emerged from 1920s-1940s during the colonial era of Beirut, Lebanon. Moreover, it discusses how classifications of what constitutes culturally significant sites can aid the process of memory capsulation of such edifices, by using a case study approach of the St. George Hotel & Bay, the first reinforced concrete modern building in Beirut city. The paper will discuss not only the interlinked relationship between modernist heritage rehabilitation and Climate-aided Design (CADe) but also how data-driven environmental design strategies can mitigate and adapt to climate change risks by utilizing advanced digital prototyping tools and regenerative design principles to enhance anti-fragile and resilient actions in Mediterranean coastal cities like Beirut.

Keywords: Urban Regeneration · Climate-aided Design · Cultural Heritage

1 Introduction

Demolishing buildings and cities, especially after crisis scenarios, can pave the way for a better future or erase certain traumatic stories. However, it is also the easiest and most effective way to obliterate the collective memory and the physical identity of the built environment. At the same time, building construction and operation is responsible for around 40 to 50% of the total energy consumption worldwide. According to the Intergovernmental Panel on Climate Change (IPCC) in their fourth Assessment Report (AR4), the increase in the atmospheric concentration of carbon dioxide (CO2), methane (CH4), and nitrous oxide (N2O) happened during the last two centuries because of human activities exceeding any precedent pre-industrial numbers [1].

D. Berkouk et al. (Eds.): CCTS 2024, PTT, pp. 66–74, 2026.
https://10.1007/978-981-95-7607-4_8

The paradigm of how to approach and conceptualize an urban intervention has been the central focus of architects, designers, and urbanists throughout the centuries. These explorations often investigate the potential for building a better future, regardless of the level of uncertainty of this future. But what if the notion of building, as in the sense of construction, becomes at the expense of destroying more buildings, especially ones that express historical and symbolic significance? According to the United Nations (UN), around 68% of the world's population will live in cities and urban areas by 2050 [2]. At the same time, new land use is becoming more disruptive to natural habitats and ecosystems, and in a future where more urbanization will be trendier, the destruction of culturally notable structures is becoming more evident. This presses the urge to reconsider not only the climatic and social challenges facing the practice of building in cities today but also to prompt mitigation and adaptation strategies to what seems to be an unavoidable climatic catastrophe.

The IPCC report concludes with the urgency to encourage integrated design solutions for buildings, which is employing both passive and active techniques in monitoring and optimizing thermal comfort, ventilation, light distribution, and visual comfort to reduce energy requirements [3]. This raises an important statement of how environmental design can be utilized as a survival tool in urban settlements for a more sustainable future, and how classification, preservation, and adaptive reuse of existing cumulative heritage buildings can be a catalyst for urban regeneration processes.

The focus of the study draws into Beirut city, Lebanon, a coastal city on the east of the Mediterranean basin, known historically for its urban complexity and diverse cultural identity. Also, Beirut is part of the Eastern Mediterranean and the Middle East (EMME) climatic zone [4], which is considered a climate change hotspot as it is highly vulnerable to shifts in both extreme and average climate conditions. This is because the EMME region is warming up twice as fast as the global average and the implications of this climatic change are already evident today in the form of intensified droughts, heatwaves, and other extreme weather events [5].

Beirut is a city with a rich and layered history, stretching back to the fifteenth century BCE. Over time, it evolved from a Phoenician port into a Roman colony, a medieval village, an Ottoman city, and later a French protectorate. The city's transformation began during the late 19th century, under Ottoman rule, with a development boom that marked the start of continuous urban change. After World War I and the fall of the Ottoman Empire, Lebanon became part of the French Mandate in 1920. This sparked the first major wave of destruction in the city, as Ottoman-era buildings were replaced by a Western "Parisian" style of urban planning in what is now known as the "Haussmannization of Beirut". More waves of destruction happened since the independence, the Lebanese civil war (1975–1990), after the war of 2006 and 2024, leaving the city in a constant state of destruction and rebuilding [6].

The chosen case is the first modern building in Beirut that has used reinforced concrete for its construction. Designed and built in 1929 during the French mandate by the Parisian architect Auguste Perret and Lebanese architect Antoine Tabet. Located on the waterfront in Beirut's downtown core, it was claimed during the 1960s as one of the best seven hotels globally. However, since the civil war, it has struggled with any form of utility and was threatened multiple times with demolition because it is not listed as part of Beirut's cultural heritage from the colonial era that is noteworthy to be saved [7]. With every wave of destruction, Beirut is subject to becoming more climatically fragile to extreme weather as well as it loses part of its cultural identity, especially the downtown of the city. Then with every wave of reconstruction, political and economic private stakes were prioritized over anti-fragile and constructive measures [8].

Therefore, the paper aims to explore how significant landmarks like the St. George Hotel, which embodies part of the city's colonial, post-independence, and post-civil war spatial memory, can become catalysts for regenerating urban landscapes. Also, to highlight the potential of Climate-Aided Design (CADe) to assist early design stages. The hypothesis is that utilizing CADe as a main driver to preserve and rehabilitate underutilized structures can be an approach to reduce the environmental impact of high carbon footprint, potential waste of energy, material resources, and maintenance by leveraging the embodied energy of the existing setting. Moreover, enhances the site's climatic performance and resilience to future environmental threats which could ultimately contribute to its preservation. The aim is to create a framework that integrates environmental de-sign practices from digital prototyping, multi-scalar environmental analysis, future scenario forecasts, and CADe to aid preservation efforts of French colonial architecture experiments during (1920–1940) in cities of the Levant, exemplified by the St. George Hotel in Beirut.

2 Methodology

The methodology embraced for this case included, first a deep historical and theoretical review that sets a foundation for allocating the significance of St. George Hotel to the spatial narrative of a city like Beirut. Second, a multi-scalar risk assessment of the current state-of-art condition and future scenario forecast of Beirut's downtown area. This multi-scalar approach (see Fig. 1) from an urban scale to the building scale would represent the site's relationship with its context and identify multi-risk factors of both the building and the downtown area. The relationship between the St. George Hotel and the downtown area was primarily studied through the writings of Lebanese historians and architectural theorists, who laid down a critical narration of the interaction between national identity conservation and the practices of destruction and expansion of the city's colonial and post-colonial periods. These include Robert Saliba, an urban historian and a Professor of urban history at the American University of Beirut, who tackled the conservation practices of colonial heritage in Beirut's central and pericentral districts since independence. Saree Makdisi, a Professor of comparative literature and colonial studies at UCLA, discussed the post-civil war aftermath of Beirut's downtown. Assem Salaam [9], a renowned Lebanese Architect and academic, criticized post-war reconstruction plans carried out during the 1990's.

On the other hand, the methodology to analyse and forecast current and future environmental and urban risks to downtown Beirut and the St. George Hotel embodied an integrated workflow starting from digital documentation from GIS data primarily obtained by Beirut Urban Lab [10], which is a collaborative, interdisciplinary research lab organised by the American University of Beirut. In addition, a collection of GIS and historical image sources like the Google Earth Engine, Arab Centre Archives (ACA) and site inspections between 2019 and 2022 was used. All maps, plans, sections, images, and Geo-referenced 3D models, obtained from open street maps, helped the digital prototyping process using Rhinoceros 3D.

Regarding current climate conditions and environmental analysis, a few key performance indicators (KPIs) were considered regarding urban and building scale levels. A complementary effort between the Autodesk Forma platform, Climate Studio plugin, and Grasshopper Ladybug plugin was utilized. Forma provided a powerful insight into running heavy urban scale simulations including sun hour analysis, wind speed and flow, comfort study, and microclimate analysis that combines thermal, wind, and shade information to visualize areas of optimal comfort within the urban context taking into account buildings orientation and shadow cast by the surrounding context. Climate Studio was useful at a site area and building-scale level where higher resolution of details is preferred to simulate primarily incident radiation analysis of the facades and roofs and the ground plane. In addition, it is effective to show an accurate perception of indoor space, direct and indirect daylight availability, direct and indirect solar radiation and glare study. Regarding the future scenario forecast, certain KPIs needed to be taken into consideration to foresee and visualize the symptoms of climate change on urban and building scale levels respectively.

This includes risks like temperature rise, the effect of the urban heat island effect (UHI), and sea level rise throughout 5, 25, and 50-year intervals. So, EPW (EnergyPlus Weather) forecast weather files, site-specific to Beirut were obtained using Meteonorm 8.0, including air temperature, relative humidity, global solar radiation, wind speed and direction, etc. This was done following Representative Concentration Pathway (RCP) 4.5 for moderate emissions scenario for 2030, 2050 and 2080. The obtained data was then plugged into a grasshopper ladybug tool to eventually visualize and compare the results of incident radiation. Ladybug enables you to view thorough climate data through interactive graphs and spatial mapping, as well as import climate files in EPW format.

Concurrently, forecast climatic data of Lebanon concerning thermal comfort, number of hot days, rise of sea level, etc. were collected from several sources, including the World Climate Research Program, the Coastal Risk Screening Tool developed by Climate Central, and Lebanon's climate fact report published by the Lebanese Red cross. These were analyzed in comparison to psychrometric charts of the key years generated by Climate Studio. Finally, sea level rise was simulated and visualized using Grasshopper Kangaroo plugin based on the data gathered from Climate Central and the Lebanese Red Cross.

3 Result

Urban landscapes in politically and economically hazardous regions like Beirut face extreme challenges to mitigate and adapt to future climate change scenarios compounded by the long-lasting struggle with the lack of adequate infrastructure, poverty, high population density, and conflicts. Eventually, they are also burdened with future scenarios of an environmental crisis. Therefore, traditional evaluation and intervention approaches must consider context, including infrastructure adequacy, socio-economic conditions, and local methods and materials.

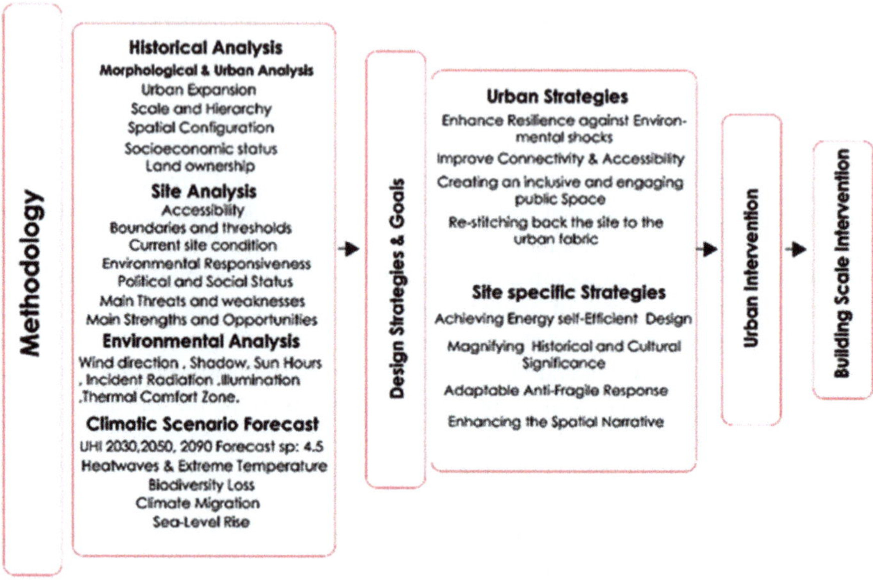

Fig. 1. Methodological Approach

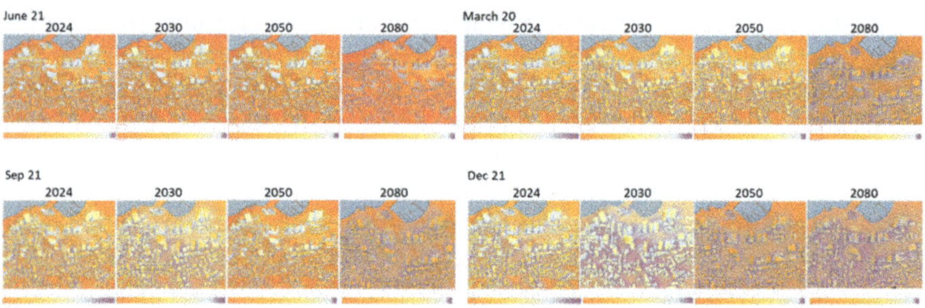

Fig. 2. Radiation Analysis and forecast of Beirut's CBD area 2024,2030,2050,2080 from left to right respectively. Top left June 21. Top right September 21. Bottom left March 20. Bottom right December 21

Since the civil-war ended (1990) The Saint George Hotel stood at the edge of Beirut's empty down. The empty promises of prosperity following the war have transformed the Central Business District (CBD) into a ghost town, alienating its connection to everyday Lebanese life. This is because 80% of buildings in the downtown have been either damaged beyond repair or demolished, two-thirds of which have been wiped out in the name of reconstruction and modernity. The area has been reserved for multi-millionaire real estate developments and seasonal tourism, pushing the majority of Beirut's residents away from its centre. The sign "STOP SOLIDERE" is a striking reminder of the ongoing battle to preserve Beirut's identity. The hotel embodies the struggle to protect the city's history from erasure, serving as a memory incubator in a city that has lost significant parts of its spatial narrative. This deeply rooted legacy raises an immediate need to visualize the future of the St. George Bay and Hotel.

From an environmental perspective, Beirut lacks adequate blue and green infrastructure, endures increasingly extreme summer hot waves, and has already experienced an annual mean temperature increase of 0.3 °C per decade since 1970, which is twice the global average. This phenomenon is forecasted to worsen even more decade after decade. This can be seen from the predictions provided by World Climate Change Program reports on Lebanon and the radiation forecast analysis carried out for the key years. Each period (2030, 2050, 2090) shows a steady increase in solar radiation levels across the entire urban landscape of the downtown, contributing to a worsening UHI effect (see Fig. 2). June 21 exhibits the highest incident radiation, with values rising to 7.5 kWh/m^2 by 2090, compared to 7.1 kWh/m^2 today. Moreover, the number of very hot days (>35 °C) will reach 11–21 days annually by 2050 and more than 25 days by 2090. Under RCP 4.5, the moderate scenario applied, mean temperature will rise by 3.1 °C, intensifying heat stress in the city.

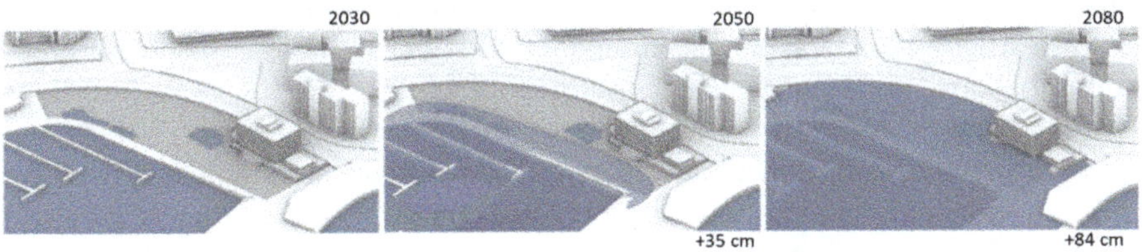

Fig. 3. Sea level rise simulation for 2030 (left), 2050 (middle), 2080 (right)

The psychrometric chart forecast analysis demonstrates an overall shift towards warmer conditions and higher humidity levels by 2090 (with ratios above 0.025 kg water/kg air). This by default decreases the number of hours that fall within both indoor and outdoor thermal comfort zones, causing more probability of extreme heat and making it harder to maintain comfort using passive strategies. According to the Lebanese Red Cross, it is projected that by 2050, the sea level in the coastal cities of Lebanon, where 90% of the population lives, will be 30–60 cm higher. Using this rate, 20 mm per year, it is expected that by 2100 sea level will rise 80–100 cm (see Fig. 3). This would not only increase the risk of coastal flooding but also cause saltwater intrusion into coastal aquifers. Concerning the site, since its construction the building morphed from a two-storey structure to triple in size to 6 storeys today.

This happened due to several reasons ranging from economic boom and expansion to war battles and conflict that have resulted in a series of restoration and reconstructions. As a result, the climatic performance of buildings was highly affected by fast and unstudied restorations. The original structure used to embed passive shading and cooling features to reach thermal comfort like wind towers, a central atrium, overhanging balconies, breeze walls etc. that were highly contextual and responsive to the climate at the time. Many of these aspects were lost and the building today finds itself again struggling to perform optimally.

4 Design Solution

The intervention aimed to represent a balanced relationship between memory preservation and addressing the pressing need to mitigate and adapt against upcoming climatic risks. This is conveyed by revitalizing the connection between the waterfront, the city, and the site as a continuously activated public space stitching it back to the city's fabric. This was done while simultaneously incorporating measures to mitigate the Urban Heat Island (UHI) effect, reaching net energy efficiency, safeguarding against sea level rise, and enhancing indoor and outdoor thermal comfort.

As a result, actions on the site involved considering Nature-Based Solutions (NBS) to transform the under-utilized outdoor area into a floodable public space that features rain gardens and water pools carefully zoned in high-radiation zones to enhance evaporative cooling. Native, drought-tolerant, and endangered plant species like Juniperus Drupacea, Quercus Calliprinos, Thymus Libanotius, Crataegus Azarolus, Cyclamen Libanoticum, Alkanna Maleolens Bornm, were deployed to reduce water consumption, combat biodiversity loss, and create shade. Also, the building's basement was re-purposed as an extension of the urban park, blurring the harsh definition between public versus private areas. This aspect is apparent throughout most of the project to achieve a better relationship between the building and its surroundings.

Moreover, at the building scale, the intervention broadens the concept of the public space by visualizing multifunctional layers stacked vertically, combining environmental performance with public accessibility. The design prioritizes two main goals: first, referring back to the original design's passive cooling strategies. Second, pulling out learnings from the local vernacular architecture, in terms of materiality, organization, and climatic responsiveness. Therefore, one of the main actions was to allow sunlight to penetrate the basement to create a chimney effect that ensures passive natural ventilation throughout the structure. This approach allows sufficient wind circulation from the basement to the roof, promoting indoor air quality and thermal comfort. A pool positioned in the basement helps in cooling hot air during the summer, thereby maximizing the building's environmental resilience. By integrating these decisions, the design combines public and private functions, optimizing usability while enhancing a sustainable and climate-responsive architectural solution.

5 Discussion

The study on the St. George Hotel employed both qualitative and quantitative methods to investigate the complementary relationship between a theoretical framework and applied research. The proposal examined a detailed assessment of Beirut's vulnerability to climate change and its potential of such under-recognized sites to be a catalyst of urban regeneration processes rather than silently wait for its dismissal.

Today, the urgency to save St. George as it is one of the few colonial heritage buildings that stand today is undeniable. Nevertheless, why does it remain unrecognized for conservation and who benefits from the current state of constant erasure of heritage combined with an urban governance framework? Moreover, While the results stress the necessity to support adaptive design intervention using advanced modeling techniques and CADe on fragile landscapes to combat future risks scenarios and aid the preservation process, it may still be a niche approach for various reasons. For instance, the economic and political atmosphere combined with the lack of the institutional agenda may be a barrier to application.

While this approach provides an essential perspective to apply CADe and assist the evolution in the field of climate risk assessment in rehabilitation practices, it is essential to acknowledge restrictions such as the reliance on limited climate models and emission scenarios that may include some inaccuracies. For instance, although climate models and projections offer valuable insight into site-specific conditions, they may not consider all microclimatic conditions. Also, the limitation of the modelling approach using multiple data sources to create close approximations. In addition, in a rapidly changing urban setting, further alterations of context structures, evolution, orientation, heights and materiality may affect the building's performance and responsiveness to adapt to future threats.

6 Conclusion

The case of the St. George Hotel re-examines the essence of what creates continuous multi-layered cities in the first place, which is the adaptability and resilience of its layers to survive the test of time, natural disasters, conflicts, and man-made destruction. The main goal of this experiment was to showcase the potential of mandating and recognizing Beirut's modernist legacy as part of the collective heritage of the city. It is time to adopt climate-resilient preservation practices as part of Lebanon's heritage conservation policy. Also, the proposal points to a framework to stimulate anti-fragile generative measures in fragile landscapes applied to other coastal cities of the Mediterranean and encourage urban policies that embrace CADe as a tool for urban regeneration practices.

References

1. Altomonte, S.: Climate change and architecture: Mitigation and adaptation strategies for a sustainable development. Sustain. Dev. **1**(1), 97–110 (2008)
2. United Nations, Department of Economic and Social Affairs, https://www.un.org/development/desa/en/news/population/2018-revision-of-world-urbanization-prospects.html (2025). Accessed 24 Feb 2025
3. IPCC: Climate change 2007: Mitigation. In: Contribution of Working Group III to the Fourth Assessment Report of the Intergovernmental Panel on Climate Change. Cambridge University Press, Cambridge, United Kingdom and New York, NY, USA (2007)
4. Lazoglou, G., et al.: Identification of climate change hotspots in the Mediterranean. Sci. Reports. **14**(1), 29817 (2024)
5. Red Cross Red Crescent Climate Centre (RCCC)., https://www.climatecentre.org/wp-content/uploads/RCCC-Country-profiles-Lebanon_2022-Final.pdf (2025). Accessed 24 Feb 2025
6. Makdisi, S.: Laying claim to Beirut: Urban narrative and spatial identity in the age of solidere. Critical Inquiry. **23**(3), 660–705 (1997)
7. Saliba, R.: Historicizing early modernity—Decolonizing heritage: Conservation design strategies in postwar Beirut. Traditional Dwellings and Settlements Rev. **25**(1), 7–24 (2013)

8. Makdisi, S.: Reconstructing history in central Beirut. Middle East Report. **203**, 23–25 (1997)
9. Salaam, A.: The reconstruction of Beirut: A lost opportunity. AA Files. **27**, 11–13 (1994)
10. Beirut Urban Lab, American University of Beirut, https://beirut-built-environment-database-bul-aub.hub.arcgis.com/maps/964194fbd53c42b197e2894497f74e3e (2025). Accessed 24 Feb 2025

An Analysis of Different Types of Gum Used in Concrete Applications: A Review

Alaa M.E.A. Saleem[1](✉), Tareg Abdalla Abdalla[2], and Abdullah Albogami[3]

[1] Department of Civil Engineering, Faculty of Engineering, University of Khartoum, Khartoum, Sudan
alaa.bow90@gmail.com

[2] Department of Civil Engineering, Faculty of Engineering Sciences, Omdurman Islamic University, Omdurman, Sudan

[3] Department of Civil Engineering, Faculty of Engineering, Al-Baha University, Al-Baha 65779, Saudi Arabia

Abstract. Given that gums are eco-friendly materials that enhance concrete's behavior, much research has recently been conducted on their use in concrete production and their influence on the properties of concrete. This review paper discusses the possible uses of different kinds of gums as environmentally friendly substitutes for traditional building materials. The properties, environmental advantages, and uses of several natural gums—such as Gum Arabic, Diutan, Welan, and others—are reviewed in concrete improvement. The results of several studies show how effective these gums are at enhancing the workability of concrete mixtures, water reducers, viscosity-modifying additives, and mechanical properties of concrete. The encouraging findings highlight the importance of using natural gums in building methods to promote ecological sustainability and economic growth, while more research is required.

Keywords: Gum · Concrete · Biopolymer · Workability · Compressive Strength

1 Introduction

Cement is the most widely used binder in construction. About 5% of greenhouse gases worldwide come from cement manufacturing, which produces one ton of cement, one ton of CO_2 is released [1, 2]. As a result, there is a need to limit the use of cement, especially given its high cost compared to natural materials [3, 4].

Considering the size and significance of the building sector, as previously mentioned, this paper tries to investigate the usage of gums as natural substitutes in the construction applications industry. Gums are affordable compared to other materials, accessible locally, and environmentally friendly. They also play a significant role in the growth of the construction industry, which consequently encourages economic growth.

Trees and shrubs release natural gums in tears, which the sun dries to produce a variety of colored, hard, crystal exudates. For example, Arabic gum is bright amber to white, tragacanth gum is white to rich brown, and karaya gum is light gray to dark brown.

D. Berkouk et al. (Eds.): CCTS 2024, PTT, pp. 75–81, 2026.
https://10.1007/978-981-95-7607-4_9

High temperatures and little moisture promote the growth of gum, and shrubs and trees may produce more if their bark is trimmed or removed. For many years, exudate gums have been utilized in food preparation for stability, thickening, and emulsification, as well as in non-food applications. In wounded trees, exudates are fluids that flow from wounds and solidify when exposed to air. Tapping (removing wood or bark to induce exudation) or unplanned or organic reasons (insect, animal, or viral attack, dryness, or wind damage) can cause injury. Although the physical and chemical characteristics of gums vary greatly, they can all be easily distinguished from balsams, resins, oleoresins, and other rubbery compounds. Gums are soluble in either benzene, chloroform, fixed oils, and other liquids that dissolve resins, but they are miscible in water. Naturally occurring gums can significantly raise the viscosity of a solution, frequently even at low concentrations [5].

2 Methodology

A comprehensive methodology is used to investigate the use of gum in construction. Various publications supplied information, including Elsevier, Springer, MDPI, Taylor & Francis, Wiley, ResearchGate, Hindawi, ScienceDirect, and other journals. The collected data was validated by Scimago Journal, country rank (Sjr). The review primarily mentioned different types of gums and their uses in construction, especially as concrete additives.

3 Results and Discussions

Gum is applied to enhance the physical properties of building materials, improve environmental resilience, and support sustainable building practices. This section discovers various construction applications of gum, summarized as follows:

3.1 Effect of Gum on General Properties of Cement and Concrete

Many researchers have investigated gums' effect on cement and concrete's properties. For instance, M.K.Rustum and K.M.Oweed [6] investigated the effects of Arabic gum (AG) using several amounts of Arabic gum (0.1%,0.3%,0.5%,0.7%, and 0.9%). The sample mortar was prepared by weighing out a 1:2 cement sand ratio and adding 0.5 water to the binder. Compressive, flexural, and workability tests were performed on each of the groups (G1, G2, G3, G4, and G5), as well as a reference group G0 that was devoid of Arabic gum. The experimental results about workability, flexural strength, and compressive strength (see Fig. 1 a, b, c) show that the use of AG in regular cement mortar at an appropriate replacement ratio improved the mortar's workability and flexural and compressive strengths; group G3 (which contained 0.5% AG) was noted as having the best results in mechanical tests when compared to other groups; the addition of AG improved the workability of ordinary cement mortar.

Neem gum was added to concrete mixtures at different weight percentages of cement content (0.1%, 0.2%, 0.4%, 0.6%, 0.8%, 1.0%, 1.2%, & 1.5%), as studied by M.G.L.

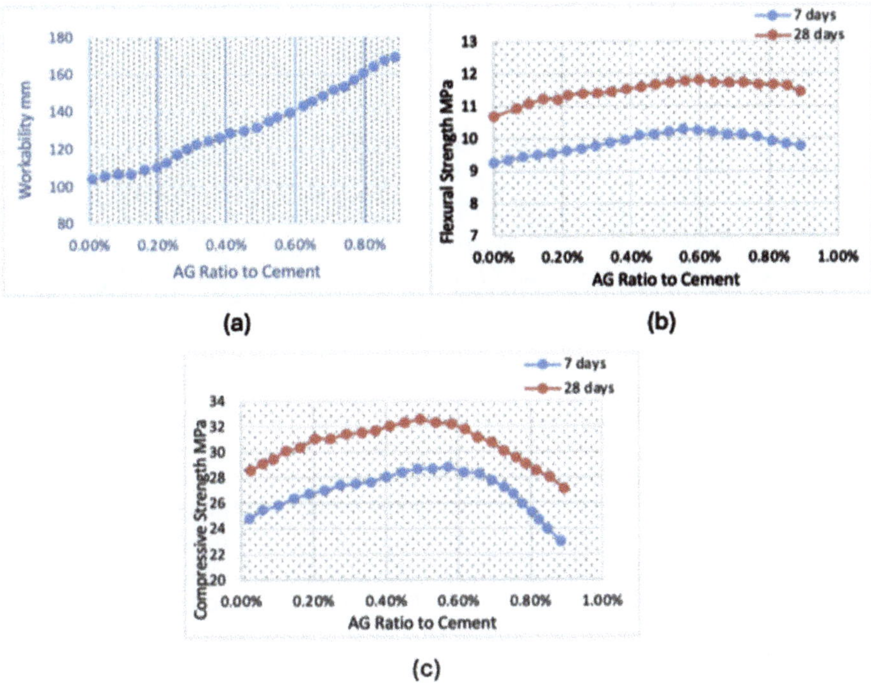

Fig. 1. Result of a) Workability, b) Flexural Strength, and c) Compressive Strength [6].

Annamalai et al. [7] The outcome demonstrated there is a noticeable difference in the concrete when neem gum is added to the mixtures. When the compressive strength was tested at 7, 21, and 28 days, it was discovered that emu gum delivers compressive strength that is noticeably higher than regular concrete, up to 1.2%. After that, it slightly falls as emu gum content in concrete mixtures increases. Concrete, with excellent workability and great compressive strength, was produced at all ratios.

To evaluate how AGB affects concrete densities and depth of carbonation and determine which percentages produce the best outcomes for each property, A. M. Mohamed et al. [8] conducted several tests of concrete samples with different AGB ratios. According to the test results, the AGB mortar's fluidity and initial setting time increased steadily as its AGB content increased, reaching an ideal 0.9% AGB weight fraction. When the quantities of AGB are increased between 0% and 0.9%, the density somewhat drops. Compared to OPC cement, the XRD findings analysis of AGB cement revealed the same hydration pattern. Increasing the amount of AGB to 0.9 results in a maximum relative improvement in carbonation of about 11%. AGB concrete's scanning electron microscopy SEM, as shown in Fig. 2, reveals fewer voids scattered throughout the microstructure of the AGB mix.

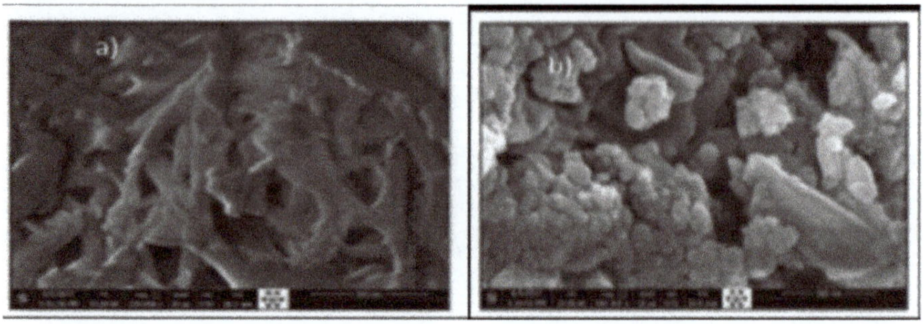

Fig. 2. SEM Structure of a) Pure Cement Material b) AGB Cement Mixes [8].

3.2 Effect of Gum as Admixture

In the building industry, AGB can be used as an addition to reduce water, which lowers the need for chemical admixtures. Developing new markets for goods would impact the economy and society [8]. Under laboratory conditions, A. U. Elinwa et al. [9] assessed the impact effects of the gum Arabic additive on the concrete's mechanical characteristics using doses ranging from 0.00% to 1.00% wt % by cement and cured for three to ninety days. Including gum Arabic, GA enhanced the qualities of gum Arabic concrete GAC and changed the GA components into minerals that affected the concrete's strength. As the dosage of GA grew, GAC's water absorption and porosity increased as well; therefore, using a water-reducing admixture with GA might be necessary. GAC has the density of regular-weight concrete. As the GA dosage was raised, GAC's compressive strength also rose. For typical concrete projects, a dosage of 0.50% to 0.75% is advised; 0.50% is ideal, as shown in Fig. 3. Linear, Natural Log, and Square Root models can all be used to represent the data produced by the studies.

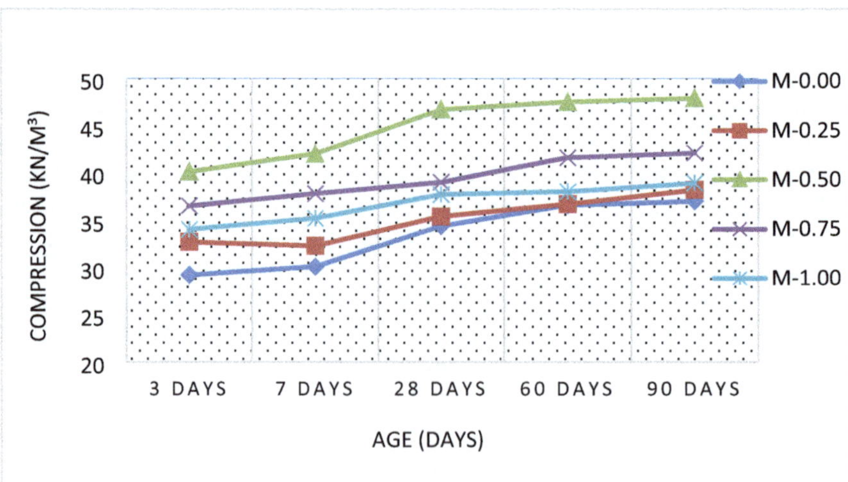

Fig. 3. Cube Compressive Strength of GA-OPC Concrete [9]

Based on the Bs EN 934–2:2009 Table (12), Y.H. Ahmed et al. [10] conclude that Acacia seyal, Taleh Gum (TG), at a dosage level of 0.8% by cement weight, has been demonstrated to be a dual-function concrete admixture, namely a setting accelerator (SA) and normal water reducer (WR). Without compromising compared to a reference mix of concrete, it lowers water content by 11.5%.

A study by R. Mbugua et al. [11] investigated using Acacia karroo (GAK) gum as a concrete water-reducing additive. Compressive strength, density, and slump tests were conducted using varying doses of GAK, with neat concrete serving as the control. According to the results, a 2% intake of GAK caused a 200% rise in slump. For samples with a 3% dosage, the water-to-binder (w/b) ratio was reduced from 0.61 to 0.48. During 180 days of curing, a dosage of 3% w/b resulted in a 37.03% improvement in compressive strength over the control. GAK in concrete also resulted in a slower rate of hydration, according to XRD measurements. It was determined that GAK may be added to the concrete as an environmentally friendly water-reducing additive, resulting in more sustainable and greener concrete.

3.3 Effect of Gum as Viscosity Modifying Agent (VMA)

Studies have investigated the effect on the effect of gum as a viscosity-modifying agent (VMA). For example, M. Sonebi [12] evaluated how the dosage of Diutan gum, a viscosity-modifying material of the second generation, affected the rheological and fluidity characteristics. Compared to Welan gum, cement-based grout. All grout was created using a water-to-binder ratio (W/B) of 0.40. In comparison to grouts manufactured using Diutan gum at 0.02, 0.04, 0.06, and 0.08% by binder weight, the new properties of control grouts made with superplasticizer (SP) and without any viscosity modifying agent (VMA) were examined. Comparable mixtures using Diutan gums and Welan gum were examined. Using Welan gum, comparable control grouts and mixes with different PFA doses were made to compare fluidity and rheological characteristics to the previous grouts made using Diutan gum. Compared to grouts containing Welan gum, the improvement in yield value of Diutan gum grouts was more significant when PFA replaced a larger percentage of the cement. The more PFA that is substituted with Welan gum grout, the lower the yield value. In contrast to welan gum, adding PFA to Diutan gum grout resulted in a minor decrease in yield value.

Y. H. Ahmed et al. [13] studied Gum Arabic (GA) as a viscosity modifying admixture (VMA) to reduce the particle content in SCC, particularly Acacia Senegal (known locally as Hashab Gum). The study's approach is predicated on BS EN 480–15 and BS EN 934–2, which provide a framework for categorizing GA as the VMA. Three SCC test mixes are experimental with varying cement contents (400, 370, and 350 kg) as powder. GA mixes when immersed at a 30% water weight concentration, and then a dose of 0.2% by weight of cement is administered. The ratio of cement to water (W/C) for all mixes remains at 0.45. The findings of the study demonstrated that the 0.2% GA dosage is suitable and satisfies every requirement for VMA specified via British Standard EN 934–2: 2009. SCC powder uses GA as a VMA of 0.2%, reducing its weight from 400 kg to 350 kg while retaining its freshness.

4 Conclusion

The building industry, which depends on conventional materials like cement, is regarded as a major sector supporting national economies. However, the manufacture of cement releases CO_2 gas, which harms the environment. Consequently, it is necessary to search for more ecologically friendly substitutes. In addition to their capacity to enhance the qualities of concrete as needed, this study focuses on gums as natural, regional, and eco-friendly materials. The following conclusion can be drawn from the study:

- The use of Arabic gum and Neem gum improves the properties of fresh and hardened concrete and workability and increases the flexural, tensile, and compressive strength of the concrete.
- Using Arabic gum, GAK, and Welan gum as water reducer admixture for concrete. In addition, Taleh Gum (TG) and Nilotica gum can be used to set accelerator admixture.
- Using Arabic gum as VMA in SCC production and self-healing concrete. Diutan and Welan gum can also be used as VMA.

It is clear from the preceding that using Gums as a building material has several advantages; nevertheless, further research is required to understand the various types and their long-term impacts fully. As a result, they will expand quickly and become a promising material in the construction industry.

Declaration of Interest Statement The authors confirm that no conflicts of interest are associated with this work.

References

1. Chang, I., Im, J., Cho, G.C.: Introduction of microbial biopolymers in soil treatment for future environmentally-friendly and sustainable geotechnical engineering. Sustainability (Switzerland). **8**(3), 251 (2016). https://doi.org/10.3390/su8030251
2. Abdalla, T.A., Omer, E.I.I., Khalifa, N.M.M., Albasheir, M.A.A., Abdulrahman, H.M.A., Abdown, A.G.M.: Effect of copper slag on workability and compressive strength of concrete. Seybold Report. **18**(July), 2657–2669 (2023). https://doi.org/10.5281/zenodo.8241087
3. Chang, I., Jeon, M., Cho, G.C.: Application of microbial biopolymers as an alternative construction binder for earth buildings in underdeveloped countries. Int J Polym Sci. **2015**, 326745 (2015). https://doi.org/10.1155/2015/326745
4. Rimbarngaye, A., Mwero, J.N., Ronoh, E.K.: Performance evaluation of compressed laterite blocks stabilised with cement and gum Arabic. Int. J. Adv. Technol. Eng. Explor. **8**(83), 1268–1278 (2021). https://doi.org/10.19101/IJATEE.2021.874536
5. Nussinovitch, A.: Plant gum exudates of the world: Sources, distribution, properties, and applications. CRC Press Taylor & Francis Group, Boca Raton, FL (2009)
6. Rustum, M.K., Oweed, K.M.: Investigation the effect of arabic gum on the physical and mechanical properties of ordinary cement mortar. J. Eng. Sustain. Dev. **24, no. Special**, 417–424 (2020). https://doi.org/10.31272/jeasd.conf.1.46
7. Annaamalai, M.G.L., Maheswaran, G., Yuvaraja, R., Jayakodi, R.: Effect of partial replacement of cement with neem gum on the strength characteristics of high performance concrete. Int. J. ChemTech Res. **8**(1), 178–183 (2015)

8. Mohamed, A.M., Osman, M.H., Smaoui, H., Mohd Ariffin, M.A.: Durability and microstructure properties of concrete with Arabic gum biopolymer admixture. Adv. Civ. Eng. **2018**, 1962832 (2018). https://doi.org/10.1155/2018/1962832
9. Elinwa, A.U., Abdulbasir, G., Abdulkadir, G.: Gum Arabic as an admixture for cement concrete production. Constr. Build. Mater. **176**, 201–212 (2018). https://doi.org/10.1016/j.conbuildmat.2018.04.160
10. Y.H. Ahmed, S. Abdullatief, and A. Satti, "USE OF GUM ARABIC (Acacia seyal) as concrete admixture," 2018. [Online]. https://www.researchgate.net/publication/330754045
11. Mbugua, R., Salim, R., Ndambuki, J.: Effect of gum arabic karroo as awater-reducing admixture in concrete. Materials. **9**(2), 80 (2016). https://doi.org/10.3390/ma9020080
12. Sonebi, M.: Rheological properties of grouts with viscosity modifying agents as diutan gum and welan gum incorporating pulverised fly ash. Cem. Concr. Res. **36**(9), 1609–1618 (2006). https://doi.org/10.1016/j.cemconres.2006.05.016
13. Y. Hummaida Ahmed, M. Ibrahim Rahamtalla, and K. Salah Eldin, Characterization of gum arabic as viscosity modifying agent (VMA) for producing self-compacting concrete (SCC)

Thermal Evaluation of Wax-Expanded Polystyrene Composite in Concrete Blocks

Aisha Hassan, Ethar Alawad, Shorouq Alsirhani, Haitham Rashed, Hassan A. Abas, and Hussain Alsadiq(✉)

Prince Mugrin University, Madinah, Saudi Arabia
H.alsadiq@upm.edu.sa

Abstract. Enhancing thermal properties of concrete blocks in hot climates is crucial for energy efficiency. In this study, the thermal properties of wax-expanded polystyrene (EPS) composite insulation were investigated to assess its effectiveness in reducing heat transfer and improving heat capacitance. Four hollow concrete blocks, measuring 10 cm × 19 cm × 40 cm were examined: three blocks were filled with varying ratios of wax and Expanded Polystyrene, and one remained hollow as a control. The specific heat capacity of the wax-Expanded Polystyrene composite was measured to evaluate its thermal storage potential. Additionally, the thermal insulation abilities of the filled blocks were assessed using a heat source and a thermal camera. The results showed that the blocks with the wax-EPS composites exhibited reduced heat transfer compared to the control block, demonstrating improved thermal performance. The block with the higher wax content additionally displayed greater specific heat capacity, achieving a value of 1825.1 (J/kg.K), indicating enhanced thermal storage capabilities. These findings highlight the potential of wax-EPS concrete blocks in improving thermal regulation, contributing to a viable and energy-efficient building material for hot climates.

Keywords: Wax-Expanded Polystyrene composite · Phase change materials · Thermal insulation

1 Introduction

Concrete blocks are the main material used for building envelopes in Saudi Arabia (KSA), due to their acceptable mechanical performance and durability in hot climates [1]. However, they has poor thermal insulation, making them sensitive to heat transfer in hot environments regions. Despite increased thermal mass, the high thermal conductivity decreases the block insulation performance, causing high energy usage [2, 3]. Integrating thermal insulation into buildings is both a technical necessity and a regulatory requirement to reduce energy consumption and support sustainability.

Expanded polystyrene is the most common insulation material used in KSA due to its lightweight properties and low thermal conductivity. Despite this, its low fire resistance leads to degradation, increasing the demand for insulation materials to improve the thermal performance of concrete blocks [4, 5].

D. Berkouk et al. (Eds.): CCTS 2024, PTT, pp. 82–90, 2026.
https://10.1007/978-981-95-7607-4_10

Changing the block's void configuration or integrating insulation materials can increase thermal performance, contributing to distinct insulating properties [6]. Wax equally functions as a phase change material (PCM) which can store and release thermal energy as it experiences phase transitions [7]. The combination of these two materials can significantly enhance thermal storage, presenting innovative insulation solutions in hollow concrete blocks [8].

The thermal performance of wax-EPS mixtures as insulation materials in concrete blocks has limited studies. Alsadiq et al. (2024) investigated Wax-EPS mixtures in insulating concrete blocks and identified their effectiveness in enhancing thermal insulation [8]. They stressed the need for further research on practical applications and long-term performance in construction. Fahrurrozi et al. (2024) analyzed the thermal behavior of paraffin wax as PCM in bricks, addressing phase change leakage and low thermal conductivity [9]. Their findings demonstrated a peak heat transfer of 2.04 W/m^2 during phase changes. Lygerakis et al. (2024) explored the integration of Paraffin wax in building materials and optimized them for maximum energy efficiency and improved thermal comfort, resulting in reduced annual energy usage by up to 12.8% [10].

This study examines the thermal properties of Wax-EPS composite insulation to evaluate its effectiveness in reducing heat transfer and determining its thermal storage capabilities through specific heat capacity. Thermal performance was evaluated using a heat source and a thermal imaging camera to measure the temperature difference through the block thickness.

2 Methodology

Four volcanic concrete blocks were obtained, each measuring 10 cm × 19 cm × 40 cm. Three blocks were filled with wax and Expanded Polystyrene (EPS) mixtures in volume ratios of 40:60, 60:40, and 80:20 (Fig. 1). One block remained hollow and served as a control to compare thermal performance.

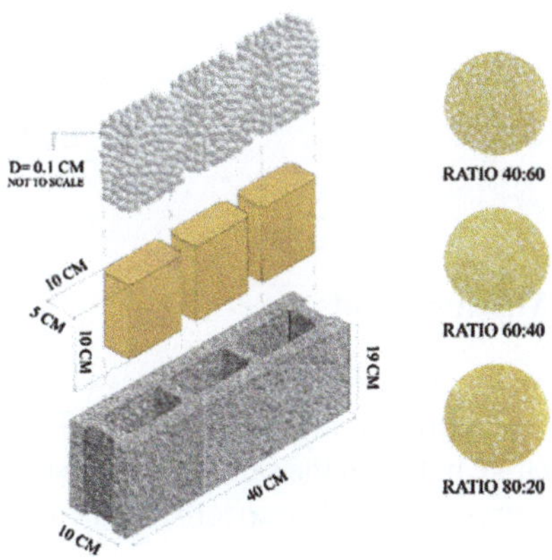

Fig. 1. A schematic of a composite concrete block infused with wax (PCM) and expanded polystyrene (EPS) in varying volume ratios of 40:60, 60:40, and 80:20.

2.1 Density Measurement of Wax-EPS Composites

The density of wax was determined using the water displacement method. This technique measured the volume of wax by recording the change in water level when submerging a known mass of wax. The mass of the wax was measured using an accuracy balance, and the density was calculated using Eq. 1:

$$\rho = \frac{m}{V} \tag{1}$$

where ρ is the density, m is the mass, and V is the measured volume.

After calculating the densities of wax (1048 kg/m^3) and EPS (23.33 kg/m^3) individually, the theoretical density of the wax-EPS mixtures was determined using the rule of mixtures using Eq. 2:

$$\rho_{mix} = \frac{\rho_{wax}V_{wax} + \rho_{EPS}V_{EPS}}{V_{wax} + V_{EPS}} \tag{2}$$

where ρ_{mix} is the density of the mixture, ρ_{wax} and ρ_{EPS} are the densities of wax and EPS, respectively, $V_{wax} + V_{EPS}$ are their respective volumes.

The experimental density values were obtained by measuring the mass and volume of prepared wax-EPS samples under controlled conditions. The calculated theoretical densities were compared with the experimental values to assess material uniformity and mixing efficiency. The results are presented in Table 1:

Table 1. Comparison of Calculated and Measured Densities of Wax-EPS Composites at Different Mixing Ratios.

Wax: EPS	Calculated density $\rho_{theo}(kg/m^3)$	Measured density $\rho_{exp}(kg/m^3)$
40%: 60%	499.8	506.4 ± 17.67
60%: 40%	638.1	607.5 ± 6.41
80%: 20%	843.1	854.4 ± 22.52

2.2 Thermal Capacity Measurements

The specific heat capacity of the wax-EPS composites was measured using a calorimeter containing water at 50 °C. The samples, initially cooled to 3 °C, were then placed into the calorimeter. The heat exchange between the sample and the water was recorded. Figure 2 shows the setup used in this experiment. The specific heat capacity was determined using Eq. 3:

$$c_{\exp.sample} = \frac{m_w C_w \left(T_{hot} - T_{final}\right)\Big)}{m_{sample}\left(T_{final} - T_{intia,sample}\right)} \tag{3}$$

where m_w and m_{sample} represent the masses of the water and the sample, respectively. C_w is the specific heat capacity of water, and T_{hot}, T_{final} and $T_{intia,\ sample}$ denote the temperatures of the hot water, the final equilibrium temperature, and the initial sample temperature, respectively.

The experimental results were compared with the theoretical specific heat capacity calculated using Eq. 4:

$$C_{theo.sample} = (V_{wax}.C_{wax}) + (V_{EPS}.C_{EPS}) \tag{4}$$

where $C_{theo.\ sample}$ is the theoretical specific heat capacity of the composite, C_{wax} and C_{EPS} are the specific heat capacities of wax and EPS, respectively, and V_{wax} and V_{EPS} represent their volume fractions.

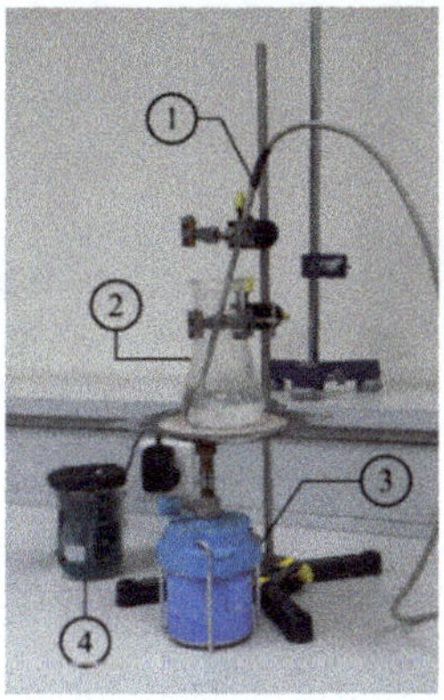

Fig. 2. Experimental Setup for Specific Heat Capacity Measurement—(1) Temperature Probe for monitoring temperature changes. (2) Beaker used to heat the water to 50–60 °C (3) Bunsen burner to heat water in the beaker. (4) Calorimeter containing heated water, where the sample (previously cooled to 3 °C) was added for heat exchange analysis.

A finite difference method (FDM) simulation was conducted using Python to estimate the specific heat capacity $C_{Num.\ sample}$ of the Wax-EPS composites. The one-dimensional transient heat conduction equation was discretised and solved over a spatially divided sample with a uniform grid. The numerical solution was implemented in Python® using NumPy™ for array calculations and Matplotlib™ for visualization. The temperature distribution was updated iteratively using explicit finite difference time-stepping, ensuring stability with a controlled time step and spatial resolution. Finally, the temperature-time data was analysed to extract specific heat capacity values using an energy balance as in equation Eq. 5:

$$Q = mC_p\Delta T \tag{5}$$

where the heat energy input Q was determined from the applied heat flux, and the mass m was calculated using the sample volume and density.

2.3 Heat Transfer Testing

The heat transfer properties of the concrete blocks were evaluated through experimental testing. An electric heat source was applied to one side of each block. The temperature variation on the opposite side was recorded using a TiS20+ thermal imager (FLUKE®, WA, US). Additional temperature measurements were taken using a Bosch UniversalTemp™ infrared thermometer (BOSCH®, Gerlingen, DE) to validate the thermal imaging results. Figure 3 shows the setup used in this experiment.

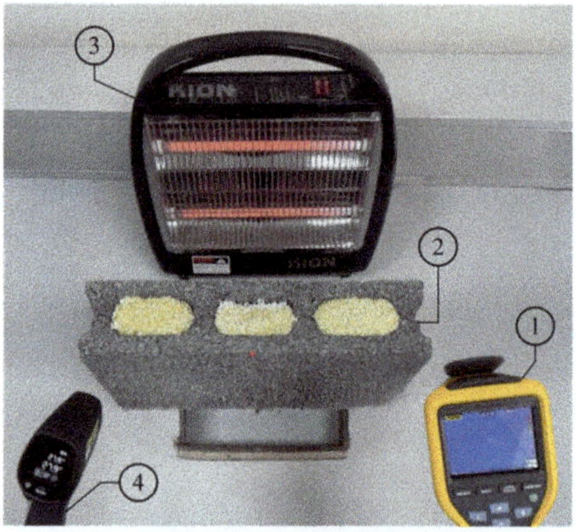

Fig. 3. Experimental Setup for Heat Transfer Testing of EPS-Wax Composite Blocks – (1) Thermal imager used to record the temperature distribution on the block's surface. (2) Tested block containing EPS-wax composite fillings, positioned for thermal evaluation. (3) Radiation heat source providing consistent heat exposure to the block. (4) Infrared thermometer used for precise spot temperature measurements.

3 Results and Discussion

The thermal behavior of wax-EPS composite samples was examined through calculating its specific heat capacity and comparing it to numerical simulation and validating it experimentally as described in the methodology. The heat transfer through the concrete blocks filled with the composite was examined.

3.1 Specific Heat Capacity of Wax-EPS Composites

The results show that the specific heat capacity increases as the wax to EPS ratio increases, as was expected from theoretical calculations and aligns with the numerical simulation. Table 2 summarizes the theoretical, empirical and numerical specific heat capacity results, and shows that the 40% Wax–60% EPS composite exhibited the lowest specific heat capacity, whereas 80% Wax–20% EPS showed the highest, which confirms the role of wax as a thermal storage medium.

Table 2. Theoretical, Numerical, and Experimental Specific Heat Capacity of Wax-EPS Composites.

Wax: EPS	Heat Capacity $(J/kg.\ K)$		
	$C_{theo.\ sample}$	$C_{nume.\ sample}$	$C_{exp.\ sample}$
40%: 60%	1600	1545	1745.6
60%: 40%	1760	1665	1762.8
80%: 20%	1920	1805	1825.1

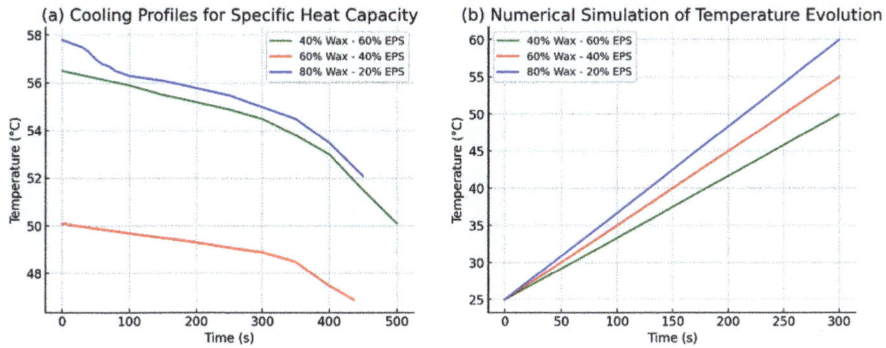

Fig. 4. a) Experimental Cooling Profiles for Specific Heat Capacity – Temperature evolution of Wax-EPS composites during the cooling phase, b) Numerical simulation of the temperature evolution over time of the Wax-EPS composite under a constant heat input over 300 s.

The experimental results show that the 80% Wax–20% EPS mixture had the highest heat capacity of 1825.1 J/kg·K, 4.5% higher than the 40% Wax–60% EPS composite (1745.6 J/kg·K). Numerical simulations closely matched the experimental values, with deviations less than 5%, confirming the model's accuracy.

The temperature trend during heating, obtained from the numerical simulation (Fig. 4b), shows that the 80%–20% Wax to EPS composite heated 13% slower than the 40%–60% Wax to EPS composite, indicating better thermal storage. The cooling curve from the experimental results shown in Fig. 4a confirms this trend, with the 80%–20% Wax to EPS composite cooling at a 10% slower rate, reducing heat loss and temperature fluctuations.

The theoretical values slightly underestimated the heat capacity for the 40% Wax - 60% EPS composition, likely due to material inconsistencies. However, the strong agreement between the experimental and simulated values supports the reliability of the results.

3.2 Heat Transfer Assessment Using Thermal Imaging

The experimental results reveal that all the blocks with wax-EPS ratios exhibited high thermal efficiency, with minimal temperature differences between the ratios, in contrast to the control block (Fig. 5). The 80:20 concrete block achieved a temperature change of 32.4 °C, 23.84% higher than the control block, which had a temperature difference of 26.24 °C, demonstrating its greater ability in reducing heat transfer (Fig. 6). These findings show that wax-EPS integrated concrete blocks have significant potential in improving thermal insulation in building materials, contributing to better energy efficiency in arid climates.

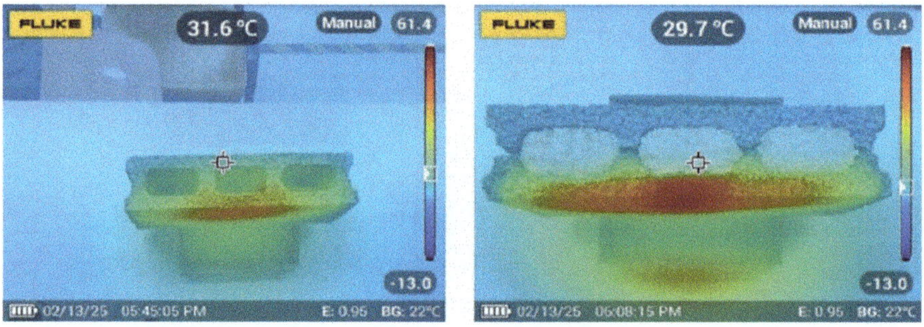

Fig. 5. Thermal image the control concrete block (left), and the 40% wax–60% EPS-filled block (Right). The color contour ranges from −13 °C (Blue) to 61.4 °C Red.

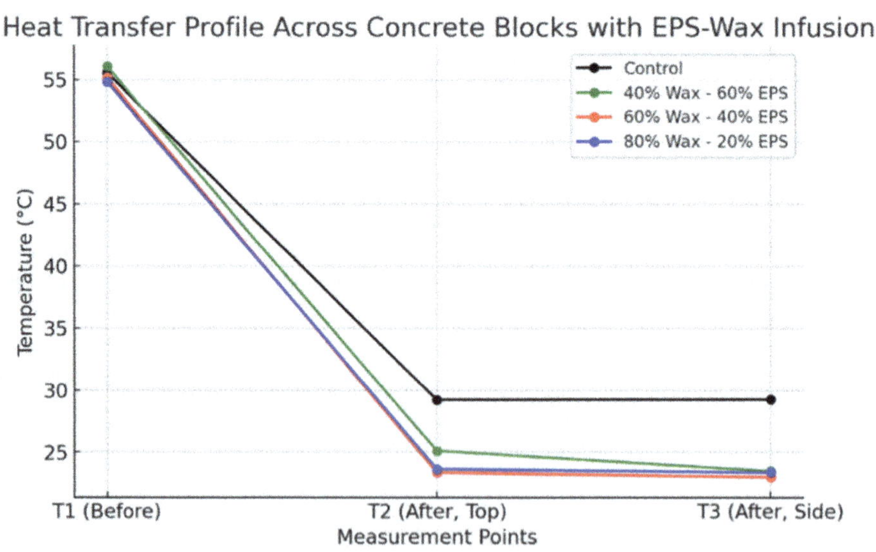

Fig. 6. Experimental heat transfer profile across concrete block for Wax-EPS different ratios.

4 Conclusion

This research demonstrated that integrating wax-EPS composites into hollow concrete blocks significantly improved their thermal insulation and heat storage capacity. The experimental results confirmed the theoretical prediction of improved thermal capacity with the wax to EPS ratio in the novel composite presented in this paper. The 80% Wax–20% EPS composite was shown to have the highest value of 1825.1 J/kg·K. Heat transfer analysis using thermal imaging have also showed the superior thermal performance of the high wax to EPS ratio composite, where the 80:20 wax to EPS filled hollow block reduced the heat transmission by up to 32.4 °C compared to the control.

Despite the promising performance of the Wax-EPS composite material, there are limitations in this study that need to be addressed in future work, such as the mechanical properties of the composites and the effect of prolonged thermal cycling loads. Also, the future work should compare the composite thermal performance to the performance of blocks filled with pure EPS.

Acknowledgments. The authors would like to express their gratitude to Eng. Nader Alharbi and Mr. Mohammed Abu-Baker for their assistance in the laboratory experiments. Special thanks are also extended to Eng. Basel Jouda for his thoughtful insights and constructive discussions.

References

1. Al-Homoud, M.S.: Performance characteristics and practical applications of common building thermal insulation materials. Build. Environ. **40**(3), 353–366 (2005). https://doi.org/10.1016/j.buildenv.2004.05.013
2. Monteiro, P.: Concrete: Microstructure, Properties, and Materials. McGraw-Hill Publishing, New York (2006)
3. Remesar, J.C., Vera, S., Lopez, M.: Assessing and understanding the interaction between mechanical and thermal properties in concrete for developing a structural and insulating material. Constr. Build. Mater. **132**, 353–364 (2017). https://doi.org/10.1016/j.conbuildmat.2016.11.116
4. Shao, X., et al.: Reduced fire hazards of expandable polystyrene building materials via intumescent flame-retardant coatings. J. Mater. Sci. **55**(17), 7555–7572 (2020). https://doi.org/10.1007/s10853-020-04548-z
5. Lu, G.: Study on the properties of rock wool for external thermal insulation of buildings under the soaking and hot & humid conditions. J. Phys. Conf. Ser. **1622**(1), 012006 (2020). https://doi.org/10.1088/1742-6596/1622/1/012006
6. Al-Mekhlafi, G.: Thermal optimization of two-way joist slabs: A comparative study using finite element analysis. Mater. Res. Proc. **48**, 181–190 (2025). https://doi.org/10.21741/978 1644903414-21
7. Trisnadewi, T., Putra, N.: Application of soy wax phase change material as thermal energy storage in wall building. J. Phys. Conf. Ser. **2739**(1), 012017 (2024). https://doi.org/10.1088/1742-6596/2739/1/012017
8. Alsadiq, H., Almekhlafi, G., Alabdullah, A., Alfatih, S., Abas, H.A.: Enhancing thermal properties in concrete blocks using Styrofoam-wax mixtures. Mater. Res. Proc. **28**, 184–192 (2025). https://doi.org/10.1007/978-981-97-8348-9_23

9. Fahrurrozi, M., Iwan, M., Gunarto, G., Fahrezi, D., Govari, M.K.: Comparative study of the effect of paraffin phase change material mixture and ice bag on temperature control in bricks. Int. J. Ind. Innov. Mech. Eng. **2**(1), 17–31 (2024). https://doi.org/10.61132/ijiime.v2i1.118
10. Lygerakis, F., Gioti, C., Gournis, D., Ioannis, V., Yentekakis, M.K., Kolokotsa, D.: Enhancing building energy efficiency with innovative paraffin-based phase change materials. Energies (Basel). **17**(16), 4155 (2024). https://doi.org/10.3390/en17164155

Sustainable Concrete Solutions: Harnessing CornCob as a Renewable Resource

Arwa Fallatah[1(✉)] and Shifana Fatima Kaafil[2]

[1] Hekma School of Engineering, Computing and Design, Dar Al-Hekma University, Jeddah, Saudi Arabia
Arwaf7@gmail.com
[2] Department of Architecture, Dar Al-Hekma University, Jeddah, Saudi Arabia
skaafil@dah.edu.sa

Abstract. The agricultural waste is disposed of either by burning it or throwing it in a landfill. In this an attempt is made to utilize the agricultural byproduct corn cob to develop a green concrete. This in turn will reduce the environmental pollution and reduce the greenhouse gas emission from conventional concrete. The used corn cob is crushed into powder and is added as additives with cement. Experiments were carried out on mortar cubes and concrete cubes with 0%, 1% and 2% of corn cob with respect to weight of the cement and studied its dry density, wet density, water absorption, compressive strength, slum value and air porosity. by using corn cob waste to make a concrete cube with different corn cob percentages. From the test results it is found that concrete with corn cob as additives of 2% with respect to weight of cement give high compressive strength and has less weight compared to conventional concrete. Corn cob is an agricultural byproduct and thereby it absorbs more water than conventional concrete, but still it increased the strength as it fills the voids between fine and coarse aggregate and is evident from its less air porosity value. It is highly recommended to use corn cob as additives in concrete but still it requires further test to use under tensile load.

Keywords: Agricultural waste · Concrete · Corn Cob · Sustainable Concrete

1 Introduction

The amount of agricultural waste in the Kingdom reached more than 1.6 million tons annually, of which more than 800 thousand tons are field crop residues. The results of current studies have confirmed that most agricultural waste in Riyadh, Dammam, and Jeddah is disposed of either by burning it or throwing it in a landfill [1–3]. Municipalities cause severe environmental damage, in addition to not benefiting from recycling these wastes, which makes them pile up in the form of large heaps near corn stores. Still, the drought conditions and the lack of animal feed in some years pushed the direction of grinding the corn cob and introducing it in a small percentage with some feed [4].

D. Berkouk et al. (Eds.): CCTS 2024, PTT, pp. 91–97, 2026.
https://10.1007/978-981-95-7607-4_11

1.1 Usage of Corn Cob in Construction

Corn cob concrete is used in non -structural applications. Pinto et al., 2012 [5] used lightweight corn cob concrete for pavement regularization layers. The primary goal of this research is to investigate the potential use of corn cob granulate (without any corn, i.e., agricultural waste) as a sustainable aggregate solution for lightweight concrete and as an alternative to commonly used products such as expanded clay, expanded polystyrene (EPS) particles, cork particles, or other lignocellulose wastes.

As an alternative natural aggregate of lightweight concrete, granulate corn cob without any corn is presented. Corn cobs are generally regarded as agricultural waste. Using it as an alternative to expanded clay, cork, and EPS, among other options, might have a significant economic and sustainable advantage. The testing findings show that corn cob concrete treated in a 6:1:1 (corn cob granulate: Portland cement: water) ratio may have acceptable material qualities. Asebiomo et al., 2024 [6] examined the effect of corncob ash as a replacement for the traditional stone dust filler and it offers ecological and economic benefits. Popoola et al., 2024 [7] and Yifru et al., 2022 [8] used corn cob ash as partial replacement of cement to stabilize soil.

Adesanya and Raheem, 2009 [9] investigated the use of corn cob ash as a pozzolan in cement manufacture to turn waste materials into usable materials for the construction sector. The research examined the chemical components of corn cob ash (CCA). The corn cob ash-blend cement was manufactured in factories by substituting CCA for ordinary Portland cement clinkers in various weight ratios ranging from 0% to 25%. The CCA tested was made by crushing dried corn cobs to a diameter of about 4.00 mm to improve combustion and decrease carbon content, which influences the pozzolanic characteristics of the CCA. The ground cobs were burned in the open air in a local blacksmith furnace powered by charcoal. The fire went continuously for almost 8 h, reaching 650 °C until the corn cobs turned to ashes.

Several research focuses on the use of corn cob ash (CCA) as a partial replacement for cement in the concrete production process to decrease cement usage and to reduce its cost. The strength of the replacement-produced concrete cubes and beams demonstrated that using corn cob ash to replace some of the cement may be used to bear some structural loads. Replacement should be limited to 10% since strength produced over this level may not be adequate to fulfill strength requirements [10–12]. It was found that the proposed high-strength concrete with CCA did not meet the compressive and split tensile strength requirements [13]. However, test findings demonstrated that using CCA as a partial replacement for cement in concrete, particularly in plain concrete works and non-load-bearing buildings, to improve the waste-to-wealth project. As a result, the application of superplasticizers is essential to improve workability [14, 15]. In earlier literature the corn cob is burned and its ash is used as partial replacement of cement and in this paper, the corn cob is ground into fine powder and is used as additive with cement.

The objective of this research is to use locally accessible resources waste reduction of corn cobs and to build sustainable green concrete by utilizing agricultural byproducts.

2 Methodology

2.1 Collection of Samples

The researcher needed to collect a certain amount of corn cob waste to start the experiment. This quantity was unavailable in Jeddah, so it was ordered from a corn farm in the eastern region, as shown in (Fig. 1a). After that, the corn was cut into small slices using an electric saw with the help of a carpenter. The last stage is grinding corn in Coffee mills. (see Fig. 1b).

2.2 Mixing and Testing of Mortar Cubes

Mortar cubes of size 5 cm × 5 cm is used for the test, the following ingredients were added: 400 g of cement, 1200 g of fine and coarse aggregate, 200 ml of water, 4 g of corn cob 1% cube, and 8 g of corn cob 2% cube (see Fig. 2a). All the dry ingredients were mixed appropriately; water was added until the mixture became homogeneous. After casting the mixture in mold (see Fig. 2b), it was left to dry in the sun for a full day. After that, the cubes were removed from the mold (see Fig. 2c). The cubes were placed in water for seven days, and their weight was measured. They were then placed in the oven at 1010C for 24 h and then the weight of cubes were measured. Then compressive strength is carried out at the laboratory (see Fig. 2d).

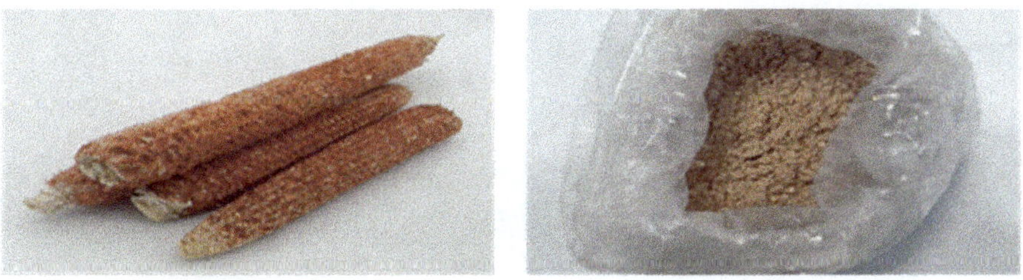

a) Raw corn cob waste   b) Corn cob after grinding

Fig. 1. Collection and Grinding of Corn Cob.

2.3 Mixing and Testing Concrete Cubes

Concrete cubes of size 150 × 150 mm are cast and tested as per ASTM standards. For the conventional cube, four cubes were cast. Two cubes were tested on day 7, and the other cubes were tested on day 28. A slump cone test and air porosity are also measured for conventional concrete cubes. Three cubes are cast for 1% corncob and 2% corncob with respect to the weight of the cubes. Two cubes were tested on day 7, and the other cubes were tested on day 28. Air porosity and workability tests were made. That test was conducted to determine how much air was in the mixture and to check how the corncob powder is easy to work with concrete (see Fig. 3). It shows the process and testing of concrete cubes with corncob.

Fig. 2. Process Involved in Making Mortar Cubes.

2.4 Water Absorbed by Corncob Powder

The powder was soaked in water at room temperature for three days to calculate the percentage of water absorbed by fibers. The initial weight of the water and corn cob powder was measured (Fig. 4). After three days, the powder was filtered and removed from the water, and then the weight was measured again.

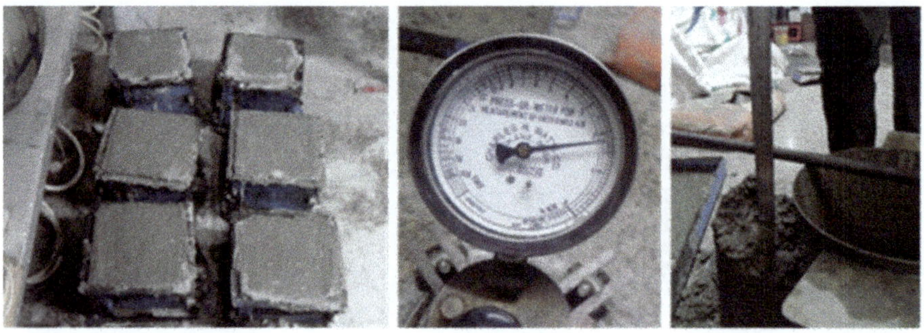

Fig. 3. Process Involved in Making and Testing Concrete Cubes.

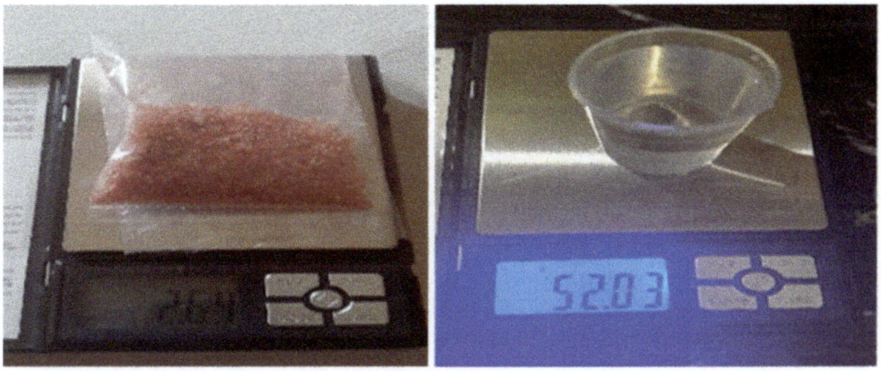

Fig. 4. Water Absorption of Corn Cob Powder.

3 Results and Discussions

3.1 Mortar Cubes Result

The results in (Table 1) show that 1% corn cob powder (CCP) wet density is less than the conventional Cube (CC) while 2% is a bit lower than the (CC) which is good. However, the dry density in 1% is so much less than the (CC), but 2% is much better than 1% and slightly lower than the (CC). Moving to the water absorption, it shows that 1% absorbed water more than the (CC), but 2% is better than 1% and a bit higher than the (CC). The last one is compressive strength. The results shown in (Figure 5) indicate that 1% is less than the (CC) because the cube was damaged, but 2% is much higher than the (CC).

Table 1. Mortar cube test results

Parameters	Wet Density (g/cc)	Dry Density (g/cc)	Water absorption (%)	Compressive Strength (N/mm2)
Conventional Cube	2.134	2.051	4.4	10
Cube with 1% corn cob	1.563	1.4492	7.8	6.8
Cube with 2% corn cob	2.053	1.960	4.7	11.4

3.2 Concrete Cube Test Results

The results in Table 2 show that 1% corn cob powder (CCP) wet density at 0 days is less than the conventional Cube (CC) while the 2% is a bit lower than the (CC) which is a good sign. Meanwhile wet density in 1% is much less than (CC) at 7 days and but 2% is slightly lower at 28 days. However, the dry density in 1% and 2% at 7 days is less than the (CC), but 1% and 2% are a bit higher than (CC) at 28 days. Moving to the water absorption, it shows that both 1% and 2% absorbed much less water than the (CC) at 7 or 28 days. Regarding the air voids, 1% is less than (CC) but 2% is much higher than (CC). The slump test shows that 1% is less than (CC) but 2% exceeds (CC) which is good. The last one is compressive strength, The results indicate that both 1% and 2% surpassed the (CC).

Table 2. Concrete cube test results

Parameters	Wet Density (g/cc)		Dry Density (g/cc)		Water absorption (%)		Compressive Strength (N/mm2)		Slump value (mm)	Air Voids (%)
No of days	7	28	7	28	7	28	7	28	0	0
Conventional Cube	2.49	2.43	2.45	2.42	2.9	3.1	17.73	22.84	145	1.8
Cube with 1% corn cob	2.42	2.45	2.40	2.436	1.7	1.8	17.82	23.82	140	1.7
Cube with 2% corn cob	2.44	2.46	2.42	2.43	1.9	2.1	19.02	25.02	150	1.44

3.3 Water Absorption of Corn Cob Powder

After soaking the corn cob powder in the water, the fiber didn't mix well with the water. It was easy to separate them from each other. The results showed that the percentage of water absorbed by corn cob powder is 10%.

4 Conclusion

It is found that concrete with 2% corn cob performs better than conventional and cube with 1% corn cob in terms of high compressive strength and workability, less density, air porosity and water absorption. It is recommended to use corn cob as additives in concrete which enhance the effective usage of agricultural by-products and to develop sustainable concrete. It is suggested to increase the percentage of corn cob and to conduct tensile and bending tests as future scope of the research.

Acknowledgments. The authors would like to acknowledge the Solutions and Mining Laboratory to conduct the experiments in the laboratory and to Mr. Abu Hussam Al-sharari, for supporting in the collection of corn cob materials.

References

1. Elegbede, J.A., Ajayi, V.A., Lateef, A.: Microbial valorization of corncob: Novel route for biotechnological products for sustainable bioeconomy. Environ. Technol. Innovation. **24**, 102073 (2021)

2. Prasad, M., et al.: Efficient transformation of agricultural waste in India. Contam. Agric. **4**, 22–28 (2020). https://doi.org/10.1007/978-3-030-41552-5_13
3. Abd Ali, B.A.: Utilization of corn cobs in the manufacturing of particleboards. Mesopotam. J. Agric. **33**, 56–61 (2005). https://doi.org/10.33899/magrj.2005.35960
4. U.S. Department of Commerce, International Trade Administration. Saudi Arabia - waste Management https://www.trade.gov/country-commercial-guides/saudi-arabia-waste-management
5. Pinto, J., et al.: Corn cob lightweight concrete for non-structural applications. Constr. Build. Mater. **34**, 346–351 (2012). https://doi.org/10.1016/j.conbuildmat.2012.02.043
6. Asebiomo, J.O., Aderinola, O.S., Ejigboye, P.O., Olusegun, V., Aderemi, P., Titiloye, O.: Evaluation of corn cob ash as mineral filler in asphalt mixture. Int. J. Pavement Res. Technol. **18**(6), 1574–1585. https://doi.org/10.1007/s42947-024-00440-0
7. Popoola, O.O., Adekanmi, J.S., Olulope, O.R.: Corn- cob ash as partial replacement of cement for stabilization of laterite soil. J. Civ. Eng. Sci. Technol. **15**(1), 7–17 (2024)
8. Yifru, W., Getu, N., Kifile, D., Mesfin, A., Sewunet, A., Tamene, M.: Effects of corn cob ash as partial replacement of cement for stabilization of an expansive clay. Adv. Civ. Eng. **2022**, e6788120 (2022)
9. Adesanya, A., Raheem, A.A.: Development of corn cob ash blended cement. Constr. Build. Mater. **23**(1), 347–352 (2009). https://doi.org/10.1016/j.conbuildmat.2007.11.013
10. Tiza, M., Augustine, A.: Partial replacement of cement with corn cob ash. Int. J. Innov. Res. Multidiscip. Field. **2**, 159–169 (2016). https://doi.org/10.13140/RG.2.2.13203.40481
11. Kamau, J., Ahmed, A., Hirst, P., Kangwa, J.: Suitability of corncob ash as a supplementary cementitious material. Int. J. Mater. Sci. Eng. **4**, 215–228 (2016)
12. Mujedu, K.A., Adebara, S.A., Lamidi, I.O.: The use of corn cob ash and saw dust ash as cement replacement in concrete works. Int. J. Eng. Sci. (IJES). **4**, 22–28 (2014)
13. Olafusi, O., Olutoge, F.: Strength properties of corn cob ash concrete. J. Emerg. Trends Eng. Appl. Sci. **3**, 297–301 (2012)
14. Abioye, A., Olutoge, F., Ofuyatan, O., Oyebisi, S.: Effect of corncob ash blended cement on the properties of lateritic interlocking blocks. Prog. Ind. Ecol. Int. J. **11**, 373 (2017). https://doi.org/10.1504/PIE.2017.10013954
15. Samson, E., Maleki, K., Zhang, T., Marchand, J.: Determination of the water diffusivity of concrete using drying/absorption test results. J. ASTM Int. **5**(7), 1–12 (2008). https://doi.org/10.1520/JAI101322

Digital Transformation Innovations for Sustainability: Advancing Workforce Training in the Digitalization Era

Azza AlGhamdi(✉) and Alia Aljindi

Faculty of Education, King Abdulaziz University, Jeddah, Saudi Arabia
aalghamdi4221@stu.kau.edu.sa, aalgnedi@kau.edu.sa

Abstract. The rapid evolution of digital technologies has led organizations to embrace innovations that align with digital transformation, aiming to improve efficiency and secure a competitive edge for the future workforce, in support of sustainable development objectives. However, the successful implementation of digital innovations and emerging technologies necessitates a thorough evaluation of various factors. Our research explores how digital transformation innovations influence workforce training and contribute to the achievement of sustainable development goals (SDGs) across three sustainability dimensions: social, economic, and environmental. Data were gathered through a survey targeting energy professionals and trainees from King Abdulaziz University's Department of Nuclear Engineering and related institutions. The findings indicate a substantial positive correlation between digital transformation innovations in workforce training and the sustainable development goals, exhibiting differing intensities across the three aspects. Social sustainability objectives, namely quality education (SDG4) and gender equality (SDG5), garnered substantial backing, although economic sustainability objectives like decent work and economic growth (SDG8) and industrial innovation and infrastructure (SDG9) also demonstrated significant support. Environmental sustainability, particularly Climate Action (SDG 13) and Life on Land (SDG 15), received widespread support, however certain SDGs had mixed reactions. Statistical analysis validated the robustness of these associations, with p-values below 0.05 and t-values signifying substantial impacts. These findings underscore the revolutionary capacity of digital advances in workforce training, aligned with the attainment of the SDGs. Subsequent study may investigate developing technologies and their extensive applications across diverse sectors.

Keywords: Digital Transformation Innovations · Workforce Training · Sustainable Development Goals (SDGs)

1 Introduction

The Sustainable development has become a top focus for institutions worldwide. The Millennium Development Goals, which included eight objectives, were introduced in 2008, and further progress was made in 2015 when 193 countries agreed to establish 17

D. Berkouk et al. (Eds.): CCTS 2024, PTT, pp. 98–106, 2026.
https://10.1007/978-981-95-7607-4_12

Sustainable Development Goals (SDGs) with associated sub-indicators, as well as the formation of a committee to track nations' progress toward these goals [1]. Saudi Arabia has made substantial contributions to the fulfillment of these objectives; Aligned with the Kingdom's Vision 2030, which aspires for comprehensive and sustainable growth, the emphasis is on meeting present societal demands while proactively planning for future generations [2]. Sustainable development is defined as growth that addresses society's current needs while ensuring its ability to fulfill future requirements; The SDGs are divided into three categories: social (SDGs 1–7, 11and 16), economic (SDGs 8, 9,10, 12 and 17), and environmental (SDGs 13, 14 and 15) [3]. This study seeks to investigate the impact of digital transformation innovations in improving workforce training, particularly in the context of sustainable development. The principal research question is: In what ways do digital transformation advances in workforce training facilitate the attainment of the Sustainable Development Goals (SDGs)? This research seeks to examine the correlation between breakthroughs in digital transformation for workforce training and the Sustainable Development Goals (SDGs). This research explores multiple dimensions of sustainability [social, economic, and environmental] by investigating the statistically significant relationships between digital transformation in training and sustainability across these dimensions.

2 Literature Review

This section analyzes literature regarding the correlation between digital transformation and sustainability, focusing on essential concepts, frameworks, and prior research that elucidate how Digital Transformation Innovations facilitate the attainment of Sustainable Development Goals (SDGs).

2.1 Sustainable Development Goals (SDGs)

Building on the remarkable success of the Sustainable Development Goals (SDGs) in the new millennium, the 2030 Agenda outlines a global strategy, this agenda encompasses 17 SDGs and 169 related targets, providing a comprehensive and unified framework to address complex global challenges [4]. The TWI2050 framework further explores the connections between socioeconomic development and environmental sustainability, employing complex systems thinking and interdisciplinary approaches. It proposes six key SDG transformations as modular components for achieving the goals, grouped into the following domains: (1) human capacity and demographics (education, gender, and inequality); (2) consumption and production (health, well-being, and demographics); (3) decarbonization and energy (energy decarbonization and sustainable industry); (4) food, biosphere, and water (sustainable food, land, water, and oceans); (5) smart cities (sustainable urban areas); and (6) the digital revolution for sustainable development [5]. The release of the QS World Rankings for Sustainable Development underscores the growing focus on achieving sustainable development goals and improving their annual implementation [6]. The rankings are built on three core pillars: social impact, economic impact, and environmental impact, along with various indicators related to the sustainable development goals and their contributions in tackling environmental and social challenges [7].

2.2 Digital Transformation Innovations

The world is undergoing swift technological progress, resulting in a transition to new horizons, the notion of digital transformation has garnered much attention lately, since it emphasizes the conversion of traditional processes into digital formats, hence enhancing the value of work [8]. The digital transformation of education represents an important opportunity to improve educational quality, increase student interaction and participation, and reflects ongoing efforts to achieve the vision of developing education while also achieving progress and innovation in the field [9]. Innovation and digital transformation have a reciprocal relationship: innovation enables digital transformation, while digital transformation develops an environment conducive to innovation. Innovative thinking is essential to any successful digital transition [10]. Underlines Schork et al. [11] the need of taking a comprehensive approach while using digital tools for sustainability. It investigates how the integration of digital transformation (DT), innovation, and sustainability (DIS) might benefit enterprises, notably by improving environmental and social practices. Based on the foregoing, our research illuminates the relationship between advancements in digital transformation innovations for workforce training and the Sustainable Development Goals (SDGs).

Main Hypothesis: There is no statistically significant link ($p < 0.05$) between Digital Transformation Innovations in workforce training and Sustainable Development Goals (SDGs).

- first Sub-H0: There is no statistically significant link between Digital Transformation Innovations in workforce training and social sustainability.
- second Sub-H0: There is no statistically significant link between Digital Transformation Innovations in workforce training and economic sustainability.
- third Sub-H0: There is no statistically significant link between Digital Transformation Innovations in workforce training and environmental sustainability.

3 Methodology

This study utilized a quantitative descriptive analytic design to examine the impact of Digital Transformation Innovations on workforce training and their contribution to achieving the Sustainable Development Goals (SDGs), emphasizing social, economic, and environmental sustainability. A survey questionnaire was employed to gather data, comprising two sections: one for demographic information and the other for evaluating the alignment of digital advances with the Sustainable Development Goals (SDGs). The second section includes 17 items derived from the SDGs, categorized into three key pillars: social, economic, and environmental. These items aim to evaluate participants' views on the topic, with responses assessed using a Likert scale to facilitate a structured analysis. A total of 150 questionnaires were distributed, receiving 87 completed responses selected using random sampling from energy Specialists and trainees affiliated with the Nuclear Engineering Department at King Abdulaziz University "20.83%", King Abdulaziz City for Science and Technology (KACST) "30.5%" and King Abdullah City for Nuclear Energy "48.6%". Out of these, 72 responses were deemed valid for statistical analysis.

Fig. 1. Participants' Views on Social Sustainability in the SDGs.

4 Results

This section presents the study's results, including statistical analyses and data visualizations, to test the relationships between digital transformation innovations in workforce training and the Sustainable Development Goals (SDGs) across social, economic, and environmental sustainability dimensions. It also evaluates the extent to which the hypotheses are accepted or rejected. The first section of the questionnaire focused on social sustainability goals, which align with nine of the Sustainable Development Goals (SDGs). Notably, SDG.4 (Quality Education), SDG.5 (Gender Equality) and SDG.16 (Peace, Justice, and Strong Institutions) received the highest levels of agreement, with 96%, 92%, and 95% of participants agreeing or strongly agreeing, respectively. However, SDG.6 (Clean Water and Sanitation) showed a more divided response, with 30% disagreeing, indicating areas that need further attention. These findings, as presented in (Fig. 1), reflect strong support for the social sustainability goals, though some goals still require more effort to achieve broader consensus. Although most participants robustly endorsed the social sustainability objectives, specific areas require enhanced attention and the implementation of digital transformation innovations in workforce training, especially within the energy sector.

The second section of the questionnaire assessed participants' perspectives on the connection between digital transformation training programs and the achievement of SDGs within the economic sustainability dimension. There was strong support for the alignment of SDGs 8 and 9 with digital transformation innovations, with 98% agreement for SDG.8 and 96% support for SDG.9. However, responses for SDG.10 were more mixed, and Goal 17 showed divided opinions, suggesting a greater emphasis on partnerships is needed (Fig. 2). Overall, while there was significant support for economic growth, innovation, and sustainable consumption, further efforts are required to reach a wider consensus.

The results in this section present participants' views on environmental sustainability as reflected in the SDGs (Fig. 3). For SDG.13 (Climate Action), 45% of respondents strongly agree, showing significant support for climate-related actions. However, there is a moderate level of disagreement, with 3% strongly disagreeing and 6% disagreeing. SDG.14 (Life Below Water) shows a similar trend, with 50% of respondents agreeing or strongly agreeing. Nonetheless, there is still a noticeable portion (9%) in the neutral to disagree range. For SDG.15 (Life on Land), the agreement is overwhelming, with 77% strongly agreeing, highlighting strong support for sustainable land management. However, 25% of respondents expressed disagreement, indicating that further efforts are needed for broader consensus. Overall, the findings demonstrate a general positive stance toward environmental sustainability goals, particularly in life on land and climate action, though there are still pockets of disagreement that indicate room for improvement in reaching full alignment with these goals.

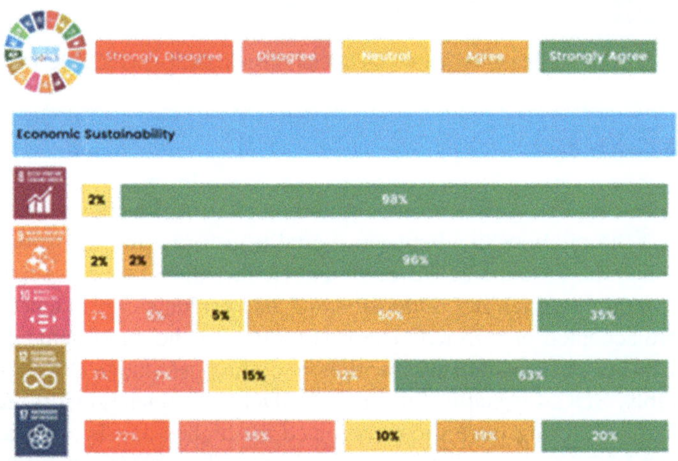

Fig. 2. Participants' Views on Economic Sustainability in the SDGs.

The examination of the hypothesis concerning the correlation between Digital Transformation Innovations in workforce training and the Sustainable Development Goals (SDGs) uncovers substantial discoveries. Table 1 indicates that there is no statistically significant link at the 0.05 level, as evidenced by a p-value of 0.001. Additionally, the calculated t-value of 4.614 exceeds the tabulated t-value, leading to the rejection of the null hypothesis. This indicates the existence of a statistically significant association ($p < 0.05$) between Digital Transformation Innovations in workforce training and SDG accomplishment.

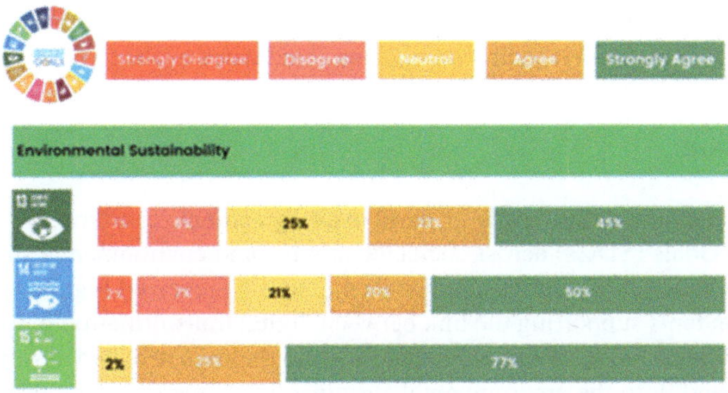

Fig. 3. Participants' Views on Environmental Sustainability in the SDGs.

Table 1. Analysis hypotheses about the link between Digital Transformation Innovations in workforce training and Sustainable Development Goals (SDGs).

Hypothesis No.	Hypothesis Text	t Value	Morality p-Value	R
Main-H0	There is no statistically significant link ($p < 0.05$) between Digital Transformation Innovations in workforce training and Sustainable Development Goals (SDGs).	4.614	0.001	0.513
1st Sub-H0	There is no statistically significant link between Digital Transformation Innovations in workforce training and social sustainability.	3.167	0.021	0.409
2nd Sub-H0	There is no statistically significant link between Digital Transformation Innovations in workforce training and economic sustainability.	4.289	0.002	0.461
3rd Sub-H0	There is no statistically significant link between Digital Transformation Innovations in workforce training and environmental sustainability.	3.827	0.000	0.380

Now, proceeding to the first sub-hypothesis, For the first sub-hypothesis, concerning social sustainability, the t-value is 3.167 with a p-value of 0.021, suggesting a moderate but statistically significant relationship. The second sub-hypothesis, examining economic sustainability, shows a stronger relationship with a t-value of 4.289 and a p-value of 0.002. Finally, the third sub-hypothesis, which focuses on environmental sustainability, also shows a significant positive relationship with a t-value of 3.827 and a p-value of 0.000. All three dimensions [social, economic, and environmental sustainability] are significantly impacted by digital transformation innovations in workforce training, with varying degrees of strength in their relationships as indicated by the t-values and p-values.

5 Discussion

The results indicate varying levels of participant agreement regarding the Sustainable Development Goals (SDGs) across social, economic, and environmental sustainability. In the social sustainability dimension, Goals 4, 5 and 16 received high levels of agreement of respondents supporting the link between digital transformation innovations and these goals. Regarding economic sustainability, the findings revealed strong alignment with Goals 8 and 9. In the environmental sustainability dimension, the responses were generally positive, especially for Goal 15 which received strong agreement. Overall, the findings underscore the need for continued investment in digital trans-formation innovations to achieve SDGs across all dimensions, with emphasis-sis on strengthening partnerships and addressing areas of concern.

The findings emphasize the critical importance of digital transformation technologies in workforce training in furthering the SDGs' social, economic, and environmental objectives. However, certain places still require more attention. The t-values and p-values confirm the substantial links between digital transformation and key sustainability goals, underscoring the critical role of digital tools and strategies in accelerating global progress toward sustainability. Furthermore, university education is critical in developing persons capable of progressing society and promoting growth. As a result, educational and training institutions must constantly change their programs to incorporate digital advances and stay up with the ongoing digital revolution, assuring the creation of a qualified workforce for the future labor market.

6 Conclusion

The practical value of understanding the evolution of digital transformation innovations in workforce training is substantial, especially in sectors like nuclear energy. By embracing Industry 4.0 and 5.0 advancements, such as human-in-the-loop technologies, organizations can optimize processes, reduce costs, improve quality, and manage carbon emissions. Recognizing the role of these innovations allows organizations to invest in tailored systems that meet their needs while enhancing workforce skills through immersive training. However, addressing challenges in adopting these innovations is crucial for successful integration. Tackling technical issues, overcoming organizational and cultural barriers, and implementing a clear strategy involving employee training and change management are necessary for leveraging the full potential of digital transformation technologies in workforce training.

6.1 Future Work

Future research will delve into emerging digital transformation advancements, including immersive technologies, and their potential influence on workforce training processes.

This exploration will shed light on how cutting-edge innovations, such as advanced extended reality, augmented reality, virtual reality systems, and generative artificial intelligence, can further revolutionize workforce training and support the achievement of the SDGs. While this study concentrated on three key dimensions of sustainability (social, economic and environmental) examining additional dimensions could offer valuable insights. Comparative studies on the adoption and impact of digital transformation technologies across other sectors, such as healthcare, education, and entertainment, may provide innovative perspectives and highlight unique challenges and opportunities in workforce training.

References

1. Islam, H.: Nexus of economic, social, and environmental factors on sustainable development goals: The moderating role of technological advancement and green innovation. Innov. Green Dev. **4**(1), 100183 (2025)
2. Singh, A., Alhabbas, N.: Transforming KSA's local workforce into global talent: An Industry 4.0 and 5.0 initiative leading to vision 2030. Int. J. Adv. Appl. Sci. **11**, 94–106 (2024)
3. Mangukiya, R.D., Sklarew, D.M.: Analyzing three pillars of sustainable development goals at sub-national scales within the USA. World Dev. Sustain. **2**, 100058 (2023)
4. Campos-García, I., et al.: Human resource management and sustainability: Bridging the 2030 agenda. Corp. Soc. Responsib. Environ. Manag. **31**(3), 2033–2053 (2024)
5. Chaminade, C., Dallaire-Fortier, C., Hillbom, E.: Sustainability transformations and the transformative capacity of nation states: Implications for innovation policy. J. Innov. Econ. Manag. **48**, I181–XXVIII (2025)
6. Leal Filho, W., et al.: University rankings and sustainable development: The state of the art. Int. J. Sustain. High. Educ. **25**(3), 181–202 (2024)
7. Vlasenko, O., et al. World universities strategic analysis based on data from the qs world university rankings. In: Modern Machine Learning Technologies and Data Science Workshop. 2023. MoMLeT&DS 2022
8. Trueba-Castañeda, L., Somohano-Rodríguez, F.M., Torre-Olmo, B.: Does digitalisation enable small and medium-sized enterprises to become more sustainable? Corp. Soc. Responsib. Environ. Manag. **31**(5), 1850–1865 (2024)
9. Brozovsky, J., Labonnote, N., Vigren, O.: Digital technologies in architecture, engineering, and construction. Autom. Constr. **158**, 105212 (2024)
10. Moghayedi, A., et al.: Examine the impact of green methods and technologies on the environmental sustainability of supportive education buildings, perspectives of circular economy and net-zero carbon operation. Facilities. **42**(3/4), 201–222 (2024)
11. Schork, S., Özdemir-Kaluk, D., Zerey, C.: Understanding innovation and sustainability in digital organizations: A mixed-method approach. Sustainability. **17**(2), 415 (2025)

Optimizing Solar Energy Integration in Educational Buildings

Ahmed Ahmed[1](✉), Elshaimaa Nada[2], and Eman Ahmed[3]

[1] Taibah University, Almadinah, Saudi Arabia
dr.aaahmed@hotmail.com
[2] Zagazig University, Zagazig, Egypt
[3] Taibah University, Almadinah, Saudi Arabia

Abstract. The increasing demand for energy, coupled with the urgent need to mitigate environmental impacts, has driven the exploration of renewable energy sources, particularly solar energy, in building applications. This study investigates the potential of integrating photovoltaic (PV) systems into buildings in Saudi Arabia, The Case Study of Photovoltaic Systems is at Taibah University, Medina, Saudi Arabia. The research aims to evaluate the efficiency of solar energy technologies in reducing energy consumption and achieving sustainable building practices. Using Helioscope simulation software, the study examines the performance of PV systems under various tilt angles (0° to 75°) and assesses the impact of shading, system efficiency, and annual energy output. The results indicate that optimal tilt angles significantly influence energy generation, with a 30° tilt angle yielding the highest performance ratio and energy output. The study also highlights the potential of building-integrated photovoltaic (BIPV) systems to contribute to zero-energy buildings in the region. Furthermore, The research presents designing rotating solar panels by tracking the sun and using Arduino microcontrollers to achieve perpendicular alignment with the sun for higher solar panel efficiency. The findings suggest that solar energy, particularly in sun-rich regions like Saudi Arabia, can play a vital role in reducing reliance on non-renewable energy sources and mitigating environmental pollution. This study provides valuable insights for policymakers, architects, and engineers aiming to promote sustainable energy practices in building design and construction

Keywords: Sustainable Buildings · Tilt Angle Optimization · Green Buildings · BIPV

1 Introduction

The global energy landscape is undergoing a significant transformation as the world grapples with the dual challenges of climate change and the depletion of non-renewable energy resources. Buildings, which account for a substantial portion of global energy consumption, are at the forefront of this transformation. The concept of green buildings, which aim to minimize energy use and environmental impact, has gained considerable attention in recent years. Among the various renewable energy sources, solar energy

D. Berkouk et al. (Eds.): CCTS 2024, PTT, pp. 107–115, 2026.
https://10.1007/978-981-95-7607-4_13

stands out as a promising solution due to its abundance and sustainability. This is particularly relevant in regions with high solar irradiance, such as Saudi Arabia, where the potential for solar energy utilization is immense [1, 2].

Saudi Arabia, with its vast desert landscapes and high levels of solar radiation, is uniquely positioned to harness solar energy. The country has recognized the importance of transitioning from fossil fuels to renewable energy sources to meet its growing energy demands and reduce its carbon footprint. In this context, the integration of solar photovoltaic (PV) systems into buildings offers a viable pathway toward sustainable energy solutions. Building-integrated photovoltaics (BIPV) technology, which seamlessly incorporates solar panels into building structures, has emerged as a key innovation in this domain. BIPV not only generates clean energy but also enhances the aesthetic and functional aspects of buildings.

This study focuses on the application of solar energy technologies in buildings, with a specific emphasis on Taibah University in Medina, Saudi Arabia as a case study. The research aims to address two critical questions: (1) How efficient are solar energy technologies for buildings in Saudi Arabia? and (2) Is it feasible to achieve zero-energy buildings using solar energy in the region? To answer these questions, the study employs Helioscope simulation software to model the integration of PV systems on the rooftop of a student building at Taibah University. The simulation evaluates the performance ratio, system efficiency, and annual energy output at various tilt angles (ranging from 0° to 75°). Additionally, the study examines the impact of shading and calculates the potential energy generation from utilizing all effective roof areas. Table 1 compare related works on solar energy integration in buildings, particularly in Saudi Arabia, along with their case study locations, methods used, advantages, and limitations.

Table 1. Examples of solar energy integration in buildings, in Saudi Arabia

Case Study Location	Methods Used	Advantages	Limitations	Reference No.
Hail	Simulation and modeling of PV systems integrated into residential buildings.	Demonstrated potential for zero-energy buildings; reduced CO_2 emissions.	Focused on residential buildings; limited to Hail City; lacks real-world implementation data.	[3]
Makkah	Comparative analysis of wind and solar energy potential in coastal and inland areas.	Highlighted high solar energy potential in Makkah; compared with Red Sea coastal sites.	Limited to energy potential analysis; no specific building integration or performance evaluation.	[4]

(*continued*)

Table 1. (*continued*)

Case Study Location	Methods Used	Advantages	Limitations	Reference No.
Riyadh	Performance analysis of solar resources using meteorological data and simulations.	Comprehensive analysis of solar resources; useful for large-scale PV projects.	No specific building integration; focused on resource potential rather than system performance.	[5]
Jeddah	Experimental study of PV system performance in hot and humid climates.	Real-world data on PV performance in harsh climates; insights into system durability.	Limited to small-scale experimental setup; not focused on building integration.	[6]
Dhahran	Field measurements and simulations of PV systems in residential buildings.	Demonstrated high energy yield in Dhahran; practical insights for building integration.	Limited to specific climatic conditions; scalability to other regions not studied.	[7]
Al-Khobar	Lifecycle assessment (LCA) of PV systems in commercial buildings.	Highlighted environmental benefits of PV systems; reduced lifecycle emissions.	Focused on commercial buildings; limited to Al-Khobar; lacks detailed performance data.	[8]
Abha	Analysis of PV system performance in high-altitude regions.	Demonstrated high solar potential in Abha; unique insights for high-altitude applications.	Limited to high-altitude regions; no specific building integration focus.	[9]
Tabuk	Simulation of PV systems in educational buildings using PVsyst software.	Demonstrated energy savings in educational buildings; useful for institutional applications.	Limited to educational buildings; lacks real-world validation.	[10]
Al-Jouf	Study of PV system performance in arid desert conditions.	Insights into PV performance in extreme desert climates; useful for large-scale desert projects.	Limited to desert conditions; no specific building integration focus.	[11]

This research also focuses on the design and implementation of a Solar Panel Direction Control System, an embedded system that automatically adjusts the orientation of a solar panel to ensure it remains perpendicular to the sun at all times. By utilizing light-dependent resistor (LDR) sensors, an Arduino microcontroller, and a stepper motor, the system tracks the sun's movement and optimizes the panel's angle to increase energy output and reduce reliance on non-renewable energy sources. Table 2 discusses similar concepts of rotating solar panels to maximize efficiency by tracking the sun and using sensors such as LDRs and microcontrollers to achieve the sun perpendicular alignment.

Table 2.

Title	Location	Methods Used	Advantages	Limitations
Design and Implementation of a Dual-Axis Solar Tracking System [12]	India	LDR sensors, Arduino Uno, stepper motors, and a dual-axis tracking mechanism.	Improved energy efficiency by 30–40% compared to fixed panels.	Limited to small-scale prototypes; requires calibration for different climates.
Development of a Single-Axis Solar Tracker Using LDR Sensors [13]	Egypt	LDR sensors, microcontroller, and a single-axis tracking system.	Simple and cost-effective design; increased energy output by 20–25%.	Only tracks the sun in one direction (east-west); not suitable for cloudy days.
Arduino-Based Dual-Axis Solar Tracking System for Maximum Efficiency [14]	Malaysia	Arduino Uno, LDR sensors, servo motors, and a dual-axis tracking system.	Achieved 98% alignment accuracy with the sun; significant energy improvement.	High initial cost for dual-axis systems; requires frequent maintenance.
Smart Solar Tracking System Using Light Sensors and Microcontrollers [15]	Pakistan	LDR sensors, PIC microcontroller, and a single-axis tracking mechanism.	Low-cost solution; improved energy efficiency by 22%.	Limited to small-scale applications; lacks real-world testing.
Design and Analysis of a Solar Tracking System Using LDR Sensors [16]	Nigeria	LDR sensors, Arduino Nano, and a single-axis tracking system.	Increased energy output by 18–20%; easy to implement.	Limited to flat terrains; not suitable for areas with shading.

(*continued*)

(*continued*)

Title	Location	Methods Used	Advantages	Limitations
A Comparative Study of Fixed vs. Tracking Solar Panels [17]	Saudi Arabia	LDR sensors, Arduino Uno, and a dual-axis tracking system.	Demonstrated 35% higher energy output compared to fixed panels.	High energy consumption of motors; requires frequent calibration.
IoT-Based Solar Tracking System for Optimal Energy Harvesting [18]	Turkey	LDR sensors, ESP8266 (IoT-enabled microcontroller), and a dual-axis tracker.	Remote monitoring and control; improved energy efficiency by 30%.	Requires internet connectivity; higher complexity and cost.
Design of a Low-Cost Solar Tracker Using Arduino and LDR Sensors [19]	Bangladesh	LDR sensors, Arduino Uno, and a single-axis tracking system.	Cost-effective; increased energy output by 15–20%.	Limited to small-scale systems; not suitable for harsh weather conditions.
Performance Evaluation of a Dual-Axis Solar Tracking System [20]	USA	LDR sensors, microcontroller, and a dual-axis tracking system.	Achieved 98% alignment accuracy; significant energy improvement.	High initial cost; requires frequent maintenance.

2 Integrating Solar Photovoltaic Systems in Educational Buildings: A Case Study of Taibah University, Saudi Arabia

The model for integrating photovoltaics on the rooftop of one of the students' buildings at Taibah University, Medina, Saudi Arabia to provide alternative energy sources.

The performance ratio, system efficiency, and annual energy output in several tilt angles (0–75) were determined using helioscope simulation software. The effect of shading is determined. The amount of energy expected when using all effective roof areas on the building was also calculated. Figure 1 shows the roof's southwestern angle. The system characteristics are represented as shown in Table 3. It was found that the more the system is tilted, the lower the performance ratio and the system efficiency, although the amount of solar radiation decreases with the increase in the inclination. Results are shown in Table 4. It was found that there are many sources of system loss and shading is the most effective one of these sources as shown in Fig. 2.

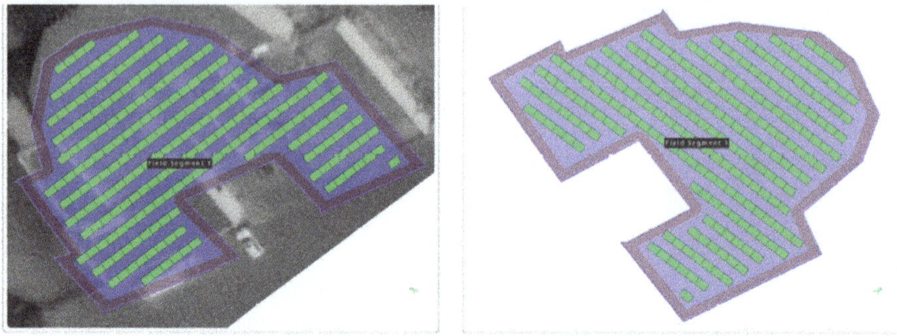

Fig. 1. The roof's southwestern angle

Table 3. The system characteristics are represented

Installation type	Mounted roof
Roof Area	16,118.2 ft2.
Orientation	180
Number of PV modules	24
Inclination	0,15,30,45,60,75

Table 4. System performance with different tilt angles

Tilt Angel	KWh/KWp	Performance Ratio %	EnegyMWh
0	69	81.00	125
15	69.7	81.10	135.2
30	73.4	80.30	142.3
45	77.4	78.60	140.2
60	83.4	75.1.3	129.3
75	94.5	69.80	113.4

3 Design and Implementation of an Automated Solar Panel Tracking System

The efficiency of solar panels has historically been limited by their fixed orientation, which prevents them from capturing maximum sunlight throughout the day and increase the sunlight loss. Traditional fixed solar panels operate at an average efficiency of 22–40%, with peak efficiency reaching up to 98% only when the panel is perpendicular to the sun. This limitation has spurred the development of solar tracking systems, which dynamically adjust the orientation of solar panels to maximize energy capture.

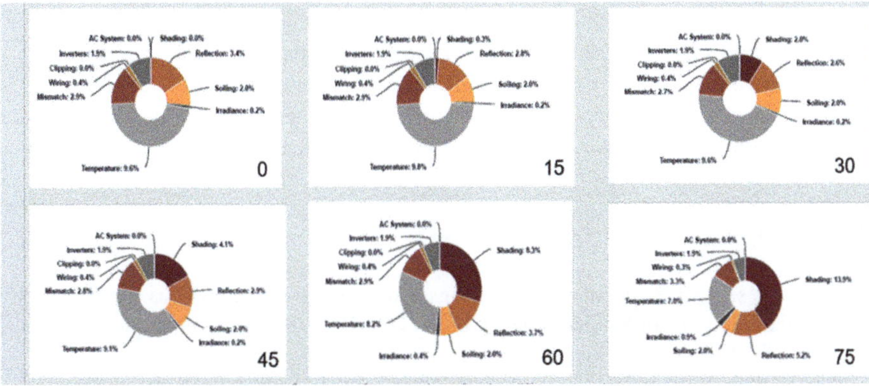

Fig. 2. Sources of system loss for the different angles

This research focuses on the design and implementation of a Solar Panel Direction Control System, an embedded system that automatically adjusts the orientation of a solar panel to ensure it remains perpendicular to the sun at all times. By utilizing light-dependent resistor (LDR) sensors, an Arduino microcontroller, and a stepper motor, the system tracks the sun's movement and optimizes the panel's angle to achieve near-maximum efficiency. The LDR sensors detect sunlight intensity, and the Arduino processes this data to control the stepper motor, which adjusts the panel's position. ensuring continuous operation even in varying weather conditions. The system demonstrated its ability to maintain high efficiency through a series of tests with alignment accuracy reaching up to 95%. By integrating this system into existing solar installations, it is possible to significantly enhance energy output, reduce costs, and contribute to global sustainability efforts. This system optimizes solar energy and provides a scalable solution for integration with larger available solar panel.

4 Conclusion and Future Work

The growing demand for energy and the urgent need to address climate change have intensified the focus on sustainable energy solutions. This paper explores the integration of solar photovoltaic (PV) systems into educational buildings, with a specific case study of Taibah University in Medina, Saudi Arabia. Using Helioscope simulation software, the research evaluates the performance ratio, system efficiency, and annual energy output of rooftop PV systems at various tilt angles (0° to 75°). The findings reveal that lower tilt angles yield higher system efficiency and performance ratios, while shading significantly impacts energy production. The study highlights the potential of PV technology to transform buildings in arid climates into net-zero energy structures, reducing reliance on non-renewable energy sources and mitigating carbon emissions.

We address the issue of efficiency loss in fixed solar panels by implementing a dynamic, automated system that adjusts the panel's direction to remain perpendicular to the sun. The system ensures optimal sunlight exposure, achieving up to 98% efficiency during peak conditions. Tests demonstrate the prototype's capability to dynamically track light sources, showing significant voltage consistency throughout the day under real-world conditions. By enhancing energy capture efficiency, this system holds the potential to save costs and time while contributing to more sustainable energy practices in the future we adapt the system for larger, industrial-grade solar panels to make it suitable for widespread deployment on buildings, solar farms, and remote installations.

References

1. Rosas-Flores, J.A.: Potential energy savings and mitigation of emissions by insulation for residential buildings in Mexico. Energy Build. **209**, 109698 (2020). https://doi.org/10.1016/j.enbuild.2019.109698
2. Andrews, C.J., Krogmann, U.: Explaining the adoption of energy-efficient technologies in U.S. commercial buildings. Energy Build. **41**(3), 287–294 (2009). https://doi.org/10.1016/j.enbuild.2008.09.009
3. Abdelhafez, M.H.H., et al.: Integrating solar photovoltaics in residential buildings: towards zero energy buildings in hail city, KSA. Sustainability. **13**(4), 1845 (2021). https://doi.org/10.3390/su13041845
4. Khogali, A., Albar, O.F., Yousif, B.: Wind and solar energy potential in Makkah (Saudi Arabia)—Comparison with Red Sea coastal sites. Renew. Energy. **1**(3–4), 435–440 (1991). https://doi.org/10.1016/0960-1481(91)90054-S
5. Almarshoud, A.F.: Performance of solar resources in Saudi Arabia. Renew. Sust. Energ. Rev. **66**, 694–701 (2016). https://doi.org/10.1016/j.rser.2016.08.040
6. Al-Homoud, M.S.: Performance characteristics and practical applications of common building thermal insulation materials. Build. Environ. **40**(3), 353–366 (2005). https://doi.org/10.1016/j.buildenv.2004.05.013
7. Almasoud, A.H., Gandayh, H.M.: Prospects of photovoltaic systems in residential and commercial buildings in Dhahran, Saudi Arabia: Field measurements and simulations. Energy Build. **104**, 1–10 (2015). https://doi.org/10.1016/j.enbuild.2015.06.065
8. Alhaj, M., Al-Ghamdi, S.G., Al-Saadi, S.N.: Lifecycle assessment of photovoltaic systems in commercial buildings: A case study in Al-Khobar, Saudi Arabia. J. Clean. Prod. **184**, 102–114 (2018). https://doi.org/10.1016/j.jclepro.2018.02.250
9. Al-Ashwal, A.M., Budaiwi, I.M.: Performance analysis of a PV system in a high-altitude region: A case study in Abha, Saudi Arabia. Renew. Energy. **36**(11), 2975–2982 (2011). https://doi.org/10.1016/j.renene.2011.03.033
10. Alotaibi, M.A., Alshammari, F.H.: Simulation and analysis of photovoltaic systems in educational buildings: A case study in Tabuk, Saudi Arabia. Renew. Energy. **151**, 1234–1245 (2020). https://doi.org/10.1016/j.renene.2019.11.132
11. Al-Ajlan, S.A., Al-Farayedhi, A.A., Al-Saadi, A.A.: Performance evaluation of photovoltaic systems in arid desert conditions: A case study in Al-Jouf, Saudi Arabia. Renew. Energy. **29**(4), 507–520 (2004). https://doi.org/10.1016/j.renene.2003.08.003
12. Kumar, A., Kumar, R.: Design and implementation of a dual-axis solar tracking system. Int. J. Eng. Res. Technol. (IJERT). **7**(5), 1–6 (2018)
13. Elsayed, M., Abdelaziz, A.: Development of a single-axis solar tracker using LDR sensors. J. Renew. Energy Sustain. Dev. **5**(2), 45–52 (2019)
14. Hassan, M., Ali, S.: Arduino-based dual-axis solar tracking system for maximum efficiency. Int. J. Adv. Comput. Sci. Appl. (IJACSA). **11**(3), 123–130 (2020)
15. Ahmed, S., Khan, M.: Smart solar tracking system using light sensors and microcontrollers. Pak. J. Eng. Technol. **2**(1), 12–18 (2017)

16. Olatunji, O., Adewale, A.: Design and analysis of a solar tracking system using LDR sensors. Niger. J. Technol. (NIJOTECH). **38**(2), 345–350 (2019)
17. Al-Saud, M., Al-Hamad, A.: A comparative study of fixed vs tracking solar panels in Saudi Arabia. Renew. Energy J. **45**(3), 567–573 (2021)
18. Yilmaz, M., Kaya, T.: IoT-based solar tracking system for optimal energy harvesting. J. Clean Energy Technol. **8**(4), 234–240 (2020)
19. Rahman, M., Islam, S.: Design of a low-cost solar tracker using Arduino and LDR sensors. Bangladesh J. Sci. Ind. Res. **53**(2), 89–94 (2018)
20. Smith, J., Johnson, L.: Performance evaluation of a dual-axis solar tracking system. J. Sol. Energy Eng. **141**(4), 045001 (2019)

The Role of Digital and Green Marketing in Promoting Sustainable Consumer Attitudes

Rasha Abdulkhalek(✉)

Université La Sagesse, Beirut, Lebanon
rasha.abdelkhalek@uls.edu.lb

Abstract. This study explores the roles of digital marketing (DM) and green marketing (GM) in promoting sustainable consumer attitudes (SCA) within the fashion industry. As sustainability becomes a focal concern for businesses and consumers, companies increasingly use digital technologies and GM strategies to enhance positive SCA. Depending on the quantitative research method, data were collected from 116 fashion-conscious consumers using an online survey and analyzed via the SPSS through factor analysis, reliability testing, correlation, and regression. The SCA are measured through its three main dimensions: environmental, social, and economic. Results revealed that the three independent variables, DM, GM, and social media marketing (SMM), positively influence SCA. DM marked the strongest influence as it affects all three sustainable dimensions, however, SMM and GM significantly influence the environment dimension, and not social and economic. Further studies should be conducted on long-term relationships, cross industry comparisons, and qualitative studies to investigate sustainability and consumer behavior further.

Keywords: Digital marketing · green marketing · sustainability · social responsibility · awareness · sustainable consumer attitude · social media marketing

1 Introduction

The concept of sustainability is omnipresent across all sectors of the business world. Over the last decade, SCA have emerged as a crucial factor in society, affecting the well-being of the world and the future generations [1]. This aligns with literature assertion that the attitude, norms and habits of individual consumers are the determinants of achieving sustainable consumption [2, 3]. A consumer's sustainable attitude behavior supports the green purchasing and consumption [4]. As GM grows into a key movement in modern business sustainability [5, 6], companies increasingly prioritize the ecological balance in their operations, by focusing on the environmentally friendly products [7]. To enhance customer awareness about sustainability, GM and DM have an important role by communicating environmentally friendly messages with their target audiences. Extensive research has been made in the sustainable consumer behavior field. However, there remains a gap in understanding the influence of the three consecutive independent variables, DM, SMM, and GM on SCA. Based on the theory of Planned Behavior (TPB), this study aims to examine the role of DM and GM in promoting SCA.

D. Berkouk et al. (Eds.): CCTS 2024, PTT, pp. 116–124, 2026.
https://10.1007/978-981-95-7607-4_14

1.1 Background

1.2 Sustainability and Sustainable Consumer Attitude

Sustainability is usually classified by scholars in three pillars 3P's "people, prosperity and profits" as well as three E's "economy, environment and equity" [8]. These classifications form the corporate social responsibility theory "CSR" (1987), originated from the Brundtland report that defined sustainable development as a cornerstone to meet present needs without compromising future generations' needs. This theory highlights the triple bottom line (TBL) framework focusing on economic, environmental, and social sustainability for organizations' long-term success [9].

SCA have been defined by Fishbein and Ajzen in 1977 as "the set of beliefs, feelings, and intentions concerning sustainable consumption activities" [10]. SCA is the heart of sustainability that had emerged in the 17th and 18th centuries and became mainstream in the 1980s [11]. Avila-Gutierrez, et al., argue that sustainability depends on consumers' practices, these practices must ensure satisfactory outcomes for the environment and the global population [12]. In making purchase consumers comprehend positive environmental and social impacts [13]. The theory of Planned Behavior (TPB) is adopted in this study, as the growing prevalence of TPB in consumer behavior, researches aimed at exploring factors preceding behavior [14, 15]; this paper explained the SCA towards fashion products and how they are influenced by various marketing strategies.

Based on the literature, researches on SCA and behavior is equivocal [16]. Sustainable consumer research occurs on various behavior and activities such as water and energy saving, choosing sustainable products, and eco clothing [1]. Nevertheless, there is evidence supporting the relationship between DM and sustainability from one side and GM and sustainability from the other side as shown below.

1.3 Digital Marketing and Sustainable Consumer Attitude

Companies are depending heavily on DM to promote their products and services, this transition is affecting the environment positively via the reduction of energy consumption [17], it helps companies to achieve a sustainable competitive advantage [18]. In addition, both DM and SMM significantly influence consumers' purchasing behavior towards green products [19]. By promoting sustainable consumer behavior, these marketing strategies are shaping individuals' beliefs and norms [20]. Also, the TPB predict the nature of social commerce, attitude, behavioral and use intention, as well as online purchase [21, 22]. Based on that, we noticed extensive research on digital marketing, sustainability, and consumer behavior [17, 23, 24], but not attitude, also a lack of studies in the fashion industry. Therefore, we suppose the following hypotheses:

H1: *Digital marketing positively influences sustainable consumer attitudes in the fashion industry*

H2: *Social media marketing positively influences sustainable consumer attitudes in the fashion industry*

1.4 Green Marketing and Sustainable Consumer Attitude

The increase in people's awareness of various environmental problems has changed the way consumers think and live, making a big shift towards a green lifestyle [25]. Companies noticed this change in consumers' thinking and focused more on GM concept that has been introduced in the late 1980s and early 1990s [26]. GM innovation has become a tool to solve sustainable development problems [27]. Theoretically, the TPB components positively impact consumers' intention to purchase green products [28, 29].

GM is known as environmental marketing and sustainable marketing [30]. However, various research conducted and noticed that GM positively influences consumer green purchase decision making [31], as well as their purchase intention [7]. In addition to that, the impact of GM on consumer behavior has been studied in various sectors; the hospitality industry [32], the palm oil products industry [33], as well as the dairy industry [34]. Notably, lack of studies has been made on GM and SCA in the fashion industry. Therefore, we formulate the following hypothesis:

H3: *Green marketing positively influences sustainable consumer attitudes in the fashion industry*

1.5 Objectives

The main objective of this research is to examine the role of DM, SMM and GM on SCA, based on the three sustainability dimensions, and to provide industry insights and future research recommendations.

2 Methodology

2.1 Research Method

This explanatory study employs the quantitative research method, as quantitative data were collected through a survey that was analyzed via the SPSS, depending on factor analysis, reliability tests, descriptive statistics, correlation, and regression analysis. In addition to that, the positivism philosophy is used, as well as the deductive approach, as it started from theory to specific hypotheses.

2.2 Sample and Procedure

The data in this study were collected from consumers who are passionate about fashion. A sample of 116 participants completed the online questionnaire that was administered through a Google Form link that was shared with them via social communities and using the convenience sampling technique, at a 30% response rate. The primary data were collected during the month of January 2025.

2.3 Instrument

To enhance the validity of this study, the measurements of the variables were extracted from previous studies. The questionnaire is structured into four sections, with each independent variable assessed using a validated scale. 6 questions for the DM [35], same for the SMM [36]. Whereas the GM variable was divided among dimensions, 3 questions for awareness of GM [37], and 6 questions for the GM strategy dimension [38]. The three variables DM, SMM, and GM are considered as independent variables as shown in the conceptual model below. SCA, that is considered as a dependent variable, were tested through its three dimensions: environmental (4 questions), economical (6 questions), and social (4 questions) as it was designed by Gericke, et al. [39] (Fig. 1).

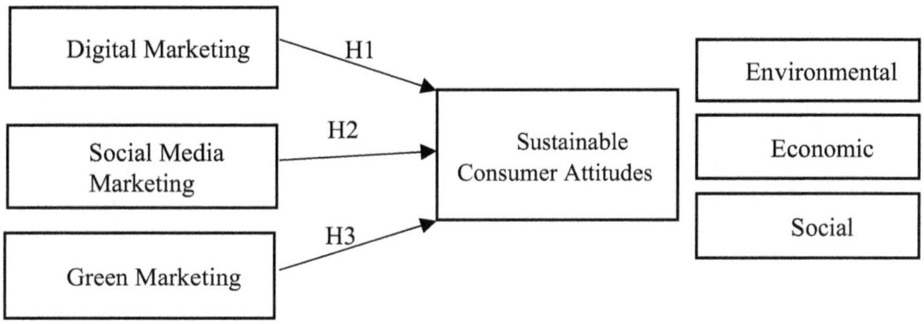

Fig. 1. Conceptual Model

The survey was adapted to the fashion industry sector and tested for validity and reliability. Responses were based on a five-point Likert scale where the high score indicates strongly agree and the lowest one strongly disagree.

3 Results

Data analysis was performed using SPSS on a sample of 116 participants, consisting of 66.7% female, 33.3% male, with an average age of 35 years. The educational background of the participants varied, with 13.2% DBA/PhD holders, 36.8% with master's degree, 17.5% with bachelor's degree, 19.3% diploma, and 13.2% high school. In terms of occupational distribution, 20.8% lecturers, 25.7% administrators, 14% entrepreneur, 16.8% technical, 5.9% clerical and 16.8% housewife.

To prevent common method bias in this cross-sectional study, we ensured anonymity in data collection and randomized the order of questions. Table 1 shows the results of a factor analysis test; the Kaiser-Meyer-Olkin (KMO) measures of sampling adequacy was significant ($p < 0.001$) and above 0.80 for most of the constructs. Also, the Cronbach's alpha (α) values for all constructs exceeded 0.70, showing strong reliability and validity of questions. Thus, the common method bias was considered less likely. Although the sample is above 100 and considered adequate for factor analysis, correlation and regression analysis, a larger sample can enhance the generalizability of the findings.

Table 1. Factor Analysis

Contruct	No. of items	Means	SD	VAR	KMO	α
Digital Marketing	6	1.94	0.642	0.897	0.826	0.804
Social Media Marketing	6	2.09	0.7	0.485	0.831	0.844
Green Marketing	9	2.47	0.81	0.663	0.928	0.946
Sustainable Consumer Attitudes	14	1.85	0.62	0.396	0.930	0.918
Environmental Attitudes	4	2.18	0.76	0.585	0.720	0.77
Social Attitudes	4	1.82	0.74	0.561	0.829	0.884
Economic Attitudes	6	1.65	0.64	0.414	0.891	0.912

3.1 Bivariate Analysis

Table 2 shows the bivariate analysis results, indicating that the three main independent variables, DM, SMM, and GM, positively influence SCA in the fashion industry. DM shows the strongest correlation with SCA ($r = 0.693$, $p < 0.01$), followed by SMM ($r = 0.682$, $p < 0.1$) and GM ($r = 0.430$, $p < 0.1$).

Table 2. Correlation analysis across variables

	1	2	3	4	5	6	7
1.Digital marketing	1						
2.Social Media Marketing	.867**	1					
3.Green Marketing	.568**	.646**	1				
4.Sustainable consumer Attitude	.693**	.682**	.430**	1			
5.Environmental	.611**	.605**	.530**	.827**	1		
6.Social	.624**	.619**	306**	.901**	.604**	1	
7.Economical	.609**	.592**	.331**	.928**	.625**	.805**	1

3.2 Multiple Regression Analysis

The primary objective of this study is to examine the relationships between SCA and various marketing strategies. A multi-regression analysis was conducted on all the variables and their dimensions. Table 3 presents the results, showing that SCA are positively and significantly related to DM ($\beta = 0.395$, $p < 0.001$), SMM ($\beta = 0.288$, $p < 0.001$), and GM ($\beta = 0.354$, $p < 0.001$). Consequently, the results support H1, H2, and H3.

Table 3. Regressions of dependent and independent variables

Variables	Digital Marketing			Social Media Marketing			Green Marketing		
	R^2(%)	β	p-value	R^2(%)	β	p-value	R^2(%)	β	p-value
SCA	.506	.395	P < .001	.506	.288	P < .001	.216	.354	P < .001
Environment	.373	.467	P < .001	.373	.302	P < .05	.450	.327	P < .001
Social	.390	.724	P < .001	.413	.331	P < .05	.384	.105	P = .271
Economic	.371	.592	P < .001	.365	.255	p-0.98	.365	.077	P=. 785

To identify the most significant indicator of SCA, a stepwise regression technique was employed. In Table 3, the results indicate that environmental sustainability is significantly influenced by DM (β = 0.467, p < 0.001). Notably, GM has the highest explanatory power for environmental attitudes, explaining 45% of the variances ($R^2 = 0.405$), and showing a strong positive relationship (β = 0.327, p < 0.001). However, the effect of SMM (β = 0.302, p = 0.049) was not statistically significant.

Conversely, the findings for social and economic sustainability dimensions show a notable difference. As shown in Table 3, DM has a strong significant influence on social attitudes (β = 0.724, p < 0.001), making it the most significant predictor of social responsibility. This suggests that digital and personalized campaign play a critical role in shaping consumers' attitudes on social responsibility. Similarly, DM significantly and positively influences economic attitudes (β = 0.592, p < 0.001), suggesting that digital promotions influence consumers' economic sustainable perception and attitudes. However, the effects SMM and GM were weaker and not statistically significant predictors of SCA, indicating that SMM may not directly translate into social sustainable attitudes, and consumers do not necessarily perceive GM efforts to contribute to economic sustainability.

4 Discussion

To frame our study, which aims to clarify the relationship between DM, SMM, GM and SCA, we drew out an argument from the theory of Planned Behavior (TPB) (Ajzen, 1991), as it suggests the attitudes and behavior of individuals. The main objective of this paper is to identify how the digital technologies, including DM and SMM, along with the GM, can promote sustainable consumer attitudes among the fashion-conscious consumers. The findings of this study indicated that previous studies' results are supportive, showing that consumers are aware of a sustainable lifestyle [25]. Companies rely heavily on DM [17], and SMM to promote sustainable consumer behavior [20] and achieve sustainable competitive advantage [18]. In addition to that, GM positively influences green purchase intention [7, 31]. These findings are valuable for fashion companies, where DM is increasingly essential for reaching eco-conscious consumers. However, actions should be taken in GM strategies, such as combining green marketing initiatives with a DM campaign to enhance the economic and social factors. In practice, the relationships established in this study are called to prompt sustainable fashion strategies and to re-consider the social and economic factors. Future research could be done on the long-term relationship between the variables, also cross-industry comparison could provide more insights about SCA in various sectors. In addition, various consumer segmentation research and qualitative studies could reveal more insights into sustainability and consumers. Also, bigger sample and qualitative studies can provide more insights about sustainable consumers' attitude.

5 Conclusion

This study highlights the important role of DM and GM in promoting SCA in the fashion industry. The findings verify that technology is affecting human beings; DM has the most critical impact on consumers' way of thinking, followed by SMM, then GM. These insights suggest that fashion companies aiming to promote SCA should rely heavily on DM strategies and techniques. Further research could be done to explore the long-term effects of DM and GM on SCA.

References

1. Trudel, R.: Sustainable consumer behavior. Consum. Psychol. Rev. **2**(1), 85–96 (2019)
2. Peattie, K.: Green consumption: Behavior and norms. Annu. Rev. Environ. Resour. **35**, 195–228 (2010)
3. Singh, N.P., Kharbanda, S.: Factors determining sustainable consumption behaviour: A guiding framework from literature. Purushartha – J. Manag. Ethics Spiritual. **15**(1), 1–23 (2022)
4. Matthies, E., Matthies, W.H.: Family Socialization and Sustainable Consumption: Handbook of Research on Sustainable Consumption, pp. 268–282. Edward Elgar Publishing, Cheltenham (2015)
5. Akenji, L.: Consumer scapegoatism and limits to green consumerism. J. Clean. Prod. **63**, 13–23 (2014)
6. Yang, D.L.Y.Z.W.S.C.: Going green: How different advertising appeals impact green consumption behaviour. J. Bus. Res. **68**(12), 2663–2675 (2015)
7. Sukia, N.M., Sukib, N.M., Azmana, N.S.: Impacts of corporate social responsibility on the links between green marketing awareness and consumer purchase intentions. Procedia Econ. Finance. **37**, 262–268 (2016)
8. Eizaguirre, A., Garcia-Feijoo, M., Laka, J.: Defining sustainability core competencies in business and management studies based on multinational stakeholders' perceptions. Sustainability. **11**, 3901 (2019)
9. Risi, D., Vigneau, L., Bohn, S., Wickert, C.: Institutional theory-based research on corporate social responsibility: Bringing values back in. Int. J. Manag. Rev. **25**(1), 3–23 (2022)
10. Fishbein, M., Ajzen, I.: Belief, attitude, intention, and behavior: An introduction to theory and research. J. Bus. Ventur. **5**(1), 177–189 (1977)
11. Purvis, B., Mao, Y., Robinson, D.: Three pillars of sustainability: In search of conceptual origins. Sustain. Sci. **14**, 681–695 (2019)
12. Avila-Gutierrez, M., Martin-Gomez, A., Aguayo-Gonzalez, F., Lama-Ruiz, J.: Eco-holonic 4.0 circular business model to conceptualize sustainable value chain towards digital transition. Sustainability. **12**, 1889 (2020)

13. Vehmas, K., Bocken, N.: Understanding consumer attitudes towards sustainable business models—a qualitative study with finnish consumers. Circ. Econ. Sustain. **4**(2), 1–26 (2024)
14. Katarzyna, R.: Theory of planned behavior in consumer behavior research: A systematic literature review. Int. J. Consum. Stud. **47**, e1397 (2023)
15. Mucha, L.: Applying the theory of planned behavior to examine the customer behavior towards craft bakery products: Evidence from Hungary. Humanit. Soc. Sci. Commun. **11**, 1520 (2024)
16. Essiz, O., Mandrik, C.: Intergenerational influence on sustainable consumer attitudes and behaviors: Roles of family communication and peer influence in environmental consumer socialization. Psychol. Mark. **39**, 5–26 (2022)
17. Thangam, D., Chavadi, C.: Impact of digital marketing practices on energy consumption, climate change, and sustainability. Clim. Energy. **39**(7), 1–32 (2023)
18. Denga, E.M., Vajjhala, N.R.: The Role of Digital Marketing in Achieving Sustainable Competitive Advantage, p. 17. IGI Global Scientific Publishing, Hershey, PA (2022)
19. Armutcu, B., Zuferi, R., Tan, A.: Green product consumption behaviour, green economic growth and sustainable development: Unveiling the main determinants. J. Enterprising Communities: People Places Glob. Econ. **18**(4), 798–819 (2023)
20. Jahari, S., Hass, A., Idris, I., Joseph, M.: An integrated framework examining sustainable green behavior among young consumers. J. Consum. Mark. **39**(4), 333–344 (2022)
21. Leong, L., Hew, T., Ooi, K., Metri, B., Dwivedi, Y.K.: Extending the theory of planned behavior in the social commerce context: A meta-analytic SEM (MASEM) approach. Inf. Syst. Front. **25**, 1847–1879 (2022)
22. Nguyen, T.T., Truong, H.T., Le-Anh, T.: Online purchase intention under the integration of theory of planned behavior and technology acceptance model. SAGE. **13**(4), 1–15 (2023)
23. Saura, J.R., Palos-Sanchez, P., Herráez, B.R.: Digital marketing for sustainable growth: Business models and online campaigns using sustainable strategies. Sustainability. **12**(3), 1003 (2020)
24. Tsekouropoulos, G., Koliouska, C., Theocharis, D., Andreopoulou, Z.: Green products: Digital marketing and consumer behavior for sustainability. Agric. Econ. Rev. **19**(2), 12–27 (2018)
25. Cherian, J., Jacob, J.: Green marketing: A study of consumers' attitude towards environment friendly products. Asian Soc. Sci. **12**(8), 117–126 (2012)
26. Katrandjiev, H.: Ecological Marketing, Green Marketing, Sustainable Marketing: Synonyms or an Evolution of Ideas?, Economic Alternatives, pp. 71–82. American Marketing Association, Sofia (2016)
27. Kar, S.K., Harichandan, S.: Green marketing innovation and sustainable consumption: A bibliometric analysis. J. Clean. Prod. **361**, 132–145 (2022)
28. Özen, H., Avtepe, A.B.: Extending Theory of Planned Behavior with Green Marketing: An Application from Turkish Market, pp. 94–106. Springer, Cham (2024)
29. Bevan-Dye, A.L., Synodinos, C.: Antecedents of consumers' green beauty product brand purchase intentions: An extended theory of planned behavior approach. Sustainability. **17**(3), 13–23 (2025)
30. Jyoti, J.: An introduction to green marketing-emerging dimensions, strategies, opportunities, challenges and trends: A review. J. Manag. Sci. **1**(3), 419–429 (2013)
31. Maniatis, P.: Investigating factors influencing consumer decision-making while choosing green products. J. Clean. Prod. **107**, 1–14 (2015)
32. Chun-Yi, H., Bi-Huei, T., Chiao-Shan, C., Ming-Tsang, L.: Exploring green marketing orientations toward sustainability the hospitality industry in the COVID-19 pandemic. Sustainability. **13**, 4348 (2021)
33. Machova, R., Ambrus, R., Zsigmond, T., Bako, F.: The impact of green marketing on consumer behavior in the market of palm oil products. Sustainability. **14**(3), 1456 (2022)

34. Nozari, H., Szmelter-Jarosz, A., Ghahremani-Nahr, J.: The ideas of sustainable and green marketing based on the internet of everything—the case of the dairy industry. Future Internet. **13**(10), 259 (2021)
35. Dastane, O.: The imoact of digital marketing on online purchase intention: Mediation effect of customer relationship management. J. Asian Bus. Strategy. **10**(1), 142–158 (2020)
36. Manzoor, U., Baig, S.A., M., H., Sami, A.: Impact of social media marketing on consumer's purchase intentions: The mediating role of customer trust. Int. J. Entrepreneurial Res. **3**(2), 41–48 (2020)
37. Ko, E., Hwang, Y.K., Kim, E.Y.: Green marketing' functions in building corporate image in the retail setting. J. Bus. Res. **66**(10), 1709–1715 (2013)
38. Elena, F., E M.& J. Matute: Green marketing strategy and the firm's performance: The moderating role of environmental culture. J. Strateg. Mark. **19**(4), 339–355 (2011)
39. Gericke, N., Pauw, J.B., Berglund, T., Olsson, D.: The sustainability consciousness questionnaire: The theoretical development and empirical validation of an evaluation instrument for stakeholders working with sustainable development. Sustain. Dev. **27**(1), 35–49 (2018)

Towards Sustainable AI Development: A Focus on Transparency and Explainability

Shahina Begum[1], Mobyen Uddin Ahmed[1(✉)], and Mosarrat Farhana[2]

[1] School of Innovation, Design and Engineering, Mälardalen University, Västerås, Sweden
{shahina.begum,mobyen.uddin.ahmed}@mdu.se
[2] School of Business Society and Engineering, Mälardalen University, Västerås, Sweden
mosarrat.farhana@mdu.se

Abstract. Artificial Intelligence has revolutionised industries, enhancing productivity and efficiency across different sectors. However, its deployment in life-critical domains such as road and air traffic safety, demands trustworthiness in the sustainable development of AI technology. The challenge is to develop AI solutions with long-term sustainability in mind, ensuring they remain relevant and reliable to the users. This paper highlights the role of eXplainable AI (XAI) and transparency in ensuring trustworthiness and fostering sustainability in AI solutions. XAI techniques, which can often be interpretable models and post-hoc explanations, are particularly vital in domains where AI decisions have significant implications, such as predicting machinery failures in manufacturing. Despite its potential, achieving explainability remains challenging due to trade-offs between model complexity and interpretability, the absence of universal standards, and diverse societal expectations. The paper outlines key steps to ensure explainability and transparency in reliable AI development supported by relevant use cases.

Keywords: Artificial Intelligence · eXplainable AI · Sustainable AI Technology

1 Introduction

Artificial Intelligence (AI) is a transformative force that reshapes industries and revolutionises problem-solving approaches [1, 2]. However, the widespread use of AI introduces complexities, where accountability and transparency go beyond decision-making [3, 4]. In situations impacting lives, maintaining a balance ensures human oversight, preventing unchecked autonomy and upholding accountability. The focus shifts from justifying AI decisions to the imperative of comprehending and explaining them, highlighting the necessity for transparency in AI development in a sustainable manner [5]. It is argued that the umbrella concept of sustainable AI intertwines two branches of research: sustainability of AI addressing the entire lifecycle of AI solutions, and AI for sustainability [25]. This concept is in its infancy and lacks clear consensus. It evolves from multidisciplinary perspectives (e.g., socio-legal and computer scientific perspectives) based on historical connotations of different aspects [26]. Though research in this domain is sporadic from a multidisciplinary perspective, it reflects core aspects

D. Berkouk et al. (Eds.): CCTS 2024, PTT, pp. 125–133, 2026.
https://10.1007/978-981-95-7607-4_15

related to sustainable development. For example, the study [8] examines how consumer packaged goods retailers can achieve sustainable growth using eXplainable AI (XAI). Again, integrating XAI into smart healthcare automation revolutionise the healthcare industry, i.e. Sustainable Development Goal (SDG) 14 [10]. Promoting biased, opaque and unethical financial decision-making AI systems can risk undermining the UN SDGs [11]. The authors in [12] emphasise the human-in-the-loop approach, accountability and data biases alongside policy recommendations to help accomplish the SDGs. The level of transparency in administrative decision-making for AI systems to ensure SDGs are discussed in [9]. Besides, the role of algorithmic transparency also has a greater impact on establishing accountability and fairness, as mentioned by the EU Commission [13]. Technical and algorithmic progress to achieve AI sustainability in various aspects including rationalisation (i.e., interpretability) in XAI methods, are presented in [23]. Thus, the development of the concept of sustainable AI encompasses different aspects such as ethics, transparency and explainability, responsibility and accountability [25–27]. In sum, it is evident that the positive impacts of AI on people and society along with its commercial benefits, are the core manifestations of this umbrella concept [13]. Having such a backdrop, in this paper, we focus on transparency and explainability in the sustainable development of AI technologies, which support making AI more socially friendly and acceptable [6, 7], in line with the key objectives for the European AI strategy from 2018 [26]. These can be leveraged in AI for sustainable efforts, enabling more efficient, data-driven solutions for global challenges as presented in the use cases.

2 Overview of AI Evolution

Before exploring the movement of sustainable AI through the scientific renewal in AI, it is useful to first examine the progress of AI. According to the Defense Advanced Research Projects Agency (DARPA) [14], the progress of AI can be broadly categorised into three waves, each leading to distinct AI paradigms as shown in Fig. 1. The first wave, starting in 1956, focused on developing AI algorithms for knowledge representation and logical reasoning but did not consider other dimensions of intelligence such as learning, adaptation, and handling uncertainty. The second wave aimed to overcome these limitations by creating value from large amounts of data using machine learning (ML) and deep learning algorithms, which were powerful in learning, characterising the problem domain and perceiving the world. However, these algorithms lacked an understanding of context and performed poorly in reasoning. This has led to unreliable outcomes in critical areas such as autonomous driving [15]. The ongoing third wave is focused on comprehending the context and enhancing reasoning to make AI algorithms more dependable. Furthermore, establishing trust is essential for sustainable AI solutions, which is a pressing need for society today. However, creating 'trustworthy AI' requires addressing multiple issues beyond just understanding the context and reasoning. Therefore, it proposes the fourth wave of "Trust" in AI, which aligns with the European perspective of AI.[1] When discussing transparency in AI, the literature often

[1] https://digital-strategy.ec.europa.eu/en/library/ethics-guidelines-trustworthy-ai

highlights explainability, which encompasses both interpretability and trust in the systems [26]. Ultimately, explainability and transparency go hand in hand to achieve the goal of trustworthy AI and promoting sustainable AI solutions.

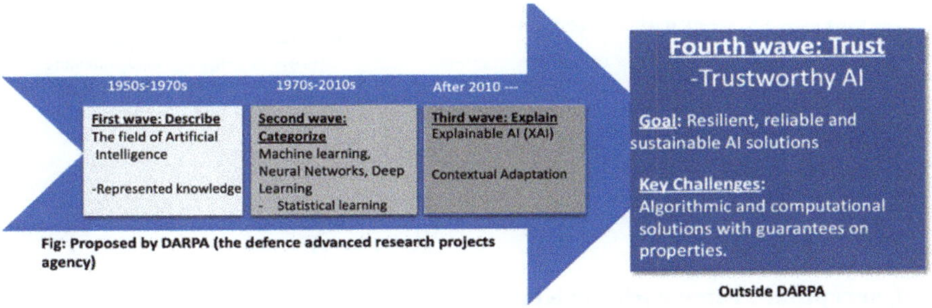

Fig. 1. Trustworthy AI presents this new era of AI in relation to the AI waves and their challenges.

3 Challenges in Ensuring Transparency and Explainability

Despite the importance of transparency and explainability, several challenges hinder their implementation in AI systems. For example, i) **Model complexity vs. interpretability**: Advanced AI models, such as deep learning, are highly complex and can have millions of parameters. While these models achieve high performance, they also become difficult to interpret. On the other hand, simpler models, such as decision trees or linear regressions, are more interpretable but may not achieve the same level of accuracy. Balancing these two factors is a significant challenge. ii) **Lack of universal standards for explainability:** Currently, there is no widely accepted framework for transparency and XAI. Different methods of explanation exist such as LIME (Local Interpretable Model-agnostic Explanations) and SHAP (Shapley Additive Explanations) offer ways to explain black-box models, but they come with limitations like inconsistency and reliability [25]. This makes it difficult for developers to choose the best approach and for stakeholders to understand the explanations provided by different AI systems [24]. Besides, there are societal and cultural differences for example, different stakeholders, such as developers, end-users, and regulatory bodies, may have different expectations of what constitutes an adequate explanation of AI decisions. Cultural, societal, and educational backgrounds can influence how explanations are perceived and understood. As such, creating universally understandable and acceptable explanations remains a significant challenge.

4 Phases to Achieving Transparency and Explainability in AI

The proposed four phases: 1) Conceptual framework for key principles, 2) Incorporating specific properties into AI models, 3) Action Plan and Road Map and 4) Evaluation Metrics, as shown in Fig. 2 are designed to ensure other properties like accountability, and fairness in the sustainable development of AI.

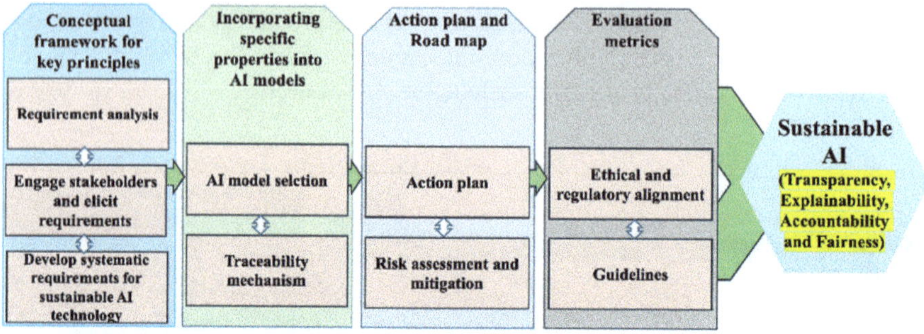

Fig. 2. Methodological steps for sustainable development of AI technologies

4.1 Conceptual Framework for Key Principles

This step started with a **Requirement analysis** to identify and analyse key principles of expected properties. This can be done through a comprehensive review of existing literature and standards, including ethical guidelines, regulatory frameworks, and industry best practices. Based on these key principles, a conceptual framework can be developed that integrates the identified principles and concepts e.g., in the context of XAI identify the definitions of "customized" and "interactive" explanations, specifying how these elements influence the AI system's trustworthiness. second step focuses on **Engaging stakeholders and eliciting requirements**, where relevant stakeholders, including end-users, domain experts, ethicists, regulatory bodies, and developers will be identified based on their roles, influence, and needs concerning AI technology. This can be done through several workshops, interviews, and surveys. The focus will be on understanding specific needs for customisation, interactivity, and trustworthiness from different stakeholder perspectives. Finally, analyse the gathered requirements based on factors such as feasibility, impact, and alignment with trustworthy AI principles. Finally, **Systematic requirements for sustainable AI technology** will translate the prioritised requirements into traceable system requirements. It defines constraints for specific AI applications and ensures a detailed specification for how the AI system will be customised and made interactive to meet user needs. It will be validated based on stakeholder feedback and expert reviews and refine the requirements to ensure feasibility.

4.2 Incorporating Specific Properties into AI Models

This step started with an approach to the **Selection of AI models. For example,** interpretable models such as decision trees, and Bayesian networks or augment complex models like deep learning with post-hoc explainability tools can be employed in AI models for explainability. Other algorithms and mechanisms to prioritise and present humanly understandable and acceptable explanations based on relevant contextual information and model data can also be investigated. Again, a **Traceability mechanism** will ensure that each requirement has a corresponding traceable component in the design and development of AI.

4.3 Action Plan and Road Map

Towards the realisation of relevant AI applications, based on the outcomes of the above steps, the **Action plan** will make a road map for the development of AI methods and algorithms, incorporating e.g., transparency, and explainability in AI techniques. Additionally, it will perform a **Risk assessment and mitigation plan and** identify potential risks associated with the requirements, particularly those that could impact the AI system's trustworthiness and develop strategies to mitigate identified risks, ensuring that these strategies are integrated into the AI system design and development process.

4.4 Evaluation Metrics

Finally, **Ethical and regulatory alignment of** existing regulations and standards for the attributes of trustworthiness, including the EU AI Act,[2] IEEE 7000 series[3,4] and Standards by ISO/IEC TR 24028:2020, ISO/IEC 23894:2023, GDPR (General Data Protection Regulation) and UNESCO AI Ethics Recommendation. This step will conduct regular ethical audits to assess the ongoing alignment of AI solutions with ethical principles and societal values and make necessary adjustments based on the findings. This step develops a reporting mechanism, where a set of lessons learned will be reported as **Guidelines** to illustrate good and bad practices encountered.

5 Explainable AI Use Cases and Sustainability Assessment

The goal of the following use cases is to contribute to the growing field of trustworthy AI, emphasising the importance of transparency and explainability in the sustainable development of AI technologies that align with societal and environmental goals.

5.1 Manufacturing

XAI has been applied to optimise production processes, identify bottlenecks, and reduce waste. For instance, AI-driven systems equipped with interpretable models help operators understand predictions related to machinery failures, enabling timely interventions and minimising downtime in our projects TRUST_GEN_Z,[5] and CPMXAI.[6]

CPMXai Cognitive predictive maintenance is essential which relies on AI models that utilise both supervised and unsupervised learning algorithms, requiring vast amounts of labelled data, including historical data, sensor data, and proprietary resources. CPMXai focuses on developing new tools through leveraging digital twin technology for creating a self-monitoring, self-learning, and self-explanatory system with automatic data labelling and XAI for predictive maintenance.

[2] https://artificialintelligenceact.eu/

[3] https://standards.ieee.org/develop/project/7000.html

[4] https://standards.ieee.org/wp-content/uploads/import/documents/other/ead1e-overview.pdf

[5] https://www.vr.se/english/swecris.html?project=2024-01402_Vinnova#/

[6] https://www.vinnova.se/en/p/cpmxaicognitive-predictive-maintenance-and-quality-assurance-using-explainable-ai-and-machine-lea/

Trust_Gen_Z Advancements in generative AI (gAI) are accelerating industrial digitalization, but challenges like biased data, opaque decisions, and adaptability hinder trust and sustainability. It [19] proposes an adaptive gAI-based multi-agent framework, featuring specialized agents (ExplainAgent, AuditAgent, RobustAgent, and AutoAgent) to enhance explainability, fairness, and robustness. By fostering human-AI collaboration and aligning with sustainable practices, it supports trustworthy AI development and promotes ethical, reliable innovation for long-term industrial growth.

5.2 Air Traffic Management

Here, XAI contributes to ensuring safety by explaining AI-generated recommendations for flight scheduling and conflict resolution, thereby aiding air traffic controllers (ATCOs) in making informed decisions in Artimation[7] and TRUSTY[8] projects.

ARTIMATION. It uses three post hoc explanation methods and conducted both quantitative and user evaluations to assess their acceptability and usability. The results showed that these methods effectively mimic the models' inference mechanisms. User evaluations revealed that user-centric explanations were more preferred by ATCOs. It highlights the potential of XAI in improving the transparency, trust, and sustainability of AI-driven decision-making for efficient air traffic operations [17]. Further, it [18] highlights the need for sustainable AI that prioritises both performance and user trust, suggesting future directions for education and training in high-stakes environments.

TRUSTY (Trustworthy Intelligent System for Remote Digital Tower), TRUSTY explores i) the development of self-explanatory and self-learning systems for critical decision-making through the utilization of transparent ML models incorporating interpretability, fairness, and accountability. It [20] investigates constructions of fair models without relying on sensitive attributes, focusing on Synthetic Minority Oversampling Technique (SMOTE)-driven methods. Further, we have [21] investigated the research issues and challenges in the computational development of trustworthy AI.

5.3 Road Safety

XAI plays a vital role in providing clear explanations for navigation and obstacle avoidance. This not only improves system reliability but also enhances user trust in projects:

SimuSafe Simulator of Behavioural Aspects for Safer Transport. Here the study [16] explores driver behaviour variations in simulators and real-world driving, focusing on risk and hurry. It highlights how spatiotemporal data relationships can lead to incorrect interpretations and impact decision-making. By using Gradient Boosted Decision Trees and SHAP for accurate decision explanations, it contributes to more reliable and sustainable AI for simulator technologies, enhancing safety and promoting environmentally conscious driving behaviour in future transport systems.

---

[7] https://www.artimation.eu/

[8] https://research.dblue.it/trusty/

FitDrive Monitoring devices for overall FITness of DRIVErs: develops an AI tool for professional drivers who are at higher risk of fatigue-related crashes, requiring law enforcement to use reliable tools for assessing driver fitness and performance on the road to enhance safety. It [22] explores two approaches: multimodal reasoning, using semi-supervised methods to label vehicular telemetry data and transfer learning, which transfers knowledge across models to reduce resource demands. These methods demonstrate the potential for sustainable, intelligent systems that optimise resources while advancing AI applications across automotive industries.

6 Conclusions

By aligning with the regulatory and sustainability goals, the methodological steps used in the showcased projects ensure that the developed AI systems are not only transparent and explainable but also ethically compliant. Ensuring transparency and explainability in sustainable AI development will promote fairness and inclusivity by avoiding discriminatory outcomes and contribute positively to society by harnessing AI for the greater social good. Establishing robust governance and accountability mechanisms, it ensures the responsible development of AI that brings long-term benefits to society.

Acknowledgement. This study is part of project 1) xBest (2024-05613) funded by the Swedish Research Council. 2) Trust_Gen_Z (2024-01402) and funded by VINNOVA, Sweden's Innovation Agency; 3) TRUSTY funded by SESAR 3 Joint Undertaking under grant agreement No 101114838; 4) CPMXai (2021-03679) funded by Sweden's Innovation Agency.

References

1. Russell, S., Norvig, P.: Artificial Intelligence: A Modern Approach, 4th edn. Pearson, Harlow (2020)
2. Chen, J., Yuan, Y., Ziabari, A.K., et al.: AI for manufacturing and healthcare: a chemistry and engineering perspective. arXiv preprint arXiv:2405.01520. (2024)
3. Ahmed M.U., et al.: Role of multi-modal machine learning. explainable AI and human-AI teaming in trusted intelligent systems for remote digital towers. In 7th Artificial Intelligence and Cloud Computing Conference (2024)
4. Degas, A., et al.: A survey on AI and eXplainable AI in air traffic management: Current trends and development with future research trajectory. Appl. Sci. **12**(21), 10834 (2022)
5. Vinuesa, R., et al.: The role of artificial intelligence in achieving the Sustainable Development Goals. Nat. Commun. **11**, 233 (2020)
6. Peñalvo, G., et al.: Safe, Transparent, and Ethical Artificial Intelligence: Keys to Quality Sustainable Education (SDG4). Int. J. Educ. Res. Innov. (IJERI). **22**, 1–21 (2024)
7. Fan, Z., Yan, Z., Wen, S.: Deep learning and artificial intelligence in sustainability: A review of SDGs, renewable energy, and environmental health. Sustainability. **15**(2), 1234 (2023)
8. Behera, R.K., Bala, P.K., Rana, N.P.: Creation of sustainable growth with explainable AI: An empirical insight from consumer packaged goods retailers. J. Clean. Prod. **398**, 136783 (2023)
9. Aldrou, K.K.A.R.: The principle of transparency in administrative decisions in light of AI for sustainable development goals. J. Lifestyle SDGs Rev. **5**(2), 1–21 (2025)

10. Shafik, W.: Sustainable development goal 14: Explainable AI for ocean health. In: AI and Edge Computing for Sustainable Ocean Health. The Springer Series in Applied. Springer, Cham (2024)
11. Truby, J.: Governing artificial intelligence to benefit the UN sustainable development goals. Sustain. Dev. **28**, 946–959 (2020)
12. Mhlanga, D.: Human-centered artificial intelligence: The superlative approach to achieve sustainable development goals in the fourth industrial revolution. Sustainability. **14**(3), 1433 (2022)
13. Larsson, S., et al.: Sustainable AI: An Inventory of the State of Knowledge of Ethical, Social, and Legal Challenges Related to Artificial Intelligence. AI Sustainability Center, Stockholm (2019)
14. Fouse, S., Cross, S., Lapin, Z.: DARPA's impact on artificial intelligence. AI Mag. **41**(2), 3–8 (2020)
15. Kumar, R., et al.: Self-driving car using machine learning. In: Abraham, A., Muda, A.K., Gandhi, J. (eds.) Emerging Technologies in Data Mining and Information Security, pp. 709–719. Springer, Singapore (2023)
16. Islam, M.R., Ahmed, M.U., Begum, S., (2023). Interpretable machine learning for modelling and explaining car drivers' behaviour: An exploratory analysis on heterogeneous data. In: 15th International Conference on Agents and Artificial Intelligence
17. Jmoona W., Explaining the unexplainable: role of XAI for flight take-off time delay prediction. In: Proceedings of the 19th International Conference on Artificial Intelligence Applications and Innovations (AIAI 2023), June 2023
18. Hurter, C., et.al., Examining decision-making in air traffic control: Enhancing transparency and decision support through machine learning, explanation, and visualization: A case study. In: 16th International Conference Agents and Artificial Intelligence (March 2024)
19. Ahmed, M.U. et al. Enhancing explainability, robustness, and autonomy: A comprehensive approach in trustworthy AI. In: Proceedings of the IEEE Symposium on Explainable, Responsible, and Trustworthy Computational Intelligence (XRT-CI 2025)
20. Kabir, M., et al. Balancing fairness: Unveiling the potential of SMOTE-driven oversampling in AI model enhancement. In: Proceedings of the International Conference on Machine Learning Technologies (ICMLT 2024)
21. Begum, S., Ahmed, M.U., Barua, S., Kabir, M.A. (2024). Research issues and challenges in the computational development of trustworthy AI. In: Proceedings of the IEEE International Conference on Artificial Intelligence in Engineering and Technology (ICAIET 2024)
22. Barua, A., Ahmed, M.U., Begum, S.: Advanced hybrid reasoning and transfer learning on multimodal data with transformers. SN Comput. Sci. **6**, 1–15 (2025)
23. Chen, Z., et al.: A Survey on AI sustainability: Emerging trends on learning algorithms and research challenges. IEEE Comput. Intell. Mag. **17**, 45–58 (2022)
24. Ahmed M.U et al. When a CBR in hand is better than twins in the bush. In: Proceedings of the Fourth Workshop on Case-Based Reasoning for the Explanation of Intelligent Systems (XCBR 2022)
25. Van Wynsberghe, A.: Sustainable AI: AI for sustainability and the sustainability of AI. AI and Ethics. **1**(3), 213–218 (2021)
26. Larsson, S., Heintz, F.: Transparency in artificial intelligence. Internet Policy Rev. **9**(2), 1–15 (2020)
27. Wilson, C., Van Der Velden, M.: Sustainable AI: An integrated model to guide public sector decision-making. Technol. Soc. **68**, 101926 (2022)

When War Threatens Sustainability: The Legal Fight to Protect Our Environment

Adnan Mahmutovic(✉)

Department of Law, Al-Yamamah University, Riyadh, Saudi Arabia
a_mahmutovic@yu.edu.sa

Abstract. Global sustainability is increasingly threatened by environmental destruction, particularly in conflict-affected regions. War accelerates ecological degradation, disrupts development, and deepens humanitarian crises. Yet, international legal frameworks remain fragmented, outdated, and difficult to enforce. This article assesses whether existing international humanitarian law (IHL) sufficiently protects the environment during armed conflicts or if further legal provisions are required. Through treaty analysis, case law examination, and comparative assessment of international legal instruments, it identifies critical gaps and inconsistencies. To enhance protections, the UN General Assembly must establish a global legal consensus, laying the foundation for a comprehensive Environmental Protection Convention. As warfare continues to inflict severe environmental harm, strengthening legal safeguards is no longer optional, it is essential for the survival of both humanity and the natural world.

Keywords: Environmental Protection · Armed Conflict · International Humanitarian Law (IHL) · Legal Frameworks · Sustainability

1 Introduction

Global sustainability is increasingly shaped by environmental factors, with every country facing unique challenges on both global and local scales. However, nations experiencing armed conflicts are particularly vulnerable, as war disrupts development and accelerates environmental degradation. Despite continuous efforts by international organizations, NGOs, and activists to curb violence, armed conflicts remain persistent. Currently, there are around 110 active wars worldwide [5], a significant rise from 2018 when 69 conflicts affected 30 states and territories. This escalation reflects a growing number of armed groups and increased foreign interventions.

The link between natural resources and conflict is well-documented. According to the United Nations (UN) and the World Bank, over 40% of internal armed conflicts in the last six decades have been tied to resource mismanagement and unequal distribution [28]. While much attention is given to the human cost of war, environmental destruction is an often-overlooked consequence that compounds the suffering of already vulnerable communities.

D. Berkouk et al. (Eds.): CCTS 2024, PTT, pp. 134–144, 2026.
https://10.1007/978-981-95-7607-4_16

Some of the most haunting images of war illustrate its environmental devastation. The napalm-scorched mangroves of Vietnam [28] and the blazing oil fields of the Persian Gulf in 1991 highlight how warfare scars the planet. The Kuwait oil fires, ignited during Iraq's invasion, provide a striking example. Over 600 oil wells burned for months, releasing massive amounts of sulfur dioxide, carbon monoxide, nitrogen oxides, and other unhealthy particulates into the air [2]. The resulting pollution caused severe respiratory illnesses and long-term ecological damage. Between six and eight million barrels of oil were discharged, devastating marine life and killing at least 30,000 seabirds [21]. The destruction of oil infrastructure not only fueled atmospheric pollution but also contaminated land and water sources, threatening biodiversity and human health.

While early predictions warned of catastrophic global consequences, studies later showed that the fires' impact remained mostly regional, with pollutants dispersing faster than expected [9]. However, localized environmental damage was severe. Hydrocarbons and heavy metals contaminated coastal, marine, and terrestrial ecosystems, leaving a lasting toxic legacy [4]. Health surveys of affected populations reported increased respiratory issues, fatigue, and psychological distress, although the full mortality impact remains uncertain [29].

Similarly, the atomic bombings of Hiroshima and Nagasaki in 1945 caused unprecedented devastation. Shock waves and extreme temperatures leveled entire city centers, while radiation exposure led to acute deaths and long-term illnesses, including cancer and genetic disorders. [38] Research by the Radiation Effects Research Foundation continues to study the health consequences among survivors, providing critical insights into radiation exposure [3]. Despite the devastation, some plant species remarkably survived the blasts, symbolizing nature's resilience in the face of extreme destruction.

More recently, the war in Ukraine has inflicted significant environmental damage. The destruction of the Nova Kakhovka Dam has led to widespread flooding, while air pollution has surged due to bombings, fuel depot fires, and industrial site attacks [10]. The country's renewable energy sector has suffered heavily, with most wind and solar stations in southern Ukraine shut down [10]. Water sources have also been contaminated by chemicals and explosive remnants, raising serious concerns about long-term sustainability. These examples illustrate how modern warfare inflicts lasting environmental damage, emphasizing the urgent need for stronger protection. Without effective safeguards, armed conflicts will continue to cause irreversible harm, jeopardizing ecosystems and long-term sustainability.

This article critically examines whether existing legal frameworks sufficiently protect the environment during armed conflicts or if additional provisions are needed. It includes treaty analysis, case law examination, and an assessment of state practice to determine the extent of legal protections. Additionally, comparative analysis of international legal instruments will highlight inconsistencies and areas for improvement. While some argue that current international laws, provide adequate protection, [28] others contend that legal gaps leave the environment vulnerable to wartime destruction [15]. If these protections are found insufficient, the article explores potential legal reforms to strengthen environmental safeguards in conflict zones.

The analysis focuses on two key areas: (1) legal rules governing environmental protection in armed conflicts, including both international and non-international armed conflicts, and (2) recommendations for legal reforms to close gaps and enhance enforcement.

2 Legal Rules Governing Environmental Protection in Armed Conflicts

International law safeguards the environment through international humanitarian law (IHL), international environmental law, and international criminal law. However, early legal efforts prioritized regulating warfare over environmental protection, focusing first on civilian safety and infrastructure security. Environmental concerns were addressed later and with less urgency. This article will focus on IHL and its role in protecting the environment during armed conflict. IHL treaty law relevant to environmental protection can be divided into three categories: provisions explicitly addressing environmental harm, general IHL principles applicable to environmental protection, and provisions that indirectly safeguard the environment.

2.1 Provisions Explicitly Addressing Environmental Harm

IHL provides limited protection for the environment during armed conflict. One of the most explicit provisions is found in AP I to the 1949 Geneva Conventions, particularly Articles 35(3) and 55(1) [12]. Article 35 prohibits combat methods expected to cause excessive, lasting, and serious environmental damage, covering both direct and indirect harm without requiring specific intent. While some argue that this strengthens environmental protections, [1] others believe it imposes rigid criteria that could, paradoxically, allow more harm than general principles such as proportionality and the prohibition of unnecessary suffering [27]. The provision applies only to foreseeable damage, leaving unintended or indirect environmental harm unregulated. Furthermore, nuclear weapons are effectively excluded, as international law does not prohibit their use outright, only restricting testing in certain environments.

Article 55 reinforces these obligations by requiring combatants to protect the environment and banning environmental reprisals. Both provisions prohibit widespread, long-term, and severe environmental damage, but their effectiveness is weakened by the cumulative requirement, meaning all three conditions must be met for a violation to occur. While "long-term" has been interpreted to mean at least ten years, the terms "widespread" and "severe" remain undefined, creating an impractically high threshold for enforcement. These limitations weaken broader principles like proportionality, potentially allowing significant environmental harm that might otherwise be restricted. Additionally, the provisions apply only to recognizable damage, excluding harm that is not immediately apparent but could lead to long-term ecological disruption.

Protection in Non-International Armed Conflicts (NIACs) is even more limited. Additional Protocol II, which governs NIACs, provides only indirect protection through Article 14, which safeguards resources essential to civilian survival, and Article 15, which protects infrastructure containing dangerous forces [12]. While these provisions restrict certain warfare tactics that could harm the environment, they are primarily designed to protect human populations, not ecosystems [6]. The Geneva and Hague

Conventions lack any direct environmental protections for NIACs, and a proposal to extend protections similar to those in Articles 35(3) and 55 of Protocol I was explicitly rejected during the Diplomatic Conference, demonstrating a lack of political will to regulate internal conflicts.

Environmental concerns in NIACs thus rely on indirect protections, such as bans on attacks against infrastructure and the use of starvation as a weapon. However, these rules are often interpreted in favor of military necessity, offering little real protection to ecosystems. Governments engaged in NIACs remain bound by peacetime environmental laws, but enforcement is nearly impossible during active conflict. Insurgent groups, on the other hand, are not bound by international treaties unless they have explicitly agreed to follow them, which is rare. This leaves environmental protection in NIACs almost entirely absent from IHL.

Beyond the Geneva Conventions, other treaties offer some protection but remain limited in scope. Protocol III of the Convention on Certain Conventional Weapons (CCW) bans incendiary weapons against forests and plant cover, but only if they are not being used for military purposes [25]. Article 55(1) of Protocol I prohibits environmental damage but ties this protection to human survival, reinforcing the anthropocentric nature of international law. Similarly, Article 55(2) bans reprisals targeting the environment, but does not prevent incidental environmental damage from military operations.

The ENMOD Convention, adopted in response to the use of defoliants during the Vietnam War, prohibits environmental modification techniques as weapons, acknowledging their long-term ecological consequences [26]. Unlike Protocol I, ENMOD sets a lower threshold for violations, banning actions that cause widespread, long-lasting, or severe damage, rather than requiring all three conditions to be met simultaneously. It also defines "long-lasting" as months rather than decades, making violations easier to prove. However, ENMOD prioritizes state interests over environmental integrity, aiming to prevent harm between nations rather than directly safeguarding ecosystems.

Other treaties provide narrow and conditional protections. The Paris Convention on Cultural and Natural Heritage offers some protection to designated sites, though its applicability in wartime remains highly debated [22]. The Geneva Protocol on Chemical and Biological Weapons restricts toxic agents, but its prohibitions focus on human harm, not direct environmental consequences [16].

2.2 General IHL Principles Applicable to Environmental Protection

The core IHL principles provide universal legal standards that support existing treaties, including those related to environmental protection. The Martens Clause, introduced in 1899, fills gaps in international law by requiring states to adhere to the principles of humanity and public conscience when no explicit legal framework exists [8]. This clause serves as a crucial safeguard, particularly in armed conflicts where environmental protection is otherwise absent.

The principle of distinction requires a clear separation between military and civilian objects, prohibiting indiscriminate attacks on civilian infrastructure. Article 52(2) of AP I defines military objectives as those that contribute to military action and whose destruction provides a definite advantage. Since most environmentally significant sites, such as protected areas, are civilian in nature, targeting them likely violates this principle. However, its application is complex. Industrial sites like power plants and chemical factories may cause severe environmental damage if attacked, but they can still be classified as legitimate military targets. The same ambiguity arises when protected areas are used to fund conflict through illegal resource extraction. These challenges expose the need for a clearer definition of military advantage, distinguishing direct strategic value from indirect economic or psychological impact.

The principle of military necessity limits military action to what is essential for achieving a legitimate military objective. Article 23(g) of the 1907 Hague Convention IV prohibits the destruction of enemy property unless strictly required by military necessity [7]. While this rule indirectly protects natural resources and ecosystems, its effectiveness is uncertain because military necessity often overrides environmental concerns.

The principle of proportionality, codified in Article 57 of Additional Protocol I, bans attacks where collateral damage outweighs military advantage. Destroying an entire forest or village to neutralize a minor target would be a clear violation. Many forms of environmental destruction, including the burning of oil fields and deliberate oil spills during the 1990–1991 Gulf War, have been widely condemned as disproportionate acts due to the overwhelming environmental devastation compared to their military benefit.

The principle of humanity prohibits unnecessary suffering, injury, and destruction, barring starvation as a method of warfare and the destruction of resources vital to civilians. Poisoning water supplies, burning agricultural land, or destroying forests, as seen in the Darfur conflict, violates this principle, as these tactics cause long-term ecological and humanitarian crises.

While these IHL principles provide some environmental protection, their indirect nature and broad interpretations leave gaps in enforcement. The Martens Clause acknowledges environmental concerns but lacks binding legal provisions, making accountability difficult. Strengthening these principles with explicit environmental protections is essential to ensuring that warfare does not continue to destroy ecosystems with impunity.

2.3 Provisions that Indirectly Safeguard the Environment

IHL provides indirect environmental protection through treaties that regulate weapons, warfare methods, and the protection of civilian property and cultural heritage. Since the Hague Conventions of 1899 and 1907, efforts have been made to limit how belligerents conduct war. Both the Hague and Geneva Conventions contain general rules that may have an environmental impact, including prohibitions on unnecessary destruction of enemy property, restrictions on attacks that cause starvation, and protections for civilian infrastructure. At first glance, these provisions might seem to provide environmental safeguards. However, closer analysis reveals significant limitations.

Many of these rules are conditional, allowing exceptions for military necessity. Others protect only civilian objects, meaning military targets can still be destroyed even if doing so causes severe environmental damage. Some apply only when attacks inflict severe civilian losses, making them difficult to invoke for purely environmental harm. Additionally, treaty obligations are often limited by the inter partes principle, meaning they only bind signatory states, and in some cases by the erga omnes clause, which requires that all parties to a conflict be bound by the treaty for it to apply. Furthermore, older Hague law applies only in times of war and does not cover armed conflicts that fall short of this definition.

Even when these rules seem to offer environmental protection, such as restrictions on starvation tactics or attacks on key infrastructure, they only do so indirectly and in limited contexts. More importantly, applying modern environmental concerns to these provisions is legally questionable, as the Hague and Geneva Conventions were drafted before ecological awareness became a global priority. The laws of war were developed in a pre-ecological era, making it difficult to retroactively interpret them as environmental protections. This raises doubts about whether such an interpretation is legally valid or enforceable.

Unlike treaties, customary international law evolves over time, making it a more flexible tool for addressing environmental harm in warfare. The Martens Clause, for instance, asserts that both civilians and combatants continue to be safeguarded by international law, which is informed by established customs, humanitarian values, and the collective conscience of society. Treaties like Additional Protocol I (AP I) to the Geneva Conventions attempted to regulate environmental protection, but key military powers, including the United States and Israel, never ratified it, while France and the UK made reservations on its environmental provisions. This limited its impact, making customary law a more viable path for strengthening environmental protections in conflicts.

In terms of indirect environmental protection, Article 56 of AP I prohibits attacks on dams, dykes, and nuclear power plants if such actions could release dangerous forces and cause severe civilian losses. Military targets near critical infrastructure cannot be attacked if doing so risks environmental catastrophe. Yet, Russia's actions in Ukraine, particularly at the Zaporizhzhia Nuclear Power Station and the destruction of the Nova Kakhovka dam, have violated this principle. The dam's collapse caused severe, long-term environmental damage across Ukraine and the Black Sea region, widely recognized as ecocide by Ukraine, several states, the Council of Europe, and the EU.

Under Ukrainian law, those responsible can be prosecuted under Articles 441 ("Ecocide") and 438 ("Violations of the laws and customs of war") [11]. Given the scale of destruction, this crime extends beyond national jurisdiction, warranting international legal action. The ICC, despite not formally recognizing ecocide as a separate crime, could prosecute it under Article 8.2(b)(iv) of the Rome Statute as a war crime involving severe environmental damage. If the ICC takes on this case, it would set a precedent for prosecuting environmental crimes in war, reinforcing international accountability for ecological destruction. Beyond treaties, UN General Assembly resolutions—though non-binding—serve as political tools for reinforcing environmental protections. The 1992 General Assembly Resolution on "Protection of the Environment in Times of Armed Conflict" was a response to large-scale environmental destruction during war, particularly Iraq's deliberate oil spills and fires during the Gulf War [23]. Though non-binding, it reaffirmed that wanton environmental destruction without military necessity violates international law. It urged states to integrate environmental protection into military protocols but did not impose new legal obligations. Despite lacking enforceability, the resolution contributed to the recognition of environmental damage as a serious concern in armed conflicts. The International Court of Justice (ICJ) later referenced it, highlighting that General Assembly resolutions can influence customary international law [13]. This principle was echoed in later legal frameworks, such as the San Remo Manual on Armed Conflicts at Sea (1994) [14].

Recognizing the need for stronger legal frameworks, the International Law Commission (ILC) developed the Draft Principles on the Protection of the Environment in Armed Conflicts [15]. These principles address environmental protection in three phases: before conflict (preventative measures), during conflict (limitations on environmental harm), and after conflict (restoration and accountability). One of the most significant provisions, Principle 13, outlines obligations to protect the environment during armed conflict. It requires states to respect and safeguard the environment in accordance with international law, prohibits methods of warfare that could cause widespread, long-term, and severe environmental damage, and bans attacks on the environment unless it is a military objective. While these principles reinforce legal norms, their effectiveness is weakened by the high threshold for prohibited environmental damage. The requirement that destruction must be widespread, long-term, and severe to qualify as a violation leaves many instances of environmental harm unaddressed. This loophole limits accountability and enforcement.

Despite its advancements, the UN 2022 resolution remains non-binding and faces resistance from major military powers unwilling to accept stronger restrictions on warfare [24]. The ILC also avoided politically sensitive topics, such as weapons of mass destruction, to ensure broader support. However, as environmental concerns grow and scientific knowledge advances, legal developments in this area are likely to continue. Strengthening accountability mechanisms and integrating these principles into binding legal frameworks will be crucial in ensuring environmental protection during armed conflicts.

International court rulings, particularly from the International Court of Justice (ICJ), help interpret these laws. While rulings apply only to the litigating parties, they clarify legal norms and expose gaps. Notable cases include New Zealand v. France (1995) on nuclear testing [19], Nicaragua v. United States (1986) on the legal weight of UN resolutions [20], and the 1996 Advisory Opinion on the legality of nuclear weapons use [18].

3 Recommendations for Legal Reforms to Close Gaps and Enhance Enforcement

International law remains inadequate in protecting the environment during armed conflict. Many treaties address wartime environmental damage ambiguously or fail to cover it entirely, reflecting a broader reluctance among states to prioritize environmental concerns over economic and military interests. Even in peacetime, environmental protections often face political resistance; during war, they are largely disregarded. Modern conflicts introduce new threats that go beyond traditional state control. The rise of terrorism and private military forces, especially those linked to nuclear powers, increases the risk of severe environmental devastation. Despite growing awareness of these risks, international legal norms remain fragmented and reactive rather than preventive.

A critical step toward stronger environmental safeguards is broader ratification and enforcement of existing treaties, such as AP I of the Geneva Conventions. The International Committee of the Red Cross has recommended incorporating environmental protection into military manuals to ensure legal obligations are upheld in battlefield practice. Another essential reform is amending AP I to classify serious environmental crimes as "grave breaches" under Article 85(3), making such violations prosecutable under international law. Existing treaties, including the ENMOD Convention and the Convention on Certain Conventional Weapons, should be revised to clarify key provisions and expand protections. Additionally, peacetime environmental treaties must be amended to prevent their suspension during armed conflicts, closing loopholes that allow environmental destruction under the pretext of military necessity.

Beyond legal adjustments, a fundamental shift in perspective is required. Environmental security should be recognized as inseparable from human security. Just as conflicts transcend national borders, so do their ecological consequences. The current human-centric legal framework must evolve to acknowledge the intrinsic value of ecosystems. The Martens Clause, which mandates adherence to humanitarian principles in legal gaps, could be interpreted to extend its protections to the environment. A precautionary approach should also be adopted, prohibiting not only known environmental damage but also actions with uncertain long-term consequences. To translate these principles into practice, the United Nations General Assembly should adopt a declaration establishing a global legal consensus on environmental protection during armed conflict. Ultimately, a comprehensive Environmental Protection Convention should be developed to consolidate existing legal obligations and close current gaps.

The current legal framework suffers from several fundamental weaknesses. Environmental protection is neither the primary focus nor the explicit aim of existing international legal frameworks, resulting in a notable absence of clear and dedicated safeguards. The relevant legal provisions are dispersed across an array of treaties, conventions, and protocols, a fragmentation that engenders inconsistencies and undermines effective enforcement. Existing treaties also fail to ensure environmental protection during large-scale hostilities, highlighting the need for stronger wartime provisions. Furthermore, the legal principles governing environmental protection in war lack broad international recognition, and the high thresholds for legal action under customary humanitarian law, such as those in Protocol II of the Geneva Conventions, limit their practical application.

The Rio Declaration's Principle 24, often cited as a customary norm for environmental protection during armed conflicts, does not impose binding obligations on states responsible for environmental destruction. Given its limitations, it should be replaced with a legally binding principle of general international law that extends beyond environmental law alone. Additionally, international law must be revised to recognize environmental protection as a shared responsibility of humanity, ensuring it becomes a global legal priority rather than a secondary concern.

This shift is necessary to address the systemic gaps in international law and to establish a legal paradigm that protects the environment as a common good of humankind. Without meaningful legal reform, the ecological consequences of armed conflict will continue to be neglected, posing long-term risks to global stability and security.

4 Conclusion

International law provides inconsistent and insufficient protection for the environment during armed conflict. While rules for IACs are stronger than those for NIACs, they remain weakly defined, difficult to enforce, and heavily constrained by military priorities. Even in IACs, existing frameworks are fragmented, outdated, and poorly enforced, leaving ecosystems vulnerable. High evidentiary burdens, vague definitions, and selective enforcement further undermine their effectiveness. Laws continue to prioritize human welfare over environmental integrity, failing to recognize the environment as a legal value in its own right.

The limited ratification of Protocol I has prevented its provisions from becoming universally binding customary international law. While some legal norms can be interpreted to protect the environment, they lack clarity and enforcement mechanisms, making accountability nearly impossible. A modern legal framework is needed—one that explicitly recognizes ecological harm as a violation of international law and imposes clear, enforceable obligations on both states and non-state actors.

To translate principles into action, the UN General Assembly must establish a global legal consensus (opinio juris) on environmental protection in armed conflict. In the long term, a comprehensive Environmental Protection Convention is necessary to consolidate legal obligations and close enforcement gaps, ensuring a more effective and binding system.

The urgency for reform cannot be overstated. Modern warfare inflicts unprecedented, long-term damage on ecosystems, with severe consequences for human populations and biodiversity. Strengthening environmental protections under international law is no longer a matter of policy preference, it is an existential necessity for both humanity and the natural world.

References

1. Bothe, M., Partsch, K.J., Solf, W.A.: New Rules for Victims of Armed Conflict. Martinus Nijhoff Publishers, The Hague (1982)
2. Cahalan, R.F.: The Kuwait oil fires as seen by Landsat. J. Geophys. Res. **97**(D14), 14565–14572 (1992)
3. Douple, E.B., et al.: Long-term radiation-related health effects in a unique human population: Lessons learned from the atomic bomb survivors of Hilton and Nagasaki. Disaster Med. Public Health Prep. **5**(Suppl 1), S122–S133 (2011)
4. Evans, J.S., Wilson, A.M., Tuomisto, J.T., Tainio, M., Cooke, R.M.: What risk assessment can tell us about the mortality impacts of the Kuwaiti oil fires. Epidemiology. **16**(5), 658–663 (2005)
5. Goldstone, J.A.: Saving the environment and political stability. Environ. Change Secur. Proj. Rep. **2**, 33–34 (1996)
6. Gună, D.A.: The analysis of the international legal framework for environmental protection in case of armed conflicts. Valahia Univ. Law Stud. **41**(1), 38–50 (2023)
7. Convention (IV) respecting the Laws and Customs of War on Land and its annex: Regulations concerning the Laws and Customs of War on Land. The Hague, 18 October 1907
8. Hague Peace Conference: Martens Clause, Preamble to the Convention (II) with Respect to the Laws and Customs of War on Land. Carnegie Endowment for International Peace, The Hague (1899)

9. Hobbs, P.V., Radke, L.F.: Airborne studies of the smoke from the Kuwait oil fires. Science. **256**(5059), 987–991 (1992)
10. Hryhorczuk, D., et al.: The environmental health impacts of Russia's war on Ukraine. J. Occup. Med. Toxicol. **19**, 50–62 (2024)
11. Hrytsai, L.: The Nova Kakhovka Dam collapse as an example of Russia's hybrid warfare in Ukraine: A threat to environmental security and beyond? Ann. Univ. Mariae Curie-Skłodowska, Sect. K – Politologia. **31**(1), 53–68 (2024)
12. International Committee of the Red Cross: Protocol Additional to the Geneva Conventions of 12 August 1949, and Relating to the Protection of Victims of International Armed Conflicts (Additional Protocol I). International Committee of the Red Cross, Geneva (1978)
13. International Court of Justice: Legality of the threat or use of nuclear weapons, Advisory Opinion. ICJ. Rep. **1996**, 254–255 (1996)
14. International Institute of Humanitarian Law: San Remo manual on international law applicable to armed conflicts at sea, https://onlinelibrary.iihl.org/wp-content/uploads/2021/04/14_SAN-REMO-EN.pdf (1994). Accessed 9 Feb 2025
15. International Law Commission: Draft principles on protection of the environment in relation to armed conflicts. https://legal.un.org/ilc/texts/instruments/english/draft_articles/8_7_2022.pdf (2022). Accessed 9 Feb 2025
16. Karna, B.K., Shivakoti, G.P., Webb, E.L.: Resilience of community forestry under conditions of armed conflict in Nepal. Environ. Conserv. **37**(2), 201–209 (2010)
17. Levine, A.S.: Staying afloat: State agencies, local communities, and international involvement in marine protected area management in Zanzibar, Tanzania. Conserv. Soc. **5**(4), 562–585 (2007)
18. International Court of Justice: Legality of the threat or use of nuclear weapons, Advisory Opinion. ICJ Rep. **1996**, 226 (1996)
19. International Court of Justice: New Zealand v. France. ICJ Rep. **1995**, 288 (1995)
20. International Court of Justice: Nicaragua v. United States of America. ICJ Rep. **1986**, 14 (1986)
21. Ogawa, M.M., Sasaki, H.: Hiroshima, an experience that can never be forgotten: Long-term follow-up of Hiroshima survivors. Int. J. Dermatol. **50**(1), 1–10 (2011)
22. United Nations Educational, Scientific and Cultural Organization: Convention Concerning the Protection of the World Cultural and Natural Heritage. UNESCO, Paris (1975)
23. United Nations General Assembly: Protection of the Environment in Times of Armed Conflict, A/RES/47/37. United Nations, New York (1992)
24. United Nations General Assembly: Protection of the Environment in Relation to Armed Conflicts, A/RES/77/104. United Nations, New York (2022)
25. United Nations: Convention on Prohibitions or Restrictions on the use of Certain Conventional Weapons Which may be Deemed to Be Excessively Injurious or to have Indiscriminate Effects (CCW). United Nations, Geneva (1983)
26. United Nations: Convention on the Prohibition of Military or any Other Hostile use of Environmental Modification Techniques (ENMOD). United Nations, Geneva (1978)
27. Verwey, W.D.: Protection of the environment in times of armed conflict: In search of a new legal perspective. Leid. J. Int. Law. **8**(1), 7–41 (1995). https://doi.org/10.1017/S0922156500003083
28. Westing, A.H.: Ecological Consequences of the Second Indochina War. Almqvist and Wiskell International, Stockholm (1976)
29. Zierler, D.: Invention of Ecocide: Agent Orange, Vietnam, and the Scientists Who Changed the Way We Think About the Environment. University of Georgia Press, Athens (2011)

Sustainable Precision Agriculture: IoT-Based Soil Monitoring and ML-Driven Crop Recommendations

Basnt Melibari(✉), Jomana Alhothly, Joud Alshareef, Lenah Alsaeedi, Arwa Alsubhi, and Sarah Al-Shareef

Computer Science & Artificial Intelligence Department, Umm Al-Qura University, Makkah, Saudi Arabia
{s443008037,s443004418,s443016585,s443005004,amsubhi, saashareef}@uqu.edu.sa

Abstract. Soil management plays a crucial role in agricultural productivity and sustainability. However, traditional soil analysis and crop recommendation methods rely on laboratory testing, which is costly, time-consuming, and inefficient, often delaying critical farming decisions. This paper introduces an IoT-enabled soil management system integrating real-time soil sensors and machine learning to provide instant soil health assessments and crop recommendations. The system employs a Random Forest model, achieving 98% accuracy in predicting optimal crops. Additionally, this work offers soil improvement suggestions, helping users optimize nutrient levels through guided fertilizer recommendations. The Streamlit-based client application provides an interactive dashboard for real-time monitoring, historical data tracking, and community-based knowledge sharing. Experimental validation confirms the system's high reliability, reducing soil analysis time from days to seconds. By eliminating reliance on expensive lab-based testing, the proposed system enhances efficiency, sustainability, and precision in agriculture, making it ideal for farmers, agronomists, and home gardeners.

Keywords: Precision agriculture · IoT-based soil monitoring · Sustainable farming technologies · Smart farming

1 Introduction

Effective soil management is essential for agricultural productivity, but traditional laboratory-based assessment is slow, costly, and inefficient, delaying critical decisions. These methods require extensive, and sometimes expensive, sampling and processing, leading to slow feedback.

Our system offers a real-time, cost-effective solution by integrating IoT-based soil sensors with machine learning. By continuously monitoring soil health and providing instant insights, farmers can make timely decisions on crop selection and management. This approach improves efficiency and sustainability and reduces reliance on costly, labor-intensive soil testing, enabling smarter, more responsive farming.

D. Berkouk et al. (Eds.): CCTS 2024, PTT, pp. 145–153, 2026.
https://10.1007/978-981-95-7607-4_17

2 Related Work

Integrating the Internet of Things (IoT), machine learning (ML), and cloud computing has significantly advanced precision agriculture, enabling real-time soil analysis, crop recommendations, and sustainable farming practices. Several recent studies have explored the role of smart agricultural technologies in optimizing soil management and improving crop yields.

One notable study by Senapaty et al. [1] proposed an IoT-enabled soil nutrient classification and crop recommendation (IoTSNA-CR) model to assist farmers in optimizing soil conditions for improved productivity. Their model utilized a sensory system equipped with soil temperature, moisture, pH, and NPK sensors, coupled with an Arduino UNO board to collect real-time soil data. The collected data was stored in Firebase cloud storage and accessed through an Android application, providing farmers with immediate insights into soil fertility levels.

A key innovation in their approach was the hybridization of ML algorithms. The study introduced the Multi-Class Support Vector Machine with Directed Acyclic Graph and Fruit Fly Optimization (MSVM-DAG-FFO) model, outperforming traditional ML models such as SVM, decision trees, and kernel-based SVMs. The proposed model achieved the highest accuracy rate of 97.3%, significantly improving crop recommendation precision. Compared to our approach, which employs Random Forest and SVMs, their model benefits from an optimized kernel selection technique using Fruit Fly Optimization, which enhances soil classification accuracy.

Other researchers have also explored IoT and machine-learning applications in agriculture. For example, Ahmed et al. [2] emphasized the importance of IoT sensors in data-driven agriculture, highlighting how cloud-based storage and real-time data acquisition can improve soil analysis and fertilizer management. Similarly, Kulkarni et al. [3] and Bouni et al. [4] developed a crop yield prediction framework that combined several models (Random Forest, Naive Bayes, SVM, etc.) to recommend the best crops based on soil and environmental parameters.

These studies align with our system's goal of integrating sensor-based real-time data collection with ML models to assist farmers in decision-making.

While existing studies have demonstrated the benefits of IoT and ML in soil analysis, challenges remain regarding scalability, cost-effectiveness, and user accessibility. Some models rely on high-cost sensors or complex algorithms, limiting their applicability to small-scale farmers or home gardeners. Our proposed system addresses these limitations by providing a low-cost, user-friendly IoT and ML-based solution that enables real-time crop recommendations, soil guidance, and historical data tracking.

Our proposed system builds on previous research by incorporating a sensor-based IoT framework, ML-driven recommendations, and a cloud-integrated user interface. By leveraging real-time soil data, our approach enhances resource efficiency, reduces reliance on traditional soil testing, and empowers farmers with actionable insights for sustainable agriculture.

3 Methodology

3.1 System Components

The proposed system is composed of three main components, as illustrated in Fig. 1:

- Sensor circuit captures real-time soil nutrient levels directly from the soil.
- The backend system processes sensor data using an ML unit to provide crop recommendations, soil health assessments, and a knowledge-sharing platform.
- Client application is a user-friendly dashboard allowing users to input soil properties and access system-generated insights manually.

These components work together to ensure accurate soil analysis and personalized agricultural recommendations.

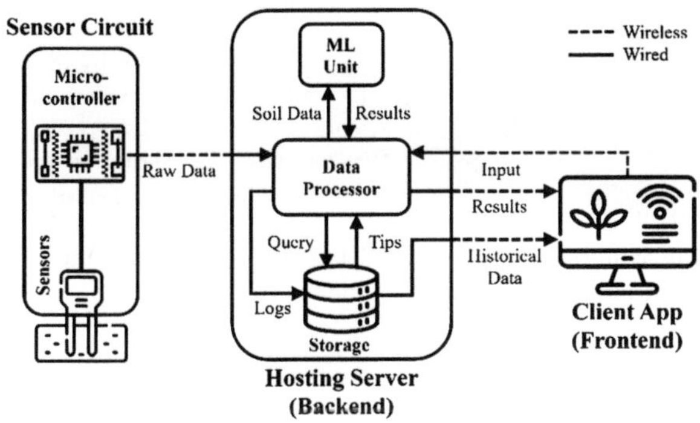

Fig. 1 The system components of the proposed IoT-based precision agriculture system. Dashed line represent wireless communications, while solid lines indicate wired communication.

3.2 Sensor Circuit Design & Integration

The Soil NPK sensor is a core component of the proposed system. It measures Nitrogen (N), Phosphorus (P), and Potassium (K) levels to assess soil fertility and provide accurate crop recommendations.

Before physical implementation, the sensor circuit was simulated using Tinker-CAD,[1] a virtual circuit design tool. This simulation ensured proper sensor connectivity and correct voltage levels, helping identify potential issues in the circuit design before assembling the real-world prototype.

The sensors used are from Luocute, which is known for its high precision and durability in agricultural environments. These sensors can measure nutrient levels with an accuracy of ±2%, ensuring precise soil condition assessments. They operate with fine resolution as high as 1 mg/kg or 1 mg/l for NPK measurements, covering the typical

[1] https://www.tinkercad.com/

nutrient levels in agricultural soils. Their sensitivity allows for the detection of minute changes in nutrient concentration, which is crucial for timely decisions.

After successful virtual testing, the physical circuit was assembled. The Soil NPK sensor was connected to an ESP32 microcontroller via an RS485 to enable stable long-distance communication. A breadboard was used for flexible and modifiable connections between the components. The circuit is powered by **dual 9 V batteries connected in series** for reliability and ease of replacement. A **power management circuit** optimizes energy efficiency, extending battery life. Figure 2 illustrates the complete sensor circuit setup, showing the integration of all components into a cohesive unit ready for deployment.

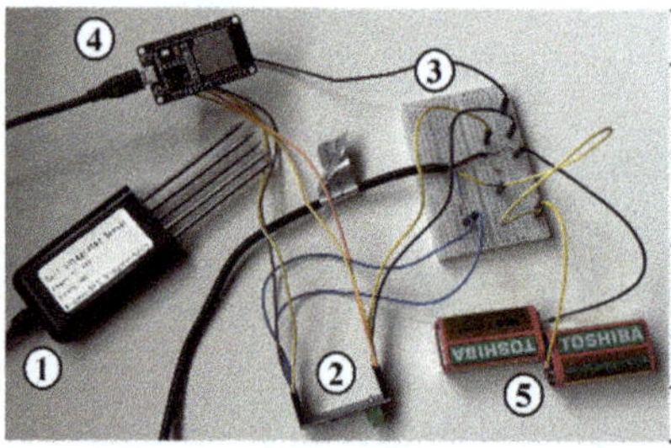

Component	Function
1. NPK Sensor	measures essential soil nutrients.
2. RS485 Modbus	communicates robustly between the sensor and the microcontroller via the RS485 protocol.
3. Breadboard	allows experimental circuit designs.
4. ESP32 Microcontroller	processes data and transmits it to the cloud-based system.
5. 2×9V Batteries	connected in series to get 18V

Fig. 2 Sensor circuit setup of the proposed IoT-based precision agriculture system with labeled components and a brief description for each of them.

3.3 Data Communication

The system implements the RS485 serial communication protocol to ensure seamless communication between the sensor and the ESP32 microcontroller.

To verify the integrity of transmitted soil data, the system employs the Modbus RTU CRC-16 verification protocol. This error-checking mechanism calculates a checksum for each transmitted data packet, ensuring that only valid data is processed.

The CRC-16 algorithm follows these steps:

1. Initialize the CRC value to 0xFFFF (a standard starting point for Modbus CRC calculations).
2. Iterate through each data byte, applying XOR operations and bit shifts.
3. Use the polynomial 0xA001 to detect transmission errors.
4. Convert the final computed CRC value into a 2-byte sequence in little-endian format for transmission.

This verification ensures that every data frame transmitted over the Modbus network includes a valid CRC checksum, allowing receivers to detect errors upon arrival.

The RS485 protocol was selected for its suitability for long-distance communication, making it ideal for agricultural settings where sensors may be distributed across large fields. Modbus RTU, a widely used industrial protocol, was chosen for its simplicity, efficiency, and reliability in handling small data packets over serial communication lines.

Data exchange between the sensor and the ESP32 microcontroller is handled via RS485 protocol with the following parameters:

- Baud Rate: Set to 9600 bits per second for stable communication.
- Parity: None (no additional parity bit used).
- Stop Bits: 1 (single stop bit for efficient transmission).
- Byte Size: 8 bits per byte.
- Timeout: 1 s, preventing the system from waiting indefinitely for a response.

Once data is received, the system checks its validity by comparing the computed CRC checksum with the received CRC value. If the values match, the NPK levels are extracted, converted into a human-readable format (mg/kg), and displayed. If the CRC check fails, the system discards the data and prompts for retransmission.

3.4 Machine Learning Unit

To develop the crop recommendation model, the system analyzes a dataset from Kaggle[2] containing 2200 soil and crop samples across 22 crop types. The dataset includes attributes like NPK levels, pH, and rainfall, but only NPK features were used for training due to sensor limitations.

To ensure high accuracy and generalization, a five-fold cross-validation approach was used. Three folds were used for training to maximize learning, and a fold was used for each validation and testing to ensure model accuracy and generalizability.

Two models were trained: Random Forest (RF) and Support Vector Machines (SVM), as they scored the best in the reviewed literature. All experiments were implemented and evaluated in Python using Google Colab's Jupyter Notebooks environment. Table 1 lists the performance of these models across all the experiments.

After extensive evaluation, the RF model was chosen due to its superior performance in agricultural predictions, with an average score of 98% in all metrics: accuracy, precision, recall, and F1-score. It efficiently handles the dataset's complexity and size, providing reliable results and crucial insights into influential soil and environmental factors.

Table 1 Average performance of the crop recommendation models: SVM and RF.

Model	Accuracy	Precision	Recall	F1-Score
RF	0.99	0.99	0.99	0.99
SVM	0.96	0.97	0.96	0.96

[2] https://www.kaggle.com/datasets/atharvaingle/crop-recommendation-dataset

3.5 Client Application

The client application is the user-facing component of the proposed system. Developed with Streamlit[3] and hosted within the Visual Studio Code environment, it offers an interactive dashboard where users can:

- Manually enter soil parameters (NPK values).
- Receive real-time crop recommendations based on the trained ML model.
- Select a specific crop and obtain customized guidelines to improve soil suitability.

The application is integrated with a Firebase database to store user queries and recommendations, ensuring data accessibility from any location.

Users can keep soil analysis records private or public. Private records are user-only, while public records allow others to search by soil properties or crop type, fostering collaborative learning and knowledge sharing on soil improvement.

3.6 System Data Flow

The system integrates sensor readings, ML, and cloud storage to deliver real-time crop recommendations, soil guidance, and historical data search with accuracy and accessibility.

Crop Recommendation. Users can input soil NPK values manually or use real-time sensor readings. The Soil NPK sensor transmits data via ESP32 and RS485 to the back-end, where CRC-16 verification ensures data integrity. Readings are updated every 2 s until stopped by the user. A pre-trained ML model then predicts the most suitable crops, ranking them based on suitability scores. Results, including confidence scores and visual nutrient levels, are displayed in the client application, with an option to store recommendations for future reference.

Soil Guidance. Users can enter NPK values or use sensor data. The system compares soil properties with ideal conditions for the selected crop, detecting nutrient deficiencies or excesses. It then retrieves fertilizing recommendations or suggests alternative crops better suited to the soil. Recommendations are displayed in the client application for informed decision-making.

Historical Data Search and Community Sharing. Users can search historical soil data by soil properties or crop type. The system queries Firebase, retrieving private or public records, sorted by date and location. Shared records help foster collaborative learning among farmers by providing insights into successful soil and crop management practices.

4 Results and Discussion

The proposed system was tested using several test cases. These test cases were grouped into two main sets: (1) to assess the sensor circuit validity and real-time update using three soil samples, (2) to assess the performance of both crop recommendation and customized guidelines to improve soil suitability.

[3] https://streamlit.io/

Table 2 Comparing the sensor circuit readings and the range of the expected value for three testing samples.

Soil Sample	Expected values (mg/kg)	Sensor Reading (mg/kg)
Normal soil	$N < 5$ $P < 7$ $K < 10$	2 3 6
Soil with natural fertilizer	$25 < N < 30$ $37 < P < 45$ $70 < K < 80$	27 39 78
Soil with natural fertilizer with excessive water	$25 < N < 30$ $37 < P < 45$ $70 < K < 80$	24 34 69

Table 2 shows the three soil samples used to test the sensor circuit: normal soil with low nutrient levels, soil with natural fertilizer, soil with natural fertilizer and excessive water. The expected outcome ranges for each sample were compared with the real-time sensor readings. The results showed that the system can provide accurate and consistent data, making it a suitable alternative to traditional, time-consuming laboratory tests.

The ML model was evaluated on 73 unseen samples and achieved an accuracy of 98% in crop recommendations based on soil conditions. Figure 3 shows the outcome of crop recommendations and soil guidance based on the sensor readings.

The system effectively reduced the time for soil analysis from several days to seconds. Users particularly appreciated graphical representations of nutrient levels and crop recommendations for their ease of understanding and immediate applicability, making the system accessible even for farmers with limited technical experience. Additionally, by reducing dependencies on traditional soil testing methods, the system cuts costs and enhances precision in crop selection, which is vital for resource conservation and sustainability in agriculture. Its ability to dynamically adjust recommendations based on changing soil and environmental conditions ensures adaptability and resilience in modern farming practices.

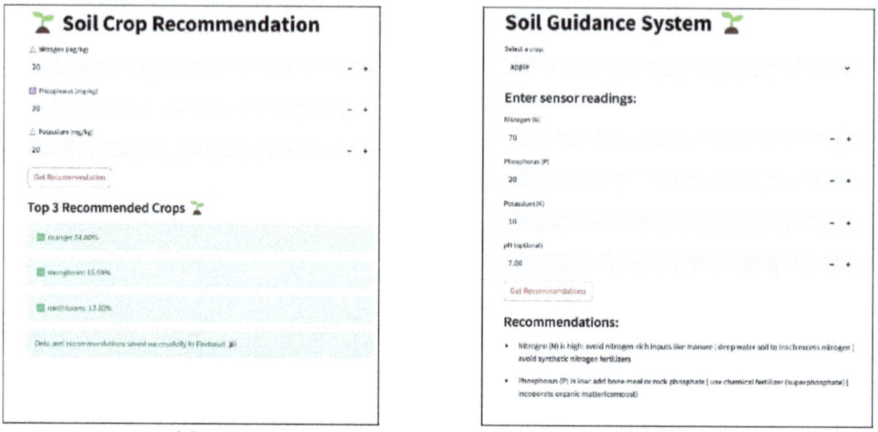

(a) (b)

Fig. 3 Interfaces for the services in the proposed system. (a) Soil recommendation service, suggesting the top three crops best suited for the given soil conditions. (b) Soil guidance service, providing customized recommendations on how to improve soil quality for a chosen crop.

5 Conclusion

This work demonstrates the potential of integrating IoT with ML to revolutionize soil management and crop recommendation practices. The system addresses critical challenges in agriculture, offering a scalable, efficient, and user-friendly solution. Future work will focus on expanding the crop database, incorporating climate variables, and enhancing the ML models to cover more diverse agricultural needs and environmental conditions.

References

1. Senapaty, M.K., Ray, A., Padhy, N.: IoT-enabled soil nutrient analysis and crop recommendation model for precision agriculture. Computers. **12**(3), 61 (2023)
2. Ahmed, U., Lin, J.C.-W., Srivastava, G., Djenouri, Y.: A nutrient recommendation system for soil fertilization based on evolutionary computation. Comput. Electron. Agric. **189**, 106407 (2021)
3. Kulkarni, N.H., Srinivasan, G., Sagar, B., Cauvery, N.: Improving Crop Productivity Through a Crop Recommendation System Using Ensembling Technique. In: 2018 3rd International Conference on Computational Systems and Information Technology for Sustainable Solutions (CSITSS), pp. 114–119. IEEE (2018).
4. Bouni, M., Hssina, B., Douzi, K., Douzi, S.: Integrated IoT approaches for crop recommendation and yield-prediction using machine-learning. IoT. **5**(4), 634–649 (2024)

Diverse Applications of Technology and Creative Solutions

CNN-Based Deep Learning for Automated Kidney Stone Classification in Medical Imaging (CT Scan)

Vishesh Tanwar[1], Bhisham Sharma[2], and Imed Ben Dhaou[3](✉)

[1] Chitkara University Institute of Engineering and Technology, Chitkara University, Rajpura 140401, Punjab, India
vishesh.tanwar@chitkara.edu.in
[2] Centre of Research Impact and Outcome, Chitkara University, Rajpura 140401, Punjab, India
[3] Department of Computer Science, Hekma School of Engineering Computing and Design, Jeddah, Saudi Arabia
Imed.bendhaou@utu.fi

Abstract. Nephrolithiasis, or kidney stones, is a common painful condition affecting millions of people around the world and has significant health and economic implications. Traditional methods of diagnosis are X-rays, ultrasound, and CT scans. These methods have several limitations: radiation exposure, operator dependency, and inconsistent results. This study covers the application of a CNN-based convolution neural network method for the classification of kidney stones on images taken using a medical imaging approach. Using a data set developed using CT and scans to enhance the model based on robust deep learning methods that potentially lead to an autonomous detection process. The performance of the CNN model was remarkable: precision, 99.33%, with a perfect recall at 100%, while its low loss value is at 0.227. It had an impressive generalization performance, so much so that precision is at an incredibly high number that will make the tool reliable for real use in the diagnostic process. Therefore, this study will open up ways for AI-based diagnostic tools to enhance early detection and reduce healthcare burdens while improving patient outcomes, paving the way for more accessible and reliable kidney stone diagnostics.

Keywords: Deep learning · Kidney stone · CNN model · Classification

1 Introduction

Kidney stones, medically known as nephrolithiasis, are solid mineral and salt deposits formed inside the kidneys as a result of the impacts of dehydration, diet, obesity, or genetic factors. The basic balance of substances in urine is disturbed in stone formation, such as crystallisation in the presence of calcium, oxalate, and phosphorus. This condition is painful and has the potential for more severe health implications, including recurrent urinary tract infections, urinary tract obstruction, and even permanent kidney damage if left untreated. The incidence of kidney stones has been steadily increasing over the past

D. Berkouk et al. (Eds.): CCTS 2024, PTT, pp. 157–167, 2026.
https://10.1007/978-981-95-7607-4_18

few decades, and studies estimate that about 12% of the world's population will suffer from this condition at some point in their lives. In some regions, such as the Middle East and South Asia, the prevalence is even higher, reaching up to 20%. This is very essential since the recurrence rate is quite high, and almost 50% of patients may have a second episode in five to ten years. Proper diagnosis of the condition in time becomes an essential element in management, thus reducing its recurrence and complications such as extreme pain infections and kidney dysfunction. Additionally, early intervention can help avoid invasive surgical procedures and improve patient outcomes [1]. Traditional methods of diagnosing kidney stones include radiography, ultrasound, and computed tomography (CT) scans. These are the most common methods, but they have many drawbacks. For example, X-rays often cannot detect smaller or radiolucent stones. By analyzing patterns and features in medical images, models of this genre can correctly diagnose abnormalities with very high accuracy, even superseding expert-level human recognition. This makes deep learning an excellent resource for quickly and accurately identifying kidney stones [2].

The variety of deep learning architectures brings up the other widely-used deep models such as ResNet, VGG, and Inception. ResNet is the model with the vanishing gradient issue for deep networks. VGG is known for being computational and having an extensive memory space; CNN, for instance, has overturned this drawback through a streamlined architecture that requires much less computational resources for tremendous accuracy. Thus, concerning the detection of kidney stones, it is well suited to limited computational applications [3]. CNN's application to kidney stone detection can change diagnostic workflows, assuring precise and reliable results much quicker than existing methods. Medical images are now processed automatically, so the workload for radiologists and the probability of human errors can be minimised. Last but not least, recent work with deep learning and disease detection in AI has provided further inspiration and motivation for better advances in medical solutions. For example, Tanwar and Lamba's work on "Classification of Multiple Maize Leaf Diseases Using a Blended Convolutional Neural Network" reflected how blending different CNN architectures can lead to enhanced classification accuracy; such results may also apply to kidney stone detection systems so that specific challenges in feature extraction can be overcome [4]. Similarly, Shweta Lamba's application of Xception for "Multiple Grapes Leaf Disease Identification" demonstrated the capability of ideal deep learning models to classify multiple diseases accurately, thus strengthening the notion of choosing the correct model architecture [5]. In addition, the "Enhancing Blood Cell Diagnosis Using Hybrid Residual and Dual Block Transformer Network" further demonstrated hybrid models incorporating CNNs and transformers, thus introducing the diverse architectures to be applied to develop high-performance medical imaging solutions [6]. In this paper, we attempt to explore the use of CNN to detect kidney stones using medical imaging data. Our objectives are to develop a robust and efficient deep learning framework based on CNN for the automated detection of kidney stones; second, to compare its performance with other SOTA deep learning models, such as ResNet, VGG, and Inception, in terms of accuracy and computational performance.

2 Literature Survey

Various studies have highlighted the accurate and early analysis of kidney stones. CT and ultrasound were considered standard diagnostic imaging methods for detecting kidney stones. These conventional imaging methods are prone to drawbacks, such as operator dependence and radiation exposure. According to [7], an automated diagnostic system must be in place to address the shortcomings associated with conventional methods and create space for AI-based systems. Recently, Deep learning has shown perfect medical imaging capabilities for fast processing of massive data sets. For example, in [8], CNN was successfully used for disease classification, obtaining excellent results in numerous medical task applications. Such research implies the potential applicability of DL in the case of kidney stone detection. The usage of CNN for processing medical images has been relatively broad. A work in [9] used ResNet to classify kidney stones, showing promising results, but some problems with its computational inefficiency and overfitting were observed. This study motivated further studies on optimized architectures. The CNN compound scaling approach has been a game changer in image classification tasks. In [10], the authors successfully used CNN to detect retinal diseases, demonstrating its superior performance to traditional CNN. Its efficiency makes it a candidate for the detection of kidney stones. Segmentation is a way through which, in medical images, kidney stones can be identified successfully. The work in [11] suggested a U-Net architecture for stone segmentation in CT images, achieving high sensitivity. However, the model experienced difficulties processing high-resolution data, as its resources were not efficient. Alternative architectures, such as CNN, would be helpful. Better performance has been achieved by combining different deep-learning architectures. The contribution in [12] proposed a hybrid architecture that integrates CNN and attention mechanisms for medical image classification. It outperforms most present SOTA anomaly detection approaches. In [13], various DL models were comprehensively compared for medical imaging. CNN was found to be superior to models such as VGG and Inception in terms of accuracy, making it very suitable for resource-limited settings. Transfer learning has been a game changer in medical imaging. The work in [14] used CNN models for kidney stone detection, reducing training time and improving model performance. This showed the possibility of CNN in actual clinical practice. There has been growing interest in automated diagnostic systems. Work in [15] proposed an end-to-end pipeline for detecting kidney stones using artificial intelligence, integrating pre-processing, segmentation, and classification. Although effective, the system relied on computationally heavy models with limited scalability. However, challenges remain, including data scarcity and model generalization [16]. proposed a framework to advance the heftiness of the model by data enhancement and multimodal learning, which can be applied to kidney stone detection systems.

3 Methodology

The methodology for detecting kidney stones using CNN is designed to ensure high accuracy, robustness, and scalability. The process involves several key stages, including the preparation of the data set, the training of the model, the assessment, and the validation. Figure 1 shows a detailed explanation of each stage.

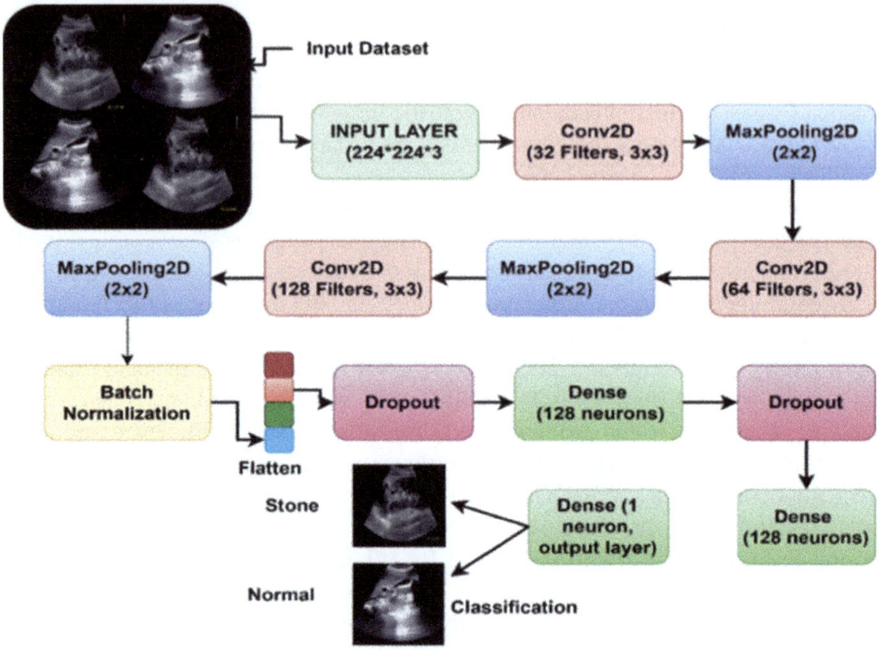

Fig. 1 Proposed methodology for kindy stone detection

3.1 Collection of Data Sets. The First Step Involves Collecting and Pre-Processing Medical Imaging Data Relevant to Kidney Stone Detection

The data set comprises computed tomography (CT) scans and ultrasound images obtained from clinical sources. Key steps in this stage include. Images are collected from hospitals, diagnostic centres, and publicly available medical imaging repositories. Radiologists perform annotations to label kidney stones on the images. The images are resized, normalised, and augmented to enhance the model's robustness. Rotation, flipping, and contrast change are used to create a rich data set. The data set is split into training, validation, and test sets, ensuring the images with and without kidney stones are properly distributed. The Kidney Stone Dataset consists of two classes, as shown in Fig. 2: Normal (4,414 samples) and Stone (5,002 samples), making it a binary classification problem [17].

3.2 CNN Model- This Model is a CNN, Tailored Explicitly for Kidney Stone Detection in Medical Imaging Such as CT or MRI Scans

The proposed model architecture is described in Table 1. It begins with an input layer accepting resized images of 224 × 224 pixels to ensure uniformity independent of the original image size. The model's heart lies in its convolutional layers, which serve as feature detectors. Layers scan the input image for patterns from simple edges and textures in the initial layers to more sophisticated shapes and features in the deeper layers. To contain computational complexity and concentrate on essential details, pooling layers

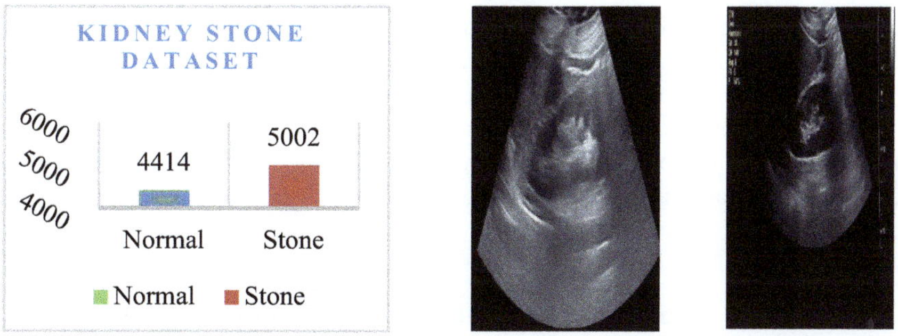

(c) (a) (b)

Fig. 2 Dataset Details in 2 (c) and sample images (a) normal (b) stone

accompany each convolutional layer, shrinking the image size while retaining crucial features. Batch normalization stabilizes and accelerates training by normalizing layer outputs. After the convolutional and pooling layers, the data is flattened into a one-dimensional vector for the fully connected layers, which undertake high-level decision-making.

Table 1 Proposed model architecture

Layer	Output	Parameters
InputLayer	(224, 224, 3)	0
Convo	(224, 224, 32)	896
Max-pool	(112, 112, 32)	0
conv2d_1	(112, 112, 64)	18,496
Max-pool	(56, 56, 64)	0
conv2d_2	(56, 56, 128)	73,856
Max-pool	(28, 28, 128)	0
Batch_Normalization	(28, 28, 128)	512
Flatten	(100352)	0
Dropout-1	(100352)	0
Dense-1	(128)	12,845,184
dropout_2	(128)	0
dense_1	(128)	16,512
dense_3	(1)	129
Total parameters	12,955,585	
Trainable parameters	12,955,329	
Non-trainable parameters	256	

These layers compile and refine all the learnt features, eventually ending with a single neurone in the last layer. This neurone emits a 0 (no kidney stone identified) or a 1 (kidney stone identified). The dropout layers are strategically located to avoid overfitting by randomly turning off some neurones during training, requiring the model to generalise better to new data. Together, the model has 12.95 million parameters, of which the majority are trainable, The model is designed to adjust and learn from the data effectively.

3.3 Training and Testing - Fine-tune the CNNB0 Trained on the Dataset for Kidney Stones

A key step involved. Loss function: Uses the binary cross-entropy loss on classification tasks. Kidney stones are present or not; categorical cross entropy might be used where multiple classes need to be found if types are distinguished. Optimiser: To use the best available convergence, the Adam Optimiser with LR Scheduler. Batch Size and Epochs: Train the model with an optimum batch size for some epochs, decided through the hyperparameter tuning for assessment. Metrics to evaluate the model: We used Accuracy, The total correctness of the predictions, to assess the performance. Recall-Evaluates true cases. Specificity: Assesses negative cases. F1-Score: The balance of precision and recall provides a complete performance metric. AUC-ROC- This score evaluates the ability of the model to distinguish between classes.

Table 2 Hyperparameters details

Parameter	Value
Activation function	ReLU
Optimizer	Adam
Learning Rate	0.0010
Batch size	16
Training Set	70%
Testing Set	30%
Loss Function	Binary Cross Entropy
Final dense Layer Activation	Sigmoid

Table 2 below depicts the primary hyperparameters and configurations used to train a binary classification model on the Kidney Stone Dataset. For hidden layers, the ReLU (Rectified Linear Unit) activation function is utilised to introduce nonlinearity so that the model can learn complex patterns efficiently. The model was optimised using Adam (Adaptive Moment Estimation), an adaptive gradient-based optimiser known for stable and efficient training. A learning rate of 0.001 ensures that weights are gradually updated to avoid fluctuating too violently but still enabling convergence. With a batch size of 16, the model updates weights after every sequence of 16 samples it reads, balancing the factors of memory use and gradient stability. The dataset is split into 70% for training and 30% for testing, ensuring ample data for model training and evaluation. For this binary classification task, the binary cross-entropy (BCE) loss function measures the alignment of predicted probabilities with actual labels. The final dense layer uses a sigmoid activation function to output a probability between 0 and 1, representing the likelihood of a sample belonging to the "Stone" class.

4 Results

Table 3 presents the performance metrics of the proposed CNN model for kidney stone classification, including Precision, Recall, F1-Score, Loss, Kappa, and Accuracy for both "Normal" and "Stone" categories. Precision measures how many predicted cases for a class were correct. A 100% precision for the "normal" class means that all predictions of "normal" kidneys were accurate. A 98% precision for the "Stone" class indicates that 2% of the predicted kidney stone cases were normal. Recall that measures how many actual cases of a category were correctly identified. The model has a 98% recall for "normal", meaning that it correctly identified 98% of normal kidneys, while the 100% recall for "stone" means it correctly detected all cases of kidney stones. Both classes have a high F1 score of 99%, indicating strong classification performance. The loss value of 0.227 represents the model error in classification. A lower loss suggests that the model makes fewer incorrect predictions. The Kappa score (98%) is a statistical measure of inter-rater agreement that considers the possibility of correct predictions by chance. A value close to 1 indicates an almost perfect agreement between predictions and actual labels. Finally, the overall accuracy (99.33%) confirms the model's outstanding performance in detecting kidney stones, making it a reliable tool for medical diagnosis.

Table 3 Performance parameters

Category	Precision (%)	Recall (%)	F1-Score (%)	Loss	Kappa (%)	Accuracy (%)
Normal	100	98	99	0.227	98	99.33
Stone	98	100	99			

Figure 3 displays a Confusion Matrix showing the model's performance on the test dataset, correctly identifying 1295 of 1324 "Normal" cases and all 1501 "Stone" cases. There are only 29 misclassifications where a normal kidney was falsely classified as having a stone, while no kidney stones were misclassified as normal. This highlights the high recall of the model for detecting kidney stones, making it particularly useful in medical diagnosis where missing a kidney stone could have significant clinical consequences.

Figure 4(a) illustrates the training and validation accuracy of the proposed model across 30 epochs. The blue curve indicates the training accuracy, while the orange curve represents the validation accuracy. Initially, both accuracies increase rapidly, with a validation accuracy of 99.33% within a few epochs, indicating fast learning and strong generalization. The training accuracy gradually improves and aligns closely with the validation

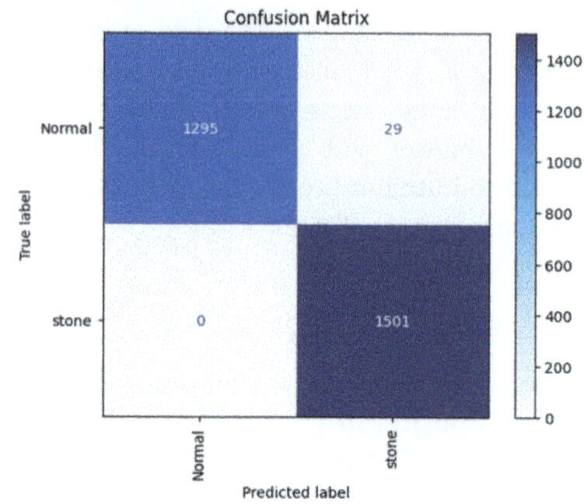

Fig. 3 Confusion matrix of the proposed model

accuracy, reflecting a stable and effective learning process. Occasional dips in validation accuracy suggest minor fluctuations in model generalization, but the overall consistently high performance highlights the robustness of the proposed model. Figure 4(b) illustrates the Training and Validation Loss over 30 epochs.

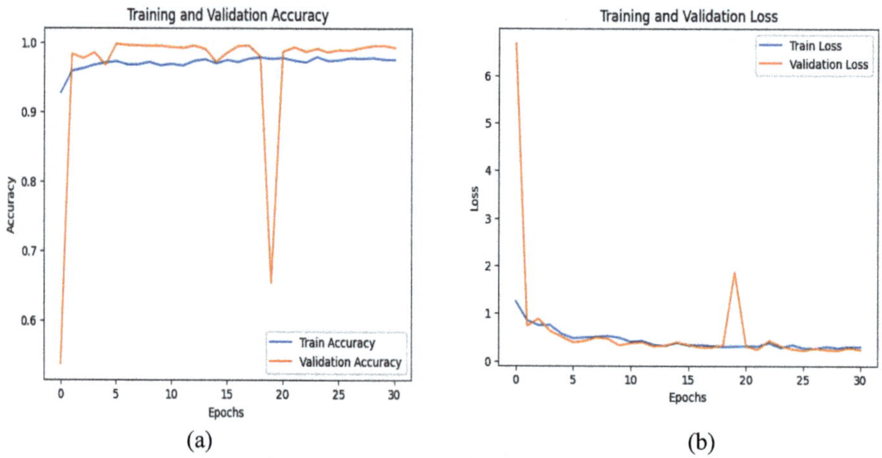

Fig. 4 Training and validation (a) accuracy and (b) loss of the proposed model

The loss indicates how well the model's predictions match the labels. The training loss decreases steadily, suggesting that the model is learning effectively. There are spikes around the 18th and 20th epochs, corresponding to the fluctuations seen in validation accuracy. These sudden increases in loss suggest that the model might have encountered complex validation samples or temporary instability. Still, the loss remains low overall, confirming the model's effectiveness in classifying kidney stones.

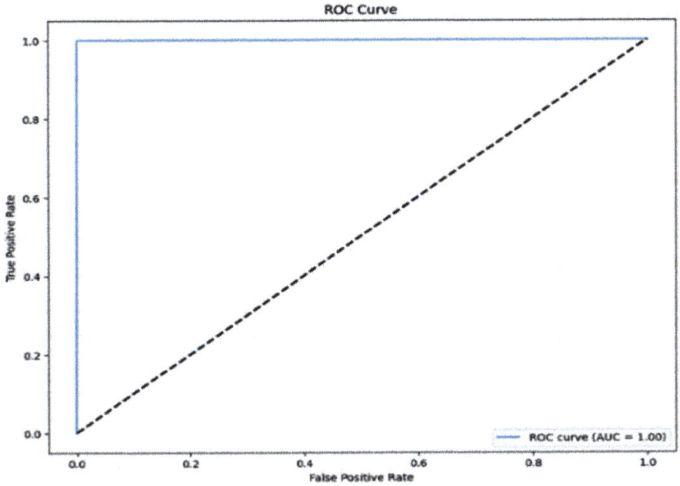

(a)

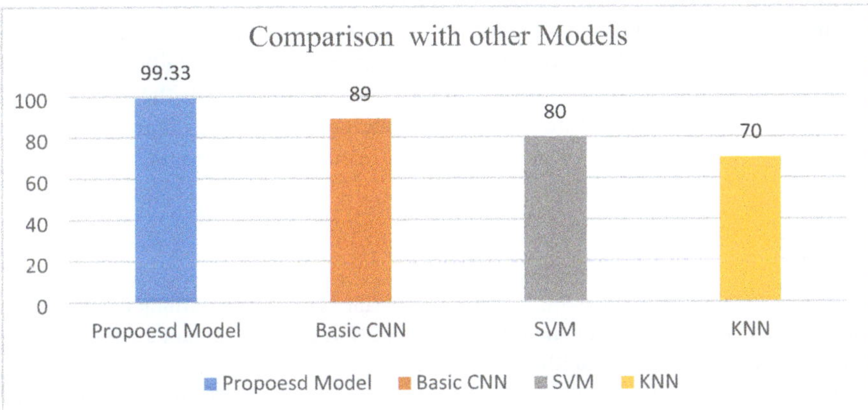

(b)

Fig. 5 (a) Roc curve of the proposed Model (b) Comparison of the proposed Model performance with the existing model on the same dataset.

Figure 5(a) displays the ROC curve for the proposed model, highlighting its excellent classification performance. It achieves a near-perfect discrimination between classes, with an AUC value of 1.00, indicating 100% accuracy in distinguishing positive and negative cases. Figure 5(b) compares the proposed model against three existing models: Basic CNN, SVM, and KNN. The proposed model achieves the highest accuracy at 99.33%, significantly outperforming the others. The Basic CNN records 89% accuracy, while SVM achieves 80%, and KNN performs the least with 70%. These results underscore the superior capabilities of the proposed model.

5 Conclusion

The emergence of kidney stones poses a significant health issue worldwide, especially in terms of increased prevalence and recurrence rates. This necessitates proper and prompt diagnosis. Traditionally, imaging-based diagnosis has proved effective but remains constrained by drawbacks like radiation exposure, operator dependency, and variability of results. In this study, it is revealed that CNN can potentially transform the practice of kidney stone detection. Our proposed CNN model, with an accuracy of 99.33%, 100% recall for kidney stone detection, and a low value of loss of 0.227, showed being profoundly reliable as a tool for medical diagnosis. It can highlight very minute patterns and features by this architecture, making it even more fit for real-world clinical applications. Integrating AI into medical imaging not only enhances the accuracy of diagnoses but also lessens the burden on radiologists, minimises human error, and provides quicker results. The current study forms part of a growing body of research on innovations in AI-driven healthcare, thus providing a scalable and efficient solution for kidney stone detection. Future work could include hybrid models, multi-modal learning, and further optimisation in CNN architectures to handle issues associated with data scarcity and generalization of models. The prospect of adopting AI technology in kidney stone diagnostics is promising for better patient outcomes, lowering healthcare costs, and advancing the field of medical imaging. Improving and expanding these technologies will help draw us closer to a future when accurate and accessible diagnostic tools can be used worldwide regardless of geographical or economic constraints.

References

1. Black, K.M., Law, H., Aldoukhi, A., Deng, J., Ghani, K.R.: Deep learning computer vision algorithm for detecting kidney stone composition: CNNs to predict kidney stone composition. BJU Int. **125**(6), 920–924 (2020)
2. Shetty, M., Shetty, S.B., Sequeria, G.P., Hegde, V.: Kidney stone detection using CNN. 2024 Second International Conference on Data Science and Information System (ICDSIS), Hassan, India, pp. 1–6 (2024). doi:https://doi.org/10.1109/ICDSIS61070.2024.10594342
3. Bhardwaj, S., H. S., S.: Modeling of a CNN architecture for kidney stone detection using image processing. In: 2022 Fourth International Conference on Emerging Research in Electronics, Computer Science and Technology (ICERECT), Mandya, India, pp. 1–5 (2022). doi:https://doi.org/10.1109/ICERECT56837.2022.10059972
4. Tanwar, V., Lamba, S.: Classification of multiple maize leaf diseases using a blended convolutional neural network. In: 2023 3rd International Conference on Intelligent Technologies (CONIT), Hubli, India, pp. 1–7 (2023). doi:https://doi.org/10.1109/CONIT59222.2023.10205797
5. Degadwala, S., Rathva, V.: A review on kidney stone detection using ML and DL techniques. Int. J. Comput. Sci. Eng. Inf. Technol., 10, 5, 101-102 (2024). doi:https://doi.org/10.32628/CSEIT2410589
6. Tanwar, V., Sharma, B., Yadav, D.P., Dwivedi, A.D.: Enhancing blood cell diagnosis using hybrid residual and dual block transformer network. Bioengineering (Basel). **12**(2), 98 (2025)
7. Menon, C.S.: Advances in kidney stone imaging techniques. J. Urol. **35**(2), 120–128 (2020)
8. Krizhevsky, A., Sutskever, I., Hinton, G.: ImageNet classification with deep convolutional neural networks. Adv. Neural Inf. Process. Syst. **25**, 1097–1105 (2012)

9. He, K., Zhang, X., Ren, S., Sun, J.: Deep residual learning for image recognition. In: Proceedings of the IEEE Conference on Computer Vision and Pattern Recognition (CVPR), pp. 770–778 (2016)
10. Tan, M., Le, Q.: EfficientNet: Rethinking model scaling for convolutional neural networks. In: Proceedings of the 36th International Conference on Machine Learning (ICML), pp. 6105–6114 (2019)
11. Ronneberger, O., Fischer, P., Brox, T.: U-Net: Convolutional networks for biomedical image segmentation. In: International Conference on Medical Image Computing and Computer-Assisted Intervention (MICCAI), pp. 234–241 (2015)
12. Xu, J., Wang, S., Zhao, H.: Attention-based hybrid models for medical image classification. IEEE Access. **9**, 85056–85067 (2021)
13. Liu, Z., et al.: Comparative analysis of deep learning models for medical imaging. IEEE Trans. Med. Imaging. **39**(6), 2231–2241 (2020)
14. Zhang, R., Wu, J., Liu, Y.: Transfer learning with CNN for kidney stone detection. Comput. Biol. Med. **128**, 104124 (2021)
15. Brown, T., et al.: Automated kidney stone detection using deep learning. Nat. Biomed. Eng. **5**, 345–355 (2022)
16. Chen, Y., et al.: Improving model robustness in medical imaging. IEEE J. Biomed. Health Inform. **25**(4), 1308–1318 (2021)
17. Kidney Stone Dataset (Binary dataset). Kaggle.com. [Online]. Available: https://www.kaggle.com/datasets/visheshtanwar26/kidneya. Accessed 1 Jan 2025.

Generative Artificial Intelligence Tools in Higher Education: Systematic Literature Review

Maha M. Hassan[1,2](✉) and Sanaa S. Askool[2,3]

[1] Civil Engineering Department, University of Prince Mugrin, Madinah, Saudi Arabia
m.modather@upm.edu.sa

[2] Dr. Hussein AlSayyed Research & Innovation Center, Dept. of SRGS, University of Prince Mugrin, Madinah, Saudi Arabia

[3] College of Computer and Cyber Sciences, University of Prince Mugrin, Madinah, Saudi Arabia

Abstract. Generative artificial intelligent tools have impacted higher education worldwide providing both opportunities and challenges. These tools offer transformative potential by enhancing teaching, learning, and administrative processes. However, the advancement in large language models with the ability to perform complex natural-language processing tasks encouraged students to use in different course work leading to risks of academic dishonesty. The effects of biases on these tools are still unclear. Moreover, use of generative AI designs and photos in art-related disciplines is another challenge. Until now, universities have been taking different approaches to integrate the use of this technology into academic policies. This paper seeks to provide an overview of research on the influence of generative AI tools on higher education through a systematic approach. 4950 publications were initially identified for the period between 2022 And 2025. This paper highlights the main findings of the review study focusing on students' perception of these tools, influence on their performance, and governing policies. It is found that these tools are reshaping the academic world, yet universities are still lagging in their integration.

Keywords: Generative artificial intelligence · Higher education · ChatGPT · Perception · Challenges

1 Introduction

Reliance on technology, specifically generative artificial intelligence (Gen AI), in higher education necessitates examining their influence on different aspects. The launch of ChatGPT in November 2022 represents a milestone where educators faced new and unprecedented challenges in addition to many opportunities. Gen AI tools have the potential to provide personalized learning support to students and immediate tutoring. However, they could be a tool for aiding academic dishonesty. Clear academic policies to deal with these challenges are still lacking [1, 2]. Moreover, hallucinations and bias shed doubts on the accuracy and reliability of generated content [3–6]. Regardless of these challenges, studies show that 43% of college students in the US use AI tools and 50% of instructors employ them to develop lessons [7].

D. Berkouk et al. (Eds.): CCTS 2024, PTT, pp. 168–175, 2026.
https://10.1007/978-981-95-7607-4_19

Generative AI refers to algorithms that can create new content, including text, images, audio, and video, based on patterns learned from existing data [8]. A surge in Gen AI tools happened since the release of ChatGPT, which lastly witnessed the release of DeepSeek. Some review studies have systematically organized current research on Gen AI in higher education [9, 10]. However, these reviews focus on aspects related to overall development trends, core issues, opportunities and risks, integration of Gen AI in instruction process, and integration strategies adopted by teachers. There is still a lack of a review that considers students' perceptions of Gen AI. This study reviews Gen AI tools usage in higher education with a focus on the following research questions:

- How do students use and perceive Gen AI tools in higher education?
- What is the influence of these tools on students' skills and understanding?
- How do different academic institutions deal with usage of Gen AI tools?
- What are the challenges for using Generative AI in education?

To answer these questions, a search methodology is employed to grasp the state of research field. Due to paper size limitations, the main research findings are summarized concisely, highlighting key results, methodologies, and implications while omitting data analysis and supplementary discussions.

2 Method

Review is performed to address specific questions based on systematic and replicable search strategy [10]. To conduct a comprehensive literature review on generative AI tools in education using ScienceDirect in February 2025, a search methodology was employed. First, relevant keywords were identified to refine the search query, including terms such as "generative AI" and "higher education". Advanced search filters were applied to limit results to peer-reviewed journal articles, conference proceedings, and publications. This initial search yielded 4950 articles. A manual screening was performed to assess the relevance of these articles to the research questions. Accordingly, irrelevant publications were removed from the database. The final dataset included 128 articles published between 2022 and 2025. The inclusion and exclusion criteria are shown in Table 1. Accordingly, findings are categorized based on broad directions and research objectives as listed below.

Table 1 Screening inclusion and exclusion criteria.

Category	Inclusion and Exclusion Criteria
Research Field	1. Articles on creating Gen AI tools for purposes other than education are excluded. 2. Articles that assess the performance of Gen AI tools in solving exams are excluded. 3. Articles dealing with AI technology generally are excluded. 4. Articles on the use of Gen AI tools in online courses are excluded.
Educational Stage	Research targeting K-12, primary and secondary schools is excluded.
Publication Type	Abstract only studies are excluded.

3 Findings

A substantial body of studies has examined Gen AI potential and applications in different subjects. Most papers are published in 2024. AI technologies have been increasingly applied to facilitate education and training in different subjects including language, STEM, and medicine [11, 12].

3.1 Students' Perception

Several studies investigate students' acceptance of Gen AI in education in different contexts. In Ghana, Salifu et al. [13] conducted a survey with 306 undergraduate participants, identifying key factors such as social influence, performance expectancy, hedonic motivation, and habitual behavior as crucial drivers of students' behavioral intentions, ultimately impacting their actual use of Gen AI. Conversely, a study in Jordan by Masadeh et al. [14], based on data from 880 students, highlighted the importance of credibility, usefulness, and ease of use in shaping students' positive attitudes toward adopting Gen AI in classrooms. Notably, student attitudes mediate the link between perceived usefulness and the intention to utilize GAI. Similarly, Rahman et al. [15] investigated Gen AI adoption among 344 students from both private and public universities in Bangladesh, emphasizing that perceived usefulness, ease of use, and informativeness significantly influence students' attitudes, which in turn shape their behavioral intentions to use GAI for learning. In Sri Lanka, research by Sabraz Nawaz et al. [16] involving 500 undergraduates across 17 government universities found that habit, performance expectancy, and perceived ease of use play a critical role in students' acceptance of AI applications. Their study underscores the need for effective implementation strategies, acknowledgment of individual differences in technology adoption, and continuous support and training to enhance student proficiency. On the other hand, a study in Malaysia by Foroughi et al. [17], which applied a hybrid analytical approach to data from 406 students, revealed that performance expectancy, effort expectancy, hedonic motivation, and learning value significantly influence students' intentions to use Gen AI, whereas social influence, facilitating conditions, and habit did not have a notable impact on Gen AI usage. Waluyo and Kusumastuti [18] examined the impact of Gen AI on English language learning in Thai higher education. Students exhibited high acceptance of generative AI (GAI) tools for English learning, except for the influence of social factors. There were no significant differences in Gen AI usage between high-performing and low-performing English language learners. Students reported improved efficiency, enhanced engagement, and increased linguistic confidence from using Gen AI tools, but also expressed concerns about over-reliance and the need for critical use. A study of 305 students' perception of usage of ChatGPT showed that they feel increased freedom and productivity in addition to strengthened academic engagement [19]. Bernabei et al. [20] focused on the use of engineering students of Gen AI tools and showed that LLM assistance did not undermine their preparation, yet many students did not rework and discuss the text produced by ChatGPT.

Students' adoption rate of Gen AI tools is faster than teachers [21]. Feelings of anxiety and discomfort have been identified as significant barriers to utilize these tools by educators [22, 23]. Results show that lecturers' and students' perception of risks and weaknesses of Gen AI tools differ. Developing and deploying efficient procedures to spread knowledge and awareness about the use of Gen AI tools in research and education could enhance teachers' attitude towards their use [24].

3.2 Adaptive Learning and Personalized Tutoring Applications

Different tools exist with the aim of customizing the learning process and creating a welcoming learning environment for students. Examples include tools targeting mathematics and chemistry, sports training, language and medical guidance [25–29]. Chan and Tsi [30] investigated the potential of Gen AI to assist or fully replace human teachers through an online survey targeting 399 university students and 184 teachers. Majority of respondents thought that human teachers are irreplaceable considering their unique qualities including critical thinking and emotions.

3.3 Impact of Gen AI on Education

Studies on impact of Gen AI on education include work by Malik et al. [31, 32] and Rajala et al. [33]. Findings showed that Gen AI positively impacted students in three Arab countries: Yemen, Jordan, and Oman [34]. Students were able to associate tools and cognitive achievement. Wang and Zhang [35] showed that Gen Z are encouraged to use Gen AI tools considering several factors including effort expectancy, price value, and hedonic motivation.

Many LMS platforms have integrated AI technologies that conduct assessment tasks and generate tailored feedback. Studies on performance evaluation by Gen AI include [36, 37] showing how these tools can be efficient in assisting instructors.

3.4 Challenges and Ethical Considerations

Gen AI tools such as ChatGPT can support the writing process of those who may need extra help [38, 39]. However, over-dependence on these tools by some students led to diminished writing ability, lack of creativity, and increased plagiarism among students [40]. Many studies address use of technology, specifically AI, in English language learning [18, 26, 41, 42]. Different reported challenges include:

- Academic Integrity: use of Gen AI tools for course work raises questions about plagiarism and originality of students' work [1, 5, 6].
- Data Privacy: use of Gen AI requires the collection and analysis of large amounts of data, raising concerns about user privacy and data security.
- Content Accuracy: Gen AI models can produce inaccurate and biased content, which could negatively impact learning outcomes [43, 44].
- Dependence on AI: these tools may hinder the development of critical thinking and problem-solving skills for both students and faculty.

3.5 Policies by Academic Institutions

Many studies have reported that there is a need for developing and enforcing policies that regulate the use of Gen AI tools [45]. A study by Ou et al. [46] shows that students possess divergent viewpoints on the ethical concerns of Gen AI tools, especially in the absence of university instructions. Absence of clear instructions combined with lack of confidence and competence perceived by educators to address the implications of Gen AI tools [47] shows a gap that needs to be addressed. Some universities banned students from using ChatGPT on assignments [48]. Use of ChatGPT is viewed as a form of plagiarism threatening pillars of contemporary education systems including accuracy of information, market value of skills, and ability to measure students' knowledge and skills [49]. Holmes and Pomsta [50] emphasized the need for developing ethical frameworks to govern the use of AI in education to minimize the potential risks. However, it is noticed that most conversations and discussions in universities are based on opinions of experts and educators as the center of the policy development process leaving out voices of students [51]. Students' viewpoints are essential to understand the impact of Gen AI on universities given they are the main users of these tools. A more inclusive approach to policy development is needed. Hence, future research should explore mechanisms for incorporating student feedback into AI governance frameworks

4 Conclusions

This study presents the findings of a literature review aimed at understanding the implications of generative artificial intelligence tools on higher education. It was found that generative artificial intelligence tools have an impact on learning processes, hold great potential and many ethical challenges. Integration of Gen AI tools in higher education requires a thoughtful approach acknowledging both benefits and limitations. Moreover, these tools provide both students and faculty with rich learning resources and help them process information across different cognitive levels. The adoption rate of this new technology is progressing with different rates between students and instructors. It is expected that the area will attract increased research efforts in the upcoming years considering the surge in reliance on Gen AI tools in different fields. Studies show that universities need to find innovative approaches to embrace these technologies and make use of their potential. Students need to be trained to be ready for employment in a job market powered by Gen AI considering the ethical, privacy and pedagogical challenges. Moreover, different strategies are needed to address challenges related to curriculum adaptation and assessment reformation.

Acknowledgement. This scientific paper is derived from a research grant funded by the Research, Development, and Innovation "Authority (RDIA) – Kingdom of Saudi Arabia – with grant number (13465-UPM-2023-UPM-R-3-1-EF-) and Dr. Husain AlSayyd Center for Research and Innovation's fund.

References

1. Hua, S., Jin, S., Jiang, S.: The limitations and ethical considerations of chatgpt. Data Intell. **6**(1), 201–239 (2024)

2. Cotton, D.R., Cotton, P.A., Shipway, J.R.: Chatting and cheating: Ensuring academic integrity in the era of ChatGPT. Innov. Educ. Teach. Int. **61**(2), 228–239 (2024)
3. Gill, S.S., et al.: Transformative effects of ChatGPT on modern education: Emerging Era of AI Chatbots. Internet Things Cyber-Phys. Syst. **4**, 19–23 (2024)
4. Motoki, F., Pinho Neto, V., Rodrigues, V.: More human than human: measuring ChatGPT political bias. Public Choice. **198**(1), 3–23 (2024)
5. Ray, P.P.: ChatGPT: A comprehensive review on background, applications, key challenges, bias, ethics, limitations and future scope. Internet Things Cyber-Phys. Syst. **3**, 121–154 (2023)
6. Rutinowski, J., Franke, S., Endendyk, J., Dormuth, I., Roidl, M., Pauly, M.: The self-perception and political biases of ChatGPT. Hum. Behav. Emerg. Technol. **1**, 7115633 (2024)
7. Businessolution.org: AI in Education Statistics. https://businessolution.org/ai-in-education-statistics/. Accessed 10 April 2023
8. Dwivedi, Y.K., et al.: Opinion Paper: "So what if ChatGPT wrote it?" Multidisciplinary perspectives on opportunities, challenges and implications of generative conversational AI for research, practice and policy. Int. J. Inf. Manag. **71**, 102642 (2023)
9. Sengar, S.S., Hasan, A.B., Kumar, S., Carroll, F.: Generative artificial intelligence: a systematic review and applications. Multimed. Tools Appl. **14**, 1–40 (2024)
10. Wang, P., Jing, Y., Shen, S.: A systematic literature review on the application of generative artificial intelligence (GAI) in teaching within higher education: Instructional contexts, process, and strategies. Internet High. Educ. **100996** (2025)
11. Perrotta, C., Selwyn, N.: Deep learning goes to school: Toward a relational understanding of AI in education. Learn. Media Technol. **45**, 251–269 (2020)
12. Ivanov, S., Soliman, M.: Game of algorithms: ChatGPT implications for the future of tourism education and research. J. Tour. Futures. **9**(2), 214–221 (2023)
13. Salifu, I., Arthur, F., Arkorful, V., Abam Nortey, S., Solomon, O.-Y.R.: Economics students' behavioural intention and usage of ChatGPT in higher education: A hybrid structural equation modelling-artificial neural network approach. Cogent Soc. Sci. **10**(1), 2300177 (2024)
14. Masadeh, R.E., et al.: Antecedents of adoption and usage of ChatGPT among Jordanian university students: Empirical study. Int. J. Data Netw. Sci. **8**(2), 1099–1110 (2024)
15. Rahman, M.S., Sabbir, M.M., Zhang, J., Moral, I.H., Hossain, G.M.: Examining students' intention to use ChatGPT: Does trust matter? Australas. J. Educ. Technol. **39**(6), 51–71 (2023)
16. Sabraz Nawaz, S., Fathima Sanjeetha, M.B., Al Murshidi, G., Mohamed Riyath, M.I., Mat Yamin, F.B., Mohamed, R.: Acceptance of ChatGPT by undergraduates in Sri Lanka: A hybrid approach of SEM-ANN. Interact. Technol. Smart Educ. **21**(4), 546–570 (2024)
17. Foroughi, B., et al.: Determinants of intention to use ChatGPT for educational purposes: Findings from PLS-SEM and fsQCA. Int. J. Hum.–Comput. Interact. **40**(17), 4501–4520 (2024)
18. Waluyo, B., Kusumastuti, S.: Generative AI in student English learning in Thai higher education: More engagement, better outcomes? Soc. Sci. Humanit. Open. **10**, 101146 (2024)
19. Rehman, A., Behera, R.K., Islam, M.S., Abbasi, F.A., Imtiaz, A.: Assessing the usage of ChatGPT on life satisfaction among higher education students: The moderating role of subjective health. Technol. Soc. **78**, 102655 (2024)
20. Bernabei, M., Colabianchi, S., Falegnami, A., Costantino, F.: Students' use of large language models in engineering education: A case study on technology acceptance, perceptions, efficacy, and detection chances. Comput. Educ.: Artif. Intell. **5**, 100172 (2023)
21. Yang, Y., Xia, Q., Liu, C., Chiu, T.K.: The impact of TPACK on teachers' willingness to integrate generative artificial intelligence (GenAI): The moderating role of negative emotions and the buffering effects of need satisfaction. Teach. Teach. Educ. **154**, 104877 (2025)
22. Kay, R.: The impact of preservice teachers' emotions on computer use: A formative analysis. J. Educ. Comput. Res. **36**(4), 455–479 (2007)

23. Şahin, F., Doğan, E., İlic, U., Şahin, Y.L.: Factors influencing instructors' intentions to use information technologies in higher education amid the pandemic. Educ. Inf. Technol. **26**, 4795–4820 (2021)
24. Ivanov, S., Soliman, M., Tuomi, A., Alkathiri, N.A., Al-Alawi, A.N.: Drivers of generative AI adoption in higher education through the lens of the theory of planned behaviour. Technol. Soc. **77**, 102521 (2024)
25. Moorhouse, B.L., Wan, Y., Wu, C., Kohnke, L., Ho, T.Y., Kwong, T.: Developing language teachers' professional generative AI competence: An intervention study in an initial language teacher education course. System. **125**, 103399 (2024)
26. Kohnke, L., Moorhouse, B.L., Zou, D.: Exploring generative artificial intelligence preparedness among university language instructors: A case study. Comput. Educ.: Artif. Intell. **5**, 100156 (2023)
27. Kohnke, L.: Exploring EAP students' perceptions of GenAI and traditional grammar-checking tools for language learning. Comput. Educ.: Artif. Intell. **7**, 100279 (2024)
28. Caccavale, F., Gargalo, C.L., Kager, J., Larsen, S., Gernaey, K.V., Krühne, U.: ChatGMP: A case of AI chatbots in chemical engineering education towards the automation of repetitive tasks. Comput. Educ.: Artif. Intell. **8**, 100354 (2025)
29. Valeri, F., Nilsson, P., Cederqvist, A.M.: Exploring students' experience of ChatGPT in STEM education. Comput. Educ.: Artif. Intell. **8**, 100360 (2025)
30. Chan, C.K., Tsi, L.H.: Will generative AI replace teachers in higher education? A study of teacher and student perceptions. Stud. Educ. Eval. **83**, 101395 (2024)
31. Malik, T., Hughes, L., Dwivedi, Y.K., Dettmer, S.: Exploring the transformative impact of generative AI on higher education. In: Conference on e-Business, e-Services and e-Society 9, pp. 69–77. Springer Nature, Cham, Switzerland (2023)
32. Malik, T., Dettmer, S., Hughes, L., Dwivedi, Y.K.: Academia and generative artificial intelligence (GenAI) SWOT analysis-higher education policy implications. In: International Working Conference on Transfer and Diffusion of IT, pp. 3–16. Springer Nature, Cham, Switzerland (2023)
33. Rajala J, Hukkanen J, Hartikainen M, Niemelä P.: Call me Kiran–ChatGPT as a Tutoring Chatbot in a Computer Science Course. In: Proceedings of the 26th International Academic Mindtrek Conference, pp. 83–94, (2023).
34. Jaboob, M., Hazaimeh, M., Al-Ansi, A.M.: Integration of generative AI techniques and applications in student behavior and cognitive achievement in Arab higher education. Int. J. Hum.–Comput. Interact. **41**(1), 353–366 (2024)
35. Wang, Y., Zhang, W.: Factors influencing the adoption of generative AI for art designing among Chinese generation Z: A structural equation modeling approach. IEEE Access. **11**, 143272–143284 (2023)
36. Currie, G., Barry, K.: ChatGPT in nuclear medicine education. J. Nucl. Med. Technol. **51**(3), 247–254 (2023)
37. Geçkin, V., Kızıltaş, E., Çınar, Ç.: Assessing second-language academic writing: AI vs. Human raters. J. Educ. Technol. Online Learn. **6**(4), 1096–1108 (2023)
38. Warschauer, M., et al.: The affordances and contradictions of AI-generated text for writers of english as a second or foreign language. J. Second. Lang. Writ. **62**, 101071 (2023)
39. Popovici, M.D.: ChatGPT in the classroom. Exploring its potential and limitations in a functional programming course. Int. J. Hum.–Comput. Interact. **40**(22), 7743–7754 (2024)
40. Draxler, F., et al.: Gender, age, and technology education influence the adoption and appropriation of LLMs. arXiv preprint arXiv:2310.06556, 10 Oct 2023
41. Moorhouse, B.L.: Beginning and first-year language teachers' readiness for the generative AI age. Comput. Educ.: Artif. Intell. **6**, 100201 (2024)
42. Tafazoli, D.: Exploring the potential of generative AI in democratizing English language education. Comput. Educ.: Artif. Intell. **7**, 100275 (2024)

43. Kohnke, L., Moorhouse, B.L., Zou, D.: ChatGPT for language teaching and learning. RELC J. **54**(2), 537–550 (2023)
44. Alm, A., Watanabe, Y.: Integrating ChatGPT in language education: A freirean perspective. Iranian J. Lang. Teach. Res. **11**, 19–30 (2023)
45. Baek, C., Tate, T., Warschauer, M.: ChatGPT seems too good to be true: College students' use and perceptions of generative AI.OSFPreprints https://osf.io/preprints/osf/6tjpk_v1. Accessed 10 April 2023
46. Ou, A.W., Stöhr, C., Malmström, H.: Academic communication with AI-powered language tools in higher education: From a post-humanist perspective. System. **121**, 103225 (2024)
47. Moorhouse, B.L., Kohnke, L.: The effects of generative AI on initial language teacher education: The perceptions of teacher educators. System. **122**, 103290 (2024)
48. Castillo, E.: These schools and colleges have banned ChatGPT and similar AI tools. BestColleges [Internet]. https://www.bestcolleges.com/news/schools-colleges-banned-chat-gpt-similar-ai-tools/. Accessed 10 April 2023
49. Steele, J.L.: To GPT or not GPT? Empowering our students to learn with AI. Comput. Educ.: Artif. Intell. **5**, 100160 (2023)
50. Holmes, W., Porayska-Pomsta, K.: The ethics of Artificial Intelligence in Education. Taylor, Routledge (2023)
51. Chiu, T.K.: Future research recommendations for transforming higher education with generative AI. Comput. Educ.: Artif. Intell. **6**, 100197 (2024)

Role of Artificial Intelligence in Redefining Modern Speech Therapy Practices in Saudi Arabia

Amjad Alsaadi(✉) and Areej Alasseri

School of Education, Health, and Behavioral Sciences, Department of Health and Behavioral Sciences, Dar Al-Hekma University, Jeddah, Saudi Arabia
Amjadalsaadi@hotmail.com

Abstract. The integration of Artificial Intelligence (AI) into speech therapy is transforming clinical practice by enhancing diagnostic precision, therapy customization, and workflow efficiency. This study aligns with Vision 2030's commitment to advancing Saudi healthcare through technology. Utilizing a mixed-methods design, this research combines a structured survey of 201 licensed Speech-Language Pathologists (SLPs) with focus group interviews to explore the role of AI tools, specifically ChatGPT, in key aspects of clinical practice. These include diagnostic applications, therapy personalization, and administrative support. Findings reveal that AI enhances treatment efficiency by automating repetitive tasks, generating innovative therapy resources, and improving accessibility in regions with limited access to speech therapy. However, significant challenges remain, including insufficient training, access limitations, and the need for customized AI solutions tailored to speech therapy contexts. Experts emphasize the critical need for specialized AI training for SLPs, ensuring they can effectively integrate AI-driven interventions into clinical practice. Beyond technical improvements, this study highlights the cultural and accessibility considerations necessary for effective AI implementation in speech therapy. The findings provide actionable insights for practitioners, educators, and policymakers, offering a strategic framework for integrating AI responsibly into clinical workflows while maintaining patient-centered care.

Keywords: Artificial intelligence · Speech therapy · Vision 2030 · Saudi Arabia

1 Introduction

Speech therapy is an essential part of health care, focused on identifying, diagnosing, and treating communication and swallowing disorders. The use of artificial intelligence (AI) in speech therapy has gained much attention over the years. It helps in diagnosis, treatment, and engagement better than usual. Natural language processing (NLP) and other artificial intelligence (AI) technologies contributed to innovation in conventional therapeutic mechanisms. For example, authors used ML algorithms to develop predictive models for speech and language disorders to aid early diagnosis and personalized intervention strategies [2, 4, 7].

D. Berkouk et al. (Eds.): CCTS 2024, PTT, pp. 176–189, 2026.
https://10.1007/978-981-95-7607-4_20

In line with Saudi Arabia's Vision 2030 focusing on technology and improving access to healthcare, the impetus is on modernizing the speech therapy practice in the Kingdom. Vision 2030 aims to use technology to improve the delivery of healthcare services, such as enhancing digital transformation and increasing accessibility of services to underserved or remote areas. The COVID-19 pandemic emphasizes the need for remote health solutions. Rapid responses to emerging health challenges through technology innovations are fundamental [9, 13, 14].

Although AI technologies are becoming widespread, little research has been done on the role of AI in the redefinition of the speech therapy process in modern-day Saudi Arabia. This study, therefore, investigates how AI can help reimagine speech therapy in Saudi Arabia. By looking into the use of AI tools in daily practice, this study aims to provide insights that could lead to processes that are more effective and efficient from a diagnostic as well as therapeutic perspective, thereby improving outcomes [4].

Moreover, this study will be beneficial to speech therapists, policy-makers, and the healthcare system. This will offer practitioners evidence-based material that they can use to advance policy and practice guidelines that facilitate the employment of AI-led methodologies within the clinical setting [5, 11, 12]. For policymakers, it is a framework to enable the usage of innovative technologies in the healthcare infrastructure; it is in line with the objectives laid down in Vision 2030. The goal of this study, then, is to raise awareness of the use of AI in speech therapy and its potential. Furthermore, the study aims to contribute to the body of evidence surrounding AI in health care.

In conclusion, artificial intelligence has the potential to transform speech therapy services by enhancing service delivery and improving patient outcomes, aligning with the objectives of Saudi Arabia's Vision 2030. This study investigates the application of AI in the delivery of healthcare and speech therapy services, with a particular focus on its implementation within the Kingdom. The findings aim to shed light on how AI-driven approaches can support and modernize therapeutic practices in Saudi Arabia's evolving healthcare landscape.

2 Methodology

This study investigates the impact of Artificial Intelligence (AI) on current speech therapy practices in Saudi Arabia. Quantitative and qualitative methods were used to analyze responses from Speech-Language Pathologists (SLPs) working in Saudi Arabia to comprehensively evaluate AI's role. The study aims to assess the effectiveness, accuracy, and impact of AI on traditional speech therapy to develop evidence-based recommendations for its integration into clinical practice.

2.1 Research Design

This study adopts a mixed-methods design that combines qualitative and quantitative data collection to ensure a comprehensive analysis. Utilizing multiple research methods enhances the depth of insight and provides a holistic perspective on AI's role in speech therapy. A descriptive cross-sectional survey was conducted to collect quantitative data from a large sample of registered SLPs. Additionally, focus group interviews were conducted to gather qualitative data, offering deeper insights into practitioners' experiences. The survey results provided a broad overview of AI adoption, while focus group discussions revealed the challenges, benefits, and real-world applications of AI in speech therapy.

2.2 Participants

This study included licensed Speech-Language Pathologists (SLPs) certified by the Saudi Commission for Health Specialties (SCFHS). A total of 200 SLPs, representing both genders working in various clinical settings across Saudi Arabia, were randomly selected to participate in a survey. The survey collected quantitative data about their perspectives regarding the use of AI in speech therapy. Furthermore, eight SLPs were selected from the survey respondents to participate in a focus group discussion. Selection criteria included work experience with AI or a demonstrated interest in its clinical applications. These discussions provided valuable insights into the advantages and challenges of integrating AI into speech therapy practice.

2.3 Data Collection Tools

To collect quantitative data on AI adoption, experience, and its perceived impact on therapy outcomes, a structured questionnaire was administered to 200 licensed Speech-Language Pathologists (SLPs). After analyzing the survey, eight interested SLPs were selected to conduct focus group interviews with a semi-structured interview guide. This approach provided flexibility, allowing for a deeper exploration of participants' experiences, challenges, and benefits related to AI in speech therapy. Participants were given two weeks to complete the survey, which was distributed online. The survey was distributed online, while focus group sessions were conducted in a controlled environment, whether in person or online, depending on the participant's preference. Each session was from 30 to 60 min in duration and was recorded, with the consent of each participant, for accurate data collection for later qualitative analysis.

2.4 Data Analysis

Numerical and response-based analyses were employed to interpret the study results. The statistical analysis was conducted using questionnaire responses processed in SPSS. Descriptive statistics were used to summarize the main features of the data. Subsequently, correlation and regression analyses were conducted to examine the relationship between artificial intelligence (AI) usage, diagnostic accuracy, and treatment efficiency. These analyses provided insights into potential associations and contributed to the structured organization of data. Simultaneously, qualitative data from the focus group discussions was transcribed and analyzed using NVivo. A thematic analysis was carried out to explore the impact of AI on speech therapy practices, identifying recurring themes and patterns.

2.5 Validity and Reliability

To ensure the validity of the data collection tools, three experts in speech therapy reviewed the survey and interview guides to verify that they comprehensively addressed all study-related topics. Their feedback was incorporated to refine the questions, ensuring better alignment with the study's objectives. Before the full implementation, a pilot study involving 15 SLPs was conducted to identify and refine any ambiguous or biased survey or interview questions. Face validity was assessed by gathering participant feedback to determine the relevance and appropriateness of the questions. For reliability, the internal consistency measured by Cronbach's Alpha was 0.78, indicating an acceptable reliability level for statistical analysis.

3 Results

3.1 Study Tool Questionnaire Analysis

Results are categorized into quantitative survey findings and qualitative interview insights. The quantitative analysis includes demographic data (Table 1), participant feedback, and emerging trends:

This study on Speech-Language Pathologists (SLPs) included a sample size of 201 participants. A demographic profile covering gender, age, education, experience, target population, disorders treated, and work settings was developed. The majority of respondents were females (84.1%), while males accounted for 15.9%. In terms of age distribution, 50.3% of participants were between 26 and 34 years old, followed by 22.9% below 25 years old and 20.9% between 35 and 44 years old. Most participants held a bachelor's degree, a few held their Master's, and a negligible number held a doctoral degree. Regarding clinical experience, 11% had less than one year, 30.3% had 1–3 years, 22.9% had 4–6 years, and 21.9% had over 10 years. The majority of participants provided therapy for school-aged children (78.2%), adolescents (67.7%), and early childhood groups (51.2%). Additionally, 48.3% worked with adults, and 15.9% treated geriatric patients. In terms of disorders treated, 78.1% of respondents worked with language disorders, 76.1% with articulation disorders, 52.2% with fluency disorders, 21.4% with voice disorders, 18.4% with swallowing disorders, and 8.5% with hearing disorders. Participants also work in diverse clinical settings, with 58.2% in private clinics, 30% in schools, and 17.9% in hospitals. Additionally, 20% engaged in telepractice or online therapy. These findings highlight the diverse backgrounds and clinical settings of SLPs, reflecting the broad scope of speech therapy practice within the sample.

3.2 Frequency Distribution for Part II: Study Items - Role of Artificial Intelligence in Enhancing Speech Therapy Practices. The following section presents a frequency analysis of the research tool

Section 1: Awareness of AI Technologies

Question 1: Are you aware/familiar with any AI technologies that can assist in diagnosing speech and language disorders? Table 2 shows the participants' knowledge of AI technologies that can help diagnose speech and language disorders. The majority

Table 1 Participant Demographics

Variable	Category	Counts	Percent
1. Gender	Females	169	84.1%
	Males	32	15.9%
	Total	201	100%
2. Age	Under 25 years	46	16.4%
	25–34 years	101	50.2%
	35–44 years	33	22.9%
	45–54 years	16	8%
	55 years and above	5	2.5%
	Total	201	100%
3. Educational Level	Bachelor's Degree	136	69.7%
	Master's Degree	53	26.4%
	Doctoral Degree	12	6%
	Total	201	100%
4. Clinical Experience	Less than 1 year	23	11.4%
	1–3 years	61	30.3%
	4–6 years	46	22.9%
	7–10 years	44	21.9%
	More than 10 years	27	13.4%
	Total	201	100%
5. Which Population(s) Do You Primarily Work With? (Select all that apply):	Early Childhood (0–5 years)	103	51.2%
	School-Age Children (6–12 years)	158	78.6%
	Adolescents (13–18 years)	135	67.2%
	Adults (19–64 years)	97	48.3%
	Geriatrics (65+ years)	32	15.9%
	Total	201	100%
6. Which Disorder(s) Do You Primarily Work With? (Select all that apply):	Articulation Disorders	133	66.2%
	Language Disorders	157	78.1%
	Voice Disorders	25	12.4%
	Fluency Disorders	105	52.2%
	Swallowing Disorders	28	13.9%

(*continued*)

Table 1 (*continued*)

Variable	Category	Counts	Percent
	Hearing Impairment	78	38.8%
	Other	1	0.5%
	Total	201	100%
7. In Which Setting(s) Do You Primarily Work?	Private Clinic	125	62.2%
	School	30	14.9%
	Hospital	36	17.9%
	Telepractice/Online Therapy	40	19.9%
	Other	2	1%
	Total	201	100%

(76.1%) of the participants reported being unaware of such technologies, while only a quarter (23.9%) said they were aware. These findings suggest that many professionals have limited knowledge of AI's role in speech therapy. The results highlight the need to increase awareness of AI-driven diagnostic tools, which could enhance the effectiveness of speech and language therapy.

Table 2 Are you aware/familiar with any AI technologies that can assist in diagnosing speech and language disorders?

Variable	Category	Counts	Percent
Q1: Are you aware/familiar with any AI technologies that can assist in diagnosing speech and language disorders?	Yes	153	76.1%
	No	48	23.9%
	Total	201	100%

Question 2: Follow-Up Question: if "Yes, "Please provide an example of an AI technology you are familiar with for diagnostic purposes. The pie chart illustrates the AI technologies participants have used or are aware of for diagnostic purposes. The most frequently mentioned tool in the survey was ChatGPT, which was used in 31.2% of the responses. Additionally, other variations such as "Currently ChatGPT" and "Chat Gpt (Other Variations)" also received another 27.1% of mentions. Notably, 41.7% of respondents reported using other AI technologies, suggesting the presence of widely varying methodologies in AI-enabled diagnostics of speech and language disorders. Although ChatGPT is the most recognized tool, this finding indicates the need to explore alternative AI technologies further for speech therapy diagnostic enhancement (Fig. 1).

Table 3 illustrates participants' familiarity with AI tools for documentation in speech therapy. The majority (73.6%) reported being unaware of such technologies. This significant gap suggests that many professionals who do not use AI for documentation lack

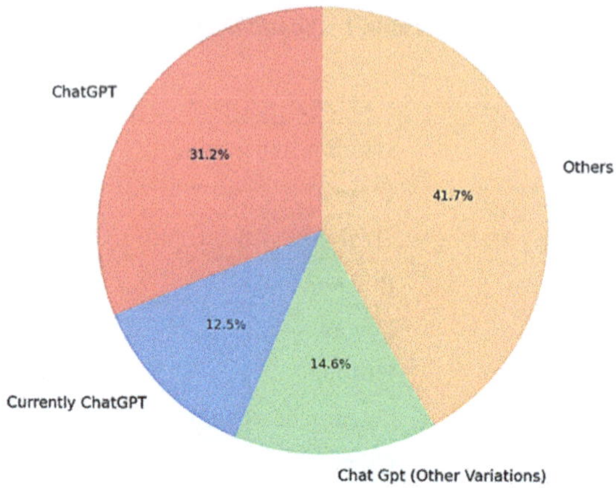

Fig. 1 Q2: Follow-Up Question: if "Yes," Please provide an example of an AI technology you are familiar with for diagnostic purposes

awareness of its potential benefits. The findings emphasize the need to raise awareness of AI-based documentation software and provide training, enabling professionals to work more efficiently and reduce their workload.

Table 3 Are you aware of any AI technologies that assist with documentation tasks in speech therapy?

Variable	Category	Counts	Percent
Q3: Are you aware of any AI technologies that assist with documentation tasks in speech therapy?	Yes	148	73.6%
	No	53	26.4%
	Total	201	100%

Question 4: Follow-Up Question: If "Yes," please provide an example of an AI technology you are aware of that assists with documentation. The pie chart illustrates participants' familiarity with AI tools used for documentation in speech therapy. ChatGPT was the most frequently reported tool, mentioned by 35.8% of participants. Additionally, variations of ChatGPT were noted in 15.1% of responses. However, nearly 49.1% of participants were categorized as "other" category, indicating they were either unaware of AI documentation tools or used lesser-known technologies such as BART and Virtual AI. Furthermore, 42.6% of participants reported actively using AI for documentation. The development of AI-based assistive technologies for speech therapy is becoming increasingly recognized. However, findings show that nearly half of the participants (49.1%) remain unfamiliar with specific AI documentation tools, highlighting the need for greater awareness and education (Fig. 2).

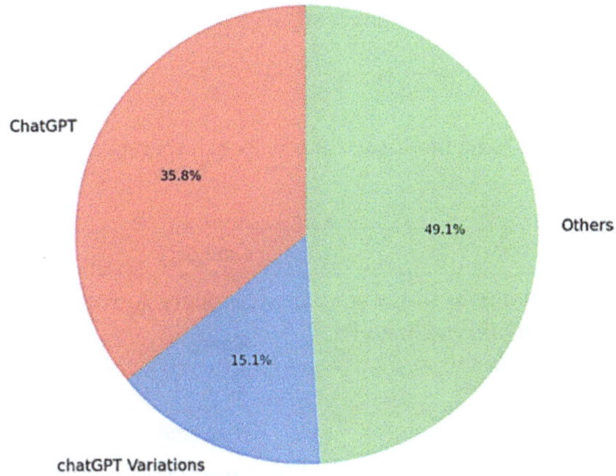

Fig. 2 Q4: Follow-Up Question: If "Yes," please provide an example of an AI technology you are aware of that assists with documentation.

As displayed in Table 4, most participants reported no experience with AI-based applications in speech therapy, such as Ambiki. Nearly 65.7% reported a level 5 familiarity, indicating a lack of exposure to these technologies. Additionally, 19.9% of participants were 'Slightly Familiar' (level 4), while moderate to high levels of familiarity were rare. Overall, these findings suggest that the use and adoption of AI tools in speech therapy remain limited. These results indicate the urgent need for greater awareness, training, and integration of AI-based applications in the field.

Table 4 Q5: On a scale of 1 to 5, how familiar are you with AI applications used in speech therapy, such as Ambiki?

Variable	Category	Counts	Percent
Q5: On a scale of 1 to 5, how familiar are you with AI applications used in speech therapy, such as Ambiki?	1- Extremely Familiar	0	0%
	2- Very Familiar	8	4%
	3-Moderately Familiar	21	10.4%
	4- Slightly Familiar	40	19.9%
	5- Not Familiar at All	132	65.7%
	Total	201	100%

Section 2: Usage and Perceived Impact of AI Technologies

Table 5 presents participants' responses to the statement: "I regularly use AI technologies in my speech therapy sessions, such as Ambiki or ChatGPT." The majority (64.7%) disagreed or strongly disagreed, indicating limited AI adoption in speech therapy. Meanwhile, 10.9% remained neutral, while 18.9% agreed and only 5.5% strongly agreed, reflecting minimal regular use of AI tools. These findings highlight the need for greater awareness, training, and accessibility to promote AI integration in speech therapy.

Table 5 Q6: I regularly use AI technologies in my speech therapy sessions, such as Ambiki or ChatGPT.

Variable	Category	Counts	Percent
Q6: I regularly use AI technologies in my speech therapy sessions, such as Ambiki or ChatGPT.	Strongly Agree	11	5.5%
	Agree	38	18.9%
	Neutral	22	10.9%
	Disagree	55	27.4%
	Strongly Disagree	75	37.3%
	Total	201	100%

Table 6 presents respondents' opinions on whether AI technologies provide detailed analysis and information for accurate diagnosis. A significant portion of respondents (43.3%) remained neutral, while 12.4% disagreed. Meanwhile, 29.4% agreed that AI can enhance diagnosis, while 8.5% strongly agreed.

Table 6 Q7: AI technologies improve the accuracy of my diagnosis by providing detailed insights and analysis.

Variable	Category	Counts	Percent
Q7: AI technologies improve the accuracy of my diagnosis by providing detailed insights and analysis.	Strongly Agree	17	8.5%
	Agree	59	29.4%
	Neutral	87	43.3%
	Disagree	25	12.4%
	Strongly Disagree	13	8.5%
	Total	201	100%

Table 7 presents respondents' views on how AI platforms simplify documentation and reporting, making speech therapy practice more efficient. The data suggest that 38.8% of participants remained neutral, while 8% disagreed about AI improving the documentation process. On the other hand, 36.8% of respondents agreed, while 10% responses strongly agreed, indicating recognition of AI's potential benefits. Meanwhile, 6.5% strongly disagreed. The study revealed mixed opinions among participants on whether AI usage will improve operations.

Q9: AI tools provide flexible options for tailoring therapy to meet diverse patient needs. Table 8 presents respondents' opinions on whether AI tools offer flexibility in customizing therapy for diverse patients. Nearly 44% of the participants disagreed, while

Table 7 Q8: AI tools help streamline documentation and reporting, saving time and increasing efficiency in my practice.

Variable	Category	Counts	Percent
Q8: AI tools help streamline documentation and reporting, saving time and increasing efficiency in my practice.	Strongly Agree	20	10.0%
	Agree	74	36.8%
	Neutral	78	38.8%
	Disagree	16	8.0%
	Strongly Disagree	13	6.5%
	Total	201	100%

8.5% strongly disagreed, reflecting doubt about AI's ability to customize therapy. Conversely, 34.8% agreed, and 8% strongly agreed, indicating AI's potential for personalized therapy. These findings suggest mixed opinions on AI's role in therapy customization. Training and practical demonstrations are necessary to illustrate how AI can enhance customized therapy.

Table 8 Q9: AI tools provide flexible options for tailoring therapy to meet diverse patient needs.

Variable	Category	Counts	Percent
Q9: AI tools provide flexible options for tailoring therapy to meet diverse patient needs	Strongly Agree	17	8.5%
	Agree	70	34.8%
	Neutral	88	43.8%
	Disagree	17	8.5%
	Strongly Disagree	9	4.5%
	Total	201	100%

Challenges and Needs for AI Integration

Q10: What challenges have you encountered in using AI technologies like ChatGPT or Ambiki in your practice? Please explain. The assessment highlights key challenges in applying AI technologies like ChatGPT and Ambiki in speech therapy. A common concern among users is a lack of knowledge about AI, leaving many unsure of how to utilize it effectively. Another key issue is limited availability, as some workplaces do not provide access to AI tools. Some participants expressed concerns about AI's accuracy, stating that its content is often too generic for individualized therapy. Furthermore, insufficient training and cost problems contribute to adoption barriers. Despite these challenges, some participants recognized AI benefits, such as enhancing creativity, facilitating communication with caregivers, and streamlining documentation. According to these findings, awareness, training, and cost-related AI tools are required to improve the use of AI in speech therapy (Figs. 3 and 4).

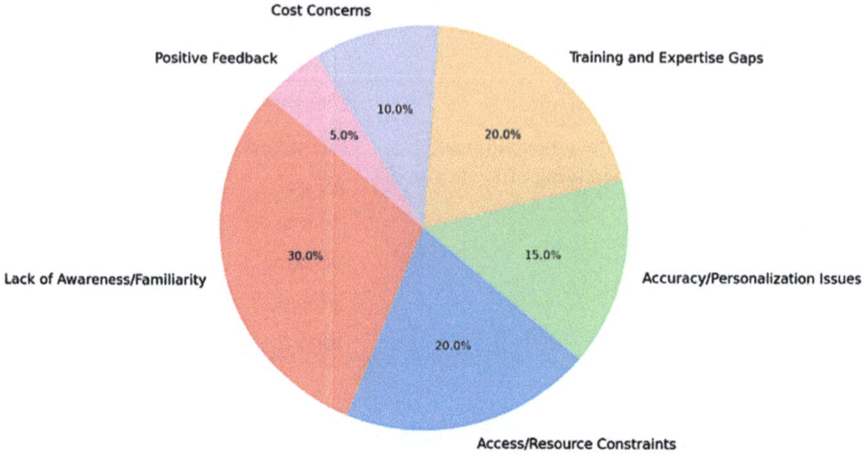

Fig. 3 Q10: What challenges have you encountered in using AI technologies like ChatGPT or Ambiki in your practice? Please explain.

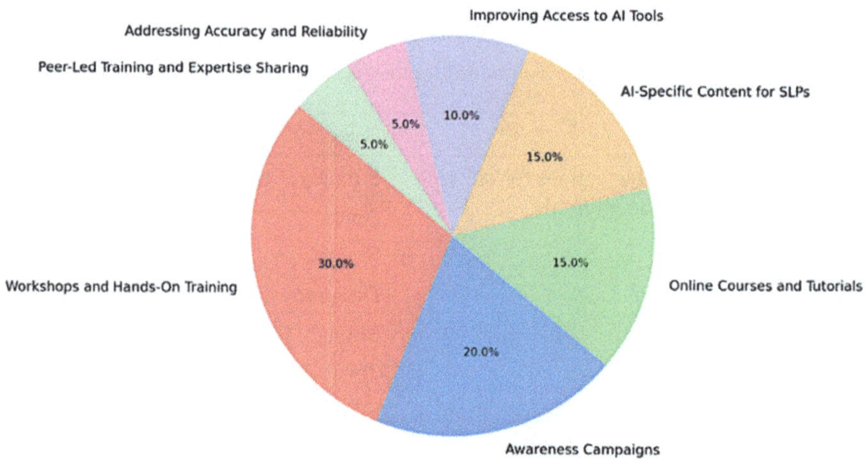

Fig. 4 Q11: What types of training or resources would help you feel more prepared to use AI effectively in therapy?

Q11: What types of training or resources would help you feel more prepared to use AI effectively in therapy? The analysis identifies key training and resources for SLPs using AI in therapy. Requests for training were the most common, highlighting the need for hands-on training and experience-based learning. Many participants suggested launching an AI learning awareness campaign, along with self-paced courses. Key concerns included improving workplace access to AI tools, as well as accuracy and reliability in AI-generated outputs. Training led by peers could be beneficial, as experienced users can provide practical insights. A well-structured, accessible training program combining awareness-building, practical instruction, and tool accessibility is needed for seamless AI implementation.

3.3 Study Tool Interview Analysis

Use of AI Technologies in Speech Therapy

Q1: What specific AI tools are you familiar with or currently use in speech therapy? The feedback indicates that participants strongly preferred ChatGPT in speech therapy for therapy planning, diagnosis, documentation, and material generation. Therapists, particularly in online and school-based settings, frequently use ChatGPT to create therapy activities and optimize workflow. Few participants mentioned Ambiki, suggesting limited awareness of alternative AI tools. The results indicate that additional training would increase speech therapists' awareness of available AI tools, enhancing their resources and clinical efficacy.

Q2: Can you describe your experience with using AI technologies in therapy? Participants reported that AI tools like ChatGPT have become essential in speech therapy, streamlining tasks, fostering creativity, and enhancing communication with clients. Many participants noted that AI significantly reduces time spent on documentation, report writing, and resource creation. AI tools can efficiently create therapy plans, provide exercises, and summarize complex disorders. AI also helps simplify communication with clients and their families by clarifying discharge instructions and therapy goals. While therapists emphasize that AI should not replace professionals, it can assist in identifying problems and developing new approaches. Overall, AI tools have been increasingly integrated into speech therapy.

Q3: Which applications or tools have you found most helpful? The responses indicate that ChatGPT is the most effective AI tool in speech therapy due to its efficiency, adaptability, and practicality. Participants emphasized that it helped with therapy ideas, writing reports, and summarizing information. Respondents added that its recommendations can be easily tailored for various clients and age groups. Currently, ChatGPT is the most well-known tool among users. Findings highlight the importance of exposure and awareness of different AI tools among professionals.

Experiences with AI in Speech Therapy

Q4: What key benefits have you noticed from using AI in therapy sessions? Responses strongly favor ChatGPT as the most effective tool in speech therapy. Respondents highlight its efficiency, adaptability, and versatility. They noted that it simplifies therapy planning, report writing, and summarization tasks. Therapists can tailor suggestions to different clients and age groups, enhancing the tool's practicality.

Q5: Have you encountered any challenges or difficulties with these tools?. The responses highlight several challenges of using ChatGPT and other AI tools in speech therapy. A major concern is determining whether AI-generated content is evidence-based. Many participants acknowledged that AI is a helpful tool but should not replace professionals. Another challenge was adapting AI content to fit hospital settings, cultural contexts, and virtual therapy.

4 Discussion

According to the findings of this research, AI in speech therapy can potentially improve efficiency and accessibility. Studies have shown that AI is beneficial for therapy-planning documentation [2, 7]. It can also facilitate therapy for users in remote locations, aligning with the findings of Franciscatto et al. [10]. However, Egaji et al. emphasized that AI cannot replace professionals [8]. Despite these benefits, challenges related to personalization, accuracy, and training still exist, as noted in [1, 2, 6]. While AI is primarily used for administrative tasks, Calvano et al.'s study highlights its emerging role in therapy sessions [4], contrasting with other applications. Furthermore, applications like Talkitt and those examined by Costanzo et al. have been used directly with patients, though they were not extensively covered in this study [6]. Overall, AI has the potential to enhance various aspects of speech therapy. However, several challenges must be addressed, including customization, training, and direct therapeutic use.

5 Conclusion

This study finds that AI is transforming speech therapy. Speech therapy is becoming more innovative, accessible, and efficient. However, AI still faces challenges regarding accuracy, personalization, and insufficient training. Building on these and previous findings, further research can improve the AI speech therapy tools to ensure that they become clinically validated, adaptable, and practical in therapeutic settings.

References

1. Benway, N.R., et al.: Reproducible speech research with the artificial intelligence–Ready PERCEPT corpora. J. Speech Lang. Hear. Res. **66**(6), 1986–2009 (2023)
2. Bhardwaj, A., Sharma, M., Kumar, S., Sharma, S., Sharma, P.C.: Transforming pediatric speech and language disorder diagnosis and therapy: The evolving role of artificial intelligence. Health Sci. Rev. **9**, 100188 (2024)
3. Brahmi, Z., Mahyoob, M., Al-Sarem, M., Algaraady, J., Bousselmi, K., Alblwi, A.: Exploring the role of machine learning in diagnosing and treating speech disorders: A systematic literature review. Psychol. Res. Behav. Manag. **17**, 2205–2232 (2024)
4. Calvano, M., Curci, A., Pagano, A., Piccinno, A.: Speech therapy supported by AI and smart assistants. In: Int. Conf. Product-Focused Softw. Process Improv, pp. 97–104. Springer, Cham (2023)
5. Chronopoulos, S.K., et al.: Exploring the speech language therapy through information communication technologies, machine learning and neural networks. In: 5th Int. Symp. Multidiscip. Stud. Innov. Technol. (ISMSIT), pp. 193–198. IEEE (2021)
6. Costanzo, F., et al.: Talkitt: Toward a new instrument based on artificial intelligence for augmentative and alternative communication in children with Down syndrome. Front. Psychol. **14**, 1176683 (2023)
7. Deka, C., Shrivastava, A., Abraham, A.K., Nautiyal, S., Chauhan, P.: AI-based automated speech therapy tools for persons with speech sound disorder: A systematic literature review. Speech Lang. Hear. **1–22** (2024)

8. Egaji, O.A., Asghar, I., Griffiths, M., Warren, W.: Digital speech therapy for the aphasia patients: Challenges, opportunities and solutions. In: *Proc. 9th Int. Conf. Inf. Commun. Manag.*, pp. 85–88 (2019)
9. Esfandbod, A., Rokhi, Z., Meghdari, A.F., Taheri, A., Alemi, M., Karimi, M.: Utilizing an emotional robot capable of lip-syncing in robot-assisted speech therapy sessions for children with language disorders. Int. J. Soc. Robot. **15**(2), 165–183 (2023)
10. Franciscatto, M.H., et al.: Towards a speech therapy support system based on phonological processes early detection. Comput. Speech Lang. **65**, 101130 (2021)
11. Kim, E.S., Shin, D.J., Cho, S.T., Chung, K.J.: Artificial intelligence-based speech analysis system for medical support. Int. Neurourol. J. **27**(2), 99 (2023)
12. Pottinger, G., Kearns, Á.: Big data and artificial intelligence in post-stroke aphasia: A mapping review. *Adv. Commun. Swallowing* (Preprint), 1–15 (2024)
13. Sabzevari, F., Amelirad, O., Moradi, Z., Habibi, M.: Artificial intelligence evaluation of COVID-19 restrictions and speech therapy effects on the autistic children's behavior. Sci. Rep. **13**(1), 4312 (2023)
14. Zhang, A., Pyon, R.E., Chen, K., Lin, A.Y.: Speech analysis of patients with cleft palate using artificial intelligence techniques: A systematic review. FACE. **4**(3), 327–337 (2023)

A Review on the Use of Artificial Intelligence for Anomaly Detection in SCADA Systems

Lujain Ali Alqahtani(✉), Sara khaed Alnatah, Maram Abdullah Alzahrani, Dalal Khalid Alkhaldi, Fatimah Saeed Aqaqah, Maryam M. Aldossary, and Mustafa Youldash

Imam Abdulrahman bin Faisal University, Dammam 34221, Saudi Arabia
{2210002740,2210002870,2210003443,2200002089,2210003507, mmsaldossary,mmzyouldash}@iau.edu.sa

Abstract. The paper fuses existing research with the contemporary debates concerning SCADA and cybersecurity augmented with machine learning. Cyber threats predominantly target SCADA systems that find application in critical infrastructure. Ordinary security tools often fail to spot anomalies in real time and thus are unable to respond quickly enough to mitigate system vulnerabilities. The infusion of machine learning into cybersecurity frameworks, however, has proved worthy during threat detection in terms of accuracy and response efficiency. Continued Internet traffic monitoring of SCADA for potential threats is needed for SCADA cybersecurity. Based on their ability to detect deviations from usual system behavior, machine learning-based anomaly detection methods can apply a proactive methodology. This paper aims, first and foremost, to examine existing ML approaches and algorithms in this field. This provides an exhaustive survey of the anomaly detection area propelled by machine learning suitable solutions thereby outlining their pros and cons in softening cyber risks.

Keywords: SCADA system · Cybersecurity · Anomaly detection · Machine learning · Operational technology · Intrusion detection systems

1 Introduction

Anomaly detection is one of the solutions used to analyze communications and maintain the security of SCADA. The anomaly is defined as a case that is different from the rest of the data in some specific ways. This misfit suggests that events of the given kind that pose a problem were produced by a mechanism entirely separate from the mechanism producing "normal" data [1]. Anomaly detection has been used for several years to detect and isolate anomalous features from datasets or data streams. There are many capacities of anomaly detection, and several methods have been devised for noticing anomalies [2].

Artificial intelligence (AI) algorithms can analyze security data from various sources like network traffic, logs, and infrastructure. This allows them to quickly locate anomalies and security threats. It can also characterize what is typical for the system and search

D. Berkouk et al. (Eds.): CCTS 2024, PTT, pp. 190–196, 2026.
https://10.1007/978-981-95-7607-4_21

for patterns that are outer from that based on existing historical security data. As such, security incidents in this case can be stopped from burdening into an active security-related event due to the real-time monitoring capability given by these solutions [3]. Our objective is to carry out a systematic review of the literature regarding SCADA anomaly detection, analyzing all approaches and their machine learning algorithms involved. This consists of evaluating the methods used, grouping them based on learning approaches employed, and their performance in detecting cyber threats. We compare these methods in order to find the best in terms of reliability and efficiency for SCADA-related security. This will further sharpen the perspective on the merits and demerits associated with these different techniques, making a commendable contribution toward enhancing cybersecurity in critical infrastructure systems.

2 Background

OT refers to the hardware and software used to control and monitor equipment, processes, and physical infrastructure. OT controls industrial operations through ICS, CNC, PLCs, SCADA systems, and robots by connecting them, tracking them, managing them, and securing them [4]. SCADA, on the other hand, is an OT system for the control of industrial processes, based on sensor data and control actions in real-time synchronization.

Anomaly detection entails the identification of points that fall outside normal expectations or of outliers [5], which can be of three types: point, contextual, or collective anomalies. ML anomaly detection allows the real-time detection of threats and thus enhances the security level of SCADA systems. These MLs can be classified as Supervised, Unsupervised, Semi-supervised, and Reinforcement learning. Supervised learning utilizes labeled datasets to train models in performing anomaly detection methods, while unsupervised learning classifies outlying behaviors without requiring labeled data, as highlighted by Christos Cholevas et al. [6]. Semi-supervised uses a mixture of supervised and unsupervised learning formats to construct classifiers using labeled and unlabeled data to provide a solution when limited labeled data is available. Lastly, reinforcement learning observes machine actions and assesses their correct decisions based on state actions and evaluation rewards.

3 Literature Review

In this literature review, we examine prior studies in SCADA security research that involve supervised, unsupervised, and semi-supervised anomaly detection techniques, considering their effectiveness, challenges, and methodologies.

3.1 Anomaly Detection using Unsupervised Learning Approaches

Researchers have investigated practical anomaly detection for industrial control systems with machine learning approaches. Sirajuddin Asjad [7] proposed an unsupervised intrusion detection model for encrypted DDS traffic using Autoencoder, Long Short-Term Memory (LSTM) Autoencoder, and One-Class SVM(OC-SVM) that reflects challenges to detect new intrusions, still it had the limitation of obfuscation. Ali Tekeoglu et al. [8] proposed real-time anomaly detection on Secure Water Treatment Testbed (SWaT) using time-series analysis and Symmetric Kullback-Leibler (KL) Divergence with an 87.11% accuracy, their approach had the weakness of using low data quality. Sajid Nazir et al. [9] used an autoencoder with Recursive Feature Elimination (RFE) and PCAa for SCADA anomaly detection by reconstruction loss that was effective on a gas pipeline dataset, yet the sensitivity of the reconstruction loss threshold impacts the balance between false positives and false negatives. All these studies emphasize the need for better feature extraction and diverse data to increase the accuracy of detection.

3.2 Anomaly Detection using Supervised Learning

Nicholas Jeffrey et al. [10] analyzed anomaly detection in Cyber-Physical Systems (CPS) through supervised learning, focusing on its precision advantages but pointing out weaknesses such as the unavailability of large, labeled datasets and high false positives. Sayawu Yakubu Diaba et al. [11] fortified SCADA security using a two-stage deep learning, showing their Genetically Seeded Flora Transformer Neural Network (GSFTNN) model outperformed Residual Neural Networks (ResNet), LSTM, and Recurrent Neural Networks (RNN) in classifying attacks, yet it has limited generalization to unseen Attacks. Nesibe Yalçın et al. [12] utilized XGBoost in conjunction with the WUSTL-IIOT-2021 dataset, which produced 99.99% accuracy in detecting cyber-attacks targeting SCADA systems; however, the method requires considerable computational power and available resources. These studies suggest that the quality of data is essential to developing effective sensors for providing cybersecurity for industrial systems, as well as more powerful models.

3.3 Anomaly Detection using Semi-Supervised Learning.

Veronika Bekbulatova et al. [13] proposed a semi-supervised anomaly detection framework by developing Deep-SAD using Federated Learning (FL) for IIoT Security. This framework allows for a secure way to train models as well as maintain significant values of privacy and scalability. The study employed the WUSTL-IIoT-2021 dataset to demonstrate the contributions of model training using FL by utilizing Federated Averaging (FedAvg) among devices and achieving high AUC-ROC and AUC-PR scores during the evaluation. However, it was also identified that the analysis was limited due to the model being targeted strictly at the analysis of Denial of Service (DoS) Attack Detection. Samson Isaac et al. [14] were able to demonstrate the effectiveness of Auto-Encoders in application to Probabilistic Bayesian Networks with the NSW-NB15_GT dataset for achieving a 98% accuracy. Chao Wang et al. [15] developed an anomaly detection mechanism based on application of Auto-Encoders in conjunction to use with One-Class Support Vector Machines (OCSVM) and a Gaussian Mixture Model (GMM) with the NF-BoT-IoT-v2 dataset and NF-CSE-CIC-IDS2018-v2 dataset framework performing tests and trained another framework to provide an accurate 98.73% accuracy rate, and demonstrate weaknesses such as upon the OCSVM results performing poorly limits the effectiveness of the GMM. These studies highlight a future improvement area in semi-supervised anomaly detection and suggest analyzing model capabilities that enhance privacy, accuracy, and value for feature extraction that improves efforts made towards supporting cybersecurity.

3.4 Comparison and Analysis

In this section, we compare different anomaly detection approaches for SCADA security, emphasizing their strengths, limitations, and key challenges in accuracy, and real-time performance (Table 1).

Table 1 Gap analysis

Study	Year	Supervised / Unsupervised /Semi-Supervised	Detection Approach	Data/Network Based Detection	Real-time	Accuracy	Used Dataset	Strength
[7]	2023	Unsupervised	Autoencoder, LSTM Autoencoder, OC-SVM, and K-Means	Network-based detection	No	91.28%	Navigation System, HMI, Hardware Monitoring System, and Alarm System datasets	Machine learning-based detection
[8]	2021	Unsupervised	(KL) Divergence	Network-based detection	Yes	87.11%	SWaT dataset	Adaptability to new threats
[9]	2021	Unsupervised	Autoencoder based detection	Data-based detection	Yes	100% (threshold =0.3) 99% (threshold =1.8)	SCADA gas pipeline dataset	Autoencoder based. High accuracy with less threshold
[10]	2023	Supervised	K-means SVM	Network-based detection	Yes	95%	ICS dataset	makes detection faster and uses less bandwidth.
[11]	2023	Supervised	GSFTNN	Network-based detection	Yes	99.12%	WUSTL-IIOT-2018 ICS	Improved Real-time Detection Capability
[12]	2024	Supervised	XGBoost	Network-based Detection	No	99.99%	WUSTL-IIOT-2018 and WUSTL-IIOT-2021	High accuracy, handles imbalanced data effectively, and minimizes false positives
[13]	2023	Semi-Supervised	Semi-supervised anomaly detection	Network-based Detection	No	100%	WUSTL-IIOT-2021	High detection Accuracy, and scalability for IIoT
[14]	2024	Semi-Supervised	combining Semi-Supervised Learning Auto-Encoders with Probabilistic Bayesian Networks (PBN)	Data-based detection	No	98%	NUSW-NB15_GT	The model effectively combines labeled and unlabeled data
[15]	2023	Semi-Supervised	AE, OC-SVM, GMM	Network-based detection	N/M	98.73% on NF-BoT-IoT-v2, 95.10% on NF-CSE-CIC-IDS2018-v2	NF-BoT-IoT-v2 and NF-CSE-CIC-IDS2018-v2	AE boosts the performance of detectors, dual anomaly detectors

Network-based anomaly detection continues to be the most commonly used mechanism for SCADA security, enabling real-time threat detection via centralized monitoring. The studies reviewed suggest that ML-based techniques gained popularity, particularly after 2020, with Autoencoders, OC-SVM, and clustering-based methods noted as representative of this trend due to their versatility and effectiveness in detecting anomalies in evolving SCADA networks. However, certain challenges are faced in the case of unsupervised methods related to encrypted traffic and real-time detection, such as Study [7].

Among supervised methods, RF, SVM, and XGBoost (Studies [10–12]) are highly accurate and require large, labeled datasets, which diminishes their scalability in cases of unseen threats. More significantly, computationally expensive algorithms, such as XGBoost, act as impediments to real-time SCADA deployment.

Semi-supervised methods: Autoencoders and OC-SVM, in particular. These methods may be applied in Studies [13–15]. They represent a middle ground between the precision of supervised and the versatility of unsupervised. Moreover, the models do not need too much labeled data to achieve high detection rates. They present a perfect fit for defining SCADA's normal operational behavior in real time, which makes them among the most promising techniques in scalability, encrypted traffic handling, and detecting anomalies in evolving SCADA networks.

4 Conclusion

SCADA systems are integral to critical infrastructure yet have a large attack surface posed by cyber threats owing to the inability of classical traditional security mechanisms to detect real-time anomalies. Enhancing anomaly detection has become a prime goal through machine learning for the detection of deviations in network behavior and improvement in response efficiency. Several studies have contributed to the detection of cyber threats in SCADA environments, which include the use of supervised, unsupervised, and semi-supervised learning approaches; each of these methods has its own strengths and weaknesses. The paper carries out a systematic review of existing research on anomaly detection based on ML for SCADA security, analyzing different approaches and algorithms. After completing a comparative review, we recognize the advantages and disadvantages of these approaches and derive lessons learned from their success. The results support the conclusion that semi-supervised methods, particularly those with autoencoders and OC-SVM, are an appropriate method to realize a trade-off between accuracy and expressiveness for real-time SCADA security. This review is a progress toward achieving greater cybersecurity for critical infrastructure.

References

1. Morrison, R., Liu, X., Lin, Z.: Anomaly detection in wind turbine SCADA data for power curve cleaning. Renew. Energy. **184**, 473–486 (2021). https://doi.org/10.1016/j.renene.2021.11.118
2. Nassif, A.B., Talib, M.A., Nasir, Q., Dakalbab, F.M.: Machine learning for anomaly detection: A systematic review. IEEE Access. **9**, 78658–78700 (2021). https://doi.org/10.1109/access.2021.3083060
3. Tatineni, S., Mustyala, A.: Advanced AI techniques for real-time anomaly detection and incident response in devops environments: Ensuring robust security and compliance. J. Comput. Intell. Robot. **2**(1), 88–121 (2022) Retrieved from https://thesciencebrigade.com/jcir/article/view/230

4. Cisco: How is OT different from IT? OT vs. IT. Cisco (n.d.). https://www.cisco.com/c/en/us/solutions/internet-of-things/what-is-ot-vs-it.html
5. Arshad, K. et al.: Deep reinforcement learning for anomaly detection: A systematic review. IEEE Access. **10**, 124017–124035 (2022). https://doi.org/10.1109/access.2022.3224023
6. Cholevas, C., Angeli, E., Sereti, Z., Mavrikos, E., Tsekouras, G.E.: Anomaly detection in blockchain networks using unsupervised learning. Algorithms. **17**(5), 201 (2024). https://doi.org/10.3390/a17050201
7. Asjad, S.: Intrusion detection and cyber attack classification for encrypted DDS communication middleware in OT networks using machine learning. University of South-Eastern Norway (2023). https://openarchive.usn.no/usn-xmlui/bitstream/handle/11250/3131438/no.usn%3awiseflow%3a6861899%3a54976491.pdf
8. Tekeoglu, A., Bekiroglu, K., Chiang, C.-F., Sengupta, S.: Unsupervised Time-Series based Anomaly Detection in ICS/SCADA Networks. In: International Symposium on Networks, Computers and Communications (ISNCC), doi:https://doi.org/10.1109/isncc52172.2021.9615827 (2021).
9. Nazir, S., Patel, S., Patel, D.: Autoencoder-based anomaly detection for SCADA networks. Int. J. Artif. Intell. Mach. Learn. **11**(2), 83–99 (2021). https://doi.org/10.4018/ijaiml.20210701.oa6
10. Jeffrey, N., Tan, Q., Villar, R.: A review of anomaly detection strategies to detect threats to cyber-physical systems. Electronics. **12**(15), 3283 (2023). https://doi.org/10.3390/electronics12153283
11. Diaba, S.Y., et al.: SCADA securing system using deep learning to prevent cyber infiltration. Neural Netw. **165**, 321–332 (2023). https://doi.org/10.1016/j.neunet.2023.05.047
12. Yalçın, N., Çakır, S., Ünaldı, S.: Attack detection using artificial intelligence methods for SCADA security. IEEE Internet Things J. **1–1** (2024). https://doi.org/10.1109/jiot.2024.3447876
13. Bekbulatova, V., Morichetta, A., Dustdar, S.: FL-SERENADE: Federated Learning for Semi-supervised Network Anomaly Detection. A Case Study. In: IEEE Intl Conf on Dependable, Autonomic and Secure Computing, Intl Conf on Pervasive Intelligence and Computing, Intl Conf on Cloud and Big Data Computing, Intl Conf on Cyber Science and Technology Congress (DASC/PiCom/CBDCom/CyberSciTech). doi:https://doi.org/10.1109/DASC/PiCom/CBDCom/Cy59711.2023.10361504, pp. 1072–1079 (2023).
14. Isaac, S., Ayodeji, D.K., Luqman, Y., Karma, S.M., Aminu, J.: Cyber security attack detection model using semi-supervised learning. FUDMA J. Sci. **8**(2), 92–100 (2024). https://doi.org/10.33003/fjs-2024-0802-2343
15. Wang, C., Sun, Y., Lv, S., Wang, C., Liu, H., Wang, B.: Intrusion detection system based on one-class support vector machine and gaussian mixture model. Electronics. **12**(4), 930 (2023). https://doi.org/10.3390/electronics12040930

Hybrid Data Augmentation model for Sequential Recommendation

Jian Gang Gao[1(✉)], Shun Zheng[1], Syed Bilal Hussain Shah[2], and Arshad Ahmed Mir[2]

[1] Tercelo Tire Group CO., LTD, Weifang, Shandong 262500, China
gaojg@tercelo.com
[2] University Centre, Leeds, United Kingdom
Bilal.syed@ucleeds.ac.uk, Arshad.Mir@leedscitycollege.ac.uk

Abstract. BERT has achieved tremendous success in the field of sequential recommendation. This bidirectional encoder, which models the capture of contextual information, can effectively capture the shifts in user interests. Although using BERT as an encoder is effective, we believe that there are still two limitations to using BERT in the sequential recommendation domain: User preferences for items may stem from multiple aspects, relying solely on the information within the user interaction sequence is difficult to capture the various features of the items and the user's emerging intentions. As an encoder that models sequences information through the Mask Language Model (MLM) task, BERT struggles to model user interaction sequences that are short in length. To address these limitations, we propose the Hybrid Data Augmentation for Sequential Recommendation (HDA4SR), which uses BERT as the encoder and, in line with BERT's characteristics, proposes two data augmentation methods from the perspectives of data-level and model-level to help the model better perform personalized user recommendations. Specifically, we propose a multi-level guided diffusion approach to introduce stable model-level data augmentation information for the target user sequence. We use other user sequences that are like the target user's interests to introduce data-level data augmentation information for the target user sequence. Additionally, we employ contrastive learning to capture the high-order features of these two types of information, enabling the model to better model user preferences. Experiments on two datasets show that HDA4SR significantly outperforms strong baselines.

Keywords: Diffusion · BERT · Data augmentation

1 Introduction

In recent years, encoder models based on self-attention, such as BERT [1], have become the focus of research in the field of sequential recommendation. The bidirectional self-attention mechanism in BERT can combine contextual information from two directions to make predictions and obtain better performance. Despite BERT is effective, we argue that such bidirectional models are sub-optimal due to the following limitations: (1) User preferences for items usually originate from multiple aspects, which may not be fully

D. Berkouk et al. (Eds.): CCTS 2024, PTT, pp. 197–205, 2026.
https://10.1007/978-981-95-7607-4_22

represented through their historical interactions. Since BERT relies solely on internal sequence information for modeling, it struggles to capture these fine-grained details. (2) Since BERT uses the MLM task to model sequences, it is difficult for BERT to capture the interest preferences of users with shorter sequence lengths.

We propose a novel Hybrid Data Augmentation model for Sequential Recommendation (HDA4SR). HDA4SR employs data-level data augmentation and model-level data augmentation tailored to BERT's characteristics to capture a more comprehensive user profile. Simultaneously, it supplements with contrastive learning to capture finer-grained user preferences from these two types of data augmentation, predicting the next item a user will prefer in the future. Specifically, we splice user sequences with similar interest preferences, enriching sequence information while mitigating the impact of the Masked Language Model (MLM) task on short interaction sequences. The improved guided diffusion technique can provide augmented information beyond the dataset for user interaction sequences from the perspective of model-level data augmentation by adding Gaussian noise and then guiding the removal of the noise. Our proposed model uses two BERTs as the underlying architecture, named the Feature Encoder and the Denoising Encoder, respectively. BERT, as a powerful encoder with denoising capabilities, can randomly mask the noise present in user interaction sequences while performing the MLM task, and its generative properties can produce more stable item features when performing diffusion tasks. These two characteristics make it an ideal choice for modeling augmented data.

2 Related Work

2.1 Self-attention

The self-attention has been proven effective in various fields such as computer vision [2]. Both the Transformer [3] and BERT models employ the self-attention mechanism. Early self-attention recommendation system models, such as SASRec [4], were the first to introduce the Transformer architecture into sequential recommendation, achieving significant results and laying the foundation for subsequent research. In the field of recommender systems, many studies are based on BERT as an encoder, such as BERT4Rec [5], which was among the first models to apply BERT to sequential recommendation. CBiT [6] fully leverages the characteristics of BERT, generates positive samples through multiple masking operations, and employs contrastive learning to enhance model performance.

In contrast to the studies, our work significantly increases the masking rate of BERT in the feature encoder and abandons the strategy of altering items during the masking process. Moreover, we introduce dynamic masking during training to enable the feature encoder to learn a more diverse set of masked samples, which plays a positive role in connecting two noisy user interaction sequences.

2.2 Uncertainty Modeling

In light of the fact that posterior distribution approximation can diminish the expressiveness of traditional VAE, ACVAE [7] introduces the adversarial variational Bayes (AVB) framework to enhance the generation capability of latent variables. Despite VAE's ability to model latent variables as distributions, it still confronts challenges with representation degradation and collapse. The diffusion model, renowned for its exceptional generative powers, has emerged as a popular choice in generative modeling, demonstrating immense potential in fields such as computer vision and natural language processing. In the domain of NLP, Diffusion-LM [8] stands as a pioneering effort that enables continuous diffusion processes in NLP tasks. DiffuSeq [9] extends the application of diffusion models to sequence-to-sequence tasks. DiffRec [10] was the first to introduce diffusion models to the field of recommender systems, ingeniously combining a VAE encoder with the diffusion model, and its effectiveness was demonstrated through extensive experimentation. DiffuRec [11] integrates the transformer into the diffusion model and designs an approximator to constrain model encoding for the acquisition of final representations, establishing a powerful baseline in the field.

Unlike their work, our method uses a multi-level guided diffusion approach to discretize the representation vectors partially generated by the diffusion process, providing direction for denoising. If the representations generated by the diffusion encoder are considered as vectors, our approach does not alter the direction of these vectors, resulting in higher accuracy.

2.3 Self-supervised Learning

Self-supervised learning is aimed at extracting contextual features from unlabeled data. In early self-supervised sequential recommendation models, S3-Rec [12] integrates contextual information into the system by maximizing mutual information. In recent years, contrastive learning has demonstrated significant application potential in the field of sequential recommendation. For instance, CL4SRec [13] employs three data augmentation strategies to generate positive samples. DuoRec [14] implements contrastive learning at the model level by combining supervised and unsupervised objectives. CBiT constructs multiple positive sample pairs using unsupervised dropout and a sliding window technique.

We use the user preferences augmented by data-level data augmentation and those augmented by model-level data augmentation as positive sample pairs for contrastive learning. This not only brings the two types of samples closer in the hyper-space, capturing finer-grained user preferences, but also pushes away the distances between user sequences with similar interests in the hyper-space, preventing the model from treating them as a single user. This enhances the discriminability between the augmented user samples.

2.4 Sequential Recommendation

Early sequential recommendation works take advantage of Markov chains to capture the dynamic transitions of user interactions, such as Fossil [15]. Later, GRU4REC [16] is the first to apply recurrent neural networks (RNNs) to model user interactions in sequential recommendation, which shows a significant improvement in performance compared to traditional methods like matrix factorization. Current cutting-edge techniques such as attention mechanisms, by achieving state-of-the-art performance in the field (including large language model-based recommendation algorithms [17]), further validate the effectiveness of sequence modeling.

As shown in Fig. 1. The key to sequential recommendation is to generate appropriate item representations. Therefore, a common approach is to shuffle the user interaction sequences through a dataloader and then input them into an encoder for encoding. Finally, the prediction layer is used to predict the next item that the user may like.

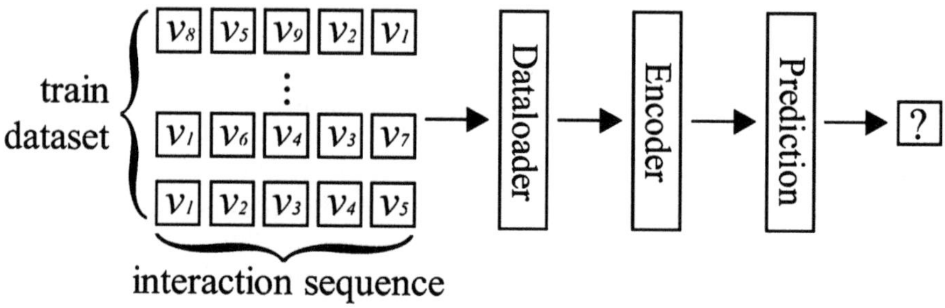

Fig. 1 The general model framework for sequential recommendation.

3 The Proposed Model

3.1 Model Framework

The framework of the proposed model is shown in Fig. 2. We first find the sequences that have the most overlapping items with the target sequence. We then concatenate these sequences, placing the target sequence at the end and the sequence of the user with similar interests at the beginning, as a means of data-level data augmentation. In the data loader, we mask the sequences, which means that the masked items in each sequence are different in each training iteration. The sequence features are sent out from the embedding layer include both item and position embeddings. Subsequently, they are fed into the feature encoder. Then the sequence features are obtained while completing the masked language modeling task. Next, we deliver these sequence features through the forward process to introduce noise and get the noisy embeddings that contain the number of noise injections. We input the noisy features into the denoising encoder for denoising. During this process, we use the sequence features of user interaction to guide the denoising encoder in its denoising work, enabling the denoising encoder to learn the sequence features corresponding to this noise. We use the sequence features obtained from the denoising encoder and the sequence features obtained from the feature encoder as positive sample pairs for contrastive learning, calculating the contrastive learning loss. We then compute the prediction loss using the feature information from these two types of features after they pass through the prediction layer. Finally, the two losses are combined by weighting and used as the ultimate loss of the model for optimization. At this point, the training phase has been completed. In the inference phase, we pass the test sequence through the embedding layer and into the feature encoder, obtaining a feature embedding. Subsequently, we sample noise embeddings from the Gaussian distribution and feed them into the denoising encoder. Guided by the feature embedding, we perform denoising after sampling, resulting in the restored user interaction feature embedding. Finally, we feed the obtained two feature embeddings into the prediction layer for prediction (Fig. 3).

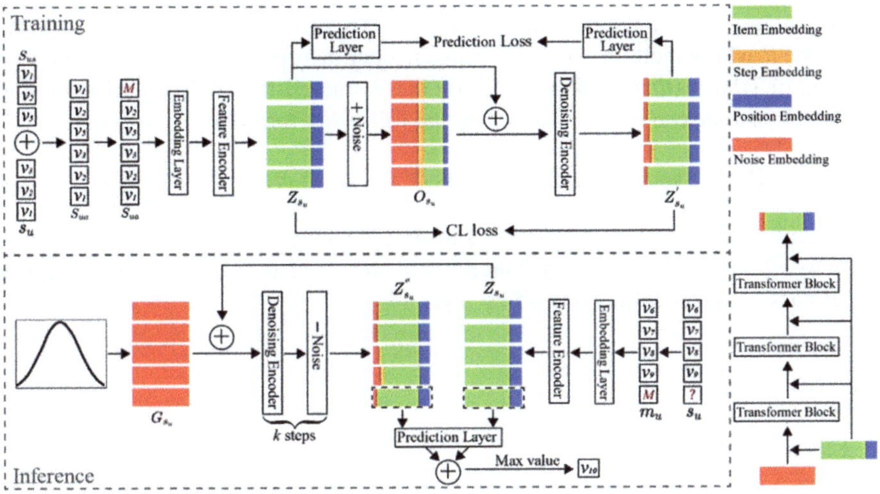

Fig. 2 The proposed model includes training and inference phases.

3.2 Problem Statement

We denote $\mathcal{U} = \{u_1, u_2, \cdots, u_{|\mathcal{U}|}\}$ as a set of users, $\mathcal{V} = \{v_1, v_2, \cdots, v_{|\mathcal{V}|}\}$ as a set of items, $s_u = \left[v_1^{(u)}, \cdots, v_t^{(u)}, \cdots v_{|s_u|}^{(u)}\right]$ as an sequence of user u sorted in chronological order, where $v_t^{(u)} \in \mathcal{V}$ represents the item that user u interacts with at t-th timestamp.

3.3 Sequence Splicing

We perform sequence splicing on the two users in the dataset who have shorter sequence lengths and the most overlapping items, with the target user's sequence spliced at the end and the sequence of the user with similar interests spliced at the front. Given the target user sequence

We perform sequence splicing on the two users in the dataset who have shorter sequence lengths and the most overlapping items, with the target user's sequence spliced at the end and the sequence of the user with similar interests spliced at the front. Given the target user sequence $s_u = \left[v_2^{(u)}, v_4^{(u)}, v_8^{(u)}\right]$ and the interaction sequence of the user with similar interests $s_{us} = \left[v_1^{(u)}, v_2^{(u)}, v_3^{(u)}, v_4^{(u)}\right]$, the augmented sequence can be shown by the following formula:

$$
\begin{aligned}
s_{ua} &= \left[v_1^{(u)}, v_2^{(u)}, v_3^{(u)}, v_4^{(u)}, v_2^{(u)}, v_4^{(u)}, v_8^{(u)}\right], \\
s_{usa} &= \left[v_2^{(u)}, v_4^{(u)}, v_8^{(u)}, v_1^{(u)}, v_2^{(u)}, v_3^{(u)}, v_4^{(u)}\right],
\end{aligned}
\tag{1}
$$

Feature Encoder and Denoising Encoder

The feature encoder retains the network architecture of BERT, with only changes made in the masking method. The denoising encoder incorporates guidance strategies at each layer of Transformer stacking based on the characteristics of diffusion technology, and its performance can be represented by the following formula:

$$
\mathbf{H}_{s_u}^{l} = \text{Activate}\left(\text{Linear}\left(Trm\left(\mathbf{H}_{s_u}^{l-1}\right) \oplus Z_{s_u}^{l}\right)\right), \forall l \in [1, \cdots, L] \tag{2}
$$

where Activate is the activation function, Linear is the linear layer, and is the Transformer layer in BERT. $\mathbf{H}_{s_u}^{l}$ are the sequence features generated from the user interaction sequence Su after passing through the embedding layer.

4 Experiments

4.1 Experimental Settings

The experiment is conducted on two datasets: Amazon [18] and MovieLens-1 M (ML-1 M) [19]. We select the Beauty domain from the Amazon dataset as the experimental dataset. The Beauty dataset have short sequence lengths. We continue to choose ML-1 M dataset, which has longer sequence lengths, as our third dataset. The statistics of the datasets are shown in Table 1.

Table 1 Statistics of the datasets

Dataset	Users	Items	Interactions	Sparsity(%)
Beauty	22,363	12,101	198,502	99.93
ML-1 M	6,040	3,416	999,611	95.16

All interactions are considered as implicit feedback. Duplicate interactions are eliminated, and the interactions of each user are arranged in chronological to form a sequence. Users with less than 5 interactions and items associated with fewer than 5 users are excluded, following previous research practices [9–11]. The evaluation strategy adopts the leave-one-out approach.

Table 2 Overall performance of different methods. (The optimal performance and the sub-optimal performance are denoted in bold and underlined fonts respectively.)

Dataset	Metric	Base on Transformer		Base on BERT		Base on Distribution			Ours	Impr (%)
		SASRec	DuoRec	BERT4Rec	CBiT	ACVAE	DiffRec	DiffuRec		
Beauty	HR@5	0.0327	0.0531	0.0213	0.0547	0.0247	0.0292	0.0558	**0.0591**	5.91
	HR@10	0.0626	0.0805	0.0372	0.0796	0.0388	0.0453	0.0791	**0.0854**	6.09
	HR@20	0.0898	0.1170	0.0579	0.1109	0.0612	0.0663	0.1111	**0.1182**	1.03
	NDCG@5	0.0240	0.0341	0.0132	0.0381	0.0256	0.0193	0.0400	**0.0419**	4.75
	NDCG@10	0.0323	0.0428	0.0183	0.0461	0.0321	0.0245	0.0475	**0.0503**	5.89
	NDCG@20	0.0366	0.0520	0.0235	0.0539	0.0376	0.0298	0.0557	**0.0586**	5.21
ML-1 M	HR@5	0.0938	0.1930	0.1364	0.2024	0.1272	0.0493	0.1797	**0.2241**	10.72
	HR@10	0.1689	0.2865	0.2057	0.2891	0.1993	0.0859	0.2626	**0.3186**	10.20
	HR@20	0.2830	0.3901	0.2995	0.3997	0.2897	0.1455	0.3679	**0.4275**	6.96
	NDCG@5	0.0532	0.1327	0.0889	0.1357	0.0823	0.0307	0.1212	**0.1552**	14.37
	NDCG@10	0.0773	0.1586	0.1113	0.1636	0.1054	0.0424	0.1479	**0.1855**	13.39
	NDCG@20	0.1059	0.1843	0.1348	0.1915	0.1282	0.0574	0.1744	**0.2130**	11.23

4.2 Experimental Results

We conduct comprehensive comparative and ablation study. The performance comparison results are shown in Table 2, and it can be observed that the experimental results of the proposed model are generally better than those of other models.

The proposed model shows good adaptability. Specifically, our model outperforms the transformer-based model on the Beauty dataset, thanks to the fact that our model introduces uncertainty in the construction of item representations, helping the model to generate more accurate predictions. For the BERT based model, which performs well on the ML-1 M dataset, the diffusion model technique used in this paper is superior to the contrastive learning (CBiT) technique. Diffusion models are first used in adversarial learning techniques of the computer vision domain where scholars utilize them as a means of data augmentation to enable models to learn more robust image representations. It is well known that contrast learning requires data augmentation to construct positive sample pairs. Data augmentation of diffusion techniques may have greater value in contrast learning. In conclusion, our model outperforms the models based on fixed vector modeling in terms of inference accuracy. For models based on distribution representation, our model may outperform DiffuRec, possibly because it uses an additional encoder to initially denoise the interaction sequences and employs a guided method to direct the denoising encoder, thereby eliminating the need for an approximator to constrain the representations trained by the model. Figure 2 is an ablation study of the model, where each component is indispensable in our model (Fig. 3).

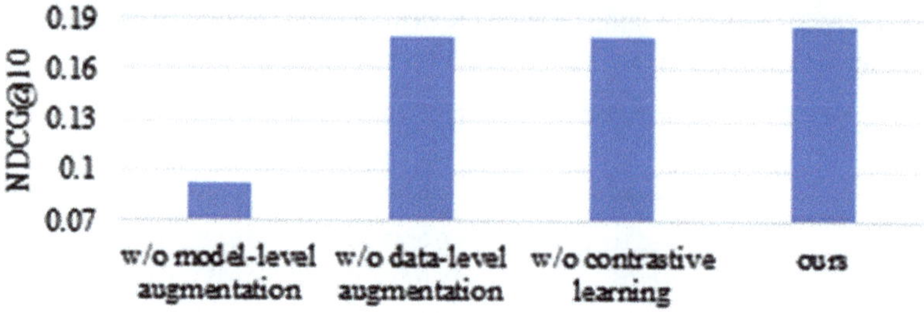

Fig. 3 Ablation study on the proposed model.

5 Conclusion

The model proposed in this paper fully utilizes two different learning methods to enable BERT to generate data with richer semantics, obtain more accurate hidden representations for each item in the sequence, and ultimately achieve more accurate semantics for the sequence. However, this study still has certain limitations. First, the model complexity is high due to the use of two separate encoders, resulting in many model parameters. Second, the inference speed of BERT is slow, which may not meet the requirements in scenarios with high real-time demands. Finally, excessive differences in user interest preferences within the dataset may adversely impact the model's recommendation performance.

References

1. Devlin, J., Chang, M., Lee, K., Toutanova, K.: BERT: Pre-training of deep bidirectional transformers for language understanding, in: NAACL-HLT 2019, Minneapolis, MN, USA, vol. 1, pp. 4171–4186 (2019).
2. Chen, Y., Fang, X., Ma, S., Bhuiyan, Z. A., Shu, L., Jing, X., TrackSpear: Attention-Guided Adversarial Patches for Probing Security in Visual Tracking. TechRxiv. 2025.
3. Vaswani, A., et al.: 2017. Attention is all you need, in: Advances in Annual Conference on Neural Information Processing Systems 2017, Long Beach, CA, USA, pp. 5998–6008 (2017).
4. Kang, W., McAuley, J.J.: Self-attentive sequential recommendation, in: ICDM 2018, Singapore, pp. 197–206 (2018).
5. Sun, F. et al.: Bert4rec: Sequential recommendation with bidirectional encoder representations from transformer, in: CIKM 2019, Beijing, China, pp. 1441–1450 (2019).
6. Du, H., et al.: Contrastive learning with bidirectional transformers for sequential recommendation, in: CIKM, Atlanta, GA, USA, pp. 396–405 (2022).
7. Xie, Z., Liu, C., Zhang, Y., Lu, H., Wang, D., Ding, Y.: Adversarial and contrastive variational autoencoder for sequential recommendation, in: WWW'21: The Web Conference 2021, Virtual Event/Ljubljana, Slovenia, ACM/IW3C2. pp. 449–459 (2021).
8. Li, X., Thickstun, J., Gulrajani, I., Liang, P., Hashimoto, T.B.: Diffusion-lm improves controllable text generation, in: NeurIPS, pp. 4328–4343 (2022).
9. Gong, S., Li, M., Feng, J., Wu, Z., Kong, L.: Diffuseq: Sequence to sequence text generation with diffusion models. CoRR abs/2210.08933 (2022).
10. Wang, W., Xu, Y., Feng, F., Lin, X., He, X., Chua, T.: Diffusion recommender model. CoRR abs/2304.04971 (2023).

11. Li, Z., Sun, A., Li, C.: Diffurec: A diffusion model for sequential recommendation. ACM Trans. Inf. Syst. **42** (2023)
12. Zhou, K., et al.: S3-rec: Self-supervised learning for sequential recommendation with mutual information maximization. In: Proceedings of the 29th ACM international conference on information & knowledge management, pp. 1893–1902 (2020).
13. Xie, X., et al.: Contrastive learning for sequential recommendation, in: ICDE 2022, Kuala Lumpur, Malaysia, pp. 1259–1273 (2022).
14. Qiu, R., Huang, Z., Yin, H., Wang, Z.: Contrastive learning for representation degeneration problem in sequential recommendation, in: WSDM 2022 Virtual Event/Tempe, AZ, USA, pp. 813–823 (2022).
15. He, R., McAuley, J.J.: Fusing similarity models with markov chains for sparse sequential recommendation, in: ICDM 2016, Barcelona, Spain, pp. 191–200 (2016).
16. Hidasi, B., Karatzoglou, A.: Recurrent neural networks with top-k gains for session-based recommendations, in: CIKM 2018, Torino, Italy, ACM. pp. 843–852 (2018).
17. Dai, S., et al.: Uncovering chatgpt's capabilities in recommender systems. Proceedings of the 17th ACM Conference on Recommender Systems. pp. 1126–1132 (2023).
18. Amazon Beauty dataset, http://snap.stanford.edu/data/amazon/productGraph/categoryFiles/ratings_Beauty.csv, last accessed 2024/06/08.
19. MovieLens-1M dataset, http://files.grouplens.org/datasets/movielens/ml-1m.zip, last accessed 2024/06/08.

Hybrid Machine Learning Models for DDoS Detection: A Performance Evaluation on the CICIDS2017 Dataset

Varkha K. Jewani[1], Pragati V. Thawani[1], Prafulla E. Ajmire[1], Mohammad Atique[1], and Imed Ben Dhaou[2](✉)

[1] Computer Science Department, SGBAU Amravati University, Mumbai, Maharashtra, India
peajmire@gmal.com

[2] Computer Science Department, School of Engineering, Computing, and Design, Dar Al-Hekma University, Jeddah, Saudi Arabia
ibendaou@dah.edu.sa

Abstract. In today's increasingly connected digital landscape, Intrusion Detection Systems (IDS) play a critical role in safeguarding networks against evolving cyber threats. Among these, Distributed Denial of Service (DDoS) attacks remain particularly challenging due to their scale and complexity. This study proposes a host-based IDS framework that dynamically detects DDoS attacks by adapting to malicious behavior patterns in local and neighboring nodes. Using the CICIDS2017 dataset—a well-established benchmark containing realistic network traffic—we evaluate the performance of various machine learning classifiers and introduce two hybrid models that combine their strengths. The analysis involves feature selection using information gain to enhance model accuracy and efficiency. Results demonstrate that the proposed hybrid models, particularly the combination of Random Forest and Decision Tree (Hybrid 2), significantly outperform individual classifiers in terms of accuracy, F1-score, and detection rate, while maintaining low false alarm rates and fast execution times. These findings highlight the potential of hybrid machine learning models in enhancing DDoS detection for real-time, resource-constrained environments such as IoT networks.

Keywords: CICIDS 2017 dataset · Feature selection · DDoS threat · Anomaly detection system

1 Introduction

Intrusion detection systems (IDS) are a critical defense against continuously evolving cyberattacks. However, anomaly-based intrusion detection algorithms face challenges due to the need for reliable datasets for training, validation, and testing [1, 2]. The two most widely used IDS types are network-based IDS (NIDS) and host-basedIDS (HIDS). Figure 1 illustrates the general structure of a NIDS.

Many machine learning algorithms are utilized in the widely used anomaly-based intrusion detection system (IDS). The ability of an intrusion detection system to accurately predict attacks is typically used to assess it.

D. Berkouk et al. (Eds.): CCTS 2024, PTT, pp. 206–214, 2026.
https://10.1007/978-981-95-7607-4_23

There are a number of possible results in the case of binary classifier IDS. This research goal is to provide an overview of CICIDS2017, an IDS dataset that includes 2, 25,745 samples from DDoS attacks. In order to identify the optimal number of flow packet feature sets for attack detection, we first analyzed the normalized data set and then applied several widely used machine learning techniques to assess it [3, 4]. Our contributions include:

- A systematic hybridization framework for IoT security.
- Validation of the models' practical applicability at the network layer [17, 18].

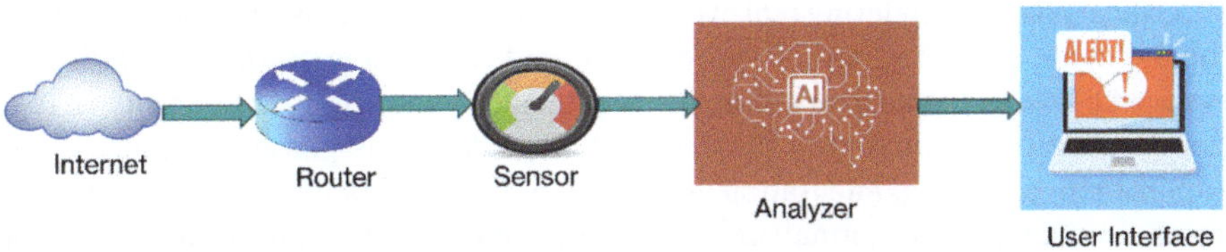

Fig. 1 Block diagram of an AI enabled network-based IDS

2 Distributed Denials of Service (DDOS)

By delivering many packets to a targeted server, service, or network, a distributed denial-of-service (DDoS) attack aims to maliciously disrupt regular traffic on that server, service, or network. Traditional rule-based or signature-based systems may be missing. The effectiveness of a DDoS attack is achieved by exploiting numerous compromised computer systems at the same time as sources of attack traffic. This completely consumes the victim's bandwidth and prevents the victim from receiving critical data from the network [4, 5].

3 Dataset

The CICIDS 2017 dataset, developed by the Canadian Institute for Cybersecurity, is a widely used benchmark for evaluating intrusion detection systems. The dataset offers a rich set of labeled network traffic that includes both benign and various attack types, such as DDoS, generated in a realistic environment. The latest features, including new attack types, are available in the dataset from the Canadian Institute of Cybersecurity's Intrusion Detection System (CICIDS 2017). This section describes the dataset containing DDoS attacks, which is used to train our algorithms [3, 6]. The dataset is processed using the openly available CICFlowMeter software from the Canadian Institute of Cybersecurity [7], which extracts 79 features, 10 of which are directly related to network traffic. These features are comprehensively labeled and calculated for all benign and malicious flows.

The CICFlowMeter generates bidirectional flows, with forward and backward directions determined by the initial packet. Statistical characteristics (such as packet length, packet count, byte count, and flow duration) are computed separately for both directions. Each flow is labeled with six initial columns: Flow ID, Source IP, Destination IP, Source Port, Destination Port, and Protocol. Notably, UDP flows terminate due to a flow timeout, whereas TCP flows typically end after a proper connection teardown. The flow timeout value is defined by the specific protocol or configuration used [7, 8].

4 Reprocessing Dataset

To enhance the precision and efficacy of subsequent machine learning algorithms, preprocessing techniques are essential to enhance data quality. These methodologies constitute a critical phase in rendering real-world data comprehensible, as such data may exhibit inconsistencies, including errors and outliers, and incomplete entries. Pre-processing plays a vital role in the analysis of network data, attributable to the variability in traffic patterns concerning format and dimensional attributes. This phase encompasses data normalization, cleansing, integration, reduction, and discretization. Numerous methods, such as decimal scaling normalisation, z-score, and min-max, are used for data normalization. The Min-Max normalization technique, shown in (1) is one of the methods used in the normalization procedure to apply numerical characteristics. It is crucial to adjust every attribute value within a specific range of [0, 1] to improve the efficacy and performance of the system. However, it does exhibit an anomalous affectability.

$$y_i = \frac{x_i - x_{min}}{x_{max} - x_{min}}, \tag{1}$$

where x_i is the data at the ith epoch, x_{min} is the minimum value of dataset, and x_{max} is the maximum value of dataset. A few variables in the CICIDS 2017 dataset are missing, which throws off the normalization method. Prior to the normalization phase, missing values were processed using different techniques like Cross-Validation, SMOTE, Heat map co-relation to enhance the accuracy of system [9].

5 Feature Selection

To improve classification performance and save memory use, feature selection is the process of identifying a subgroup of important features from the initial collection of datasets and discarding duplicate, unnecessary features from the dataset. Feature selection improves learning accuracy, lessens the impact of the "curse of dimensionality," reduces computation requirements, and helps identify which features might be pertinent to a given problem. Many techniques for supervised feature selection fall broadly within the categories of embedding models, wrapper, and filter. Information gain is one of the most extensively used filter model techniques in feature selection [10, 11]. It calculates each attribute's information gain by assessing its value in relation to the class using entropy. The characteristics with greater information richness have higher entropy." From the 79 characteristics in the CICIDS2017 dataset the 10 features that are extracted with good value are-, Idle.Min, Idle.Max, Active.Mean, Active.Max, Active.Min, Fwd.IAT.Max, Flow.IAT.Std, Fwd.IAT.Mean, Fwd.IAT.Total, Flow.IAT.Max [12].

6 Machine Learning Algorithms

In this paper, Linear Regression, Random Forest, and Decision Tree algorithms were utilized [13, 14]. Building on their results, we developed two novel hybrid machine learning models to enhance DDoS attack detection in IoT networks:

- Hybrid 1 (Linear Regression + Decision Tree): Combines interpretability with the ability to model non-linear relationships.
- Hybrid 2 (Random Forest + Decision Tree): Optimizes computational efficiency while maintaining high detection accuracy.

We conducted a comparative analysis of these models to identify the most effective approach [15, 16]. Experimental evaluation using the CICIDS 2017 dataset demonstrated that both hybrid models outperformed standalone classifiers. Notably, Hybrid 2 achieved a 10% improvement in F1-score and reduced false positives by 15% compared to traditional Decision Trees.

7 Experimental Analyses

To distinguish between BENIGN and DDoS attacks, an evaluation was conducted on 225,745 samples from the CICIDS 2017 dataset, which were split into 80% for training and 20% for testing. R Studio facilitated the experiments with its numerous libraries for machine learning techniques. A confusion matrix is utilized to summarize the performance of a classification algorithm by showcasing both correct and incorrect classification results. This matrix aids in understanding the analysis and the effectiveness of classification models. The categorization events are outlined as follows [18]: True Negatives (TN) and True Positives (TP) represent accurate classifications. A false positive (FP) occurs when the outcome is wrongly predicted as positive when it is actually negative. Conversely, a false negative (FN) happens when the outcome is mistakenly predicted as negative when it is actually positive. To enhance results, machine learning algorithms have been used to assess the performance and accuracy of the selected features. Table 1 presents the confusion matrix outcomes for the LR, DT, RF, hybrid1, and hybrid2 algorithms.

Table 1 The confusion matrix for ML classifiers

Random Forest		Predicted Class	
Actual Class		BENIGN	DDoS
	BENIGN	19,499	190
	DDoS	160	25,300
Hybrid 1 (Linear Regression + Decision Tree)		Predicted Class	
Actual Class		BENIGN	DDoS
	BENIGN	19,350	338

(*continued*)

Table 1 (*continued*)

Random Forest		Predicted Class	
	DDoS	220	25,241
Hybrid 2 (Random Forest + Decision Tree)		Predicted Class	
Actual Class		BENIN	DDoS
	BENIGN	19,428	260
	DDoS	114	25,347

8 Performance Evaluation

The performance of the IDS has been determined using the following standard metrics Correctness (Pr) or Accuracy refers to the ratio of accurately identified attack flows (TP) compared to the total number of classified flows (TP + FP), which is also known as the predictive value. Suggestion (S) or Memory (Rc) is defined as the ratio of correctly identified attack flows (TP) to all flows produced (TP + FN). The detection rate indicates the likelihood that actual events will also be predicted to happen.

Table 2 The performance examination results

ML Classifiers	Accuracy	Precision	Recall	F1-Score	Time (in sec)	Detection-Rate	False-Alarm-Rate
Random Forest	0.9528	0.9621	0.9514	0.9558	0.0306	0.5365	0.0453
Hybrid 1 (DT + LR)	0.9491	0.9637	0.9419	0.9512	0.0237	0.5312	0.0416
Hybrid 2 (DT + RF)	0.9541	0.9611	0.9548	0.9571	0.0353	0.5384	0.0467

The performance examination results for the commonly used machine learning algorithms—Random Forest (RF), Hybrid1 and Hybrid 2 are displayed in Table 2 presents the evaluation measures as a weighted average. The Hybrid2 (DT + RF) classifiers outperformed the others with an average accuracy of 95.41%., with good recall value-95.48%, F1-Score- 95.71% respectively. The detection rate of hybrid2 is 0.5384 respectively and false alarm rate is 0.0467. All parameters were evaluated on 10 features of the original dataset. Finally, the average values of all parameters were considered for comparison. The graphical output is represented using Figs. 2 and 3.

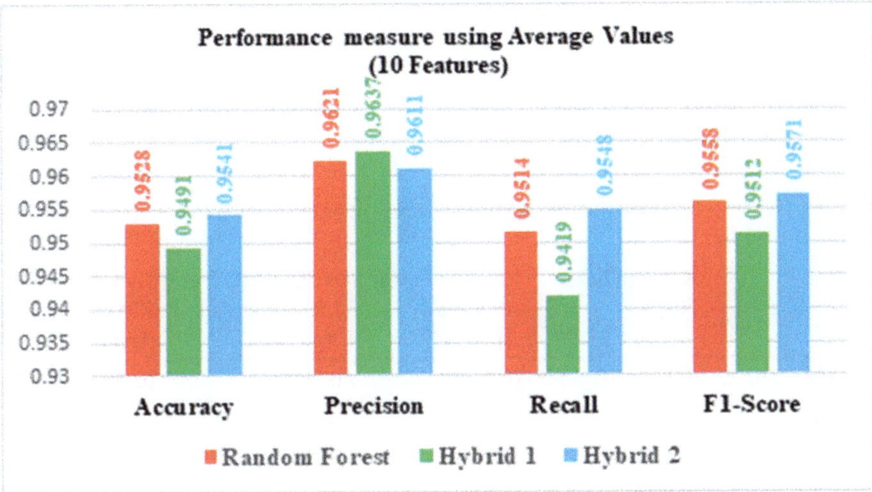

Fig. 2 Graphical representation of average values of ML classifiers

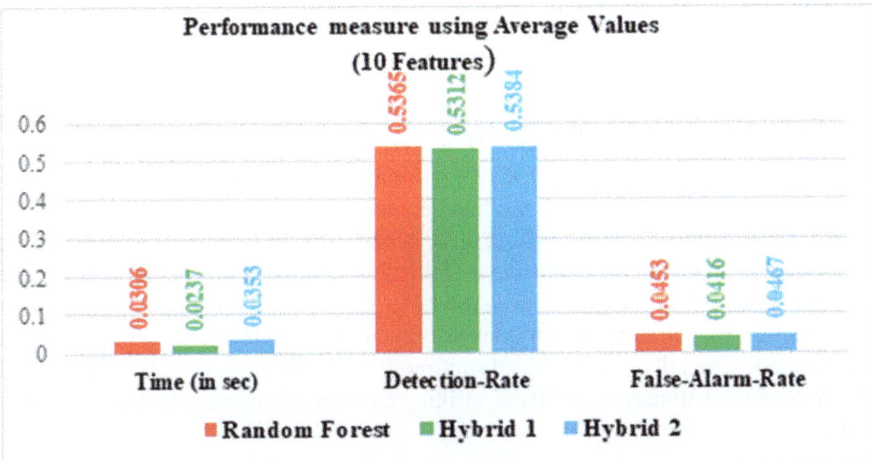

Fig. 3 Graphical representation of average values of ML classifiers

Table 3 shows comparisons with previous research [19]. The comparative analysis between the performance metrics in Table 2 and Table 3 reveals a trade-off between accuracy and computational efficiency. Models in Table 3, particularly Random Forest and Decision Tree, demonstrate superior performance in terms of accuracy (up to 99.98%), precision, recall, and F1 score, making them highly reliable for detecting anomalies. However, they are computationally more intensive, with execution times ranging from 35 s to over 4 min. On the other hand, the hybrid models from Table 2 especially Hybrid 2 (DT + RF), offer a more balanced approach by achieving competitive performance (accuracy of 95.41% and F1 score of 0.9571) while maintaining exceptionally low execution times (as low as 0.0237 s). This makes them more suitable for real-time IoT network environments where speed and efficiency are critical. Additionally, the inclusion of detection and false alarm rates in the models provides further insight into their practical applicability in security-sensitive scenarios.

Table 3 Comparison overview of ML classifiers

Metric	Random Forest [19]	Random Forest (Proposed Approach)	Hybrid 1 (DT + LR) (Proposed Approach)	Hybrid 2 (DT + RF) (Proposed Approach)
Accuracy	0.9998	0.9528	0.9491	0.9541
Precision	0.9835	0.9621	0.9637	0.9611
Recall	0.9851	0.9514	0.9419	0.9548
F1 Score	0.9842	0.9558	0.9512	0.9571
Time	4.48 min (~268.8 s)	0.0306 s	0.0237 s	0.0353 s
Detection Rate	—	0.5365	0.5312	0.5384
False Alarm Rate	—	0.0453	0.0416	0.0467

The comparative analysis reveals distinct performance trade-offs: Models in Table 3 achieve superior accuracy (over 99%) compared to Table 2 (around 95%), with higher precision, recall, and F1 scores overall. However, Hybrid 2 (DT + RF) in Table 2 stands out for its exceptional F1 score (0.9571), demonstrating the strongest balance among proposed methods. While Table 3 models demand longer execution times (seconds to minutes), Table 2 models operate within milliseconds, prioritizing speed. Notably, detection and false alarm rates are exclusive to Table 2, where Hybrid 2 achieves the highest detection rate despite a marginally elevated false alarm rate. Though Hybrid 2's ensemble architecture increases computational resource demands, it enhances robustness and generalization. Crucially, its optimized runtime (~0.035 s) remains viable for near real-time intrusion detection systems (IDS), striking a pragmatic balance between speed and efficacy.

9 Discussion and Conclusion

This study presented a performance evaluation of machine learning-based intrusion detection models for identifying Distributed Denial of Service (DDoS) attacks using the CICIDS2017 dataset. Recognizing the limitations of individual classifiers in real-time and resource-constrained environments, we proposed two hybrid models—Hybrid 1 (Linear Regression + Decision Tree) and Hybrid 2 (Random Forest + Decision Tree)—to enhance detection accuracy and computational efficiency.

Among all models evaluated, Hybrid 2 consistently demonstrated superior performance, achieving high accuracy (95.41%), strong F1-score (0.9571), and a favorable balance between detection rate and false alarm rate. These results affirm the advantage of combining ensemble learning with decision-tree-based approaches in enhancing robustness against complex, high-volume attacks such as DDoS.

Furthermore, this work highlights the importance of effective feature selection. By leveraging information gain for selecting the top contributing features from the dataset, we significantly improved model performance and reduced training overhead.

Future work will focus on extending this approach by incorporating deep learning architectures (e.g., LSTM, CNN) and advanced ensemble methods such as stacking and boosting. Additionally, applying hyperparameter optimization and regularization techniques could further improve generalization across dynamic and heterogeneous IoT environments. Evaluating these models in real-world deployment scenarios, including software-defined networks (SDN), will also be a critical step toward developing scalable, intelligent, and adaptive intrusion detection systems.

References

1. Prem Sankar, G., Di Francesco, M., Taleb, T.: Edge computing for the internet of things: A case study. IEEE Internet Things J. **5**(2), 1275–1284 (2018)
2. Madni, H., Javed, M., Arshad, M.J.: An overview of intrusion detection system (IDS) along with its commonly used techniques and classifications. Int. J. Comput. Sci. Telecommun. **5**(2) (2014)
3. Lashkari, A.H., Draper-Gil, G., Mamun, M.S.I., Ghorbani, A.A.: Characterization of Tor traffic using time-based features. In: ICISSP 2017, pp. 253–262. SciTePress, Porto (2017)
4. Duda, R.O., Hart, P.E., Stork, D.G.: Pattern Classification, 2nd edn. Wiley, New York (2000)
5. Patel, J., Panchal, K.: Effective intrusion detection system using data mining technique. J. Emerg. Technol. Innov. Res. **2**(6), 1869–1878 (2015)
6. Tayyab, M., Belaton, B., Anbar, M.: ICMPv6-based DoS and DDoS attacks detection using machine learning techniques, open challenges, and blockchain applicability: A review. IEEE Access. **8**, 170529–170547 (2020)
7. Ahmed, S.: A study of ML algorithms for DDoS detection. Int. J. Res. Appl. Sci. Eng. Technol. (2021)
8. Hacıbeyoğlu, M., Ibrahim, M.H.: Comparison of the effect of unsupervised and supervised discretization methods on classification process. Int. J. Intell. Syst. Appl. Eng. **4**(Special Issue-1), 105–108 (2016)
9. Said, E.M., Le-Khac, N.A., Albahar, M.A., Jurcut, A.: A novel hybrid model for intrusion detection systems in SDNs based on CNN and a new regularization technique. J. Netw. Comput. Appl. **191**, 103160 (2021)
10. Elejla, O.E., et al.: Labeled flow-based dataset of ICMPv6-based DDoS attacks. Neural Comput. & Applic. **31**(12), 3629–3646 (2019)
11. Tang, T.A., Mhamdi, L., McLernon, D., Zaidi, S.A.R., Ghogho, M., El Moussa, F.: DeepIDS: Deep learning approach for intrusion detection in software defined networking. Electronics. **9**(9), 1533 (2020)
12. Singh, J., Behal, S.: Detection and mitigation of DDoS attacks in SDN: a comprehensive review, research challenges and future directions. Comput Sci Rev. **37**, 100279 (2020)
13. Khashab, F., Moubarak, J., Feghali, A., Bassil, C.: DDoS attack detection and mitigation in SDN using machine learning. In: 2021 IEEE 7th International Conference on Network Softwarization (NetSoft), pp. 395–401. IEEE, Tokyo (2021)

14. Santos, R., Souza, D., Santo, W., Ribeiro, A., Moreno, E.: Machine learning algorithms to detect DDoS attacks in SDN. Concurrency Computat. Pract. Exp. **32**, e5402 (2020)
15. Bahashwan, A.A., Anbar, M., Hanshi, S.M.: Overview of IPv6-based DDoS and DoS attacks detection mechanisms. In: Anbar, M., Abdullah, N., Manickam, S. (eds.) *Advances in Cyber Security*, CCIS, vol. 1132, pp. 123–134. Springer, Singapore (2020)
16. Fernández, A., Garcia, S., Herrera, F., Chawla, N.V.: SMOTE for learning from imbalanced data: progress and challenges, marking the 15-year anniversary. J. Artif. Intell. Res. **61**, 863–905 (2018)
17. Zubaydi, H.D., Anbar, M., Wey, C.Y.: Review on detection techniques against DDoS attacks on a software-defined networking controller. In: Proceedings of the 2017 Palestinian International Conference on Information and Communication Technology (PICICT), pp. 10–16. IEEE, Gaza (2017)
18. Mansoor, A., Anbar, M., Bahashwan, A.A., Alabsi, B.A., Rihan, S.D.A.: Deep learning-based approach for detecting DDoS attack on software-defined networking controller. Systems. **11**, 296 (2023)
19. Gharkan, D.K., Abdulrahman, A.A.: Construct an efficient distributed denial of service attack detection system based on data mining techniques. Indones. J. Electr. Eng. Comput. Sci. **29**(1), 591–597 (2023)

Innovation-Driven Education Reforming Engineering Education in the AI Era

Jouni Isoaho[1], Hannu Tenhunen[2,3], Imed Ben Dhaou[3](✉), Xing Guo[4], and Zhuo Zou[4]

[1] University of Turku, Turku, Finland
jouni.isoaho@utu.fi
[2] KTH-Royal Institute of Technology, Stockholm, Sweden
hannu@kth.se
[3] Dar Al-Hekma University, Jeddah, Saudi Arabia
ibendhaou@dah.edu.sa
[4] Fudan University, Shanghai, China
{xingg,zhuo}@fudan.edu.cn

Abstract. Currently, the global business and innovation environment is rapidly evolving due to the rise of the platform economy, digital transformation, and the shift towards remote and hybrid work. This shift emphasizes personal innovation and entrepreneurship, leading to an increase in self-employment and consulting roles over traditional employment. Simultaneously, key technological drivers such as artificial intelligence and advancements in system integration technologies, like semiconductors and advanced packaging, are significantly impacting on the expertise required for business. In this paper, we first identify key factors and analyze their impact on the future business and innovation environment in the era of artificial intelligence on education. We then introduce the concept of innovation-driven education and its impact on the ICT education framework. Our baseline is the European Institute of Innovation & Technology's (EIT) educational framework and quality measures. An new key themes include innovation entrepreneurship, multifaceted innovation processes, multicultural business ecosystems, and the platform economy. The future engineering profession will be characterized by 1) innovation entrepreneurship, 2) communication, collaboration, and workflow management with AI, and 3) a holistic and diverse understanding of entireties.

Keywords: Innovation-driven education · Digital transformation · Education 4.0 · Engineering profession · Education transformation · Innovation entrepreneurship · Artificial intelligence

1 Introduction

The global business landscape and innovation environment are undergoing a fundamental transformation, driven by the rise of the platform economy, e-commerce, rapid digital transformation, and the shift toward remote and hybrid work models [1]. The COVID-19 pandemic has accelerated these trends, leading to the adoption of advanced ICT infrastructure to support remote work and collaboration.

D. Berkouk et al. (Eds.): CCTS 2024, PTT, pp. 215–222, 2026.
https://10.1007/978-981-95-7607-4_24

These shifts are changing traditional business models, enhancing digital ecosystem, and introducing new opportunities and challenges for organizations worldwide. The growing emphasis on individual innovation and innovation entrepreneurship has led to an expansion of self-employment and consulting work. This signals a departure from traditional employment structures. At the same time, technological advancments in artificial intelligence (AI) and system integration technologies are redefining engineering skills sets in the ICT sector. To address these new demands, leading universities are redefining curriculum aligned with Education 4.0 [2, 3]. This paradigm focuses on technical skills and with transferable critical soft skills such as creativity, design thinking, communication, and emotional intelligence. These skills are essential for adapting to a technology-driven, interconnected world.

As highlighted by the IEEE Future Directions group, three key technological megatrends are shaping the future of business and innovation [4]. Firstly, the growing focus on sustainability drives businesses to adopt renewable energy, circular economy principles, and eco-friendly products. Secondly, digital transformation continues to reshape societal operations and job markets, creating demand for new engineering professions and fostering remote work models. Thirdly, advancements in Artificial General Intelligence (AGI) promise breakthroughs in productivity and innovation, enhancing multiple engineering fields and human-machine communication.

This study examines the transformation of ICT engineering education toward innovation-driven models in the era of AI. It provides insights into rethinking educational frameworks to meet emerging challenges. The primary contributions of this work are:

- Analyzing the key factors reshaping the future engineering education
- Introducing the concept of innovation-driven education
- Identifying key aspects reformulating next generation engineering profession

2 Global Business and Innovation Environment

The worldwide business scene and innovation setting are experiencing immense transformation. This shift is mainly fueled by the emergence of the platform economy, digital transition, and the transition towards remote and hybrid work arrangements [5]. As per the World Economic Forum's Future of Jobs Report 2025, the top 10 skills needed by 2030 include: (1) AI and big data; (2) analytical thinking; (3) creative thinking; (4) resilience, flexibility, and agility; (5) technological literacy; (6) leadership and social influence; (7) curiosity and lifelong learning; (8) systems thinking; (9) talent management; and (10) motivation and self-awareness [18]. These changing demands not only transform job roles and skill sets but also significantly impact education and training.

For instance, the platform economy depends on essential technologies like cloud computing, blockchain, cybersecurity, data analytics, IoT, AI, mobile technology, and GPS, which necessitates a heightened focus on digital competencies. These relations are shown in Fig. 1. As these platforms redefine industry borders, educational institutions must modify their curricula to prepare students with the skills needed for success in a digitally powered innovative environment. In addition to technical expertise, a comprehensive understanding of network economics and platform-based business models is essential for navigating this new environment.

Interdisciplinary learning becomes crucial as platforms integrate technology, business, and social networks. Education systems need to encourage cross-disciplinary collaboration and creative problem solving. The students can link theoretical knowledge with practical applications by emphasizing project-based learning, and real-world case studies. This will prepare them to tackle complex challenges in platform-centric industries. Aligning education with industry demands requires frequent updates to curricular content and teaching methodologies [6]. The quadruple-helix model fosters collaboration among academia, industry, government, and society. This will play a pivotal role in ensuring that graduates have not only knowledge, but also practical skills needed.

Digital transformation [7] shift allows companies to allocate resources more effectively and focus efforts towards innovation and creativity. Advanced data analytics will provide businesses with deeper insight and enable more informed and strategic decision-making processes. This transformation will enhance collaboration and communications for efficient and cohesive organizational workflows. The cultural shift provided by digital transformation encourages innovation and agility. The curricula need to include digital and data analysis skills, and prepares students to excel in dynamic innovation and co-creation environments.

The shift towards remote and hybrid work models [8], primarily stems from advancements in communication technologies, virtual presence technologies, and evolving engineers' preferences. Increased flexibility reduced commuting time, and access to a wider talent pool are the key benefits of remote work. The challenges are related to collaboration, communication, and company working culture. The shift towards remote and hybrid work catalyzes the need to adjust teaching methodologies for preparing students to working life. This creates a need to develop skills on virtual collaboration tools and co-creation.

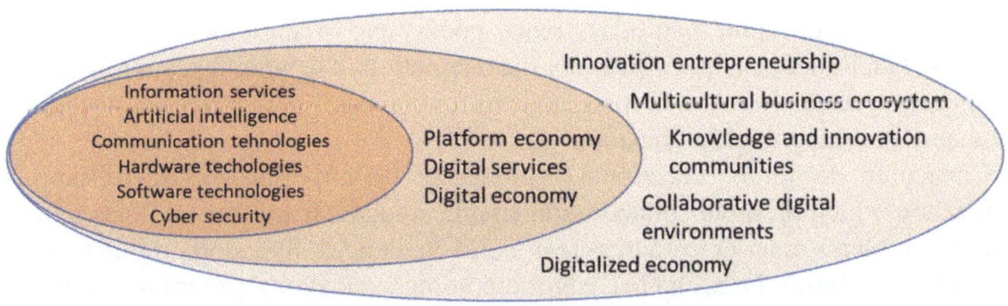

Fig. 1 Platforms connect technologies to business and innovation.

3 Impact of Technological Drivers

Key technological drivers, such as AI and system integration technologies are fundamentally changing the expertise required in business. Artificial intelligence [9] will dramatically reshape how we work, innovate, and interact socially by automating complex tasks and enhancing human capabilities [10]. This allows engineers to concentrate on strategic thinking and creative and innovative tasks. Universities must incorporate AI skills into their curricula to prepare students for an evolving job market, which increasingly relies on technology to achieve unprecedented productivity levels [11]. Equipping students with hands-on experience empowers them to harness these tools for personal and professional growth. AI's ability to analyze vast datasets and recognize patterns can lead to significant breakthroughs in complex problems across various industrial sectors. By exposing students to real-world scenarios and data-intensive analyses, educational systems can cultivate a new generation of problem-solvers. They are capable of leveraging AI to create meaningful and innovative solutions.

Advances in system integration technologies [12] have significantly enhanced computational power and empowered the development of energy-efficient processing capacity, which is essential for AI, big data, and other data-intensive applications. The progress in AI and system integration technologies goes hand in hand, like two sides of a coin. Cost reductions driven by technological advancements and mass production have made semiconductors and advanced technologies more accessible, broadening their availability across the whole of society. This democratization facilitates widespread digital transformation within education and operations of educational institutions. This provides the necessary computational backbone for innovative applications and enriched learning experiences. Educational goals should focus on digital literacy, technology integration, and ethical technology usage, aligned with system integration technologies and AI advancements. This prepares students to thoughtfully engage with emerging technologies as responsible, innovative contributors to a technology-driven future.

4 Re-Envision Educational Models

The rapid evolution of the business and innovation environment has forced the leading universities to re-envision their educational models and to align them with Education 4.0 goals [13]. Education 4.0 emphasizes the integration of technology, interdisciplinary learning, and practical skills to prepare students for the demands of the modern workforce in the new emerging world of business.

Curriculum design in Education 4.0 emphasizes the importance of interdisciplinary approaches, which promote collaboration among students from various fields to tackle complex problems in order to address real-world challenges. By incorporating technologies such as AI, blockchain, and IoT/edge computing, educational programs will provide students with hands-on experience through labs, coding workshops, and hackathons. This pragmatical engagement allows students to apply theoretical knowledge and develop skills that are essential for today's technology-driven industries. Industry partnerships [14] play a crucial role in bridging the gap between academic knowledge and practical application. Co-locating industry and university education and R&D activities will enforce the mutual benefits, as described above. Co-locating education programs are particularly effective to allow students to work within an industry while completing their studies, thus acquiring firsthand insights into real-world business.

5 Innovation-Driven Education

Innovation-driven education is a new paradigm, which emphasizes creativity, problem-solving, and innovation entrepreneurship. It is built on the European Institute of Innovation & Technology (EIT) educational framework and quality measures [15]. The EIT framework has been successful in integrating academic learning with innovation and entrepreneurship. This is achieved by embedding technology and entrepreneurship in educational settings, promoting a culture of exploration and creativity in co-location center environments.

The rise of the platform economy and digital transformation has increasingly shifted focus towards personal innovation and entrepreneurship. This provides individuals opportunities to start businesses, offer consulting services, and work as freelancers.

Three fundamental changes are currently shaping the future of education, particularly in addressing technological content challenges:

Firstly, the shifting student demographics are notable. The student population is transitioning from the digital/internet-savvy Generation Z to the AI-savvy Generation Alpha. Both generations have grown up with advanced technology, resulting in distinct expectations and learning preferences. To effectively engage and inspire these tech-savvy students, universities must adapt their teaching methods to integrate more interactive, technology-driven learning strategies, embracing tools and platforms. As student populations become increasingly varied, based on a mix of cultural, socioeconomic, and educational backgrounds, educational institutions need to prioritize inclusive and equitable education practices. These can significantly differ from the ones of the past.

Secondly, the evolving nature of work is progressing from the industrial age to a knowledge and intelligence age. This transition, reminiscent of the shift from steam to electrical power, is now transferring us to an intelligence age. Here AI and robotics are beginning to supplant both blue- and white-collar jobs, and providing more time to creativity and to innovations of the whole work force. As these technologies become integral to various sectors, they are not only reshaping business practices but also fundamentally transforming societal structures, future of jobs, and employment landscapes.

Thirdly, the global landscape is undergoing a fast shift towards the geopolitically multi-polar world. Global South will play a crucial role in re-shaping innovations, technologies, and business practices. This will result to significantly growing economic impact to world's economy and trade. Understanding the dynamics of this new multipolar environment is essential for all stakeholders to navigate successfully in a rapidly changing global world. Education plays a pivotal role in these global shifts by preparing students to engage with, and to contribute to this multipolar world. By creating and fostering an education environment that promotes international collaboration and free flow of ideas, education can provide students the new needed skills.

6 The Innovation-Driven Education Framework

The restructuring of the educational framework focuses on transitioning traditional approaches like the Conceiving-Designing-Implementing-Operating (CDIO) model to more dynamic, challenge-driven methodologies. The CDIO framework, known for its emphasis on hands-on, project-based learning, is being evolved to address real- world problems specifically [16]. Its strategic element is the capstone projects based on complex teamwork in order to solve a specific engineering complex problem. Challenge-driven education (CDE) [17] further builds on this by immersing students in projects that target pressing global challenges such as sustainability, healthcare, and urbanization. These projects foster interdisciplinary collaboration, requiring students to work with industry partners and community organizations, and with students from other disciplines.

Innovation-driven education (IDE) advances CDE by embedding innovation and entrepreneurship directly into the curriculum. This approach teaches students to not only identify and solve existing problems but also to conceive innovative solutions and effectively bring them to market. By integrating entrepreneurial elements, IDE prepares students to become proactive contributors to the economy, ready to navigate the challenges of contemporary business landscapes and platform-based economies. Innovation entrepreneurship is a core component in this transformation, because future engineers are expected to not only have technical expertise, but also to have fore- sight to identify and pursue new business opportunities. Furthermore, engineers are required to have a holistic and diverse understanding of entire systems, considering the social, economic, and environmental impacts of their work. A systems thinking approach is essential, with an awareness of ethical and sustainability considerations. The innovation-driven ICT engineering Education 4.0 framework is illustrated in Fig. 2.

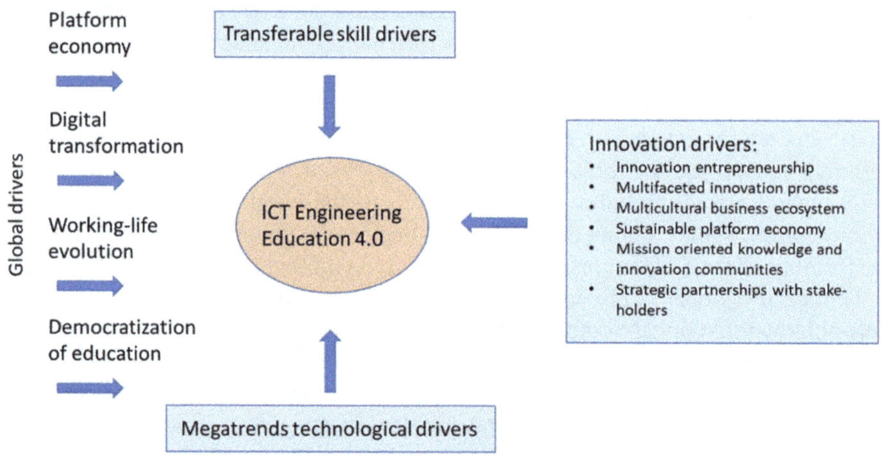

Fig. 2 Innovation-driven ICT Engineering Education 4.0

7 Future Engineering Profession

The future engineering profession will be characterized by 1) innovation entrepreneurship, 2) communication, collaboration, and workflow management with AI, and 3) a holistic and diverse understanding of the whole. Innovation entrepreneurship requires engineers to recognize and capitalize on new business opportunities. The purpose is to increase their chances of launching successful startups, bringing innovative products to market, and selling their own innovative ability as freelancers. AI enables engineers to concentrate on strategic and visionary thinking. The key focus is in holistic and multi-faced understanding of the whole, instead of concentrating detailed engineering. This new skill is becoming one of the core competences for engineering professionals.

Ethical considerations and sustainability are increasingly vital elements in modern engineering profession. According to industrial leaders, important individual transferable skills for our Education 4.0 approach include:

- Collaboration Skills: Building partnerships within and outside the team, creating trust.
- Integrity and Trust: Forming open and honest relationships across the team and stakeholders.
- Innovation Skills: Being open-minded and creative to seize innovation opportunities.
- Personal Growth: Pursuing continuous learning and development.
- Passion for Excellence: Striving for the best and continuously improving.

8 Conclusion and Future Vision

In this paper we analyzed key factors reshaping the future engineering education. Futher we introduced the concept of innovation-driven education and the key aspects reformulating next generation strategic engineering profession. Our vision is to equip students with the skills required to thrive in a rapidly evolving world through an innovation-driven education. Our commitment extends to preparing students to navigate and address challenges caused by emerging global megatrends. By integrating education, research, and innovation, we aim to empower students to drive societal progress and environmental impact positively. By collaborating effectively, we believe in creating a dynamic educational environment that cultivates future leaders and innovators.

References

1. Kamble, M.A.S., Gunasekaran, S., Sharma, S.: A survey of the digital plat- form economy: Opportunities and challenges for small and medium enterprises. Int. J. Prod. Econ. **211**, 68–85 (2019)
2. World Economic Forum: Catalysing Education 4.0: Investing in the Future of Learning for a Human-Centric Recovery. Geneva: World Economic Forum. https://www3.wefrum.org/docs/WEF_Catalysing_Education_4.0_2022.pdf (2022). Accessed 20 Apr 2026
3. Supriya, Y., et al.: Industry 5.0 in smart education: Concepts, applications, challenges, opportunities, and future direc- tions. IEEE Access. **12**, 81938–81967 (2024) https://doi.org/10.1109/ACCESS. 2024.3401473
4. IEEE: 2024 Technology Megatrends Committee. https://engage.ieee.org/FD-Megatrends-2024.html (2024)
5. Gawer, A.: Digital platforms and ecosystems: remarks on the dominant organizational forms of the digital age. Innovation. **24**(1), 110–124 (2022)
6. Dhaou, I.B.: Aligning engineering education with industrial needs through specialized courses. In: 2022 IEEE Global Engineering Education Conference (EDUCON), pp. 1664–1669 (2022)

7. Mihu, C., Pitic, A.G., Bayraktar, D.: Drivers of digital transformation and their impact on organizational management. Stud. Bus. Econ. **18**(1), 149–170 (2023)
8. Milton, S.A.: Rethinking hybrid and remote work in higher education. Am. J. STEM Educ. **1** (2024). https://doi.org/10.32674/8wzp3g35
9. World Economic Forum: Harnessing the AI Revolution in Industrial Operations:A Guidebook. https://www.weforum.org/publications/harnessing-the-ai-revolution-in-industrial-operations-a-guidebook/ (2023) Accessed 31 Dec 2024
10. World Economic Forum: AI for Impact: The Role of Artificial Intelligence in Social Innovation, https://www3.weforum.org/ (2024) Accessed 31 2024
11. Southworth, J. et al.: Developing a model for ai across the curriculum: Transforming the higher education landscape via innovation in ai literacy. Comput. Educ.: Artif. Intell. **4**, 100127 (2023). https://doi.org/10.1016/j.caeai.2023.100127
12. IEEE. International Roadmap for Devices and Systems (IRDS™) 2023 Update. https://irds.ieee.org/editions/2023 (2023). Accessed 20 Apr 2026
13. World Economic Forum. Shaping the future of learning: The role of AI in Education 4.0. https://www.weforum.org/publications/shaping-the-future-of-learning-the-role-of-ai-in-education-4-0 (2024)
14. Forum, W.E.: Digital Transition Framework: An action plan for public-private collaboration. https://www3.weforum.org
15. European Institute of Innovation & Technology: Quality for learning: EIT Label Handbook for Quality system for non-degree education and training and EIT Competence Certification Model, 1st edn. https://eit.europa.eu/ (2022). Accessed 27 Dec 2024
16. Malmqvist, J., Edström, K., Rosen, A.: CDIO standards 3.0–updates to the core CDIO standards. In: 16th International CDIO Conference, Online, pp. 60–76. Chalmers University of Technology (2020)
17. Magnell, M., Högfeldt, A.-K.: Guide to Challenge-Driven Education. https://www.kth.se/social/group/guide-to-challenge-d/. Accessed 27 Dec 2024
18. World Economic Forum: The Future of Jobs Report 2025. World Economic Forum. https://www.weforum.org/publications/the-future-of-jobs-report-2025/ (2025). Accessed 20 Apr 2026

The Impact of Intensive Training on General Education Teachers for Including Students with Autism Spectrum Disorder (ASD) in Regular Classrooms in Saudi Arabia

Mona Al-Haddad(✉)

Dar Al-Hekma University, Jeddah, Saudi Arabia
malhaddad@dah.edu.sa

Abstract. This study aimed to develop an autism training program for general education teachers in inclusive classrooms with students diagnosed with autism spectrum disorder (ASD) and to evaluate its impact on teaching skills and student performance. Despite ongoing training efforts, many teachers report feeling unprepared to effectively teach students with special education needs. A mixed-methods approach was used for data collection, involving eight general education teachers, each with at least one student with ASD in their classroom. The intervention included a training package featuring evidence-based practices (EBPs) such as Functional Behavioral Assessment (FBA), reinforcement strategies, differential reinforcement (DR), modeling and prompts, peer-mediated strategies, behavior contracts, self-management, visual supports (e.g., activity schedules, task analysis, signs, scripts), and post-implementation performance feedback. An introduction to ASD was also provided. Data collected from classroom observations and teacher interviews showed that the training had a significant positive impact on both teachers' instructional skills and the performance of students with ASD. The results highlighted a strong need for autism-specific training among general education teachers to effectively support inclusive education. Based on the findings, it is recommended that educational institutions implement autism training programs to enhance teacher preparedness and support the successful sustainable inclusion of students with ASD in general education settings.

Keywords: Inclusive education · Special education · Teacher training · Autism

1 Introduction

The successful inclusion of students with autism spectrum disorder (ASD) in general education classrooms requires well-prepared teachers (Morrier et al., 2010). Effective teacher training should encompass course content, delivery methods, long-term reinforcement, and the use of evidence-based practices (EBPs) (Scheuermann et al., 2003). Despite various efforts, many teachers still feel unprepared to educate students with special education needs (Lee et al., 2015). In Saudi Arabia, the prevalence of ASD is rising,

D. Berkouk et al. (Eds.): CCTS 2024, PTT, pp. 223–230, 2026.
https://10.1007/978-981-95-7607-4_25

though its exact rate remains unclear (Khan et al., 2020). As this trend continues, it is critical for general education teachers to be equipped with research-backed strategies for inclusive learning. However, there is limited research in Saudi Arabia on the impact of teacher training, particularly that which includes in-situ observation and performance feedback for ASD inclusion. Thus, academic and professional development are key to improving teacher competency and ensuring meaningful inclusion of students with ASD in mainstream classrooms.

This study explored the effects of teacher training in evidence-based strategies on both teachers' skills and their acceptance of students with autism spectrum disorder (ASD) in inclusive classrooms. It also assessed the academic and social outcomes of students with ASD. Using a mixed-methods approach, the study focused on four main areas: (1) effective training strategies for supporting students with ASD in inclusive environments, (2) the benefits of inclusion for students with ASD, (3) essential components of successful teacher training, and (4) key elements of best practices in ASD inclusion.

2 Methodology

The study was conducted in three phases. Phase 1 identified key challenges, selected participants, and assessed the classroom environment. Phase 2 involved teacher training using the evidence-based Behavioral Skills Training (BST) method, which consists of six steps: skill description, demonstration, practice, and performance feedback until mastery is achieved. Phase 3 evaluated the effectiveness of the training on teachers and students with ASD. The primary research approach was a mixed-methods approach, incorporating questionnaires, interviews, and observations.

2.1 Phase 1: Identification of Challenges, Participants and the Current Context

A questionnaire was distributed to general education teachers in Jeddah, Saudi Arabia, to assess challenges in including students with autism spectrum disorder (ASD). It covered topics such as the prevalence of ASD, classroom inclusion levels, teachers' ASD training, their perceptions of inclusion, preparedness, and the obstacles they face.

The study included 150 elementary school teachers from both public and private schools in North Jeddah, Saudi Arabia, representing inclusive and non-inclusive environments. Questionnaires were distributed with the assistance of a local university's special education department, receiving approval from school principals and support from school supervisors for participant recruitment.

In Phase 1, the questionnaire consisted of 10 multiple-choice questions, 2 rating scale questions, and 2 open-ended questions. The multiple-choice questions gathered basic demographic information, such as teachers' experience, role, qualifications, and familiarity with students with ASD. Participants also identified potential interventions for supporting ASD students. The open-ended questions allowed teachers to describe the challenges they face and suggest preferred training strategies for interventions. Out of 150 questionnaires distributed, 121 were returned, and 82 were completed. Pre-interview results highlighted three key themes: 1) Prerequisite Skills for Inclusion, 2) Challenges in Teaching Students with ASD, and 3) Teaching Strategies and Training Needs.

2.2 Phase 2: Teacher Training Intervention

In this phase, a 14-week teacher training intervention was implemented for selected teachers who met specific inclusion criteria. The training program, which included two hours of training per week, aimed to equip teachers from two experimental schools with evidence-based strategies for supporting students with ASD in inclusive classrooms. Eligibility requires teachers to have an education degree, at least two years of teaching experience, and experience teaching formally diagnosed ASD students. Teachers with special education degrees, prior ASD training, less than one year of experience, or no diagnosed ASD students were excluded. The training duration aligned with research suggesting effective inclusive education training lasts between 3 days and 14 weeks (Kurniawati et al., 2014).

The study involved three schools that met specific criteria, including eight teachers and eight students with ASD. Two schools were assigned to the experimental group (six teachers and six students), while one served as the control group (two teachers and two students). All teachers were female, averaging 29.7 years of age with 4.87 years of experience. The student group included seven boys and one girl, with an average age of 8.5 years. Phase 2 began following the completion of participant recruitment.

In Phase 2, two experimental groups, each consisting of three female teachers and three students with ASD, participated in a 14-week training program on inclusive practices. Training sessions were conducted weekly in the afternoons using various instructional tools. Post-training evaluations were carried out in the final week, with data collected through interviews, observations, and feedback. The control group, which did not receive training, only participated in observations and interviews.

2.3 Phase 3: Evaluation (Post -Training)

The final phase of the study assessed the effectiveness of the training intervention for teachers working with ASD students in inclusive classrooms. Semi-structured interviews were conducted with all eight participants from both experimental and control groups to validate the training phase's findings. These interviews aimed to emphasize the importance of teacher training in inclusive education and compare the outcomes for the experimental group to the control group.

3 Results

3.1 Intervention

Pre-intervention classroom observations were conducted, in week 2, at both experimental and control schools using the Teacher Observational Skills Checklist (TOSC). This checklist focused on six key skills, with data recorded through researcher notes in open-ended fields. A key theme emerged on how teachers applied these skills when working with students with ASD. For each skill, the researcher marked "yes" if the teacher demonstrated it, or "no" if they did not.

Teachers Rarely Implemented Effective Instructional Skills with Their ASD Students. Before the intervention, observational data showed that none of the eight teachers, including those in the experimental group, effectively applied key instructional skills with students with ASD. Using the TOSC, six essential skills were assessed, revealing issues such as reinforcement of negative behaviors but not positive ones, lack of peer mediation attempts, ineffective engagement and motivation strategies, and inconsistent organizational skills. Both the experimental and control groups demonstrated similar skill levels prior to the training.

By mid-intervention, teachers showed significant improvement in applying strategies for ASD students (weeks 4–14). Initially struggling with techniques like differential reinforcement and peer mediation, they progressed with feedback and implemented these strategies by the fourth observation. By week 8, most teachers achieved 100% accuracy in addressing ASD students' needs, using visual supports, reinforcement, peer collaboration, and sensory adaptations. Teachers in experimental schools were notably more effective, while the control group showed minimal change.

3.2 Training Evaluation

The final phase assessed the training program and teachers' perceptions through a questionnaire and rating scale. The questionnaire included 10 multiple-choice and two short-answer questions on ASD and evidence-based practices, while the rating scale had 11 statements scored from 0 to 4.

Results of the Questionnaire on Pedagogical Practices (Theoretical Knowledge of ASD and EBPs). The training program significantly enhanced teachers' knowledge of ASD, pedagogical practices, and evidence-based strategies (EBPs), improving their understanding of autistic students' needs. Teachers demonstrated a solid grasp of techniques such as reinforcement strategies, differential reinforcement (DR), DSM-5 criteria, Functional Behavioral Assessment (FBA), activity schedules, and visual supports. The inclusion of FBA was especially helpful, with all teachers correctly defining the A-B-C procedure, though only three initially identified its function accurately. Teachers also showed strong knowledge of token economy, social skills techniques, and peer-mediated strategies. Despite some misunderstandings, such as misclassifying token economy and peer-mediated strategies, the overall results indicated significant growth in teachers' knowledge and ability to implement ASD-specific strategies in the classroom.

3.3 Training Evaluation Form/Rating Scale

At the end of the training, teachers evaluated the program using a scale from 0 to 4. The highest-rated aspects were the effectiveness of role-playing ($M = 3.89$), feedback sessions with classroom observations ($M = 3.67$), and content organization and clarity ($M = 3.67$). Teachers also valued participation and training materials ($M = 3.50$). The lowest-rated aspects were training time ($M = 1.67$) and facility adequacy ($M = 2.67$). Overall, the training significantly enhanced teachers' knowledge of evidence-based practices, with all participants rating it as highly effective.

4 Discussion

The findings from this study, which aimed to investigate four key objectives related to the inclusion of students with ASD in general education classrooms, are discussed below:

4.1 Developing a Teacher Training Program for Inclusive Classrooms

The study aimed to develop an evidence-based teacher training program for general education teachers working with students with ASD in inclusive classrooms. The program addressed key challenges such as out-of-seat behaviors, aggression, and inappropriate vocalizations. Content was developed based on research and teacher feedback, incorporating strategies like functional behavior assessments, direct instruction, visual supports, and self-management techniques (Brennan, 2018). Teachers found these strategies practical, and post-interviews confirmed their effectiveness in classrooms.

4.2 Benefits of Inclusion for Students with ASD

Inclusion provided academic and social benefits for both students with ASD and their neurotypical peers. For students with ASD, it improved social interactions, communication, and positive behavioral imitation (Anthony et al., 2019). Neurotypical students acted as peer buddies, enhancing their acceptance and tolerance of differences. Teachers noted reduced stigma and improved social harmony in inclusive settings, aligning with findings by Choi et al. (2016).

4.3 Factors for Successful Teacher Training

Effective teacher training for inclusive classrooms involves five key elements: foundational knowledge of ASD and its impact on learning and behavior (Güleç-Aslan, 2013), school-based strategies like peer-mediated interaction and self-monitoring to foster engagement (Loftin et al., 2008), and expert-led instruction to ensure accurate strategy use (Simonsmeier et al., 2021). Practical approaches, such as BST, including video modeling, role-playing, and feedback, have proven highly effective (Güleç-Aslan, 2013). Ongoing support through collaboration and expert guidance is crucial to maintaining positive outcomes (Lauderdale-Littin and Brennan, 2018).

4.4 Prerequisites for Successful Inclusion

The study identified three key factors for successful inclusion of students with ASD. First, comprehensive pre-service teacher training is essential for equipping educators with the necessary knowledge and skills for inclusive classrooms (Khalil et al., 2020). Second, students with ASD should possess foundational skills, such as adaptive behavior and language abilities, to succeed in these environments (Fava et al., 2012). Third, schools must establish flexible and supportive environments that cater to the diverse needs of ASD students, fostering both academic and social inclusion (Anthony et al., 2019). Together, these factors enhance the effectiveness and sustainability of inclusion practices.

5 Conclusion

Equipping teachers with training in EBPs is essential for sustaining inclusive education for students with ASD. Inclusion benefits all students by promoting academic growth and tolerance (Anthony et al., 2019), but the rising prevalence of autism (1 in 59; CDC,

2018) poses challenges, especially in Saudi Arabia, where many teachers lack proper training (Khalil et al., 2020). Few studies have focused on EBPs in Saudi Arabian classrooms (Alnemary et al., 2016; Khalil et al., 2020). The Autism Teacher Training Program (ATP) addressed this gap, improving teachers' skills, confidence, and ability to implement EBPs, enhancing inclusion outcomes. Despite progress in inclusive education (Vision 2030, 2016; MoE, 2016), public schools in Saudi Arabia are still underprepared, with most ASD students attending private institutions. The researcher plans to expand the ATP annually to align with national inclusion goals.

General education teachers require effective strategies to support students with ASD, and evidence-based practices (EBPs) are essential for fostering their academic and social development. Successful autism training programs should incorporate five key elements: (1) foundational knowledge of ASD, (2) school-based interventions, (3) expert-led instruction, (4) hands-on training with feedback, and (5) ongoing support through materials and consultations.

The study demonstrated that mentoring, classroom observation, and feedback significantly improved teachers' use of evidence-based practices (EBPs) for students with ASD. The training was expanded to include a fourth follow-up phase to support long-term retention of EBPs. The study recommends a four-phase model for teacher training: BST training, hands-on coaching, evaluation, and follow-up. It also suggests using the train-the-trainer model to increase efficiency, where experienced trainers teach a few key individuals, who then train other staff (Fig. 1).

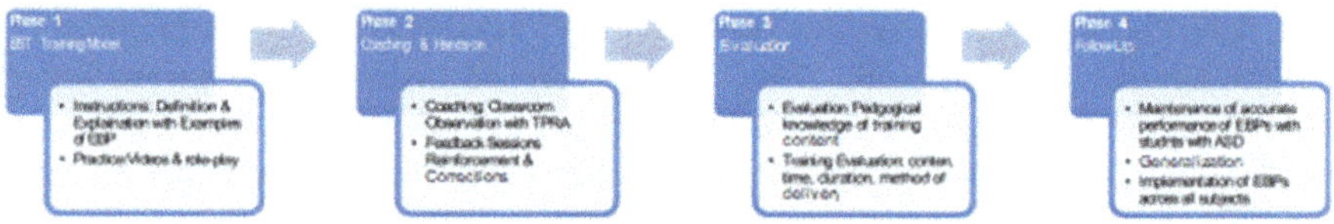

Fig. 1 Four-phase model of teacher training

The training equipped the teachers with a strong foundation in understanding ASD, as well as practical strategies such as functional behavior assessments, visual supports, and self-management techniques. This enhanced the teachers' capacity to address the unique learning and behavioural needs of students with ASD, fostering their academic and social inclusion within the mainstream setting. Sustainable inclusion requires a systematic approach that prioritizes the professional development of teachers. The findings of this study highlight the critical role of specialized teacher training in establishing and maintaining effective inclusive education for students with ASD, ultimately contributing to their overall growth and integration within the school community.

This study highlights the need for improved teacher training for general education teachers working with ASD students in inclusive classrooms. It recommends that Saudi universities create specialized autism training programs with practical components for both pre-service and in-service teachers. The Ministry of Education (MoE) should mandate such training, along with hands-on experience and clear inclusive education policies. Additionally, teachers implementing evidence-based practices (EBPs) should be provided with professional development opportunities, incentives, and promotions.

It addressed all research questions but had several limitations, including participant selection being influenced by school principals and supervisors, which limited the sample size. All participating schools were private, potentially limiting the generalizability to public schools. Additionally, due to COVID-19 disruptions and school lockdowns in Saudi Arabia, the researcher was unable to assess retention rates or conduct follow-up interviews after week 15. Future research should focus on barriers to inclusion in public schools and assess effective training methods and measurement tools.

References

Ainscow, M., Slee, R., Best, M.: Editorial: The salamanca statement: 25 years on. Int. J. Incl. Educ. **23**(7–8), 671–676 (2019). https://doi.org/10.1080/13603116.2019.1622800

Alnemary, F.M., Aldhalaan, H.M., Simon-Cereijido, G., Alnemary, F.M.: Services for children with autism in the Kingdom of Saudi Arabia. Autism. **21**(5), 592–602 (2016). https://doi.org/10.1177/1362361316664868

American Psychiatric Association: Diagnostic and Statistical Manual of Mental Disorders: DSM-5, 5th edn. American Psychiatric Association Publishing, Washington DC (2013). https://doi.org/10.1176/appi.books.9780890425596

Anthony, B.J. et al.: Increasing autism acceptance: The impact of the sesame street "See amazing in all children" initiative. Autism. **24**(1), 95–108 (2019). https://doi.org/10.1177/1362361319847927

CDC estimate on autism prevalence increases by nearly 10 percent, to 1 in 54 children in the U.S. Autism Speaks, from https://www.autismspeaks.org/press-release/cdc-estimate-autism-prevalence-increases-nearly-10-percent-1-54-children-us. (2018). Retrieved October 18, 2021.

Choi, J., Meisenheimer, J., McCart, A., Sailor, W.: Improving learning for all students through equity-based inclusive reform practices. Remedial Spec. Educ. **38**(1), 28–41 (2016). https://doi.org/10.1177/0741932516644054

Fava, L., Vicari, S., Valeri, G., D'Elia, L., Arima, S., Strauss, K.: Intensive behavioral intervention for school-aged children with autism: Una breccia nel muro (UBM)—A comprehensive behavioral model. Res. Autism Spectr. Disord. **6**(4), 1273–1288 (2012). https://doi.org/10.1016/j.rasd.2012.03.011

Güleç-Aslan, Y.: A training programme for a teacher working with a student with ASD: An action research. Educ. Sci. Theory Pract. **13**(4), 1–18 (2013) https://doi.org/10.12738/estp.2013.4.1738

Khalil, A.I., Salman, A., Helabi, R., Khalid, M.: Teachers' knowledge and opinions toward integrating children with autism spectrum disorder in mainstream primary school in Jeddah, Saudi Arabia. Saudi J. Humanit. Soc. Sci. **5**(6), 282–293 (2020) https://doi.org/10.36348/sjhss.2020.v05i06.004

Khan, A., AlGhadeer, H., Al-Qassimi, A., Al-Jubran, T., Al-Momen, H., Al-Nazzal, M.: Autism in Saudi Arabia, a challenge to Saudi families: A cross-sectional study. Int. J. Med. Dev. Ctries. **4**(9), 1453–1458 (2020) https://doi.org/10.24911/ijmdc.51-1595277794

Kurniawati, F., de Boer, A.A., Minnaert, A.E.M.G., Mangunsong, F.: Characteristics of primary teacher training programmes on inclusion: A literature focus. Educ. Rev. **66**(2), 235–256 (2014)

Lauderdale-Littin, S., Brennan, M.: Evidence-based practices in the public school: The role of preservice teacher training. Int. Electron. J. Elem. Educ. **10**(3), 369–375 (2018) https://doi.org/10.26822/iejee.2018336195

Lee, F.L., Yeung, A.S., Tracey, D., Barker, K.: Inclusion of children with special needs in early childhood education. Top. Early Child. Spec. Educ. **35**(2), 79–88 (2015). https://doi.org/10.1177/0271121414566014

Loftin, R.L., Odom, S.L., Lantz, J.F.: Social interaction and repetitive motor behaviors. J. Autism Dev. Disord. **38**, 1124–1135 (2008). https://doi.org/10.1007/s10803-007-0499-5
Ministry of Education: Procedural Guide for Special Education. Special Education Department (2016). Retrieved July 10, 2021, from https://departments.moe.gov.sa/SPED/Pages/Evidence.aspx
Morrier, M. J., Rice, C. E., Baio, J., Wagner, S. J., & Nicholas, J.: Changes in educational placements for children with autism spectrum disorders (ASDs) in a population-based sample in a US metropolitan area. In: Int. Meet. Autism Res., Philadelphia, PA, USA, March 21 (2010).
Parsons, M.B., Rollyson, J.H., Reid, D.H.: Evidence-based staff training: a guide for practitioners. Behav. Anal. Pract. **5**(2), 2–11 (2012). https://doi.org/10.1007/bf03391819
Scheuermann, B., Webber, J., Boutot, E.A., Goodwin, M.: Problems with personnel preparation in autism spectrum disorders. Focus Autism Other Dev. Disabil. **18**(3), 197–206 (2003). https://doi.org/10.1177/10883576030180030801
Simonsmeier, B.A., Flaig, M., Deiglmayr, A., Schalk, L., Schneider, M.: Domain-specific prior knowledge and learning: A meta-analysis. Educ. Psychol. **1–24** (2021). https://doi.org/10.1080/00461520.2021.1939700

VitalTwin: A Predictive and Management Prototype Digital Twin-Based System for Health and Wellbeing of Hajj and Umrah Pilgrims

Fatema Sabeen Shaikh, Rayhanah Jassim Aldolaim(✉), Abrar Saleh AlGhamdi, Aisha Saleh AlKatheer, Reem Zaki AlMahfoud, Solaf Mohammed Alhotailah, Rawan Hamed Alghamdi, and Abdullah Ali M. Alqahtani

Department of Computer Information Systems, Imam Abdulrahman Bin Faisal University, P.O. Box 1982, Dammam 31441, Saudi Arabia
{fsshaikh,2210003177,2210002657,2210002819,2210003119, 2210003159,rhalghamdi,aamqahtani}@iau.edu.sa

Abstract. Every year, millions of pilgrims perform the Hajj and Umrah (H&U), which present unique physical challenges due to intense heat, crowding, and physical exertion. Elderly pilgrims and those with preexisting comorbidities are at a higher risk of increased severity in their health issues. H&U authorities have a very short time window to identify pilgrims with higher health risks and treat them before their condition worsens. To address these problems, in this paper, we propose the conceptual idea of VitalTwin, a digital twin-based health monitoring system designed to enhance pilgrims' quality of life. VitalTwin takes a variety of different factors into account, such as environmental factors, including temperature and humidity, a pilgrim's physical activity, location and vital statistics, and health history to forecast potential health risks and notify medical professionals in real time. By enabling early medical interventions, we hope to improve pilgrims' health outcomes and make their pilgrimage a more pleasant experience. This proactive approach to healthcare, which integrates digital technologies, aims to not only improve the healthcare provided to pilgrims but also offers remote health professionals located in different parts of the Kingdom of Saudi Arabia an excellent opportunity for community service by providing their expertise during busy Hajj and Umrah periods, hence meeting the objectives of the Kingdom of Saudi Arabia's Vision 2030.

Keywords: Digital twin · Predictive system · Hajj and umrah · Diabetes · Cardiovascular · Respiratory · Hypertension · Heatstroke

1 Introduction

Digital twin (DT) technology in healthcare provides a virtual replica of people or real-life systems using real-time data and advanced analytics to monitor, analyze, and optimize the performance of the quality of healthcare [8].

D. Berkouk et al. (Eds.): CCTS 2024, PTT, pp. 231–237, 2026.
https://10.1007/978-981-95-7607-4_26

Due to the sheer scale of the millions of pilgrims that gather for Hajj and Umrah, the pilgrimages present unique management challenges to government authorities, public health professionals, medical professionals, and volunteers.

On the healthcare front, pilgrims vary widely in age, health, and physical capabilities, and environmental factors like extreme heat, overcrowded conditions, and physical exertion increase the probability of health risks, including dehydration, cardiovascular strain, and heat-related illnesses [9]. Tragic incidents like the 2015 Hajj stampede, which claimed over 2,400 lives, emphasize the urgent need for innovative technologies to ensure the safety and well-being of those performing the pilgrimage [10].

In this paper, we propose a digital twin system to monitor the health of pilgrims during Hajj and Umrah and facilitate the strategic management of health resources. The proposed system can proactively predict potential medium to high-risk cases and health risks such as heatstroke and dehydration [11] and, together with the location of pilgrims, can allow for the strategic deployment of emergency services system-generated automatic alerts, would enable timely notification of potential health risks to the pilgrims, healthcare providers, and public health professionals and enable proactive responses before a health emergency arises.

The layout of the paper is as follows: Sect. 2 discusses the methodology applied in the development of VitalTwin, Sect. 3 lists the related projects and existing applications, Sect. 4 proposes VitalTwin system architecture, Sect. 5 summaries the paper and the new solution, Sect. 6 lists the used references.

2 Methodology

In this paper we mainly present our idea of the VitalTwin. This project is still in the early stage of implementation, and we hope to publish developments in reputable journals in the near future. For our literature review, we applied Google Scholar to search for related work on implementing DT applications which are mentioned in Sect. 3. We first searched for related applications and research focusing on the use of digital twins in the medical field and specific comorbidities which are diabetes, hypertension, cardiovascular, respiratory and heatstroke involving physical exercise and fitness. Next, we studied smart devices used to monitor different vital signs such as glucose level, blood pressure, heart rate, oxygen level and fitness as shown in Table 1.

In addition, During the implementation phase, we will use HTML, CSS and JavaScript as front-end, Python as back-end and SQLite as the main database technology. Our datasets consist of a wide variety of different information including comorbidities like diabetes, cardio, respiratory, blood pressure and heatstroke, weather and crowd data. These datasets all underwent preprocessing and preparation for analysis using the Pandas library in Python. The hyperlink at [14] presents the excel sheet that summarizes our datasets, the sample sizes and the preprocessing techniques that were applied to them. The cleaned data will then be applied to DT which consists of different types of machine learning (ML) models which are chosen based on several criteria like model's accuracy, efficiency and suitability for data nature. These models include Random Forest (RF) which is optimal for dealing with high-dimensional and complex data. Additionally, it provides high-accuracy analysis and classification of health conditions. Another ML model which is Linear Regression (LR) that accurately estimates and provides a clear vision of the health trends and changes over time. The process of building the models will be done by dividing the data into two parts. 70% of the data will be as training data, while 30% of data will be as testing data. To ensure the model's accuracy Cross-Validation Techniques like Leave-One-Out Cross-Validation (LOOCV) will be applied. On the user side, all users, which include pilgrims, healthcare professionals and public health professionals, will interact with the digital twin using web-based interactive dashboards which will consist of interactive maps, advanced analytics and graphs, and alerts and notifications.

3 Literature Review

In the first part of this section, we discuss different DT studies in literature and specify how our proposed DT solution adopts a more comprehensive approach. We believe our proposed idea is the first that is fully customized to address the unique challenges presented in the H&U context. In the second part, we present the results of a survey we conducted to explore the features available in IoT-based smart health devices for measuring vital statistics.

Several studies proposed DT models to monitor and predict the health of diabetes patients. One study [1] developed a DT model for mellitus diabetes (DM) management built using Virtual Reality (VR), AI, and deep machine learning (DL) technologies to create and provide the best decision and feedback to manage the disease in different stages. Another study proposed a healthy lifestyle system based on DT technology for patients who have Type 2 Diabetes (T2D) and Metabolic Dysfunction-Associated Fatty Liver Disease involving food, activities, and sleep habits [2]. Many DT programs were developed to simulate the cardiovascular system and discover diseases; some of these projects the "Heart Living" [3] program and "Heartmodel" program [4] simulate the physical human heart to enable the users to discover heart issues and functionality, develop new medical tools, suggest correct equipment, evaluate the safety of drugs and customize surgery treatments. DT technology was used during the COVID-19 pandemic to effectively monitor the spread of COVID-19 by detecting the areas that have a high chance of COVID-19 spreading and providing real-time alerts to the visitors [3]. In terms of blood pressure (BP), the "Twin Precision Nutrition (TPN)" [5] program aims to develop a whole-body model for T2D and concomitant hypertension patients to assess the resistance of insulin (HOMA2-IR) and systemic hypertension (Systolic BP, Diastolic BP). Yet another study [6] developed a DT model for T2D patients that provides improvements recommendations in terms of food, activities, sleep and BP drugs commitments. In the field of fitness and obesity control, the "SmartFit" framework [5] developed a model the athletes' body to analyze and evaluate their behavior, future actions and diet. Within the Kingdom, the Ministry of Health launched their own digital twin model [13] application that captures the medical history of a person and builds a virtual DT model reflecting the person's medical details. However, this application is limited to displaying the current medical state of the person and does not provide proactive treatment plans.

In all the above examples, most DT solutions in literature have focused on one specific health issue and simulated its effect on the human body. All these proposed approaches serve their individual purposes; however, our proposed DT model is different for the following reasons:

1. It couples different types of information about the pilgrims' external context with the pilgrim's health data to simulate the effects in real-time on the pilgrim, and it aims to do this simultaneously for a large number of pilgrims. This information is not only beneficial to the healthcare professionals, but also to the H&U authorities and public health professionals to detect the formation of problem hotspots early on e.g. increased dehydration symptoms among groups of pilgrims in a particular area and deploy preventative measures such as medical teams to address the arising concerns before the situation worsens.
2. The VitalTwin System is built around the idea of community service, putting people first. The system allows specialized healthcare professionals from around the kingdom to volunteer during the busy Hajj and Umrah season to actively contribute towards enhancing the quality of health care offered to pilgrims by monitoring health data, offering tailored advice, and guiding on-site healthcare providers during critical situations. This approach eases the pressure on on-site healthcare providers, while ensuring that pilgrims receive the medical attention, they need by supporting them in real time.
3. Through interactive dashboards and advanced analytics, the proposed system will not only predict, monitor, and manage health risks, but also help to tackle challenges like the detection of potential problem hotspots due to extreme heat, overcrowding, and physical exertion.
4. The proposed VitalTwin System acts as a direct point of contact for medium to high-risk pilgrims to seek medical assistance virtually. This will reduce the burden on already crowded medical centers in the Hajj precincts and enhance the quality of healthcare offered to pilgrims of all ages. This is in line with the ambitious Vision 2030 objectives of the Kingdom of Saudi Arabia.

Table 1 presents the different IoT smart devices available today that measure the vital statistics for a human body. The columns present the measurable vital statistics, and the rows the individual solutions. Currently, Apple and Samsung smartwatches offer the largest variety of measurements. The Ministry of Hajj has also proposed the Smart Hajj bracelet [13] which holds the pilgrim's medical data and allows for the measurement of heart rate and blood oxygen levels. Our proposed solution will benefit from such existing technology and advance it further through the application of the virtual digital twin.

Table 1 Examples of current smart devices that measure vital statistics

Device	Features							
	Blood Sugar	BP	ECG	SpO2	Temperature	Fitness	BMI	Alerts
Apple Watch Series 9 [7]	√		√	√	√	√	√	√
Samsung Galaxy Watch Ultra [7]	√	√	√	√	√	√	√	√

(*continued*)

Table 1 (*continued*)

Device	Features							
	Blood Sugar	BP	ECG	SpO2	Temperature	Fitness	BMI	Alerts
Fitbit Sense 2 [7]	√		√	√	√	√		√
Huawei Watch D [7]		√	√	√	√	√		√
Garmin Vivoactive 4 [7]			√	√		√		√
Xiaomi Redmi Watch 5 [7]			√	√		√		√
Spire Health Tag [12]			√	√		√		√
Smart Hajj bracelet [13]			√	√				√

4 Results – The Proposed Solution

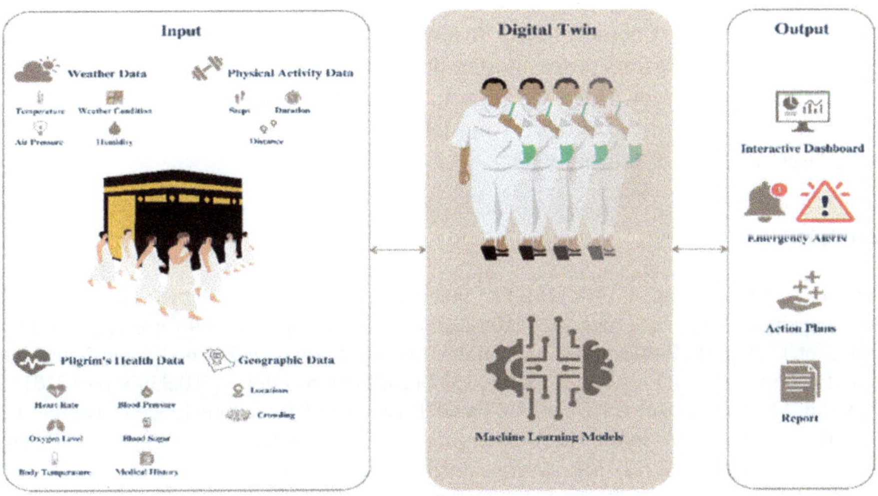

Fig. 1 VitalTwin system diagram

Compared to the existing systems, VitalTwin simulates the Hajj and Umrah environments next to the main 5 vital signs (blood glucose and pressure, heart rate, oxygen level and body temperature) then it helps to monitor and provide real-time analysis and alerts of medium to high-risk situations for both pilgrims and healthcare providers. Figure 1 is designed to represent the architecture of the system as a combination of three main parts which are input, process unit and output.

A huge amount of data coming from several sources will be as an input into the system. These data will be climate data which includes weather conditions (cloudy, sunny, rainy), temperature, air pressure and humidity, activities data which include the steps count, distance and duration, the pilgrim's health data which involves medical records, heart rate, blood sugar and pressure levels, SpO2 and body temperature which gathered using smart devices, and the geographic data which includes the pilgrim locations and crowding percentage. Finally, the output of the system will be in the form of an interactive dashboard that represents the analytics, emergency alerts and notifications about the health or environmental high-risk cases, action plans as guidance to handle the condition, and reports that summarize the situations.

5 Conclusion

this paper outlines the most important points about the VitalTwin DT project. VitalTwin system is designed to support the health and well-being of Hajj and Umrah pilgrims, which predicts the potential health risks during Hajj and Umrah through real-time analysis and alerts using advanced AI and ML. Furthermore, this project enables community services at the national level by allowing multiple healthcare specialists to support and help pilgrims remotely around the Kingdom. Our proposed solution could also be generalized further in future to monitor other large-scale events. Some key limitations of the VitalTwin is that we are currently using existing datasets available freely on the Internet so our results at times may vary from reality. Another limitation is that we currently limit our assessment of comorbidities to five health conditions. The unavailability of real-time data from pilgrims poses another problem and may require us to build our DT using normal patient data which we hope to collect in the near furture. Due to limited processing resources, our working DT model will be a protoype and may not be able to handle large amounts of pilgrim data.

References

1. Chu, Y., Li, S., Tang, J., Wu, H.: The potential of the Medical Digital Twin in diabetes management: a review. Front. Med. **10** (2023). https://doi.org/10.3389/fmed.2023.1178912
2. Meijer, C., Uh, H., Bouhaddani, S.E.: Digital twins in healthcare: Methodological challenges and opportunities. J. Pers. Med. **13**(10), 1522 (2023). https://doi.org/10.3390/jpm13101522
3. Aydın, Ö., Karaarslan, E.: OpenAI ChatGPT Generated Literature Review: Digital Twin in Healthcare. In: Ö. Aydın (Ed.), *Emerging Computer Technologies 2* (pp. 22–31). İzmir Akademi Dernegi. (2022). https://ssrn.com/abstract=4308687.
4. Armeni, P., Polat, I., De Rossi, L.M., Diaferia, L., Meregalli, S., Gatti, A.: Digital twins in healthcare: Is it the beginning of a new era of Evidence-Based Medicine? A critical review. J. Pers. Med. **12**(8), 1255 (2022). https://doi.org/10.3390/jpm12081255
5. Lin, Y., et al.: Human digital twin: a survey. J Cloud Comp. **13**(1), 131 (2024). https://doi.org/10.1186/s13677-024-00691-z
6. Tibrewala, A., Itchhaporia, D.: Revolutionizing hypertension management in Type 2 diabetes. *JACC*. Advances. **3(9), 101173** (2024). https://doi.org/10.1016/j.jacadv.2024.101173
7. Tony, L.: Top 10 best smart watch manufacturers in Saudi Arabia. IWO. (2023) https://www.iwosmartwatch.com/top-10-best-smart-watch-manufacturers-in-saudi-arabia/. Access 28 Jan 2025

8. Attaran, M., Celik, B.G.: Digital Twin: Benefits, use cases, challenges, and opportunities. Decis. Anal. J. **6**, 100165 (2023). https://doi.org/10.1016/j.dajour.2023.100165
9. Kshetri, N.: The Economics of Digital Twins. Computer. **54**(4), 86–90 (2021). https://doi.org/10.1109/mc.2021.3055683
10. Botín-Sanabria, D.M., et al.: Digital twin technology challenges and applications: A comprehensive review. Remote Sens. **14**(6), 1335 (2022). https://doi.org/10.3390/rs14061335
11. Abuzinadah, N. E.: Enhancing Pilgrim Safety During Hajj: A Smart Healthcare Solution with MYPARAMEDIC App and Vital Sign Monitoring Bracelet. In: Lecture notes in networks and systems (pp. 13–34) (2024) doi:https://doi.org/10.1007/978-3-031-62281-6_2.
12. Smart Scale Market Size, Share, Trends, Growth 2030. Zion Market Research. (2023). https://www.zionmarketresearch.com/report/smart-scale-market. Accessed 29 Jan 2025.
13. Smart Hajj. *Saudipedia* https://saudipedia.com/en/article/1230/religion/hajj-and-umrah/smart-hajj. Accessed 20 Mar 2025.
14. Datasets. https://udksa-my.sharepoint.com/:x:/g/personal/fsshaikh_iau_edu_sa/EUk-n_LDxsVCvwwsnGn4ePUBtGm7D4Yzx8BtUdJ5VciC_g?e=UVSvDi. Accessed 20 Mar 2025.

Workplace Toxicity and Its Shadow: Investigating Inner Resignation Through Employee Well-Being and Creativity in Saudi Academia

Komal Khalid(✉)

King Abdulaziz University, Jeddah, Saudi Arabia
kmuddassar@kau.edu.sa

Abstract. This study explores the impact of workplace toxicity on employees' well-being, creativity, and internal resignation in higher education institutes in Saudi Arabia, considering its relevance to the ambitions of Vision 2030. Based on a three-wave data collection, survey responses from 474 employees, including faculty members and administrative staff from top-ranking institutions, were collated. Partial least squares-structural equation modeling (PLS-SEM) was used for hypothesis testing. The research presents that workplace toxicity significantly reduces employee's well-being and creativity, which mediates their influence on inner resignation. This study also offers unique insights by highlighting these association within the context of Saudi higher education. It demonstrates the requirement for inclusive and psychologically safe environments policies to decrease workplace toxicity and inner resignation. These findings have implications for institutional leaders and policymakers to support employees' well-being and employee creativity in achieving sustainable innovation and organizational resilience aligning with the Vision 2030 goals of Saudi Arabia.

Keywords: Workplace toxicity · Inner resignation · Well-being · Creativity

1 Introduction

A workplace where creativity is suppressed, well-being is corroded, and employees silently withdraw from their roles, disengaging from the goals they once aspired to achieve is an environment often defined by harassment, bullying, and favoritism. This environment is not only corrosive to employee well-being (EW) but also a deep-seated threat to organizational sustainability [1]. In the HEI context, where innovative thinking and intellectual engagement are a prime requirement, the impact of workplace toxicity (WT) stretches further. This is particularly important in the Higher Education Institutions (HEIs), where intellectual creativity and employee well-being (EW) are key to nurturing innovation, advancing knowledge, and preparing future leaders [2]. Faculty members, the core of knowledge creation and dissemination, face unique pressures in dealing with workplace toxicity (WT) that distort their mental health and reduces their

D. Berkouk et al. (Eds.): CCTS 2024, PTT, pp. 238–246, 2026.
https://10.1007/978-981-95-7607-4_27

creative potential. Workplace toxicity (WT), characterized by bullying, harassment, and ostracism [3], stifles collaboration and innovation, leading to emotional withdrawal or "inner resignation (IR)", where employees disengage from organizational goals.

The necessity to address WT is heightened within the context of Saudi Arabia's HEI, which plays a pivotal role in achieving the socio-economic transformation envisioned in Vision 2030. This ambitious national strategy seeks to reduce dependency on oil and foster a knowledge-based economy through investments in human capital, quality education, and innovation [4]. Despite their strategic importance, HEIs have been paid little attention in research about the impact of WT on critical outcomes like EC, EW, and internal resignation (IR), especially in Saudi Arabia. Addressing this gap is relevant to enabling HEIs to play their transformative role in national development.

Workplace toxicity (WT) is a pressing organizational issue characterized by negativity and vicious behaviors encompassing harassment, bullying, violence, and ostracism [2]. In the aggregate, these behaviors tend to destroy trust, hinder team collaboration, and consequently effecting EW [2, 5]. These destructive behaviors can seep into the culture of an organization and form an environment that is not conducive to innovation and hurts organizational relationships [5]. EW is defined as physical, psychological, and emotional health. The concept is important in building resilience against workplace stressors and in improving job performance [6]. EW initiatives reduce stress and increase engagement, therefore buffering the harmful effects of WT [7]. Toxicity erodes trust, respect, and fairness; it violates psychological safety and triggers stress, anxiety, and burnout [2].

H_1. Workplace toxicity has a significant influence on employee well-being

Creativity is one of the most crucial determinants of organizational success and a source of individual fulfillment for employees to exercise their skills, talent, and unique perspectives [1]. In toxic environments, EC falls due to stress, fear, and lack of psychological safety, hence underlining the critical need for enabling organizational cultures to drive innovation and creativity [8]. WT harms creativity by creating an atmosphere of fear, distrust, and emotional instability. Toxic behaviors impede cognitive capacities, which in turn block creative thought processes [1, 8].

H_2. Workplace toxicity significantly affects employee creativity.

IR is a form of psychological withdrawal where employees disengage both emotionally and cognitively while maintaining physical presence. This state results from continued exposure to stressors such as toxicity and mirrors a broken psychological contract [9]. IR is characterized by reduced motivation, organizational commitment, and productivity. IR attribute it to feelings of helplessness and inequity in workplace exchanges [10]. The sustained exposure to WT mostly results in IR as employees withdraw psychologically to protect themselves against the constant negativity. Adverse environments engender discrepancies in exchange relationships, characterized by a lack of equitable reciprocity and support [11].

H_3. Workplace toxicity significantly impacts inner resignation.

EW is an antecedent of engagement and productivity. When well-being is threatened, emotional exhaustion and lowered motivation lead to inner resignation [12]. Employees with low levels of well-being tend to experience difficulties in staying focused, enthused, and purposeful in their work. Lack of reciprocation of effort with sufficient support and recognition breeds psychological withdrawal [13]. Organizations that disregard EW risk creating a context where IR becomes the norm, compromising individual and group performance [9].

$H_{4.}$ Employee well-being influences inner resignation.

EC alleviates IR through its provision of means for overcoming challenges while holding onto meaning [1]. Conversely, stifled creativity due to WT brings about frustration and disengagement, which heightens IR [3]. Employees not permitted to realize their creative potential often feel undervalued and demotivated, which strengthens their IR [14]. Thus, the cultivation of creativity turns out to be a critical factor in the prevention of IR and the facilitation of culture of engagement and innovation [10].

$H_{5.}$ Employee creativity significantly influences inner resignation.

WT deteriorate EW by fostering negative emotions and reducing access to support [7]. SET posits that diminished EW intensifies perceptions of inequitable exchanges, prompting withdrawal [15]. WT is significantly related to decreased EW, promoting IR [16]. This agrees with SET's proposition that behaviors correspond with the perceived quality of exchange relationships. A focus on improving EW will offset the harmfulness of WT, re-establishing a balance in workplace exchanges that reduces IR [12].

$H_{6.}$ Employee well-being significantly mediates the relationship between toxicity and inner resignation.

WT leads to the suppression of creativity because it increases fear and greatly reduces autonomy [14]. Creativity requires a supportive environment where there is a balance in workplace exchanges and positive outcomes [3]. Promoting creativity enables employees to stay engaged and develop novel solutions to the stressors in the workplace, which decreases the chances of IR [10]. Organizations that value EC as a core value can be effective in neutralizing the negative impact of WT [15]. As represented in (Fig. 1).

$H_{7.}$ Employee creativity significantly mediates the relationship between toxicity and inner resignation.

2 Methodology

2.1 Measurement

The study used standardized scales to measure the constructs of interest, which were WT, EW, EC, and IR. Each construct was measured using a 5-point Likert scale, with responses ranging from 1 (strongly disagree) to 5 (strongly agree). The survey was prepared in both the Arabic and English languages to reflect the linguistic diversity of the target population within Saudi Arabia's HEI. Translation was conducted using the back-to-back method to ensure accuracy and uniformity. WT was assessed by using an 8-item scale [2]. This scale measures three dimensions of WT: harassment (i.e., assessed

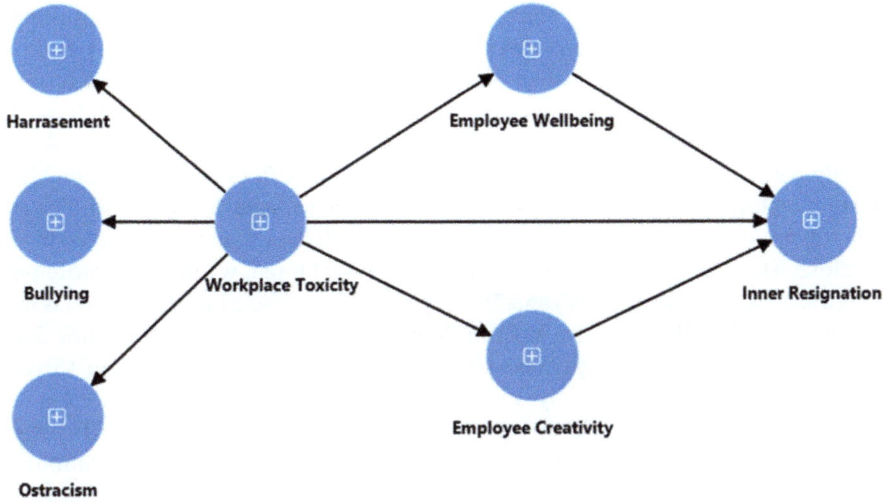

Fig. 1 Theoretical Framework

by 3 items), bullying (i.e., assessed by 3 items), and ostracism (i.e., assessed by 2 items). The EW was assessed with a 5-item scale [6]. EC was assessed using a 5-item scale [17], whereas IR was assessed by using a 3-item scale [9]. This IR scale measured the degree of respondents' disengagement, withdrawal, and reduced motivation in response to negative work environment conditions.

2.2 Sampling and Procedure

To analyze that how a WT affects EW, EC, and IR, the study adopts a three-wave data collection approach. The sample of the study was hired from the 15 top HEI of Saudi Arabia according to the 2024 ranking of the Times Higher Education World university. The study included both administrative and academic members of HEI. To draw a comprehensive insight into effects of WT dynamics on diverse duties within a HEI, academic and administrative personnel were included in the study. Given the sensitive nature of the research topic and the effectiveness of snowball sampling, this sampling technique was adopted to recruit participants. Survey forms were distributed by using respondent's email and WhatsApp. Participants were ensured for their anonymity and confidentiality. Moreover, they were also informed regarding the voluntary nature of the survey.

2.3 Data Collection

Data were collected in three waves over four months (September 2024 – December 2024), each focusing on different variables. This process was selected to reduce common method bias. In the first wave, almost 900 survey forms were distributed, to gather data regarding demographic (e.g., age, gender, education, experience, position, and marital status) and IR were collected, yielding 760 responses. Data regarding EW and EC was collected during the second phase from the respondents of first phase, resulting in 689 responses. In the third phase, the data were collected regarding the WT from the respondents of second phase, which produced 492 responses. This reduced response rate was because some of the respondents lost interest, did not respond, or shift the job. After removing 18 outliers, sample size was 474. The demographic characteristics of the study participants demonstrate that sample population from Saudi Arabia's higher education sector is in line with strategic objectives such as Saudi Vision 2030. The gender distribution is balanced, with 45.9% males, 50.8% females, and 3.6% preferring not to disclose. The age demographics reflect a diverse workforce, where most participants fall between 36–45 years (28.9%) and 25–35 years (27%). The educational level is high, with 58.9% holding PhDs, 26.2% holding Master's degrees, and 11.4% holding Bachelor's degrees. Professional experiences span less than one year to more than 15 years, with 7.6% and 12.2%, respectively; moreover, a large fraction has experience between 6–10 years, with 30% falling in this group. The dominant groupings by staff category in this personnel profile are faculty, 61.4%, which indicates academic orientation, and 38.6% were administrative. These findings underscore higher education's role in fostering a sustainable, inclusive, and research-oriented system that aligns with national innovation and development goals.

2.4 Data Analysis

This study employed partial least squares-structural equation modeling (PLS-SEM) to test theoretical models. As PLS-SEM combines multiple latent variables, it can remedy certain measurement deficiencies inherent in social sciences research [18]. SEM estimates and refines observed variables in an attempt to reduce ambiguities with regard to latent constructs while concurrently explaining causal relationships among latent and observed variables [19]. The data were collected from a single source, to assess the common method variance, both procedural and statistical controls were made [20].

2.5 Measurement Assessment Model

For the measurement and assessment of structural model, this study followed the guidelines given by [18]. The study assesses reflective measurement and composite models for harassment, bullying, ostracism, EW, EC, IR, and WT. Then, three reflective dimensions of WT were extended, namely harassment, bullying, and ostracism. The second step from [19] shows that WT is a second-order composite construct based on the score of its associated dimensions and that the related criteria were applied to assess WT as a composite second-order construct for the data. The reflective constructs reveal strong psychometric properties; that is, both CA and Rho_A values exceed the threshold of 0.70, indicating internal consistency. AVE values also exceed the minimum recommended for all constructs of 0.5, thus confirming convergent validity.

3 Results

3.1 Descriptive and Pearson Correlation Analyses.

The mean and standard deviation values of the WT and IR are 1.545 ± 0.484 and 1.541 ± 0.473 respectively, indicating this points to moderate consistency in perception among respondents about WT and IR. EC and EW have relatively low variability, with mean and standard deviations of 4.521 ± 0.403 and 4.486 ± 0.440, signaling relatively homogeneous distribution of responses. The Pearson correlation analysis demonstrate the strong relationships between WT, IR, EC, and EW. Moreover, strong positive correlation ($r = 0.689$, $p < 0.01$) between the WT and IR, moderate negative correlation between a WT and EW ($r = -0.528$, $p < 0.01$), and a strong negative relationship between WT and EC ($r = -0.714$, $p < 0.01$) was found.

3.2 Hypothesis Testing (via Structural Equation Modeling).

The direct effects in the structural model represent the direct relationships between the variables (Table 1). WT has a significant negative effect on EW ($\beta = -0.531$, $t = 14.637$, $p < 0.001$), indicating that a WT directly damages the EW (H_1 supported). Similarly, WT has a negative effect on EC ($\beta = -0.718$, $t = 28.19$, $p < 0.001$), indicating that WT diminishes EC (H_2 supported). Moreover, WT directly influences IR positively ($\beta = 0.247$, $t = 5.547$, $p < 0.001$) (H_3 supported). Additionally, EW and EC have a direct influence on IR ($\beta = -0.185$, $t = 5.137$, $p < 0.001$; $\beta = -0.484$, $t = 9.688$, $p < 0.001$, respectively), indicating that the lower the EW and EC, the higher the IR (H_4 and H_5 supported). The indirect effects in the SEM represents the mediating roles of EW and EC in the relationship between WT and IR. The results show that WT indirectly increases IR through its negative influence on EW ($\beta = 0.298$, $t = 4.653$, $p < 0.001$) and EC ($\beta = 0.347$, $t = 8.461$, $p < 0.001$). These findings suggest that a WT first reduces EW and EC, which then leads to greater psychological withdrawal among employees. Both hypotheses H_6 and H_7 relating to the indirect effects of EW and EC on the relationship of WT and IR are statistically supported given the significant p-values and confidence intervals.

Table 1. Results: Structural Equation Modeling (SEM).

Hypothesis	Direct/Indirect Effects	β Value	T statistics	P values	BCCI	
					ULCI	LLCI
H_1	WT -> EW	-0.531	14.637	0.000	-0.597	-0.457
H_2	WT -> EC	-0.718	28.19	0.000	-0.765	-0.665
H_3	WT -> IR	0.247	5.547	0.000	0.161	0.335
H_4	EW -> IR	-0.185	5.137	0.000	-0.255	-0.115
H_5	EC -> IR	-0.484	9.688	0.000	-0.585	-0.39
H_6	WT -> EW -> IR	0.298	4.653	0.000	0.261	0.342
H_7	WT -> EC -> IR	0.347	8.461	0.000	0.272	0.432

4 Discussion and Implications

The findings of the current study shed significant light on the interaction between WT, EW, EC, and IR, aligned with previous research while also bringing forth new perspectives. In line with the research by [2, 7, 8], this analysis confirms that WT significantly diminishes both EW and EC. Consistent with these findings, the negative correlation between WT and EW observed here aligns with [15], which highlighted the adverse impact of toxic behaviors on psychological safety and stress levels. Moreover, the results verify [1, 14], emphasizing that WT suppresses EC by eroding psychological safety and increasing stress, ultimately repressing creativity. The strong mediating role of EC and EW in the WT–IR relationship adds to prior research by demonstrating how these variables serve as mechanisms through which toxicity influences IR (i.e., a novel contribution to the literature). The present study builds on the previous findings of [10, 15] by further showing that the promotion of EW and EC could act as a protective factor against IR; in fact, this provides actionable insights on how to mitigate adverse effects of WT. However, contrary to previous studies, this study places these relationships within the context of HEI in Saudi Arabia. It puts forward the significant role of institutional culture in WT resilience.

The findings show some imperative steps that policymakers and the management of HEIs need to take to curb WT and its adverse effects on the EC, EW, and IR. Policymakers can build a framework for inclusive and supportive environment policies in line with Vision 2030 to improve education and innovation. Part of the framework should include clear policies against harassment and bullying, along with strict measures of enforcement, to significantly reduce WT. Addressing WT and fostering EW and EC, HEIs can contribute to the national goals of economic diversification, knowledge-based growth, and sustainable development. Policymakers should incorporate anti-toxicity frameworks into broader institutional guidelines, emphasizing transparency, fairness, and psychological safety. This will go in tandem with Vision 2030's objectives of improving the quality of education and decent work environments (i.e., SDG 4 & 8).

Management should give high priority to initiatives like EW by mental health programs, stress management and workshops that can act as a buffer to the negative consequences of WT. Moreover, management must cultivate a culture of creativity by facilitating interdisciplinary collaboration, rewarding innovative solutions, and supporting academic freedom. Regular leadership training in emotional intelligence and conflict resolution should be implemented to instill empathy and inclusivity. These efforts reinforce HEIs as hubs of creativity and leadership development, crucial for Vision 2030's socio-economic transformation. Periodic assessments of organizational culture through anonymous surveys may help to identify and correct behaviors that are harmful. Further investment in leadership development programs focused on empathy and emotional intelligence will only serve to further solidify a positive workplace culture.

5 Conclusion

This study was about the impact of WT on EW, EC, and IR in a HEI in Saudi Arabia. The outcome of this present study indicates that WT diminishes EW significantly, as does EC, further leading to disengagement among its staff members (IR). The mediating functions of EW and EC reveal how supportive, inclusive workplace environments can cushion the unwanted effects of WT. These results are aligned with the goals of Vision 2030, which aim at promoting innovation, performance, and employee engagement. Effective policies and leadership practices addressing WT are important in creating resilient and sustainable institutions that support national socio-economic transformation goals.

References

1. Arubayi, D.O.: Workplace toxicity and employee performance of manufacturing firms: Evidence from Nigeria. Int. J. Manag. Entrep. Res. **5**(2), 143–157 (2023)
2. Rasool, S.F., Wang, M., Tang, M., Saeed, A., Iqbal, J.: How toxic workplace environment effects the employee engagement: The mediating role of organizational support and employee wellbeing. Int. J. Environ. Res. Public Health. **18**(5), 1–17 (2021)
3. Khalid, K.: Workplace dynamics and individual outcomes: Role of workplace ostracism in shaping inner resignation. Amazonia Investiga. **12**(69), 120–131 (2023)
4. Mohiuddin, K., Nasr, O.A., Miladi, M.N., Fatima, H., Shahwar, S., Naveed, N.Q.: Potentialities and priorities for higher educational development in Saudi Arabia for the next decade: Critical reflections of the Vision 2030 framework. Heliyon. **9**(5), 1–16 (2023)
5. Sari, R.D., Dudija, N.: The impact of toxic workplace environments on employee productivity: A systematic literature review. Int. J. Sci. Technol. Manag. **5**(4), 878–882 (2024)
6. Ahmed, M., Zehou, S., Raza, S.A., Qureshi, M.A., Yousufi, S.Q.: Impact of CSR and environmental triggers on employee green behavior: The mediating effect of employee well-being. Corp. Soc. Responsib. Environ. Manag. **27**(5), 2225–2239 (2020)
7. Ahmed, A.K., Atta, M.H.R., El-Monshed, A.H., Mohamed, A.I.: The effect of toxic leadership on workplace deviance: The mediating effect of emotional exhaustion, and the moderating effect of organizational cynicism. BMC Nurs. **23**(1), 1–14 (2024)
8. Ruanggoon, J.: Organizational toxicity: A painful situation and impacts that occur to high-performing sales employee in the workplace. NIDA Case Res. J. **16**(2), 1–25 (2024)
9. Tiefenthaler, H.M.: Supervisors' power to deal with employees' inner resignation: How perceived power of the organization and the supervisor relate to employees' voluntary and enforced work behavior. Eur. Manag. J. **39**(2), 260–269 (2021)
10. Mohammadi, H., Ahmadi, H., Rajabpour, E.: Impact of toxic work environment on quiet quitting with psychological capital mediation. J. Strateg. Hum. Resour. Manag. **13**(1), 29–42 (2024)
11. Ilyas, A., Gulraiz, M., Zubair, Z., Khan, A.H., Munir, W.: Unraveling the impact of abusive supervision on emotional exhaustion: A serial mediation model of stress, relational conflict, and toxic workplaces. Soc. Sci. Rev. Arch. **3**(1), 2109–2130 (2025)
12. Pevec, N.: The concept of identifying factors of quiet quitting in organizations: An integrative literature review. Chall. Future. **8**(2), 128–147 (2023)
13. Smith, N., Lowman, F.I.: Conflict in the workplace: A 10-year review of toxic leadership in higher education. Int. J. Leadersh. Educ. **23**(5), 538–551 (2020)
14. Octavian, S.M.: About the impact and effects of toxic leadership on employees and organizations. Eur. Rev. Appl. Sociol. **16**(27), 87–93 (2023)
15. Batiste, H.: Management in times of crisis: A qualitative exploration of the great resignation from a social exchange perspective. Compens. Benefits Rev. **56**(3), 138–160 (2024)
16. Dahlan, M.A., Omar, R., Kamarudin, S.: Influence of toxic leadership behaviour on employee performance in higher educational institutions in Saudi Arabia. Int. J. Organ. Leadersh. **13**(1), 79–101 (2024)

17. Coelho, F., Augusto, M.: Job characteristics and the creativity of frontline service employees. J. Serv. Res. **13**(4), 426–438 (2010)
18. Hair, J.F., Sarstedt, M., Ringle, C.M., Sharma, P.N., Liengaard, B.D.: Going beyond the untold facts in PLS–SEM and moving forward. Eur. J. Mark. **58**(13), 81–106 (2024)
19. Gannon, M., Rasoolimanesh, S.M., Taheri, B.: Assessing the mediating role of residents' perceptions toward tourism development. J. Travel Res. **60**(1), 149–171 (2021)
20. Podsakoff, P.M., MacKenzie, S.B., Podsakoff, N.P.: Sources of method bias in social science research and recommendations on how to control it. Annu. Rev. Psychol. **63**(1), 539–569 (2012)

Exploring the Impact of Instant Gratification Factors on Long-Term Sustainable Financial Planning Among Saudi Youth

Fatima Balfaqih(✉) and Samar Altarteer

Dar Al-Hekma University, Jeddah, Kingdom of Saudi Arabia
faabalfaqih@gmail.com, starteer@dah.edu.sa

Abstract. This study explores the impact of instant gratification factors on long-term financial planning among Saudi youth and assesses suitable visual communication solutions to promote long-term financial planning and control instant gratification. To address this, a mixed-methods research approach was used, encompassing surveys completed by 270 participants to gather data on Saudi youth financial practices, alongside visual research to assess existing financial digital tools. The results indicate that the majority of Saudi youth lack engagement in structured financial planning, frequently prioritizing short-term satisfaction over long-term goals due to impulsive financial decisions. Moreover, there was a clear relationship between instant gratification and a lack of financial planning. This study identifies a gap in understanding the financial behavior of Saudi youth, particularly regarding instant gratification and financial planning. Findings show that many Saudi youth lack financial plans and tend to engage in impulsive spending. Visual research reveals that while existing financial apps emphasize clear visuals and tools such as budgeting and retirement planning, they often neglect the psychological factors and engaging elements that influence financial behavior. To address this, the proposed solution is a mobile app that integrates features such as spending charts, financial health indicators, goal-setting tools, and gamified rewards with social challenges, encouraging immediate behavioral changes and supporting long-term financial planning, thus providing a more holistic approach to personal finance management.

Keywords: Instant gratification · Sustainable financial planning · Saudi youth · impulsive spending

1 Introduction

1.1 Overview of Topic

Managing finances can be challenging for young people, who often prioritize immediate rewards over long-term financial planning [1]. In Saudi Arabia, this issue is especially significant due to the focus on future financial security under Vision 2030 [2]. Moreover, during the COVID-19 pandemic, the widespread availability of quick financial

D. Berkouk et al. (Eds.): CCTS 2024, PTT, pp. 247–255, 2026.
https://10.1007/978-981-95-7607-4_28

transactions has amplified the desire for instant gratification, making impulsive financial choices more common [3]. Although several studies address financial literacy and short-term behavior, the impact of instant gratification on long-term financial planning among Saudi youth remains underexplored [4]. Few studies also consider how technological and visual communication solutions can enhance financial understanding and encourage long-term planning [5]. Therefore, this study explores the impact of instant gratification factors on long-term financial planning among Saudi youth and assesses suitable technological approaches to promote long-term financial planning and control instant gratification. This research aligns with Saudi Arabia's Vision 2030 by emphasizing the new generation's financial well-being and contributing to the field of sustainable behavioral finance. It also identifies key influences on Saudi youth's financial decisions and proposes strategies, including visual communication methods, to mitigate impulsive spending.

1.2 Literature Review

Saudi young adults' financial planning behavior is increasing, though they face various challenges. For instance, 45% of young Saudis do not save and over 80% lack financial plans, indicating a focus on short-term survival [4]. Furthermore, COVID-19 pandemic-related economic stress led many young adults to prioritize immediate needs [6]. Additionally, impulsive buying is driven by psychological triggers and the pursuit of immediate gratification [7]. Also, the influence of social and economic factors on financial decision-making was emphasized, as well as the role of family and peers [8]. Similarly, social influences also encourage short-term planning [4].

Higher financial literacy reduces the tendency for instant gratification among Saudi youth [4]. Likewise, poor budgeting skills correlate with impulsive spending, which can lead to financial instability [9]. Moreover, stress can spur financial impulsivity, prompting immediate rewards to alleviate stress [10]. Additionally, FinTech offers convenience and fosters the desire for instant gratification, as digital financial services simplify credit access [11].

On the other hand, self-determination in financial planning improves debt management and reduces impulsive buying [11]. Furthermore, religious and moral values impact the financial decisions of Muslim youth, discouraging impulsive consumption by guiding individuals toward more prudent spending [12]. Additionally, real-time spending visibility in budgeting apps helps counter impulsive purchases [8].

To address these challenges, researchers developed a three-dimensional financial environment that allows participants to interact with their financial data in real-time, making complex financial data more accessible and intuitive [13]. The findings revealed improved long-term engagement and understanding of complex financial concepts. Furthermore, participants were shown current and future-aged avatars of themselves, indicating a desire to increase financial knowledge and confidence, as well as an interest in attending retirement-focused financial planning workshops [1]. In another study, a suite of smartphone apps collectively known as "Money Matters" was assessed to help youth gain more financial knowledge. These apps included tools for expenditure tracking, these tools helped them resist unnecessary spending [14]. Moreover, achievement- and progression-oriented elements, such as rewards and challenges, fostered user engagement, satisfied psychological needs, and influenced behavior [15]. Existing studies have shown improvements in financial knowledge but overlook psychological factors like instant gratification. This study contributes by proposing a mobile app that integrates financial tools with interactive features, such as gamification and virtual characters, to promote long-term financial planning and address these psychological barriers.

2 Methodology

2.1 Method One

The survey was administered to gather common practices and perspectives among Saudi youth, aiming to provide a clearer understanding of the relationship between instant gratification and long-term financial planning. Moreover, this method shows percentage data and findings that contribute to a comprehensive understanding of a large population of Saudi youth in a cost-effective manner. In addition, mixed-method data, including open-ended responses, yield insights on the correlation between instant gratification and long-term planning behaviors. The survey was distributed using Microsoft Forms through online platforms, including WhatsApp and Telegram, which are the most used apps among the target audience. To ensure proper sampling, a Purposive sampling was utilized, consisting of young Saudi adults aged 18 to 22 who have a source of income, allowance, or stipend. The age and the requirements were stated in both the survey preface and the consent form. The survey consisted of a total of ten questions, incorporating Likert scales, multiple-choice questions, and short-answer questions, and it was completed by 270 participants. Regarding ethical considerations, participants were informed of the research purpose, all responses were anonymous, participation was voluntary, and they had the option to withdraw at any time.

2.2 Method Two

This section examines existing visual communication solutions designed to manage instant gratification or to support long-term planning. The objective is to analyze critical elements for a suitable design solution using Compositional and Semiotics Interpretation. Two representative applications, "Spendee [16]: Money & Budget Planner" and "Empower [17]," were selected for their popularity, robust features, and capacity to address impulsivit. Spendee, developed by Cleevio s.r.o., focuses on budgeting, expense tracking, and collaboration through shared wallets, using color-coded indicators and playful visuals to engage users. Empower, developed by Empower Retirement, provides an all-in-one platform for budgeting, investments, and goal setting, with a professional interface, consolidated financial tools, and interactive graphs. Both apps guide decision-making with alerts, progress bars, and real-time feedback, though Spendee adopts a more vibrant, gamified style while Empower emphasizes a streamlined, corporate aesthetic. Additionally, both solutions rely on analytics to highlight spending patterns, encouraging long-term thinking. By comparing these solutions, essential design factors emerge, including targeted features, intuitive navigation, and clear visual cues that foster sustainable financial behaviors. This approach underscores how effectively designed tools can mitigate impulsivity, thus informing the direction for a tailored solution aimed at Saudi youth.

3 Results

The first section of the survey aimed to gather insights into financial planning. The survey found that (73%) of participants do not engage in financial planning. The primary reasons cited were uncertainty about where to begin (38%) and a desire for immediate pleasure (21%). Nearly half of the respondents (49%) preferred short-term financial planning, and only (10%) engaged in long-term planning. Regarding motivation to start planning, personal goals (41%) emerged as the most common motivator for beginning to plan financially (see Fig. 1).

Fig. 1. What type of financial planning do you follow? Pie chart

The next section of the survey aimed to gather insights about instant gratification. More than half (54.6%) are inclined to make impulsive financial decisions. Furthermore, to understand the factors that lead to instant gratification, respondents identified multiple influences, including social influence (20%) and psychological stress (30%).

In terms of managing impulsive spending. Among the strategies employed by participants, self-discipline and mindfulness were relied upon by (40%), value-based influences (such as religious or moral values) by (23%), and technological tools (e.g., budgeting apps) by (12%), highlighting a multifaceted approach to managing impulsive behaviors. To examine how visual communication solutions can influence financial behavior, there was an optional short-answer question about the tools participants use. The responses were categorized into five main themes: budgeting and tracking tools, investment platforms, banking and savings apps, simple tracking methods, and no tools (see Table. 1).

The results of the survey indicate that social and psychological factors drive instant gratification, prompting impulsive spending and short-term planning. Individuals inclined toward immediate rewards underuse digital tools, struggle with consistent long-term plans, and favor short-term decisions over sustainable financial habits (see Fig. 2).

For the visual research findings on Spendee and Empower, both offer clear interfaces, goal-setting features, and real-time progress tracking. Spendee stands out with its Budget Planner and shared wallet for goal setting and spending monitoring, yet it is too simple and basic, lacking interactive elements to address impulsivity. Empower

Table 1. (see Table. 1).

Budgeting and Tracking Tools	"Excel sheet to manage finances" "Masaareef app" "WalletApp and STC Pay"
Investment Platforms	"Tamra financial app" "Investment in stocks" "Bank investment funds"
Banking and Savings Apps	"Al Rajhi Capital app" "Bank savings programs"
Simple Tracking Methods	"Using notes on my phone" "Manual tracking through paper" "Record expenses in a notebook"
No Tools	"I don't use any" "Nothing specific"

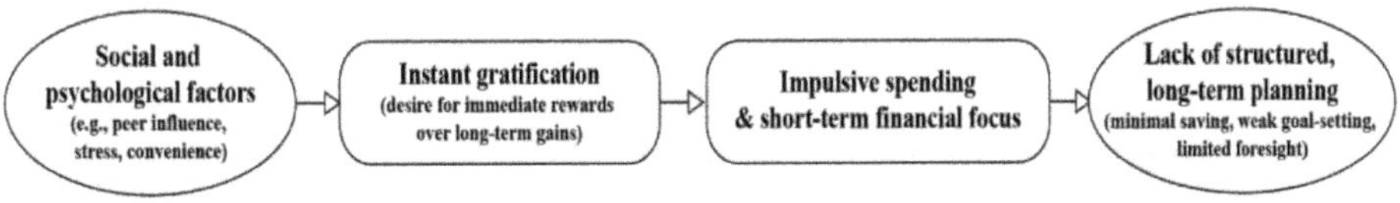

Fig. 2. correlation flowchart

offers comprehensive investment tools and a Retirement Planner, but it is too advanced for beginners, not user-friendly, and not culturally relevant. While both prioritize visual hierarchy and seamless UX, they do not fully integrate behavioral insights—such as personalized financial health indicators and social challenges—needed to mitigate impulsivity. This gap highlights the need for a new solution tailored to Saudi youth's financial decision-making.

4 Discussion

Regarding financial engagement and preferences, the findings reveal that (73%) of participants reported not engaging in financial planning. These results align with [4], which found that (45%) of young Saudis do not save and over (80%) lack structured financial plans. The primary reason cited for not having a financial plan was uncertainty about where to begin (38%). Notably, nearly half of the respondents (49%) preferred short-term financial planning, consistent with [4], which observed that young Saudis often focus on immediate needs rather than long-term strategies. In terms of motivation, personal goals (41%) were the most common impetus for beginning financial planning, while (21%) identified a desire for immediate pleasure as a barrier. Similarly, [7] highlighted that impulsive buying behaviors, driven by psychological triggers such as instant gratification, are a common obstacle to financial planning. Furthermore, (11%) of participants attributed their lack of financial planning to influences from their surroundings, aligning with [4], which identified social influences as a key factor in promoting short-term planning. Regarding the factors contributing to instant gratification in financial decision-making, the findings indicate multiple influences. Psychological stress was reported by (30%) of participants; in a similar vein, [10] identified financial stress as a key factor that increases impulsive financial behaviors. Additionally, (24%) of participants highlighted the convenience of technology as a driver of instant gratification, consistent with [8], who emphasized that the accessibility provided by FinTech fosters immediate rewards. Overall, the survey suggests that individuals strongly influenced by instant gratification tend to gravitate toward short-term decisions. Finally, regarding the strategies, (40%) of participants rely on self-discipline and mindfulness to limit impulsive spending. These findings align with [11]'s research on self-determination in managing debt. Additionally, (23%) cited value-based influences, such as religious and moral values, which align with [12]. Furthermore, (12%) indicated that technological tools serve as a strategy for resisting impulsive behaviors. This supports [8], who noted that budgeting apps offering real-time spending visibility can foster disciplined financial planning. Current digital solutions identified by the survey include budgeting apps, psychological features, and savings or investment platforms that help users resist instant gratification. These findings align with the visual research on Spendee and Empower. Spendee offers customizable budget planners, shared wallets, and some elements (e.g., progress bars and achievements), which parallels respondents' mention of the 50–30-20 rule for income allocation. Empower provides portfolio management, net worth tracking, emergency fund planning, and investment tools aimed at long-term financial goals. Both applications emphasize clear visual hierarchies, progress indicators, and color-coding for quick assessments of financial status. Spendee's user-friendly interface boosts engagement, while Empower's Retirement Planner and Investment Checkup simplify complex data for informed decision-making.

Fig.3. First section "Financial health"

4.1 Proposed Technological Solution Prototype

The proposed prototype is a mobile application conceived as an interactive personal finance companion, designed to guide users through complex financial decisions, curb

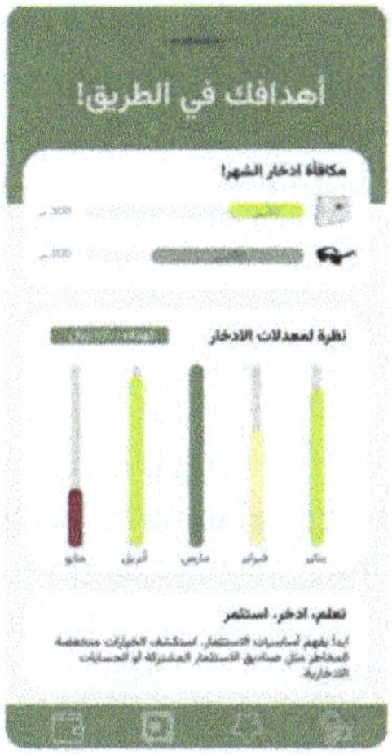

Fig. 4. Second Screen "financial planning"

Fig. 5. Third Screen "Gamification elements"

impulsive spending, and cultivate more sustainable financial behaviors. It integrates three core objectives across distinct screens in the app interface. The first section focuses on impulsivity, offering spending charts and a virtual character that visually reflects the user's financial health (see Fig. 3). The second section fosters strategic financial planning by providing goal-setting frameworks and predictive tools for more complex concepts (see Fig. 4). The third counters the human desire for instant gratification by incorporating gamified and rewarding elements such as achievement milestones, community challenges, and social sharing features (see Fig. 5). Notably, this integrated design not only encourages immediate behavioral changes but also supports evolving financial needs over time. Furthermore, this interface represents an alpha version of the application. To ensure its efficacy, a heuristic evaluation will be conducted with specialists in finance and users to assess its usability and overall effectiveness. In addition, future studies will examine the system with a wider target audience to further validate and refine its features.

5 Conclusion and Future Works

This thesis investigated the impact of instant gratification on long-term financial planning among Saudi youth and proposed a visual communication solution to promote healthier financial habits. Drawing on literature, surveys, and visual research, the study found that many Saudi youth prioritize short-term financial needs due to uncertainty about where to begin and a preference for immediate rewards, with psychological stress and social influences further driving impulsivity. Digital tools like Spendee and Empower show promise in promoting financial literacy through features such as budgeting frameworks, saving planners, goal-setting, and real-time feedback. Based on these insights, a mobile app prototype was developed to address impulsive spending and a lack of strategic planning with supportive psychological features and engaging elements. Limitations of the study include its exclusive focus on Saudi youth, which may not capture the full spectrum of financial behaviors across different age groups or cultural contexts. Future research should explore its applicability in diverse cultural and socioeconomic contexts, incorporating beta testing with the target audience to ensure scalability and address implementation challenges, ultimately helping the proposed approach effectively guide Saudi youth—and broader communities—toward more sustainable financial habits.

References

1. Sims, T., Pham, P., Jorgensen, B.L.: Long-term financial planning with future-aged avatars: Impacts on youth's retirement planning attitudes. J. Financ. Plan. **33**(6), 56–63 (2020)
2. Kingdom of Saudi Arabia (KSA): Vision 2030: A strategic roadmap for the future. Government of Saudi Arabia (2016)
3. Chandak, A.: The role of digital tools in financial literacy. Financ. Technol. Today. **10**(3), 66–82 (2022)
4. Alshebami, A.S., Aldhyani, T.H.H.: The interplay of social influence, financial literacy, and saving behaviour among Saudi youth and the moderating effect of self-control. Sustainability. **14**(14), 8780 (2022)
5. Mueangpud, A., Phakdee, W.: Enhancing financial literacy through digital motion graphics. J. Technol. Des. **22**(3), 40–52 (2019)
6. Simanungkalit, E., Nainggolan, Y.A., Hasibuan, R.H., Nababan, D.J.: Financial stress and its impact on financial behavior among young adults during the COVID-19 pandemic. J. Econ. Bus. **5**(2), 78–89 (2022)
7. Alqutub, M.: Impulse buying behavior and financial planning among Saudi youth. J. Consum. Aff. **57**(2), 231–247 (2023)
8. Ali, A., Alyami, M.: Financial technology (FinTech) and its impact on financial decision-making among Saudi youth. J. Financ. Technol. Innov. **2**(3), 25–39 (2023)
9. Htet, K., Wongsunopparat, W.: The role of financial literacy and social influences in personal financial management: Evidence from Thailand. Int. J. Econ. Fin. Issues. **11**(4), 17–24 (2021)

10. Al Balawi, K., Al Ghamdi, A., Al Harbi, N.: Financial difficulties and instant gratification among young adults. J. Behav. Financ. **20**(3), 177–186 (2019)
11. Park, N.: Self-determination in financial planning: A longitudinal study of youth financial resilience. J. Youth Adolesc. **50**(6), 1234–1245 (2021)
12. Ashraf, M.: The influence of religious and moral values on youth financial behavior. J. Islam. Finance. **34**(2), 85–94 (2022)
13. Kojic, M., Knudsen, K., Grytnes, J.: Exploring financial behavior through VR: The impact of virtual reality on youth financial literacy. J. Exp. Psychol. Appl. **26**(3), 309–321 (2020)
14. French, M.T., McCarty, D., Hastings, D.J.: Money Matters: Financial planning applications and their impact on financial literacy and behavior. J. Financ. Couns. Plan. **30**(1), 50–63 (2019)
15. Bitrián, P., Buil, I., Catalán, S.: Gamification in personal finance apps: A framework for improving financial habits. Comput. Hum. Behav. **120**, 106767 (2021)
16. Spendee. (n.d.). Money manager & budget planner. https://www.spendee.com/
17. Empower. (n.d.). Empower personal dashboard. https://www.empower.com/

The Use of Zeta Technology to Modify Vermiculite Clay for the Adsorptive Removal of Anionic Detergent Residues from Wastewater

Saad Mohamed Elsaid[1,2] and Ali ben Saleh Albarak[3](✉)

[1] ECRi, Environment and Climate change Research Institute, NWRC, Cairo, Egypt
saad4034m@gmail.com
[2] Onaizah Private Colleges, Onaizah, Saudi Arabia
[3] Medical Applied College, Qassim university, Buraidah, Saudi Arabia
alialbarrak@hotmail.com

Abstract. Wastewater from anionic detergents pollutes the environment, emphasizing the importance of developing low-cost, high-efficiency adsorbents for their removal. Water contaminated with anionic detergents is made by dissolving alkyl benzene sulfonate, one of the most often used detergents, in an appropriate volume of distilled water. Vermiculite is coated with positive charges after being activated with a saturated potassium hydroxide solution in a cationic alkaline media. Zeta potential technique has been studied in order to maximise detergent removal efficiency by utilising the features of vermiculite. The zeta potential (zp) of vermiculite clay varies with its type, mineral makeup, and, most critically, treatment procedure. Zeta potential tests have revealed that vermiculite has a positive charge when treated with 2 M calcium hydroxide at all pH levels, making it appropriate for eliminating negatively charged detergent molecules. Clay was further enhanced by treating alkali vermiculite with a stabiliser containing sodium dodecyl sulphate, a surfactant, which increased pore volume and surface area. Alkali vermiculite treated with 20% stabiliser has a greater detergent removal efficiency (98.6%) than alkali vermiculite (65%). Scanning electron microscopy (SEM) measurements were used to show how the stabiliser affected the pore volume and surface area of the clay. 20% sodium dodecyl sulphate was discovered to be the most effective solution for creating clay with increased surface area, pore volume, and efficiency. Batch tests were conducted with sorbent contact times ranging from 1 to 60 min, initial detergent concentrations ranging from 2 to 50 mg/L, clay dosages ranging from 0.01 g to 1 g, and pH ranging from 2 to 11. Sulfonate adsorption increased with increasing contact time, clay weight, and sorbent concentration, but decreased with increasing solution pH. Batch studies were conducted with a contact time of 20 min, vermiculite weight of 0.1 g/L, and pH 4. The adsorption effectiveness (qe) of detergent molecules was 281 mg/g. The Freundlich model of dynamic multilayer chemistry provided the most effective explanation for the elimination trend.

Keywords: Adsorption · Detergent -Alkali Vermiculite · Batch Adsorption

D. Berkouk et al. (Eds.): CCTS 2024, PTT, pp. 256–263, 2026.
https://10.1007/978-981-95-7607-4_29

1 Introduction

Water pollution from industrial detergents has increased significantly over the past few decades. In 2014, the Egyptian market consumed 306,000 tons of detergents and soaps, compared to 1,471,000 tons in 2015 [1]. The value of the cleaning and detergent products market in Egypt was valued at USD 8.2 billion in 2020 and is expected to grow to USD 11.9 billion by 2026, at a CAGR of 5.1% between 2020 and 2023 [1]. Vermiculite is considered one of the most important natural and environmentally friendly absorbent materials [2]. Vermiculite has recently been used to filter contaminated water due to its outstanding electrical properties. Water pollutants are divided into two categories based on their electrical potential: negatively charged organic compounds (such as pesticides and detergents) and positively charged ones (such as heavy metals). To remove pollutants via adsorption, vermiculite clay must have an opposite charge to the pollutant to enhance affinity, resulting in rapid removal and effective water purification. Researchers have experimented with zeta potential technology to achieve this goal. Alkaline activation can increase the adsorption sites of natural clay minerals, enhancing their efficiency [3]. A detergent can be considered to have a more general structure, such as R-SO4-, where R denotes an extended alkyl group. Detergents, such as soap, are negatively charged. Detergents have higher solubility in hard water than soap due to the distinct behavior of their chemical components [4]. Adsorption is a popular approach because it is simple, effective, and inexpensive [5, 6]. Clay minerals have received significant attention as promising materials for heavy metal removal from wastewater due to their strong adsorption capacity and natural abundance [7]. In this work, vermiculite was first treated with calcium hydroxides to enhance its positive charge, making it more suitable for the adsorption of negatively charged detergents. Zeta potential studies were performed on the adsorbent to confirm the clay charge. The effectiveness of the alkaline clay was also improved by increasing its surface area and pore volume after treatment with a surfactant-stabilizing solution. These results were confirmed using the SEM approach.

2 Methodology

(1) Vermiculite clay was made and collected from the Al-Khubaibiyah location in Buraidah, Al-Qassim, Saudi Arabia. The clay samples were cleaned, activated, and ready for use in studies to remove industrial detergent residues from effluent water.
(2) Vermiculite was treated with calcium hydroxide at concentrations ranging from 1.0 to 2.0 M to produce a positive zeta potential sufficient to remove the negative detergent residues.
(3) Sodium dodecyl sulfate stabilizing agents were added at 5 to 15% concentrations to increase the porosity and surface area sufficient to remove detergent contaminants from the contaminated water [8, 9]. The surface morphologies of the raw vermiculite and the alkaline properties of the adsorbents were examined using a scanning electron microscope (Nova SEM 230, USA). The hydrostatic charges of the four clay forms were also estimated by measuring their zeta potential using a Zeta-Esser (SZ 100 HORIBA, USA).
(4) Brunauer-Eimmelt-Teller (BET) measurements were used to calculate pore volume, surface area, and pore volume.

(5) Batch experiments were performed according to [10].
(6) Batch technology was used to study the adsorption of negative detergents onto positive vermiculite at varying doses (from 2 to 20 ppm). Isothermal equilibrium modeling was used to collect quantitative and qualitative data on detergent adsorption onto vermiculite, and the results were modeled using the well-known Freundlich equations.

3 Results

3.1 Zeta Potential

Raw clays often carry a negative surface charge due to their inherent electronegativity [11]. As pH increases, the electronegativity of the adsorbent increases, due to the deprotonation of functional groups [12]. The enhanced electronegativity of clay not only repels positively charged impurities but also prevents the adsorption of negatively charged detergents [13]. As demonstrated in this study, water is frequently contaminated with industrial detergents, the most common of which is alkyl benzene sulfonate (ABS), an anionic detergent. Applying the zeta potential to both raw and alkali-activated vermiculite clays provides a comprehensive view of the adsorption properties of detergents used in activated vermiculite clays. Figure 1 shows the zeta potential values of raw vermiculite, ranging from +40 to −50 mV for pH values from 2 to 13. Zeta potential studies of raw vermiculite yielded negative values ranging from 0 to −50 at pH values from 5 to 12, indicating that negatively charged detergents cannot be adsorbed at these pH levels. Positive zeta potential measurements indicated that raw vermiculite could only remove negative detergent ions within a limited pH range from 2 to 5, where the raw clay is positively charged. Figure 2 shows the change in vermiculite charge after alkaline activation with calcium hydroxide solutions of varying concentrations, which gave the clay a positive charge. As the concentration increased from 1 M to 2 M, the clay charge increased until it acquired a positive charge across all pH levels when an alkaline solution of 1.5 M or higher was used. When raw vermiculite was treated with a calcium cation solution, the clay exhibited positive zeta potential values across the pH range from 2 to 11. This observed behavior results from the dissociation of calcium ions on the surfaces and edges of clay particles in water, which is affected by pH and increases the tendency of oxygen ions (H+) to migrate from the clay into the solution. This study found that measuring the zeta potential distribution is a useful technique for investigating adsorption events in complex colloidal systems. It also showed that the zeta potential of alkali-activated vermiculite increases as the pH decreases from 13 to 2. It was also discovered that the zeta potential of vermiculite is highly sensitive to the valence of calcium ions [14].

3.2 Adsorbent Characterization

Figure 3 shows microscopic images of raw and alkali-activated vermiculite at different concentrations of stabilizer sodium dodecyl sulfate. Scanning electron microscopy (SEM) revealed an increase in the number of pores distributed on the clay surface. As

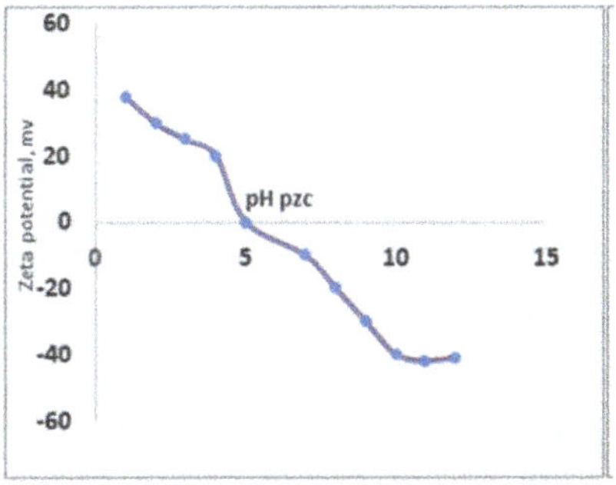

Fig. 1. zeta potential of raw vermiculite clay

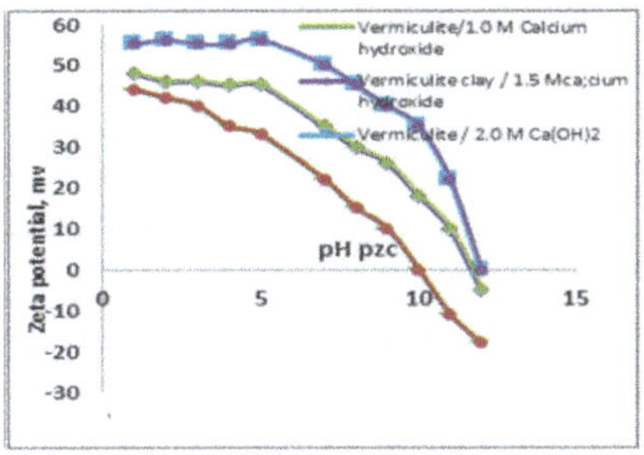

Fig. 2. zeta potential of alkali activated vermiculite clay at different alkali ratio

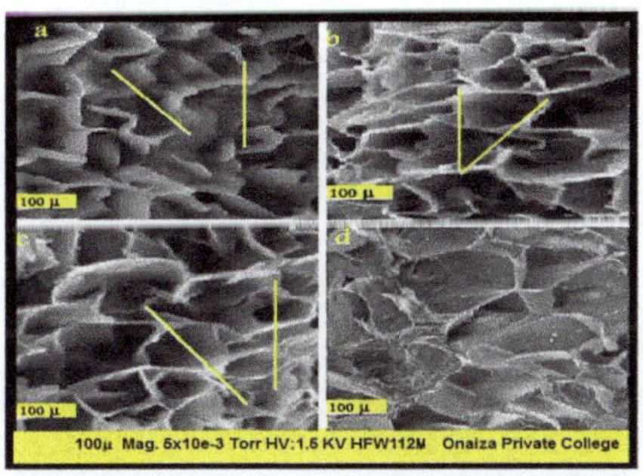

Fig. 3. a, Raw vermiculite clay, b, 5%, c, 10% and d, 15% stabilizer sodium dodecyl sulfate

the stabilizer concentration increased from 5 to 20%, the surface area of the clay also increased with increasing pore size [15].

BET studies were conducted to measure the surface area and pore volume of raw vermiculite and activated clay at different alkali and stabilizer concentrations. As shown in Table 1, the clay activated with 2 M Ca(OH)2 and 20% stabilizers had the best surface and pore properties (Table 2).

Table 1. (BET) characteristics of different shapes of vermiculite clay

Clay mineral	Pore size nm	Pore volume cm3/g	Surface area m2/g
Raw vermiculite clay	4.13	0.62	85.35
1.0 M (2CaOH)2vermiculite clay	4.32	0.65	88.1
1.5 M (2CaOH)2vermiculite clay	4.61	0.66	89.32
2.0 M (2CaOH)2vermiculite clay	4.82	0.68	89.8
2.0 M (2CaOH)2vermiculite clay/ 5%syabilizers	6.25	2.31	165.62
2.0 M (2CaOH)2vermiculite clay/ 5%syabilizers	7.23	3.98	183.52
2.0 M (2CaOH)2vermiculite clay/ 5%syabilizers	7.35	4.22	201.32

Table 2. BET of raw vermiculite and activated clay

Pore Volume /cm3/g	Pore Size /nm	Surface Area /m2/g
105.63	4.31	0.52
182.62	6.05	0.65
223.52	6.35	0.76
225.36	6.32	0.74

3.3 Batch Techniques

As shown in Fig. 4a, b, and c, the objectives for batch application focused on the ideal clay weight, the duration of the experiment, and the solution pH that yielded the highest adsorption rate. The ideal pH was 4, which yielded 99.2% adsorption, a contact time of 30 min, and 1.2 g of activated clay per liter of solution. To analyze the balanced Freundlich equation, shown in Fig. 5, the effect of detergent concentration on adsorption rate was also investigated. Freundlich coefficients and correlation coefficients ($R \geq 99.5$) could be calculated from the graph, demonstrating that the balanced Freundlich equation was suitable for the experiment.

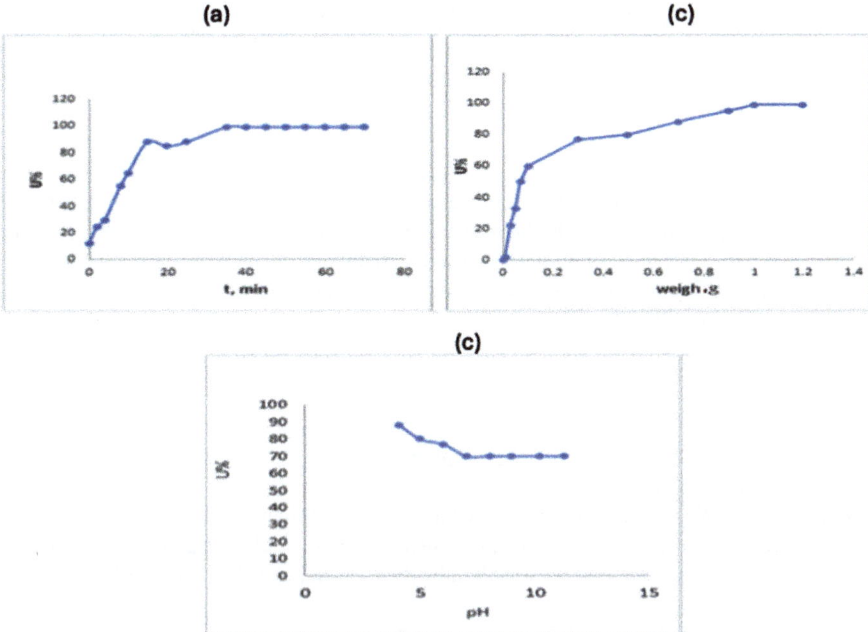

Fig. 4. (A) U% vs. t for the adsorption of detergents of detergents onto alkali activated vermiculite, (b) U% vs. clay weight for the adsorption of anionic onto alkali activated clay, (4) U% vs. pH for the adsorption of detergents of detergents onto alkali activated vermiculite clay

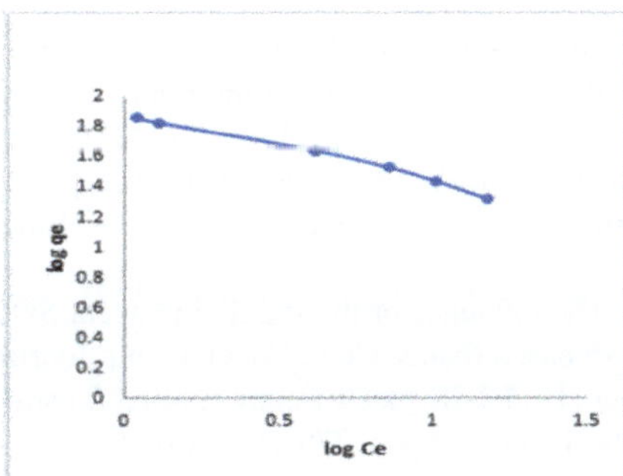

Fig. 5. Freundlich isotherm for the adsorption of anionic onto alkali vermiculite clay

4 Discussion

Using electronegativity through zeta potential and ionic radius determinations, BET and SEM techniques were used to study the adsorption efficiency of an anionic detergent, the negative dodecyl sulfate, using a base-activated vermiculite clay, which was stabilized to give it a positive charge. Furthermore, the batch adsorption process ensured satisfactory results. The results showed that when alkali-activated vermiculite clay was used with 2 M Ca OH 2 and 20% stabilizers, the maximum adsorption rate was achieved after 30 min, at pH 3.5, and using 1.2 g of adsorbent.

Acknowledgement. We, the aforementioned researchers, would like to express our heartfelt gratitude and appreciation to the officials of the Second Technical College in Buraidah, Qassim, Saudi Arabia, for providing us with chemicals and equipment and allowing us to carry out all experiments in their college laboratories. We also like to thank and appreciate the Unaizah Private Colleges in Qassim, Saudi Arabia, for their financial and logistical help.

References

1. Egyptian Research Market Forecast: (Anionic Detergents, Product Code: ETC017712) Product Type: Report Publisher, pp. 1–70 (2023)
2. Stawiński, W., Węgrzyn, A., Freitas, O., Chmielarz, L., Mordarski, G., Figueiredo, S.: Simultaneous removal of dyes and metal cations using an acid, acid-base and base modified vermiculite as a sustainable and recyclable adsorbent. J. Sci. Total Environ. **576**(1), 398–408 (2016)
3. Chen, X., Tong, D., Fang, Z., Gao, Z., Yu, W.: Acid leaching vermiculite: a multi-functional solid catalyst with a strongly electrostatic field and brönsted acid for depolymerization of cellulose in water. J. Molecules. **27**(10), 3149–3172 (2022)
4. Ahmer, A., Muhammad, R., Shamsuddin, L.: A review on recent developments in the adsorption of surfactants from, wastewater. J. Environ. Manag. **254**(15), 109–127 (2020)
5. Ersan, G., Dos Santos, A.J., Lanza, M.R., Perreault, F., Garcia, S.: Enhancing the selective ciprofloxacin adsorption in urine matrices through the metal-doping of carbon sorbents. J. Environ. Manag. **348**(1), 119–138 (2023)
6. Assad, H., Lone, I., Kumar, A., Kumar, A.: Unveiling the contemporary progress of graphene-based nanomaterials with a particular focus on the removal of contaminants from water: a comprehensive review. Front. Chem. **12**(1), 134–157 (2024)
7. Khurana, P., Pulicharla, R., Brar, S.: Imipenem-metal complexes: Computational analysis and toxicity studies with wastewater model microorganisms. Environ. Res. **239**(1), 117–135 (2023)
8. Korat, L., Ducman, V.: The influence of the stabilizing agent SDS on porosity development in alkali-activated fly-ash based foams. Cem. Concr. Comp Journal. **80**(1), 168–174 (2017)
9. Djomgoue, P., Njopwouo, D.: FT-IR spectroscopy applied for surface clays characterization. J. Surf. Eng. Mater. Adv. Technol. **3**(04), 275–293 (2013)
10. Pawar, R., Bajaj, H., Lee, S.: Activated bentonite as a low-cost adsorbent for the removal of Cu (II) and Pb (II) from aqueous solutions: Batch and column studies. J. Ind. Eng. Chem. **34**(1), 213–223 (2016)
11. Chen, T., Yuan, Y., Zhao, Y., Rao, F., Song, S.: Effect of layer charges on exfoliation of montmorillonite in aqueous solutions. Colloids Surf. A Physicochem. Eng. Asp. **548**(1), 92–97 (2018)
12. Zhang, J., Wang, S., Liu, W.: Anisotropic wetting surfaces with reversed pH-responsive wetting directions. Adv. Funct. Mater. **28**(1), 180–201 (2018)
13. Paoli, L., Vannini, A., Monaci, F., Loppi, S.: Competition between heavy metal ions for binding sites in lichens: Implications for biomonitoring studies. Chemosphere. **199**(2), 655–660 (2018)

14. Yukselen, Abidin, K.: The zeta potential of a mixed mineral clay in the presence of cations yeliz. J. Eng. Earth Sci. **1**(1), 14–21 (2016)
15. Yorgun, S., Yıldız, D.: Preparation and characterization of activated carbons from Paulownia wood by chemical activation with H3PO4. J. Taiwan Inst. Chem. Eng. **53**(1), 122–131 (2015)

Assessing the Combined Effects of Climate Change and Urbanization on Water Resources in Arid Environments: A Comprehensive Machine Learning and Data Analysis Approach—Case Study Saudi Arabia

Heba Fathi(✉)

College of Arts and Humanities, Jazan University, Jazan 45142, Saudi Arabia
hamansour@jazanu.edu.sa

Abstract. The two most significant worldwide challenges affecting ecosystems, particularly in dry regions with limited water supplies, are urbanization and climate change. Due to rapid urbanization and socioeconomic growth, several Middle Eastern nations rely heavily on desalination to meet their water needs. This study highlights the urgent importance of sustainable water management techniques by examining the effects of urbanization and climate change on Saudi Arabia's water resources. The study examines the relationship between population growth and a number of water pollution indicators from 2017 to 2023 using Pearson's correlation coefficient. It finds that dissolved oxygen (DO), biochemical oxygen demand (BOD_5), and total suspended solids (TSS) have significant positive correlations. With a high prediction accuracy ($R = 0.98$), a neural network model demonstrated its ability to impact decisions on water quality management. The findings demonstrate the importance of ongoing monitoring and adaptable management in addressing water quality issues, particularly in view of expanding population demands. Future research should focus on expanding data sources, improving modeling techniques, and promoting public involvement to ensure sustainable water resource management in the region.

Keywords: sustainable water management · environmental monitoring · water quality indices · Saudi Arabia · ANN

1 Introduction

The two most prominent worldwide events of the twenty-first century that are influencing the ecosystem's natural balance are urbanization and climate change. Most vulnerable to such shifts impacting the delicate balance in the essentially restricted water supplies are arid places [1]. In dry and semi-arid regions, particularly in many Middle Eastern countries, water is a scarce and restricted resource. Many of these countries depend on desalination to satisfy their water demands since they lack sufficient freshwater resources. In addition to their fast socioeconomic growth and urbanization, which

D. Berkouk et al. (Eds.): CCTS 2024, PTT, pp. 264–273, 2026.
https://10.1007/978-981-95-7607-4_30

has a significant impact on their water budget [2]. Understanding how urbanization and climate change interact to alter water resources is crucial for developing sustainable plans for managing them and ensuring socioeconomic growth and social stability in the region. The susceptibility of hydrologic systems to climatic fluctuations and the historical shifts in urbanizing watersheds are highlighted in Batra [3], which offers a fundamental knowledge of how climate change affects these systems. The analysis by Keller et al. [4] emphasized how urgently the global water problem must be addressed as a security concern. Otherwise, Saikia et al. and Daloğlu Çetinkaya et al. [5, 6] emphasized the tensions that exist between ecological sustainability and urban development, emphasizing the value of adaptive management techniques in attaining water security in the face of obstacles brought on by climatic unpredictability and urban expansion. Sherif et al. [7] focused on the need for research that is in line with the Sustainable Development Goals of the UN while examining the wider effects of climate change on water resources in the Gulf Cooperation Council nations. By examining new developments in hydrology and using topic modeling to pinpoint important research topics, Korkut et al. [8] highlighted the need for continued research into techniques for managing water resources by illustrating the dynamic nature of hydrological research in response to urbanization and climate change. Machine learning (ML) has a highly effective ability to classify the degrees of water contamination [9, 10]. A landscape full of opportunities for further research and real-world applications was revealed by numerous studies that demonstrated how ML frameworks could improve environmental monitoring and management practices. These studies also highlighted the use of ML in analyzing microbial community dynamics, which is crucial for guaranteeing safe drinking water and informing decision-making processes in water treatment and quality management [11, 12]. While Castrillo & García [13] advanced an approach by concentrating on the use of machine learning techniques to estimate nutrient concentrations from water quality surrogates. In order to better understand the intricate relationships for a case study in Saudi Arabia, the combined impact of urbanization and climate change on water resources could be investigated utilizing a machine learning and data analysis technique. The literature on Saudi Arabia's water resources shows a complicated interaction of issues and solutions meant to alleviate the lack of this essential resource in dry conditions. Aqel et al.'s study [14] advocates for proactive action and ongoing monitoring to protect water quality, with a particular focus on the Wadi Hanifah region in Saudi Arabia. Their suggestions for more stringent pollution control methods and more public awareness show that authorities are increasingly realizing the importance of working together to solve problems with water quality that are made worse by climate change.

Overall, the literature mentioned above demonstrates a deep understanding of the complex interactions between climate change and water resources, highlighting the urgent need for innovative, integrated approaches to water management and sustainable practices that can adjust to the changing environmental landscape. To evaluate past, present, and future changes in water resources in Saudi Arabia's arid environment, this study offers a thorough approach to the development and application of combined machine learning and data analysis methodologies. It focuses on major contamination indicators that indicate pollution caused by domestic, industrial, and agricultural disposal in the study area by measuring the collected water samples (from the year 2017 to

2023) for seven parameters, including pH, biological oxygen demand (BOD_5), chemical oxygen demand (COD), total suspended solids (TSS), dissolved oxygen (DO), nitrogen (N), and phosphorus (P), which were monitored by 16 monitoring stations distributed throughout the Kingdom of Saudi Arabia. The main objective of this study is to create methods for evaluating water resources to manage them sustainably and make decisions in light of the region's increasing demand brought on by population growth and subsequent urbanization. Based on an examination of the application of artificial neural networks (ANNs) to water contamination modeling, the technique offers a simple algorithm for determining the optimal ANN configuration that best reflects the complex behavior of water contamination. Decision-makers need this information in order to include the region in plans for sustainable development.

2 Study Area

Saudi Arabia is the largest country in the Middle East (with an area of more than 2,250,000 km^2), covering most of the Arabian Peninsula. It lies between latitudes 16° and 32°N and longitudes 34° and 56°E, as shown in Fig. 1. The Red Sea and the Arabian Gulf coastlines extend 1,760 km and 650 km, respectively, and the land is generally situated at an elevation of about 600 m above sea level [15, 16]. In the west, the country is characterized by mountains and hills, which steadily decline to lower lands in the eastern part where the Arabian Gulf is located. Mountainous areas have significant elevations, with the highest one being Jabal Sawda (3,015 m) [17].

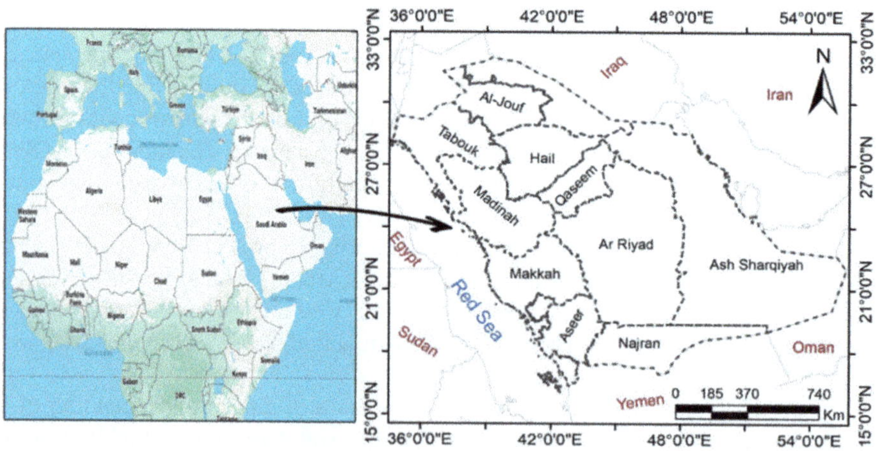

Fig. 1. Map of the study area

The climate is primarily downstream from the large landmasses to the north and is extreme in its aridity, with virtually no rainfall during the summer months. Virtually the entire period from May through September is rainless. According to the National Center for Meteorology, Ministry of Environment, Water, and Agriculture, the rainfall during the winter months in 2023 is erratic and varies from month to month, within the range of

15 to 32 cm [18]. The country gets most of its fresh water from desalination plants and underground water wells. Due to the growing population and increase in urban areas, the natural water reservoir has been severely affected [19].

3 Data Used and Methods

Based on the database of the General Authority for Statistics in Saudi Arabia, the population is increasing exponentially by 2.5% every year. The increase in construction, agricultural, and industrial areas has a direct impact on the available water resources [20]. In this regard, it is very important to analyze the impacts of spatial and temporal changes in the climatic conditions on the available water resources. Water shortage is worsened by rising demand and declining supplies, especially in dry countries. According to the Regional Initiative for the Assessment of Climate Change Impacts on Water Resources and Socio-Economic Vulnerability in the Arab Region (RICCAR) and the Intergovernmental Panel on Climate Change (IPCC), certain areas of Saudi Arabia will face elevated climate risks in the ensuing decades, particularly the following: In terms of temperature, Saudi Arabia's temperatures rose by 0.5 °C between 1975 and 2000, which was more than the global average of 0.3 °C. The annual temperature rate will rise to around 0.6 °C per ten years by 2030 [21].

The data used for this study are taken from the General Authority for Statistics in Saudi Arabia, which collects national and governmental data from several official institutions. As for environmental data, the main source is the Ministry of Environment, Water and Agriculture, Saudi Water Authority. In order to study the extent to which water sources are affected by pollutants resulting from population growth and the resulting increase in urbanization, the focus was on several criteria for water pollution indicators, which reflect the degree of organic contamination, often from sewage, industrial waste, or agricultural runoff. To guarantee compliance with water quality criteria for human consumption and ecological balance, it is essential to identify the source and types of pollution (organic, chemical, and nutrient-related) using these parameters. An overall indicator of water pollution that renders the environment inhospitable for fish and other aquatic plants and creatures is the rise in BOD_5, COD, TSS, and DO. Similarly, an increase in pH levels can damage organisms and change the toxicity of other substances in water. On the other hand, too much nitrogen and phosphorus (from fertilizers, detergents, and untreated wastewater) cause eutrophication and algal blooms, which lower oxygen levels and harm aquatic life [22, 23]. The above-mentioned pollution indicator data were monitored from 2017 to 2023, except for 2021 due to the Corona pandemic and the difficulty of taking samples and conducting the required analyses as shown in Fig. 2. The data for the seven water pollution indicators mentioned above were monitored from 2017 to 2023, except for 2021, due to the Corona pandemic and the difficulty of taking samples and conducting the required analyses. The monitoring was carried out through 16 monitoring stations distributed throughout Saudi Arabia, as follows: Ras Al Khair Plant, Jubail station, Al Khobar station, Haql station, Duba station, Al Wajh station, Amlaj station, Jeddah station, Fursan Station, Yunbu station, Shuqaiq Station, Al Qunfudhah Station, Al lith station, Al-Shaiba Station, Al Aziziyah station, and Rabigh station.

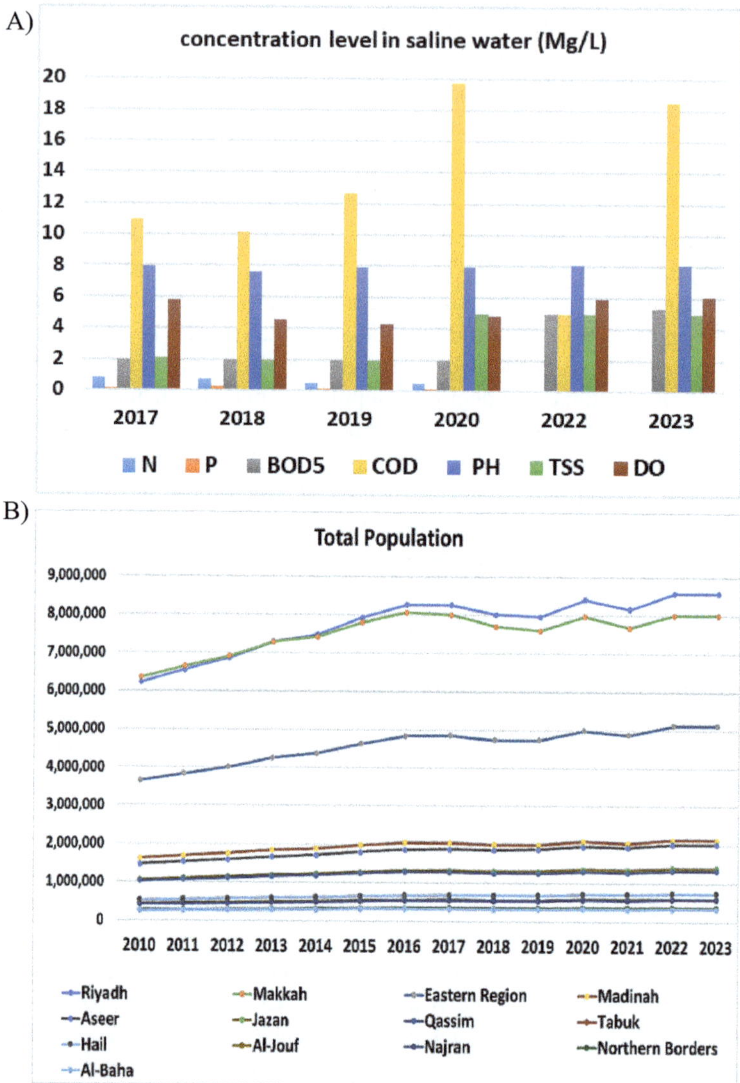

Fig. 2. (A), (B) The increase in population and the pollutants resulting from this increase in the last decade; Source: the General Authority for Statistics [20]

4 Results and Discussion

Statistical analysis was carried out by calculating Pearson's correlation coefficient (r) value for the contamination indicators (parameters) mentioned above using XLSTAT. Table 1 shows the Karl-Pearson correlation matrix for the parameters and the population from 2017 to 2023. The matrix declares positive correlation for DO ($r = 0.637$, $p < 0.001$), BOD_5 ($r = 0.797$, $p < 0.001$), and TSS ($r = 0.911$, $p < 0.001$). This study's results indicate that the relationship among the three parameters examined is strong and very reliable compared to other studies where the p-value was below 0.01 and the correlation coefficients were lower [13, 24]. It is clear that TSS and BOD_5 are significantly increasing with population growth, as an increased population typically generates more wastewater. Since more organic pollution usually means less dissolved oxygen, the positive relationship between population and dissolved oxygen could mean that the water has enough natural aeration or that the wastewater is being treated well. Furthermore, we can observe that phosphorus and pH have a strong negative correlation, meaning that more acidic conditions follow a decrease in pH that corresponds with an increase in phosphorus levels. These results are consistent with Aqel et al.'s [14] results in the Riyadh region.

Table 1. Correlation matrix for the parameters and the population from 2017 to 2023

Parameters	N	P	BOD_5	COD	PH	TSS	DO	Population
N	**1**							
P	0.362	**1**						
BOD_5	−0.802	−0.258	**1**					
COD	−0.032	−0.165	−0.051	**1**				
PH	−0.217	−0.777	0.170	0.143	**1**			
TSS	−0.684	−0.277	0.683	0.260	0.173	**1**		
DO	−0.430	−0.313	0.581	−0.036	0.315	0.410	**1**	
Population	−0.691	−0.279	**0.797**	0.115	0.199	**0.911**	**0.637**	**1**

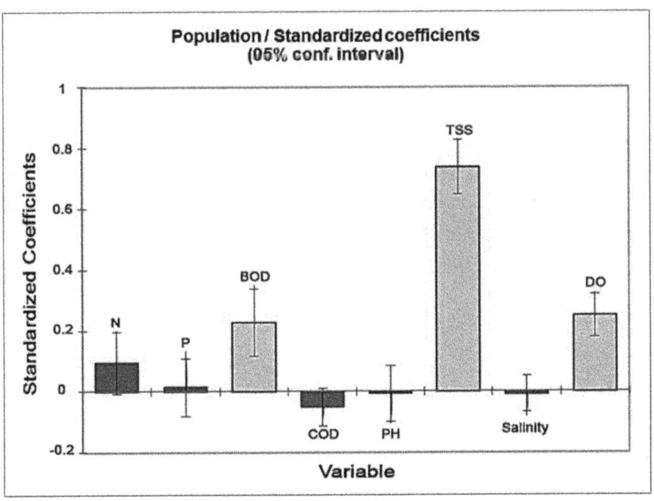

Fig. 3. Boxplot of the affected parameters and the population

Figure 3 shows the directional relationship between the increase in some contamination concentrations and the population increase, which is evident and calls for attention since the TSS is the most significant measure, and most likely because of its connection to habitat deterioration, water turbidity, and pollution transfer. The significance of oxygen dynamics and organic pollution is further supported by DO and BOD_5. Aquatic photosynthesis can be hampered, light penetration can be decreased, and aquatic ecosystems might be disturbed by high TSS. Given its powerful effect, monitoring and sediment control ought to be top priorities when it comes to enhancing water quality.

The neural network model's capacity to manage deep relationships, forecast continuous events with accuracy, and adjust to different data scenarios makes it significant. Its durability, scalability, and adaptability make it an effective tool in today's data-driven environment, facilitating improved insights and decision-making across various applications. Data preparation, model construction, training, and assessment are the phases involved in developing a neural network model for continuous response prediction. Here is a thorough analysis of the main positively connected parameters, DO, BOD_5, and TSS, utilizing the data and illustrations used in this study: 1) Training Set: This is where the model is trained. 2) Validation Set: Prevents overfitting by adjusting hyperparameters, and the observed values from 2017 to 2022 were utilized. 3) Test Set: This was used to assess how well the model performed on data that had not yet been used; in this case, 2023 data. Because Levenberg-Marquardt is effective at converging solutions, it was employed for training in regression problems. However, the model's performance was assessed using the Mean Squared Error (MSE), which calculates the average of the squares of the errors between the actual and projected values. Correlation coefficients, or R-values, show how closely the model's predictions match the actual target values. After several examinations, the optimal model's outcomes were (training R: 0.97888, validation R: 0.99343, test R: 0.98867, and overall R: 0.9816) as shown in Fig. 4.

The higher R-values in Fig. 5 indicate a stronger correlation between the expected and actual values. The data points in the neural network prediction model show different clusters instead of an equal distribution. This data shows that in the investigated water system, the water quality parameters (DO, BOD_5, and TSS) typically fall within specific ranges or combinations. With an R-value of 0.96441, the model covers most of the variation in the actual data. For environmental factors, the outcome is considered outstanding performance, as it often shows complicated multidimensional interactions. We will be able to create and test a neural network model that can generate continuous response predictions, assist individuals in making better judgments depending on data, simplify monitoring, solve environmental concerns, and promote sustainability.

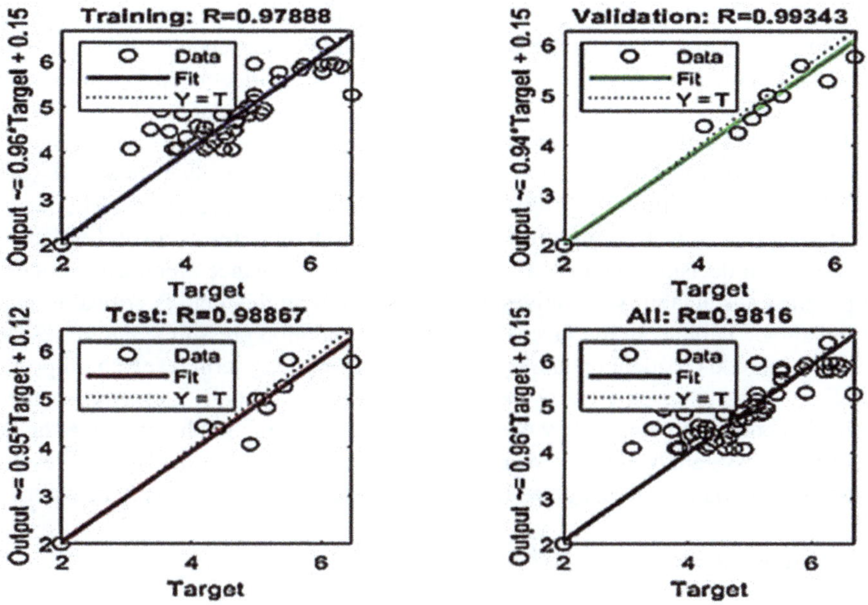

Fig. 4. The best ANN network regression

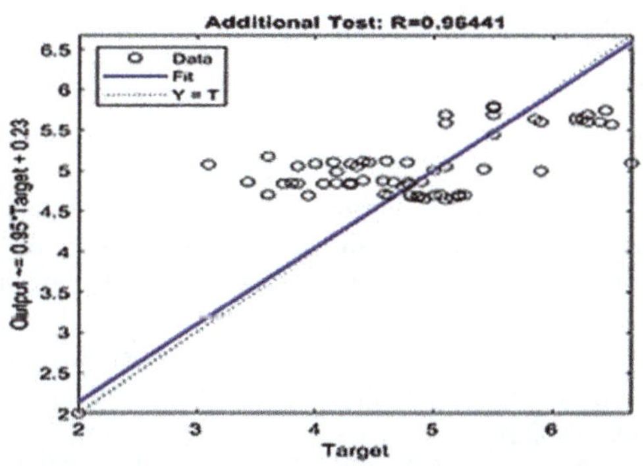

Fig. 5. Measured and the neural network prediction of DO, BOD_5, and TSS concentration in 2023

5 Conclusion and Future Research Directions

Saudi Arabia faces major water resource issues due to its dry environment and a 2.5% annual population increase, which adds further strain on limited water supplies. As climate change accelerates temperature rises and affects precipitation patterns, proactive water management methods are critical. Addressing pollution and increasing conservation measures will be critical to safeguarding Saudi Arabia's long-term water supply in the face of continuous environmental difficulties. This study used Pearson's correlation coefficient to examine the link between pollution indicators and population growth rates from 2017 to 2023. Significant positive correlations were discovered for dissolved oxygen (DO), biochemical oxygen demand (BOD_5), and total suspended solids (TSS) with increasing population levels ($r > 0.50$), emphasizing the critical need to address water quality concerns as population density increases. TSS, in particular, has been associated with habitat destruction and decreased aquatic health, highlighting the need for monitoring sediment levels. The neural network model performed effectively, with high R-values ($R = 0.98$), showing that it can reliably predict environmental reactions. The integration of analytical statistics and machine learning modeling creates a solid foundation for data-driven decision-making, which aids attempts to enhance monitoring strategies and progress sustainability initiatives. Continued study in this field is critical to maintaining aquatic habitats in the face of increasing population pressures. Future research directions include broadening data sources, undertaking long-term research, improving modeling methodologies, creating real-time monitoring systems, including socioeconomic elements, analyzing policy consequences, encouraging public participation, and measuring ecosystem health. These activities are critical for sustainable water resource management in Saudi Arabia, given the country's persistent environmental concerns.

References

1. Urban, M.C., Alberti, M., De Meester, L., Zhou, Y., Verrelli, B.C., Szulkin, M.: Interactions between climate change and urbanization will shape the future of biodiversity. Nat. Clim. Chang. **14**(5), 436–447 (2024)
2. El-Rawy, M., Fathi, H., Wahba, M., Negm, A.: An overview of water resource challenges and managed aquifer recharge (MAR) in the MENA region. In: El-Rawy, M., Negm, A. (eds.) Managed Aquifer Recharge in MENA Countries: Developments, Applications, Challenges, Strategies, and Sustainability, pp. 1–27. Springer International Publishing, Cham (2024)
3. Batra, N.: Seasonal Hydrologic Dynamics Under Changing Climate, Land Use-Land Cover and Human Influence. University of Illinois at Urbana-Champaign (2010)
4. Keller, A.A., Garner, K.L., Rao, N., Knipping, E., Thomas, J.: Downscaling approaches of climate change projections for watershed modeling: Review of theoretical and practical considerations. PLoS Water. **1**(9), e0000046 (2022)
5. Saikia, P., Beane, G., Garriga, R.G., Avello, P., Ellis, L., Fisher, S., et al.: City Water Resilience Framework; A governance based planning tool to enhance urban water resilience. Sustain. Cities Soc. **77**, 103497 (2022)
6. Daloğlu Çetinkaya, I., Yazar, M., Kılınç, S., Güven, B.: Urban climate resilience and water insecurity; future scenarios of water supply and demand in Istanbul. Urban Water J. **20**(10), 1336–1347 (2023)
7. Sherif, M., Abrar, M., Baig, F., Kabeer, S.: Gulf cooperation council countries' water and climate research to strengthen UN's SDGs 6 and 13. Heliyon. **9**(3), e14584 (2023)
8. Korkut SO, Kaymak OO, Onan A, Ulker E, Yalcin F. :A Roadmap of Emerging Trends Discovery in Hydrology: A Topic Modeling Approach, arXiv preprint arXiv:231015943(2023).
9. Xu X, Lai T, Jahan S, Farid F.: Water and Sediment Analyse Using Predictive Models. arXiv preprint arXiv (2022).
10. Mahajna, A., Dinkla, I.J., Euverink, G.J.W., Keesman, K.J., Jayawardhana, B.: Clean and safe drinking water systems via metagenomics data and artificial intelligence: State-of-the-art and future perspective. Front. Microbiol. **13**, 832452 (2022)
11. Papacharalampous, G., Tyralis, H.: A review of machine learning concepts and methods for addressing challenges in probabilistic hydrological post-processing and forecasting. Front. Water. **4**, 961954 (2022)

12. Kovačević, M., Jabbarian Amiri, B., Lozančić, S., Hadzima-Nyarko, M., Radu, D., Nyarko, E.K.: Application of machine learning in modeling the relationship between catchment attributes and instream water quality in data-scarce regions. Toxics. **11**(12), 996 (2023)
13. Castrillo, M., García, Á.L.: Estimation of high frequency nutrient concentrations from water quality surrogates using machine learning methods. Water Res. **172**, 115490 (2020)
14. Aqel, H., Sannan, N., Al-Hunaiti, A., Fodah, R.: Integrated water quality dynamics in Wadi Hanifah: Physical, chemical, and biological perspectives. PLoS One. **19**(2), e0298200 (2024)
15. The World Bank, Data about Saudi Arabia., https://www.worldbank.org/ext/en/home (2025), Accessed 2025.
16. MEWA, Coastlines and Marine Resources., https://www.mewa.gov.sa/ar/Pages/default.aspx (2025). Accessed 2025.
17. SGS, Topography of Saudi Arabia., https://sgs.gov.sa/en (2025). Accessed 2025.
18. NCM, National Center for Meteorology., https://ncm.gov.sa/Ar/Pages/default.aspx (2025). Accessed 2025.
19. El-Rawy, M., Fathi, H.: Groundwater pollution sources and its quality in the kingdom of Saudi Arabia: State of the art. In: Ali, S., Negm, A. (eds.) Groundwater Quality and Geochemistry in Arid and Semi-Arid Regions, pp. 215–235. Springer Nature Switzerland, Cham (2024)
20. GAS, General Authority for Statistics., https://www.stats.gov.sa (2025). Accessed 2025.
21. ESCWA.: Regional climate modelling outputs for Saudi Arabia: Key findings. The United Nations Economic and Social Commission for Western Asia (2023).
22. Giao, N.T., Nhien, H.T.H., Anh, P.K., Thuptimdang, P.: Combination of water quality, pollution indices, and multivariate statistical techniques for evaluating the surface water quality variation in Can Tho City, Vietnam. Environ. Monit. Assess. **194**(11), 844 (2022)
23. Razif, M.: BOD, COD, and TSS predictions from DO measurement results for the Surabaya River, Indonesia. J. Civ. Eng. Plan. Des. **1**(1), 1–7 (2022)
24. Patela, J.Y. and M.V. Vaghanib: Correlation Study for Assessment of Water Quality and Its Parameters of Par River Valsad, Gujarat, India (2015).

A Review of Capture the Flag (CTF) Platforms for OT/ICS Security Training

Maryam M. Aldossary(✉), Sara F. Aldossary, Zahra A. Albeesh, Rahaf A. Al-Ghamdi, Shujon A. Alshehri, Dalal M. Al-Thawadi, Khalid A. Alissa, and Abdulrahman M. Alharby

Imam Abdulrahman bin Faisal University, Dammam, Saudi Arabia
{mmsaldossary,2210002798,2210002639,2210002952,2210002923, 2210002445,kaalissa,aalharby}@iau.edu.sa

Abstract. Operational technology plays a vital role in managing and monitoring critical infrastructure by the help of industrial control systems (ICS). However, the interconnection of OT environments with enterprise networks exposes these systems to critical threats that could result in catastrophic damage and disrupt operational continuity. OT security is an emerging field within cybersecurity, however, due to its novelty, organizations face challenges in securing these systems due to the lack of skilled professionals and hands-on training platforms. Capture the Flag (CTF) platforms have proven to be useful for cybersecurity training. However, their applicability to OT/ICS-specific training remains limited. This paper reviews the existing literature on OT/ICS-focused cybersecurity education, security measures and CTF platforms. The analysis highlights a gap in the availability of realistic, permanent and scalable OT/ICS-focused CTF platforms. This paper emphasizes the need for developing specialized training solutions to better equip cybersecurity professionals with the practical skills required to protect critical industrial systems.

Keywords: Operational Technology (OT) · Industrial Control Systems (ICS) · Capture the Flag (CTF) · OT Security Training · Hands-on Cybersecurity Training

1 Introduction and Background

Infrastructure technology (IT)-operational technology (OT) convergence has increased the cyber threats to critical infrastructure in sectors like manufacturing, energy, and transportation. Once independent OT/Industrial Control System (ICS) systems based on protocols like Modbus and DNP3, they have been exposed to sophisticated attacks following IT-OT convergence. Convergence is beneficial to operations but widens the attack vector and creates enormous challenges.

One of the major challenges is the lack of skills, as IT professionals often lack OT knowledge, and OT engineers lack cybersecurity knowledge—and therefore, a lot of organizations are still ill-equipped [1]. These OT/ICS systems also carry vulnerabilities

D. Berkouk et al. (Eds.): CCTS 2024, PTT, pp. 274–281, 2026.
https://10.1007/978-981-95-7607-4_31

in the form of insecure design, weak encryption, and outdated firmware [2]. Perpetrators of attacks like Oldsmar and by malware like TRITON and Industroyer.

To counter these issues, experiential training is required. Capture-the-Flag (CTF) platforms are interactive, game-like training environments in which students can practice OT/ICS security [3]. This paper presents a review of existing CTF platforms and research, gaps—particularly the lack of OT/ICS content—and underscores the need to develop realistic, scenario-driven OT/ICS training solutions.

2 Literature Review

This section reviews related works on CTF platform development, educational approaches in OT/ICS, and security vulnerabilities. It identifies contributions and case studies to support gap analysis and highlights the need for this project's approach.

2.1 CTF Platforms Development and Implementation

S. Karagiannis et al. [3] present CTF challenges as a successful instrument in cybersecurity education because they help bridge the gap between theoretical knowledge and practical application. As a reaction to these challenges, authors present a systematic approach to integrating CTF challenges into curricula that blends gamification, storytelling, and collaborative learning, following the ARCS (Attention, Relevance, Confidence, Satisfaction) model. The study demonstrates that students who take part in CTF-based learning environments are more self-assured, improve their problem-solving ability, and become more effective learners of cybersecurity principles. By using pre-experiment questionnaires and observations, the study demonstrates CTF's effectiveness. Despite limitations such as small sample size, the study shows that CTF challenges can readily be adopted in formal learning settings.

T. Williams et al. [4] addressed the need for practical, hands-on cybersecurity training for ICS. They present RADICL CTF, a gamified expansion of the picoCTF framework that focusses on ICS cybersecurity and aims to teach adversarial thinking. By integrating open-source software and low-cost hardware, RADICL CTF allows students to engage with ICS network simulations in a simulated operational environment at a low cost. The platform lets students design, assess, and attack systems pertaining to ICS by utilizing the Purdue Reference Architecture, a well-known paradigm for ICS network segmentation. It aims to bridge the gap between theoretical education and practical skills. The project also focuses on scalability, ease of management, and multi-user access. The outcomes demonstrate how well RADICL CTF was implemented, how the current picoCTF architecture was improved, and how future plans are to expand the system to address a wider range of cybersecurity concerns.

D. Antonioli et al. [5] proposed a solution in the field of the ICS challenges which is ICS testbed. This challenge causes a lack of understanding of the real ICS and inability to test on it new attacks and countermeasures. They build a gamifying CTF based on secure water treatment environment called S3. The gamification idea provides access to real and simulated ICS to give opportunity for beginners and experts from different communities (academic and industry). The CTF consists of two phases the first phase conducted online in a jeopardy-style as a training session, the second phase conducted live as an attack-defense style. The first phase categories into MiniCPS, Trivia, Forensics, PLC and Misc. There are different attacks conducted during live phase such as SYN flooding (DoS) PLC, DoS Layer 1 network, Tank level sensor tampering and Chemical dosing pump manipulation. This paper represents a powerful reference to build similar events in the future.

I. Ortiz-Garces et al. [6] addresses emergent needs, which are induced by digital menaces and a lack of skilled professionals for interactive education in cybersecurity. The proposed platform is designed to provide training with gamified CTF competitions that ensure practical experience in finding and correcting security vulnerabilities. Much emphasis will be given to users' privacy while fostering enhanced learning and skill development on the proposed platform. Given the sensitive nature of cybersecurity issues, measures should be taken so that participant information, decisions, and results are well-protected against unauthorized access. As these kinds of platforms integrate learning in artificial hacking environments, privacy issues are at stake; therefore, user data protection should be well-ensured.

S. Karagiannis et al. [7] have presented a deep comparison of four popular open-source CTF platforms, including FBCTF, CTFd, Mellivora, and Root the Box. The effectiveness of these frameworks as a learning tool in cybersecurity training has formed the base for evaluating the frameworks. In this work, considering important features such as functionality, extensibility, and teaching presence that enhance learning, the study has been performed. In an empirical study, various interviews are conducted by a researcher with the help of undergraduate computer science students to attain results about usability, educational value, and gamification elements on the platform. Results showed that CTFd and Root the Box are more suitable for situations because they are built on well-organized challenge systems and have broad reporting capabilities, while event-based CTF competitions are recommended to be performed by FBCTF due to its very large graphical environment.

S. Karagiannis et al. [8] introduce PocketCTF, a platform tailored to the realization of cybersecurity training activities within educational contexts. The platform offers full support for offensive and defensive exercises: students can conduct DDoS attacks and network reconnaissance, among other things, and practice defensive tasks such as SIEM configuration and intrusion detection. The architecture consists of LXC containers running Docker instances, providing each trainee with an isolated environment while minimizing the complexity of the setup and the consumption of system resources. With low-cost, scalable architecture, PocketCTF is more accessible and adaptable for educational use.

2.2 Educational Approaches in OT/ICS

K. Karampidis et al. [9] motioned that improved cybersecurity training is desperately needed in Industry 4.0. The industry is facing more cyber hazards as a result of the growth of networked industrial systems, but there is a severe lack of qualified personnel to handle these risks. The European Union-funded DICYSTECH project aims to close this gap by providing open-source cybersecurity education through a blend of theoretical modules and hands-on training conducted in virtual labs. Gamified CTF challenges are the focal point of the project's training methodology, wherein users participate in simulated cybersecurity scenarios such as SQL injections and forensic analysis. In this hands-on way, students get to practice their skills in a risk-free environment. The CTF challenges, supported through the platform CTFd, promote competitive and active learning to improve cybersecurity skillset of trainees. Results were favorable during a pilot stage at the Hellenic Mediterranean University, forty-one students said they felt more confident about cybersecurity because of the course and wanted to learn more. The virtual labs, where participants may practice actual scenarios, were extremely well received. The project plans, based on the results of this pilot, to further enhance the content and level of interactivity soon.

R. Raman et al. [10] described how CTFs can be extremely valuable to improve cybersecurity training by their practical hands-on approach - a key factor of real-world learning. Most of the options available for CTFs do not include formal training. Because of this, it might present some sort of difficulty or a barrier for those who have low experience, but at the same time, it is an excellent teaching tool. The aim of this paper is to describe a methodological approach that tries to fill this gap by presenting a formal approach toward assessing and ranking CTF competitions, considering factors such as problem difficulty, solvability, scope of effort, rewards, and educational benefit of prevent Training. The framework aims at guiding competitors and organizers toward more successful events, which should be more congruent with the demands of the industry and allows them to identify where the development needs lie.

V. Švábenský et al. [11] emphasized the significance of CTF challenges in fostering practical skills essential for learning cybersecurity. Additionally, participants can acquire a variety of skills through these challenges and games, which encompass diverse formats such as cryptography, reverse engineering, and service exploitation. Furthermore, further study advises examining non-technical issues like phishing and studying additional writeups to increase accuracy. The article's conclusion stresses that although CTFs can be effective at teaching technical skills, they don't give enough attention to non-technical topics, which are essential for fending off sophisticated cyberattacks.

2.3 OT/ICS Vulnerabilities

G. M. Makrakis et al. [12] provided an extensive review of the cybersecurity threats facing ICS and CI. With increased integration with IT, these systems become exposed to various types of cyber-attacks. Key vulnerabilities at the level of ICS architectures, devices, and communication protocols are analyzed in this paper, with the discussion centered on how such insecurities can lead to catastrophic consequences on physical processes. The authors categorize the threats by adversary motives and techniques. They observe that most ICS systems are lacking in adequate security controls since they are complex and outdated. Then, some real-world incidents such as Stuxnet and BlackEnergy are taken into consideration to study how vulnerabilities in the ICS communication protocols and devices could be exploited. It mentions that critical infrastructure attacks gain momentum, given the tools increasingly available for adversaries. The paper underlines, in particular, the need for awareness of operational differences between IT and OT networks which further complicates applying standard cybersecurity measures in ICS environments. The authors emphasize the need for more focused approach in the security solutions particular to ICSs when it comes to preventing wide-scale disruption and damage. This review provides wide coverage necessary for understanding the status of security in ICS and sheds light on how to mitigate the vulnerabilities that affect these systems.

SCADA vulnerabilities are underlined by G. Yadav and K. Paul [13] to show their crucial role in the supervision and control of critical infrastructure, such as industrial processes, water systems, and power grids. However, these systems are increasingly more vulnerable to cyberattacks due to their integration with networks connected to the internet. Events such as the attack on the Ukrainian power grid and the Stuxnet attack on Iranian nuclear facilities give relevance to just how damaging these breaches can be. SCADA vulnerabilities arise due to three critical issues: buffer overflows, outdated systems, and poor communication protocols. While many other threats exist, availability remains the most worrisome concern, as 89% of attacks exploit network vulnerabilities, often due to security measures that become outdated frequently. Although the Common Vulnerability Scoring System (CVSS) is in use almost everywhere, the sophisticated dependencies and priorities of the SCADA system limit its usage. Therefore, a domain-based vulnerability scoring system is proposed by researchers to improve threat mitigation and management techniques in such critical systems.

3 Comparison and Analysis

This section presents a comparative analysis of existing CTF platforms, identifying their shortcomings based on the following parameters:

- Low Cost: Outlines the cost of using the platform (e.g., free, low cost, high cost).
- OT/ICS Coverage: Indicates whether the platform is focused on IT, OT, or both.
- Permanent Platform: Examines of whether the platform is temporary for a specific event or permanent.
- Diversity of OT/ICS Challenge Scenarios: Explores the range of attack methods included in challenges such as SQL injection, DoS, and Modbus exploitation.
- Protocol-Specific Vulnerability Exploits: Determines if challenges address vulnerabilities related to specific protocols such as TCP, UDP, and Modbus.
- Functionality: Examines how well the platform ensures user-friendly interaction, remains easy to deploy and resource-efficient.
- Scalability: Determines how the platform can support increasing demand to accommodate more users and resources while maintaining performance.

Results from Table 1 show that the majority of current CTF platforms are IT-focused, with only a limited number targeting OT environments. However, those OT-focused platforms have several shortcomings that reduce their effectiveness. The CTF competitions described in [4, 5, 9] run as temporary, event-based platforms rather than permanent resources used for training. In addition, they do not cover essential OT-specific topics, such as PLC logic manipulation and firmware reverse engineering.

As cyber threats are now targeting industrial systems more than ever, there is a pressing need for a platform that can address OT-specific challenges and provide hands-on experience in securing critical infrastructure. Moreover, most CTFs lack realistic

Table 1. Gap Analysis on Existing CTF Platforms

Ref.	Low cost	OT/ICS Coverage	Permanent platform	Diversity of OT/ICS challenges Scenarios	Protocol-Specific Vulnerability Exploits	Functionality	Scalability
[3]	√					√	
[4]	√	√		√	√	√	√
[5]		√		√	√	√	
[6]	√		√			√	√
[7]	√				√	√	√
[8]	√		√			√	√
[9]		√		√	√	√	√
[10]	N/A				N/A	√	
[11]					√		

a N/A indicates data is not available from the source, ✓ indicates the feature is supported; a blank indicates the feature is not supported.

OT/ICS environments. Only [5] utilizes physical OT hardware, but due to its high cost, the use of physical hardware in OT-focused CTFs is highly limited. Additionally, most CTFs prioritize general cybersecurity skills rather than OT-specific ones, such as OT forensics, incident response, and protocol analysis. Developing a balanced OT CTF training platform would help ensure these essential skills are covered properly, thereby equipping professionals to manage the unique challenges of OT security.

Functionality and scalability are also crucial to the creation of an effective CTF platform that simulates a real-world training environment. A functional platform facilitates the seamless simulation of realistic OT/ICS attacks with less setup overhead, leading to a productive learning process. Scalability enhances this by accommodating additional players, introducing diverse types of challenges, and handling resources dynamically without excessive manual effort. With modular add-ons and dynamic resource management, a robust CTF platform remains efficient, responsive, and capable of keeping pace with the evolving landscape of OT security training.

This analysis identifies substantial gaps in existing CTF platforms, particularly in terms of OT/ICS-specific coverage, diversity of OT/ICS challenge scenarios and some platforms have industrial cybersecurity components; however, they lack full scenarios, protocol-specific attacks and a persistent infrastructure. An effective OT CTF platform should focus on realism, scalability and functionality to better prepare security professionals for real-world industrial threats.

4 Results & Discussion

G. M. Makrakis et al. [12] and G. Yadav and K. Paul [13] identify critical ICS vulnerabilities such as legacy systems and poor security controls, SCADA vulnerabilities, and Stuxnet and Tri-ton attacks demonstrate their devastating impact. These attacks underscore the need for beefed-up defenses as a matter of urgency. Effective training, such as CTF-based training, might have mitigated some of these attacks by elevating awareness of such vulnerabilities and response readiness. However, as shown in our analysis, most of the available CTF platforms are IT-focused with few provisions for OT-specific exercises. While there are OT-specific platforms available, they are ephemeral, event-driven, and lack coverage of key OT topics like PLC logic manipulation and firmware reverse engineering. In addition, most of the platforms do not simulate realistic OT/ICS setups, which are essential for hands-on practice in industrial system security. As seen from S. Karagiannis et al. [3], Training based on CTF enhances cybersecurity skills and could have reduced ICS attack danger by providing real-world, hands-on experiences. Addressing the exposed deficiencies, future research needs to focus on developing long-term, large-scale OT-specific CTF platforms with diverse and realistic ICS challenges.

5 Conclusion

In conclusion, this paper provides an extensive review of the literature, covering five major areas: CTF Platforms Development and Implementation, Educational Approaches in OT/ICS, OT/ICS vulnerabilities, Security Measures for OT/ICS systems and Contributions and Case Studies in OT/ICS. From this review, it has become evident that all these fields are still a field of ongoing effort, and several challenges have been identified. It was followed by a gap analysis that showed significant gaps regarding the availability and sustainability of permanent knowledge platforms in the field of OT/ICS training. Even as awareness increases in the importance of hardy cybersecurity training, the lack of such dedicated long-term platforms creates a barrier in addressing the emerging security challenges within OT/ICS environments. This underscores the need for further development and deployment of such platforms to better prepare professionals and students for the critical task of safeguarding these systems.

References

1. Schwartz, S.: ICS security investments blocked by management confusion. Cybersecurity Dive, Nov. 10 (2021). https://www.cybersecuritydive.com/news/dragos-ponemon-ics-operational-tech-cyber-security/609739/
2. Anonymous: Knowing vulnerabilities in OT systems can help cybersecurity efforts. Food Eng. **94**(8), 25–25 (2022) https://www.foodengineeringmag.com/articles/100512-knowing-vulnerabilities-in-ot-systems-can-help-cybersecurity-efforts
3. Karagiannis, S., Magkos, E.: Adapting CTF challenges into virtual cybersecurity learning environments. Inf. Comput. Secur. **29**(1), 105–132 (2021). https://doi.org/10.1108/ICS-04-2019-0050
4. Williams, T., Fuhrmann, T., Haney, M.: Radicl CTF. J. Colloq. Inf. Syst. Secur. Educ. **10**(1), 10 (Mar. 2023). https://doi.org/10.53735/cisse.v10i1.167

5. D. Antonioli et al.: "Gamifying education and research on ICS security: Design, implementation and results of S3," Cornell University Library., arXiv.org, Ithaca (2017). doi: https://doi.org/10.48550/arxiv.1702.03067.
6. Ortiz-Garces, I., Gutierrez, R., Guerra, D., Sanchez-Viteri, S., Villegas-Ch, W.: Development of a platform for learning cybersecurity using capturing the flag competitions. Electronics. **12**(7), 1753 (Apr. 2023). https://doi.org/10.3390/electronics12071753
7. Karagiannis, S., Maragkos-Belmpas, E., Magkos, E.: An Analysis and Evaluation of Open Source Capture the Flag Platforms as Cybersecurity e-Learning Tools. Inf. Secur. Educ. Inf. Secur. Action, 61–77 (2020). https://doi.org/10.1007/978-3-030-59291-2_5
8. Karagiannis, S., Ntantogian, C., Magkos, E., Ribeiro, L.L., Campos, L.: PocketCTF: A fully featured approach for hosting portable attack and defense cybersecurity exercises. Information. **12**(8), 318 (2021)
9. Karampidis, K., Panagiotakis, S., Vasilakis, M., Lamari, A.T., Markakis, E., Papadourakis, G.: Digital training for cybersecurity in industrial fields via virtual labs and capture-the-flag challenges. IEEE Xplore. (Jun. 2023). https://doi.org/10.23919/eaeeie55804.2023.10181644
10. Raman, R., Sunny, S., Pavithran, V., Achuthan, K.: Framework for evaluating Capture the Flag (CTF) security competitions. IEEE Xplore. (Apr. 2014). https://doi.org/10.1109/i2ct.2014.7092098
11. Švábenský, V., Čeleda, P., Vykopal, J., Brišáková, S.: Cybersecurity knowledge and skills taught in capture the flag challenges. Comput. Secur. **102**, 102154 (2021). https://doi.org/10.1016/j.cose.2020.102154
12. G. M. Makrakis, C. Kolias, G. Kambourakis, C. Rieger, and J. Benjamin: "Vulnerabilities and attacks against industrial control systems and critical infrastructures," arXiv preprint arXiv:2109.03945 (2021). https://arxiv.org/abs/2109.03945.
13. Yadav, G., Paul, K.: Assessment of SCADA system vulnerabilities. IEEE Xplore. (Sep. 01, 2019) DOI: https://ieeexplore.ieee.org/document/8869541/keywords#keywords

SWOT Analysis of MIS in Higher Education: A Case Study of KKU

Nouf S. Aldahwan(✉)

Department of Information System, King Khalid University, Abha, Saudi Arabia
naldhwan@kku.edu.sa

Abstract. The introduction of an emerging technology is a vital phase on its own, understanding and use of technology are important well. Usage of information technology has been a primary dependent variable in research into management information systems (MIS). Universities use MIS to facilitate a wide variety of administrative, teaching, and learning operations, including attendance tracking, evaluation reports, reporting, financial planning, and preparation of course materials, and allocation of resources and staff. Such programs were developed by universities to improve process efficiency and to coordinate the delivery and allocation of training services. Thus, the purpose of this study was to provide systematic Literature Review (SLR) of MIS adoption and to examine strengths, weaknesses, opportunities, and threats (SWOT) connected to the incorporation of management information systems into university learning environments by enabling communication between instructors and students through online lectures, and to explore applications of various MIS in education work in this area. The process involves conducting surveys of Community College teachers and students to gather statistical data for a later assessment using the SWOT analysis. This research outcome will reflect the overall MIS adoption in higher education institutions and how students and teachers perceive SWOT terms for LMS and how they might use them.

Keywords: Higher Education Institutions · Management Information System · SWOT Analysis

1 Introduction

MIS is a very significant component in universities' reliable and active information management systems that help to meet the needs of internal and external clients. Students, teachers, program director, manager, and rector are expected to introduce educational activities in universities through the information system [1].

During the teaching of MIS courses, using a "task-driven" approach will effectively re- solve management students' fear of computer training difficulties, completely mobilizing their excitement and initiation [2]. As for producing the academic information network, all colleges are checked by the university information management department and have a high-quality information system department [3].

D. Berkouk et al. (Eds.): CCTS 2024, PTT, pp. 282–294, 2026.
https://10.1007/978-981-95-7607-4_32

The goal of this research, To provide Systematic Literature Review of existing MIS in education, and through surveys, is to analyze the opinions of students and their lecturers for further assessment through the SWOT analysis This paper is split into seven sections. The first section is an introduction. The second section contains the related study contains overview of IS for higher education, it explained the management information system in education and MIES in e-learning. Third, is the methodology. The fourth section is data collection. The fifth results and discussion. Sixth, is the conclusion. Finally, limitations and future work.

2 Literature Review

2.1 Information System in Education

Data, information, and knowledge principles form the basis of information systems (ITs). Data is used for knowledge production and can be represented in numerical, alphabetical, symbolic, and graphical ways. Once reel event results data are converted and made more relevant for users, information is created [4]. The study provides a cloud-based online education management system that is effective and may be a useful online education assignment. It enhances the conventional information management process and information categorization algorithm [5].

This paper presented detailed information on the evaluation is significant in that it provides information on the weaknesses and strengths of an EMIS system. This will furnish a more complete picture of the status of the EMIS. The scope of small and medium-sized education is restricted to basic and secondary education, including formal education in public and private schools [6].

The study aims to discover out the degree of satisfaction of decision-making for the use the student management System to handle classrooms in institutions of higher education. The results of this study, the quality of information is the most powerful predictor on satisfaction in decision making [7].

2.2 Education Management Information System (EMIS)

There is a growing increase in MIS in the field of education. The following summarize the research in this area:

Applications for Management Information System. The research analyzes and implements a artificial intelligent system. This system includes external processing management tasks and routine teacher training [8]. Another study focused on outlining what matters most in attaining a functional education management information system. The aim of this research shows that the management information system can use the data it produces to upgrade the quality of education as well as the efficiency of operations in schools [9]. Research has examined the significance of information systems in enhancing business performance, emphasizing the necessity of precise, dependable, and prompt data [1]. Another research was performed since the authors believed that the information technology plays a big part in the education of Pakistan and investigate the issues in the 21st century [10]. A systematic review study is conducted with published literature regarding self- management education on diabetes which focuses on high and low mortality rates in the developed world. This study discovered that self-managing diabetes has positive outcomes for short-term improvement in the countries [11]. Another study conceived an online platform for handling postgraduate education with emerging web technologies. The purpose of this design is to improve the collaboration and communication of advisors, the student, the institute, and the department [12]. The researcher presented the paper on the evaluation satisfaction of students' to use of LMS in higher education system. The research evaluated different factors of the most utilized and commercially available state-of-the-art LMS systems [13]. Another study indicated that the use Online Writing Instruction (OWI) with more basic features are preferred promotes engagement and collaboration in LMS. This study can help create a future LMS that incorporates the experiences and perspectives of both students and faculty. It can also help in developing writing materials to satisfy writing outcome [14].

Success Related to the use of MIS in Education. The researcher evaluated the success factors influence EMIS adoption. It focused on comprehending the EMIS use by students and its advantages [15]. Another study offers insight into the body of knowledge and practitioners while examining the strategy of character education deployment at the level of grassroots Islamic educational institutions [16]. Another study analyzes the effects of the model and technology on organizational performance while developing a cloud-based MIS for small and medium-sized enterprises (SMEs) in Ethiopia [17]. Another study investigated the behavioral intentions of Saudi Arabian university students to utilize a mobile learning management system (m-LMS)—such as the Blackboard application—by conducting an empirical test of an expanded UTAUT model. The literature on technology acceptance in general and m-learning adoption in particular is enhanced by the results, which benefit academia [18].

Impact of MIS in Educational. This study showed that the implementation of Artificial Intelligent (AI) in management information systems within the enterprise has inherent advantages. The users of the program were largely pleased with the management information systems introduced as the systems had helped to increase the efficiency and performance of duties. The program also reduced administrative mistakes and helped take decisions [19].

The research aimed at highlighting the development of IT growth, especially in the management of EMIS processes, with the key factors (timely and accurate data productivity, data integration & knowledge sharing, and successful data utilization) required for the successful growth and functionality of the EMIS for effective planning [20]. Another study to develop cloud computing MIS models in academic work, and to evaluate cloud computing MIS models and assess the system structure in educational work [21].

2.3 Education Management Information System in e-Learning

Critical Factors for EMIS Using E-Learning. A study conducted on a conventional LMS system, which is typical for all learners. Learners are different in the sense of learning and understanding skills. The significance and adaptations required to convert the LMS to different learners analyzed in this paper. This paper explores the impact of learning styles by incorporating integrated courses to match the learner's patterns within the LMS [22]. In addition to the crucial factors, analysis on LMS had been carried out using different modules of motivational factors that make effectiveness to the learners [23]. The effect of mobile learning in students had been analyzed by conducting a study. Factors impacting positively or negatively on students' learning are found initially [24].

Challenges of LMS Environment. The author conducted a study at two Norwegian Universities among the students and instructors who use Learning Management Systems (LMSs) and Social Networking Technologies (SNTs). The survey conducted was conducted as two parts seeking challenges and weaknesses of LMS followed by the second part seeking the merits and demerits of SNT [25]. Learning Analytics is an emerging field of e-learning. LA uses various methods like visual data analysis, social networking analysis, prediction, clustering, human review analysis, etc. And these would result in the improvement of both the instructor and learners [26].

MIS Models in E-Learning. Institutions and learners are converting into e-learning systems. Since the material supplied to all learners is the same and follows the same mode of learning for all, the authors implementing user intelligent adaptive learning to LMS, which helps the platform to deliver suitable material for the learners [27]. Web, Database, and Video servers required for creating an e-learning platform. The creation of this system seeks security issues, web threats [28]. The goal of the study is to increase present understanding of the status of the online learning challenge. The process involves employing surveys of teachers and students to gather statistical data for a later evaluation using SWOT analysis [29].

Intelligent Techniques of Implemented MIS in E-Learning. An intelligent system was created as part of data mining on LMS and implemented at the University of Kragujevac. Knowing the behavioral pattern of learners and adjusting the course structure, using custom modules. PDCA (Plan, Do, Check, Act) method had implemented [30]. Task-oriented, User-centered, multi-method evaluation (TUME) technique is used in LMS to find the usability and accessibility for Blind and Visually Impaired BIV students. The problems hide in web-based applications, and the evaluation and design level analysis carried in that to find out recommended corrections [31].

2.4 SWOT Analysis

Any firm can be examined using a SWOT framework to determine its opportunities, threats, weaknesses, and strengths [32]. The study of the advantages and disadvantages is done using the organization's internal environment. The external environment or environmental elements that must be considered in strategic planning are related to the opportunities and risks. Ideally, SWOT analysis aids in comprehending the following four crucial elements: [32], an organization's internal components that help it achieve its objectives are referred to as its strengths, whilst its internal components that obstruct organizational achievement are referred to as weaknesses. Opportunities are external factors that aid a company in achieving its objectives. These factors include favorable environmental features as well as chances to close gaps and start new projects. Threats, on the other hand, are elements of the external environment of the organization that act as obstacles or could act as obstacles to the achievement of its objectives.

3 Methodology

The process involves employing surveys of Community College teachers and students to gather statistical data for an assessment using the SWOT analysis. The survey was created and approved by experts in the field before being distributed to the intended audience.

3.1 Systematic Literature Review (SLR)

The survey searches for the various aspects attached to the EMIS. Several studies that have made identifiable contributions to EMIS models and implementation in the last five years have been selected and then, reviewed. The study presents an IS in education. The application, impact and success related to MIS. In addition, EMIS in e-learning that explains the challenges, factors, technique, and models in MIS was discussed.

3.2 Design of Questionnaires

There were four components of the questionnaire. These comprised the following: (a) general data about the respondent. (b) a SWOT analysis of the viewpoints of IT management. As research tools, closed questions were employed.

3.3 Research Procedure

A qualitative research design was chosen with the aim of assessing the general state of management information system in educational. Data from the closed questions were categorized using the qualitative content analysis, and they were then positioned within the SWOT analysis matrix. The inductive category development [33] divides the content analysis into five steps: (1) To define the categories, researchers looked over the materials that included responses to the closed SWOT questions. (2) Materials were used to create inductive categories. (3) The categories underwent revisions before being narrowed down to the primary categories. (4) The last thorough examination of all the materials was completed. (5) Quantitative analysis was done on the frequencies of coded categories.

3.4 Data Collection

Sample Size. The primary data for statistical analysis was selected to be gathered via questionnaire. All 120 of the target public communities applied to college higher education institutions received the questionnaire (Table 1).

Table 1. General information of respondents

Gender	No.	Percent %	Mean	Std.Dev
Male	51	42.5	1.78	0.65
Female	69	57.5		

Reliability and Validity. One common technique for pretesting questionnaires is an expert review. There is a range in the organization and rigor of expert evaluations. At the other end of the scale, a survey methodologist or subject matter expert with experience examines a draft questionnaire to find problems with the administration or wording of the questions that could cause measurement error [34]. When the study tool was applied to staff and students who were randomly selected at random from outside the study sample at an interval, the stability of accuracy was confirmed using Crombach's Alpha. Alpha to determine the reliability coefficient. The questionnaire's testability was determined to be (0.77), one of the high reliability coefficients that indicates that answering the questions will help the study be completed.

4 Results and Discussion

MIS is a central subject taught in University Management courses. This is a detailed theoretical and practical course and is also as a 'bridge' between management science and IT, cross-discipline and the learning approach is a combination of management science, computer science, IS and the concepts and methods of other topics. MIS is an organization's mechanism for streamlining the total knowledge used to achieve organizational objectives.

(1) **SWOT analysis: strength.** The purpose of this study was to examine, from the perspective of the staff and students, the strengths, weaknesses, opportunities, and threats associated with utilizing the MIS in higher education. Table 2 presents the findings of the descriptive statistical analysis. It indicates that the averages ranging from 6 to 26. As displayed in Fig. 2 the distribution for replies for ICT infrastructure covers all buildings, with a strong agree and 67.5% and Data Protection are stressful with low response by of 2.5% with strongly disagree.

(2) **SWOT analysis: weakness.** Table 3 presents the findings of the descriptive statistical analysis. It indicates that the averages ranging from 8 to 25. As displayed in Fig. 2 the distribution for replies for ICT infrastructure covers all buildings, with strong agreement and 45.8% and the time-limited assignments are stressful with low response by 2.5% with strongly disagree (Fig. 1).

(3) **SWOT analysis: opportunity.** Table 4 presents the findings of the descriptive statistical analysis. It indicates that the averages ranging from 8 to 28. As displayed in Fig. 3 the distribution for replies for develop adequate cloud computing options with

Table 2. SWOT analysis: Strength

No	Items	Mean	Std.Dev
1	Data availability.	23	16.7
2	Daily reports according to daily data loading.	6	20.1
3	Data recovery mechanism.	14	22.0
4	Structured and customized e-learning instruction	27	18.9
5	Institutions have policies and guidelines to continuously improve IT.	15	18.5
6	Learning process analysis	8	21.8
7	Web reporting add-on feature.	18	19.8
8	Data Protection	23	15.1
9	Improved communication.	9	18.7
10	ICT infrastructure covers all buildings	9	25.6
11	Senior management places importance on IT systems.	26	23.9

Table 3. SWOT analysis: Weakness

No	Items	Mean	Std.Dev
1	Many source system managers.	22	10.4
2	Shortage of personnel with knowledge, skills and IT expertise.	24	10.8
3	Absence of a personalized strategy	25	9.5
4	Connectivity problems.	8	16.3
5	Personnel do not realize or become aware of the importance of IT adoption in practice	25	15.8
6	Improvement and adaptation are very expensive	21	16.3
7	Insufficient hands-on training	18	17.3
8	Improvement and adaptation are very lengthy processes.	19	16.2
9	Reports lack flexibility.	22	13.9
10	Time-limited assignments are stressful	23	13.6
11	Incompatibility of hardware and software.	22	10.4

strongly agree and 53.3% and personnel do not realize or aware of the importance by of 0.3% with strongly disagree.

(4) **SWOT analysis: threat.** Table 5 presents the findings of the descriptive statistical analysis. It indicates that the averages ranging from 99 to 29. As displayed in the Fig. 4 the distribution for replies for less physical interaction between lecturers and

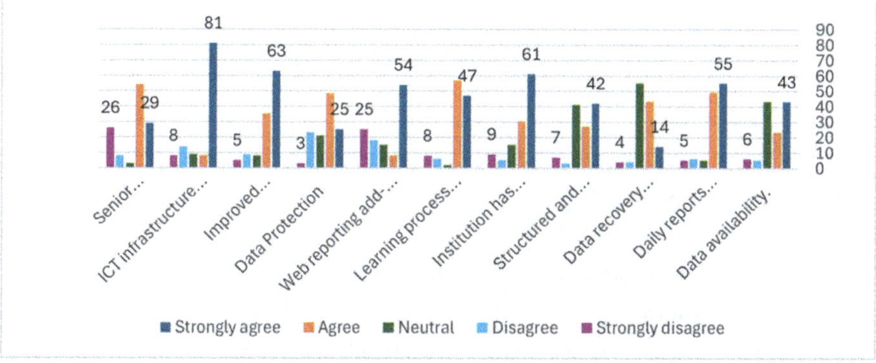

Fig. 1. Strengths

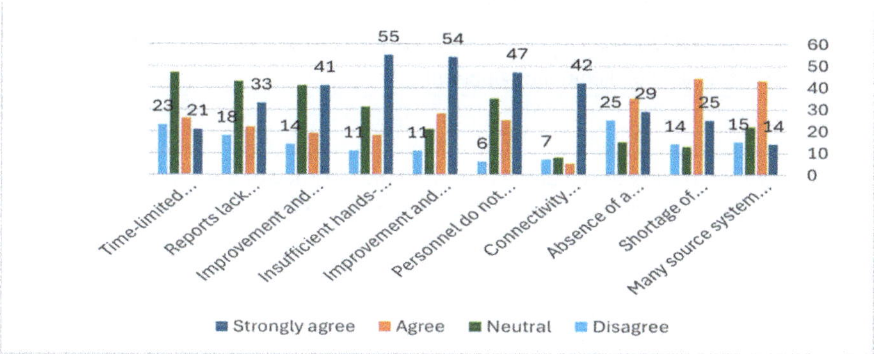

Fig. 2. Weaknesses

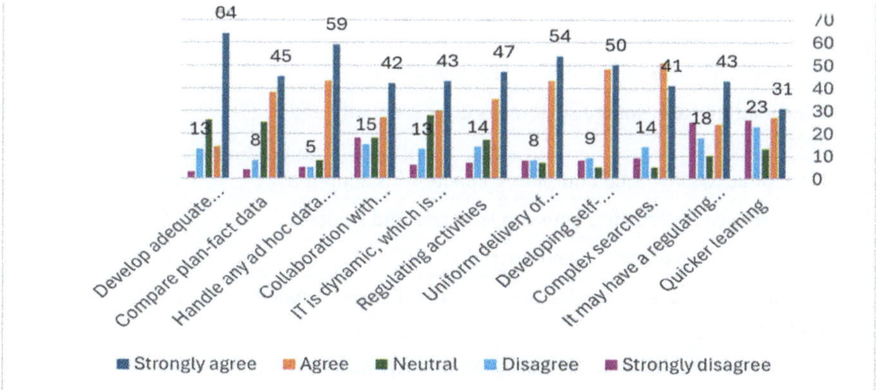

Fig. 3. Opportunity

Table 4. SWOT analysis: Opportunity

No	Items	Mean	Std.Dev
1	Quicker learning	26	6.0
2	It may have a regulating effect on the source systems' data purity.	24	8.8
3	Complex searches.	14	15.1
4	Developing self-evaluation abilities.	9	19.4
5	Uniform delivery of content handles faculty shortage	8	20.3
6	Regulating activities	17	17.7
7	IT is dynamic, which is an opportunity to constantly develop new or improved IT systems	28	13.9
8	Collaboration with outside agencies (public and private sectors).	18	11.5
9	Handle any ad hoc data retrieval needs.	8	17.4
10	Compare plan-fact data	25	19.6
11	Develop adequate cloud computing options	14	18.8

students with strongly agree and 47.5% and the absence of a socio-academic setting that fosters student growth with low response by 0.3% with strongly disagree.

Table 5. SWOT analysis: Threat

No	Items	Mean	Std.Dev
1	Leaders remain less driven while using MIS daily.	23	20.5
2	Reliability and efficacy of online instruction and learning for students or learners.	21	19.8
3	Excessive reliance on IT service providers	9	21.0
4	Data loading from source systems is done manually and sporadically.	25	20.4
5	Lack of adequate cloud computing capabilities adoptions	22	17.6
6	Less physical interaction between lecturers and students	15	18.3
7	The costs of development and operation.	18	16.8
8	The absence of a semiacademic setting that fosters a student's growth	21	15.3
9	Technology and demand are changing rapidly, which results in high-frequency investment	29	15.2
10	Insufficient technical support.	25	13.2
11	The absence of a socio-academic setting that fosters student growth.	24	14.8

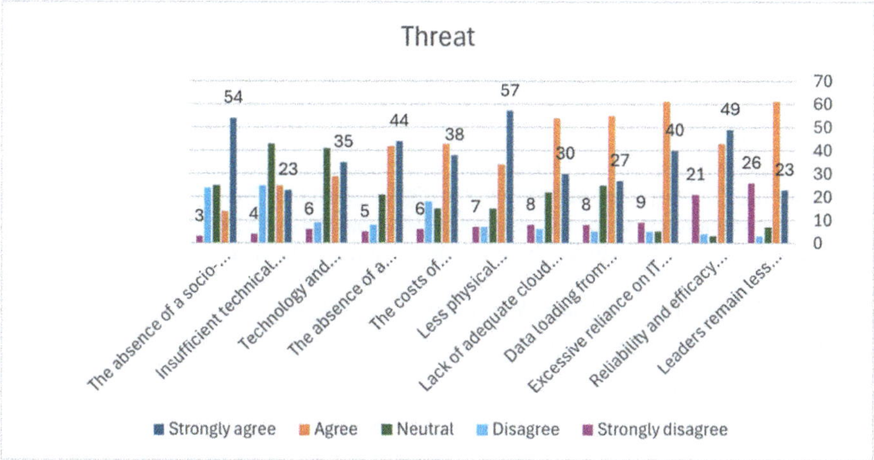

Fig. 4. Threat

5 Conclusion

MIS implementation is important for the current management of the education systems in any higher learning institution. Computer systems, technology and the database help collect, use, and distribute information and data. Developing an efficient and productive MIS therefore requires a clear vision of which products to use and which company to participate in. MIS implementation is not restricted to the establishment of a data and information system but, more specifically, to the establishment of a new culture of information exchange management.

The survey results point to several studies focus on explaining the EMIS; it presents an IS in education, a model that states that the educational community must ensure that their MIS are of great quality, with the same time keeping students happy with the system and engaging in constant use. Management information system that facilitates communication between students and instructors for use in university education contexts. A SWOT study of MIS utilization looked at its strengths, weaknesses, opportunities, and threats. The researcher employed multiple techniques and protocols, then employed qualitative content analysis to categorize and arrange the information.

While the current work offers some useful insights into the general view of the information system for education management, there are some limitations which should be recognized from our perspective. The mature ICT used in MIS in higher education is recommended for further research. In addition, is to explain the role of Artificial Intelligent in MIS. Hopefully this work will be able to clarify the capacity of the information management systems to serve institutional vision and mission.

Acknowledgments. The authors extend their appreciation to the Deanship of Research and Graduate Studies at King Khalid University for funding this work through small group research under grant number RGP1/309/45.

References

1. Taherdoost, H.: The role of different types of management information system applications in business development: Concepts, and limitations. Cloud Comput. Data Sci. (2023)
2. Calavia, M.B., Blanco, T., Casas, R.: Fostering creativity as a problem-solving competence through design: Think-Create-Learn, a tool for teacher. Think. Skills Creat. **39**, 100761 (2021)
3. Alyoussef, I.Y.: Acceptance of e-learning in higher education: The role of task-technology fit with the information systems success model. Heliyon. **9**(3) (2023)
4. Recker, J.: Scientific Research in Information Systems: A Beginner's Guide. Springer Nature (2021)
5. Li, K., Jing, M., Tao, X., Duan, Y.: Research on online management system of network ideological and political education of college students. Int. J. Electr. Eng. Educ. **60**(2_suppl), 377–388 (2023)
6. Meas, S., No, F., Sok, S.: Basic Education in Cambodia. In: International Handbook on Education in South-east Asia, pp. 1–33. Springer Nature Singapore, Singapore (2023)
7. Gupta, P., & Yadav, S. "A TAM-based study on the ICT usage by the academicians in higher educational institutions of Delhi NCR," In Congress on Intelligent Systems: Proceedings of CIS 2021*, Volume 2*, pp. 329–353. Singapore: Springer Nature Singapore. (2021).
8. Liu, T., Gao, Z., Guan, H.: Educational information system optimization for artificial intelligence teaching strategies. Complexity, 1–13 (2021)
9. Habib, M.N., Jamal, W., Khalil, U., Khan, Z.: Transforming universities in interactive digital platform: case of city university of science and information technology. Educ. Inf. Technol. **26**, 517–541 (2021)
10. Khahro, S.H., Javed, Y.: Key challenges in 21st century learning: a way forward towards sustainable higher educational institutions. Sustainability. **14**(23), 16080 (2022)
11. Camargo-Plazas, P., Robertson, M., Alvarado, B., Paré, G.C., Costa, I.G., Duhn, L.: Diabetes self-management education (DSME) for older persons in Western countries: A scoping review. PLoS One. **18**(8), e0288797 (2023)
12. Fang, M., Jandigulov, A., Snezhko, Z., Volkov, L., Dudnik, O.: New technologies in educational solutions in the field of STEM: The use of online communication services to manage teamwork in project-based learning activities. Int. J. Emerg. Technol. Learn. **16**(24), 4–22 (2021)
13. Toring, H., et al.: Evaluation of students' satisfaction toward an adopted learning management system at Indiana Aerospace University: A structural equation modelling approach. Asia Pac. Manag. Rev. **28**(3), 336–346 (2023)
14. Febriyanti, R. H., Usman, H., Lustyantie, N., Iasha, V., & Setiawan, B. "Utilizing learning management system in online writing instruction in higher education: Indonesian faculty member perspectives," J. High. Educ. Theory Pract., 22 (10, 2022).
15. Stamenkov, G., Zhaku-Hani, R.: Perceived benefits and post-adoption usage of education management information system. Libr. Hi Tech. **41**(4), 1063–1083 (2023)
16. Arif, A. M., Nurdin, N., & Elya, E. "Character education management at islamic grassroot education: The integration of local social and wisdom values," Al-Tanzim: J. Manag. Pendidik. Islam, 7(2), 435–450, (2023).
17. Jain, R., Palaniappan, D., Bekuma, Y., Parmar, K.: Cloud-based MIS model for medium and small enterprises in Ethiopia. Int. Res. J. Multidiscip. Scope. **5**(1), 766–780 (2024)
18. Alfalah, A.A.: Factors influencing students' adoption and use of mobile learning management systems (m-LMSs): A quantitative study of Saudi Arabia. Int. J. Inf. Manag. Data Insights. **3**(1), 100143 (2023)
19. Suharyat, Y.: Artificial intelligent: Positive and negative role in education management. In: International Conference on Education, pp. 349–357 (2023)

20. Bhatti, S.A., Adnan, A.: Challenges in education management information system in develpoing countries. In: 2010 International Conference on Information and Emerging Technologies, pp. 1–6 (2010)
21. Odmunkong, T., Wannapiroon, P., Nilsook, P.: The challenges of cloud computing management information system in academic work. Int. J. Signal Process. Syst. **2**(2), 160–165 (2014)
22. Lim, E.H., Ahmad, W.F.W., Hashim, A.S.: Enhancement of learning management system by integrating learning styles and adaptive courses. In: Phon-Amnuaisuk, S., Au, T.W., Omar, S. (eds.) Computational Intelligence in Information Systems. CIIS 2016. Advances in Intelligent Systems and Computing, vol. 532, pp. 211–218. Springer, Cham (2017)
23. M. Rau, C. H¨osel, C. Roschke, R. Thomanek, and M. Ritter, "Impact of motivational factors on the learning process in the use of learning management systems: An empirical study based on learners' experiences," HCII, 278–283, (2019).
24. Hamidi, H., Jahanshaheefard, M.: Essential factors for the application of education information system using mobile learning: A case study of students of the university of technology. Telemat. Inform. **38**, 207–224 (2019)
25. Hustad, E., Arntzen, A.A.A.: Facilitating teaching and learning capabilities in social learning management systems: challenges, issues, and implications for design. J. Integr. Des. Process. Sci. **17**(1), 17–35 (2013)
26. Avella, J.T., Kebritchi, M., Nunn, S.G., Kanai, T.: Learning analytics methods, benefits, and challenges in higher education: A systematic literature review. Online Learning. **20**(2) (2016)
27. Sivakumarand, S., Venkataraman, S.: A user-intelligent adaptive learning model for learning management system using data mining and artificial intelligence. Int. J. Innov. Res. Sci. Technol., IJIRST. **1**(10), 78–81 (2015)
28. Jung, S., Huh, J.: An efficient LMS platform and its test bed. J. Electron. MDPI. (2019)
29. A. Safonov, M., S. Usov, S., & V. Arkhipov, S. "E-learning application effectiveness in higher education. General research based on SWOT analysis," In Proceedings of the 5th International Conference on Education and Multimedia Technology, 07–212, (2021).
30. Blagojevic, M., Micic, Z.: A web-based intelligent report e-learning system using data mining techniques. J. Comput. Electr. Eng. **39**(2), 465–474 (2031)
31. Babu, R., Singh, R.: Enhancing learning management systems utility for blind students: A task-oriented, user-centered, multi-method evaluation technique. J. Inf. Technol. Educ. Res. **12**(1), 1–32 (2013)
32. Benzaghta, M.A., Elwalda, A., Mousa, M.M., Erkan, I., Rahman, M.: SWOT analysis applications: An integrative literature review. J. Glob. Bus. Insights. **6**(1), 54–72 (2021)
33. Nalintippayawong, S., Atchariyachanvanich, K.: IT management status in public higher education institutions in Thailand. In: 2016 IEEE/ACIS 15th International Conference on Computer and Information Science (ICIS), pp. 1–5, Okayama, Japan (2016)
34. Ikart, E.M.: Survey questionnaire survey pretesting method: An evaluation of survey questionnaire via expert reviews technique. Asian J. Soc. Sci. Stud. **4**(2), 1 (2019)

A Study on the Global Hype of Metaverse for Sustainable Growth in the Future

Saima Abdullah[1(✉)], Umara Zahid[1], and Zahid Rashid[2]

[1] Department of Computer Science, International Islamic University, Islamabad, Pakistan
{Saima.abdullah,umara.zahid}@iiu.edu.pk

[2] Technology Management Economics and Policy Program, College of Engineering, Seoul National University, 1 Gwanak-Ro, Gwanak-Gu, 08826 Seoul, South Korea
rashidzahid@snu.ac.kr

Abstract. Metaverse is expanding the real world by adding an immersive next-generation virtual world. The users, by using their avatars, can smoothly interact within real and virtual worlds. The metaverse is expected to have a high impact on how people live, travel, shop, and interact with each other. As the technologies related to the metaverse are advancing, researchers are increasingly analyzing the effects of the metaverse in different fields. At this stage, it is imperative to assess the acceptance of the metaverse by various user communities. Further, it is also essential to explore the researchers' interest across the globe for sustainable growth of the emerging metaverse. To fill this gap, in this study, we analyze the interest in the metaverse of three communities i.e., end users, media, and, researchers across six continents of the world. For this purpose, internet search traffic, research articles' frequency, and news frequency are analyzed. The results of this study showed that user's hype for metaverse was more in Europe and Asia as compare to other regions. Also the surge and decline patterns in news were found almost similar to that in search traffic. Starting at the end of 2021, researchers have shown great interest in the field of metaverse and there were huge number of research articles published in the domain onwards. The results of this research are likely to provide intuition for the steady success of metaverse evolutions and connectivity of the physical world and virtual world through storage, computing, and networking convergence. The research's approach and findings should shed light on the steady metaverse transformation success and the coexistence of the real and virtual worlds through hyper-connection and hyper-convergence. This study not only explores the acceptance of the metaverse but also provides valuable insights into the sustainable growth and transformation dynamics of this emerging virtual world.

Keywords: Metaverse · Sustainable Growth · Search Traffic · Technology Adoption Life cycle · Technology Forecasting · Topic Modeling · Metaverse Hype

D. Berkouk et al. (Eds.): CCTS 2024, PTT, pp. 295–303, 2026.
https://10.1007/978-981-95-7607-4_33

1 Introduction

There is a digital technology paradigm shift every few years from PCs to mobile phones and then to the smartphone era. Currently, the digital technology paradigm is shifting toward the "Metaverse". The concept of metaverse is not new and was first coined in the early nineties in a science fiction novel [1]. The author describes a virtual reality-based successor to the internet, which he called the "metaverse". "Ready Player One" by "Ernest Cline" [2] was another novel that made the idea of the metaverse more popular. The metaverse is considered to be a new iteration of the internet that makes use of Virtual Reality (VR) headsets, blockchain technology and avatars within a new synthesis of the real and virtual worlds [3].

With the development of 5G [4], there is a huge rise in research and development on the Metaverse in both business and education. Ever since "Mark Zuckerberg" gave the new name "Meta" to the parent company that owns Facebook [5], the metaverse has been mentioned often in investor presentations [6]. There is common consent that the Metaverse technology will embody physical world activities into this "new" virtual world in many different ways [7]. Considering the importance of this new technology, there is a need to know about people's adoption trends for this new technology. Already study has been conducted to know the people, media, and researcher's interests but it is limited to one country [18].

Given the previous work of researchers and coverage of metaverse hype from the media, in this research work, user search trends, researcher's interest, and media interest are analyzed and compared focusing on countries originating from the six continents of the world (Asia, Africa, North America, South America, Europe, and Australia). This research work aims to provide details of metaverse adoption for sustainable growth of this technology.

The outline of the rest of the paper is as follows: Sect. 2 discusses in detail the related work in the area of the Metaverse. Sect. 3 presents the research methodology used in this research work. The research findings and discussions are presented in Sect. 4. Finally, Sect. 5 concludes the research with a summary of the work presented and its limitations along with directions for future research.

2 Related Work

The word "Metaverse" was first used in future-focused fiction and till now it does not have any single scientifically grounded definition [1]. In [8], the author has defined the Metaverse as a layer between the user and reality. The Metaverse platform has gained the popularity in last few years as many people are shifting their activities online, especially during the coronavirus pandemic [9].

As of, October 2021, according to Google Trend reports, there is a great number of searches on Metaverse in the Google search engine [10] which is more than previous trends. As the metaverse is an environment that is virtual and participants interact with each other through avatars [11], in its different technologies, Virtual worlds have developed from VR, AR, and mixed reality. VR presents an artificial world and lets you be part of it. AR adds virtual objects to the physical world while Mixed Reality (MR) combines AR and VR [12].

New technologies, mostly make a life cycle through the process of social acceptance. So, there is a need for new technologies to analyze the acceptance process and know the development process [13]. According to E. M. Rogers, users can be divided into five categories in the technology adoption life cycle that are innovators, early adopters, early majority, late majority, and laggards [14]. In [15], Moore identified that when it comes to the rapid spread of new technology, there is a static phenomenon in Roger's presented life cycle. Moore defined this static period as a chasm. This chasm occurs between early adopters and the early majority. M. Blosch and J. Fenn presented a model of the hype cycle that identifies the gap between consumer's interests and acceptance decisions [16]. The user's enthusiasm for new technology determines whether they succeed or fail. Therefore, it's important to control the hype and make appropriate recommendations [13]. According to Guihua Zhang, there is much evidence that the current popularity of the metaverse is not motivated by technical factors, but rather related to the public hype [17]. In [18], a study has been presented that analyzes the hype of the metaverse and analyzes the user interests, media interests, and researcher's interests in the metaverse. The limitation of this research study is that it analyzes the hype of metaverse only in South Korea. Also, it is considering only the Scopus database for analyzing metaverse research trends.

In this research work, the hype of metaverse is being analyzed by considering six continents. Also for analyzing research trends, three databases are considered including IEEE, Scopus and Elsevier.

3 Research Methodology

The objective of this research is to assess the hype of metaverse for viable growth of this emerging technology (Fig. 1).

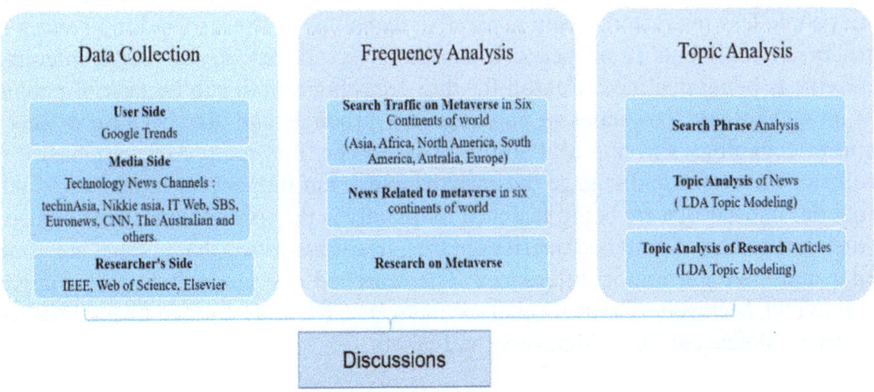

Fig. 1. Research Workflow

3.1 Search Traffic Trends Analysis of Metaverse

Data for search traffic is gathered to analyze the user's interests as search traffic is related to the level of potential adopters' interest. In this study, search traffic on google is analyzed and Google Trends tool is used for this purpose.

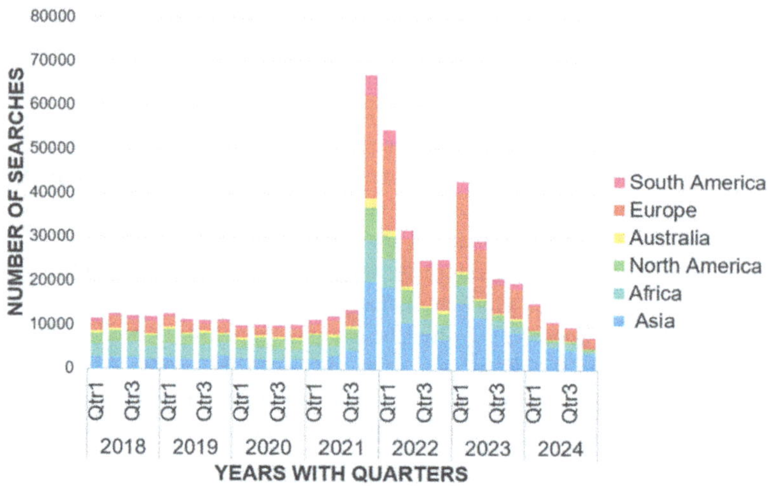

Fig. 2. Metaverse search trends in six continents – A comparison

Figure 2 shows a comparative summary of the search trends of the continents. Data is collected for the past years (2018-2024) and frequency analysis is performed. Also search traffic for the keyword metaverse in each continent is compared during specified time period. For South America, poverty and persistent inflation can be the reasons that made people less interested in any innovative technology. The rise and fall pattern are same for all continents in all years. It can be observed that in year 2024, interest in Metaverse is being declined. Reason for this declining search can be lack of people's trust on technology companies to purchase new products and also because of cost of this new technology and non-availability of resources. It is required to emphasize the benefits of metaverse and engage the early adopters and influencers in this work. Also the queries phrases that are being searched frequently in the six continents are compared. On the comparison of searched queries phrases, it was identified that phrases are almost similar and have got minor differences. The searched queries include Non Fungible Token (NFT), Metaverse Crypto, Sandbox metaverse, Metaverse green token, Facebook Metaverse, Metaverse Coin, Metaverse and others.

3.2 News Media Analysis of Metaverse

News data is collected from different technology related news channels such as: for Asia (techinasia, and NIKKEI Asia), for Africa (IT Web, techmetroafrica, and Africa press), for Australia (SBS, and the Australian), for Europe (Euronews, tech, and techday), for

North America (CNN, and IT world Canada), and for South America (consultancy, and techcrunch) using the key word "metaverse". Metaverse related news from January, 2018 to December, 2022 are searched and collected for all six continents. The news channels that are included are the ones where most of the Metaverse related news were available in the considered years. The year-wise frequency of news mentioning the "metaverse" is analyzed. The news frequencies related to metaverse during 2018-2022 for above mentioned six continents is shown in Fig. 3.

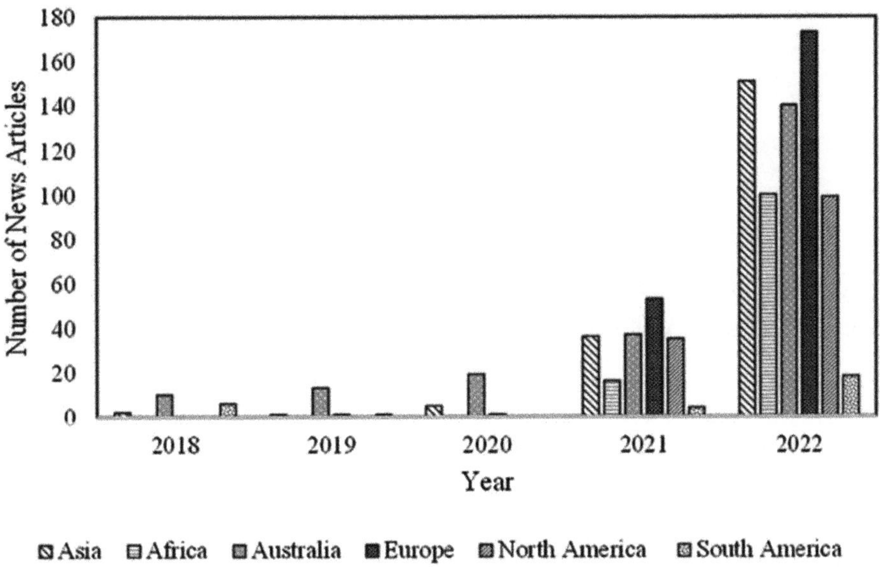

Fig. 3. Yearly graph of number of metaverse related news in news channels of continents

From the results, it can be confirmed that currently metaverse is getting hype in news media. Also, due to COVID-19, metaverse media hype has got immense surge. Availability of money with the big companies can be observed as another reason for this surge. It can be confirmed that as awareness of this technology is increased, media's interest is also increasing. The topics analysis was also performed and it was observed that during the mentioned years, metaverse Facebook, virtual reality, metaverse blockchain were main topics of news including gaming.

News Analysis results can be improved by adding more news channels from different continents and also most recent data can be included in future.

3.3 Analysis of Research on Metaverse

To know about researcher's interests in metaverse, research papers are collected from academic databases IEEE, Web of Science, and Elsevier.

In this study, data has been collected from 2000 to 2024. Although the term metaverse seems to be new but researchers have already used this term long time ago in their work.

Figure 4. Shows the frequency of research papers in three most authentic academic databases.

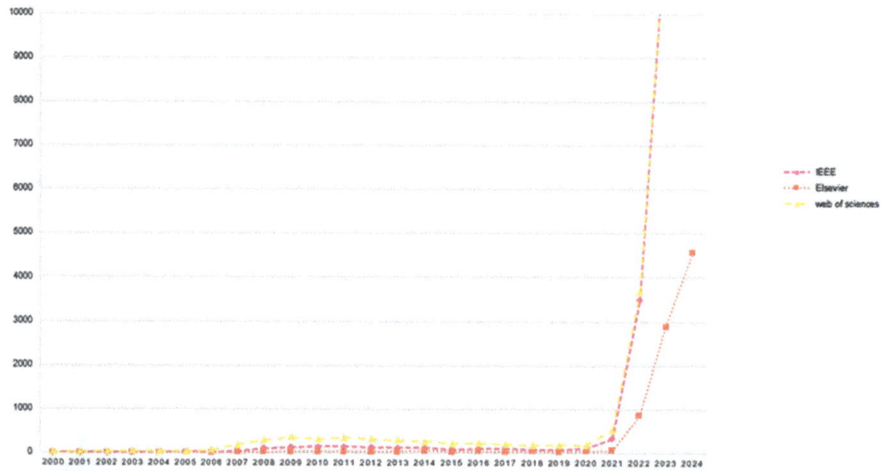

Fig. 4. Annual frequency of research papers in IEEE, Web of Science and Elsevier on metaverse (2005-2024)

As shown, researcher's interest in metaverse was increasing in years after 2007. After 2011, there was a decline in number of related publications. The reason of this decline could be non-availability of infrastructure and cost of metaverse. Then after 2020, metaverse got researcher's attention once again and their interest was escalated. With the availability of money and interest of big companies and brands in metaverse, researchers also found it interesting to work on metaverse. It can be confirmed that current time is apex of metaverse research.

Also topic analysis is performed for researches. Abstracts of research papers were collected from IEEE, Elsevier, Web of science databases to know about topics that were of interest of researchers.

4 Research Findings and Discussions

In this research, search traffic is analyzed to know about interests of users. There is rise in search in last quarter of 2021 in all continents under consideration. Then there was a decline after first quarter of 2022. In last quarter of 2022, again there was surge in search traffic in Europe but not in other continents. In 2024 there was decline in searches for Metaverse. It can be confirmed that user's hype for metaverse was more in Europe and Asia as compare to other regions. Most frequently searched phrases with metaverse were Metaverse NFT, Crypto, metaverse facebook, real estate, metaverse games and some others. This can take us to conclusion that people of ages 30 or more who are economically stable are also interested in metaverse and can use this technology for their business. The news frequency and topics of different technology related news channels for all six continents were analyzed to know about media interest in this new technology. Surge and decline in news were found almost similar to that in search traffic. Difference was for Year 2022, where news number was escalated. Researcher's interest was increasing from 2007 till 2011, then there was slight decline and then since end of 2021 till now there is huge rise in researcher's attention to metaverse. These results are similar to news analysis results. Both shows increasing interests after 2021. Researchers are of the opinion that the next-generation platform would be metaverse that would take the role of online portals, e-learning, and e-commerce.. Metaverse has the ability to effect everyday life. It can create better ways to interact with people who are not closer, have physical disability. Considering the role of Metaverse in future for different fields, there are some probable effects of the Metaverse on medicine in the future. It can be used for remote patient monitoring. Healthcare professionals could schedule virtual consultations and telemedicine appointments in lifelike virtual settings. Through the use of avatars, interactions between patients and doctors could be more individualized and interesting. It can provide users with all the ability to perform any everyday task like shopping, access to doctor, work, play, fashion etc. Despite the fact that the Metaverse is only recently been used in the physical world, there are numerous opportunities for future research with multidisciplinary research teams. It may significantly affect social life. This is the basis for current field research. Also, for this research, we have exclusively used the "Metaverse" as search term. The choice of this keyword is based on its increasing importance in academics and industry discourse. For future research, wider range of keywords like "virtual world" and "extended reality" should be incorporated to include a diverse set of relevant literature studies about this evolving domain.

5 Conclusion and Future Recommendations

Considering the importance of managing the hype of the metaverse, in this research work, the user, media, and researcher's interests are analyzed so that possible users of the metaverse can be provided details about the sustainability of the metaverse. Based on the analysis of this research work, some recommendations are proposed for the viable growth of the metaverse. First, it is required to get the metaverse as a completely different world that is interactive with the physical world. For this purpose, there is a need to develop new procedures. Secondly, studies should be conducted about new users of the metaverse and their satisfactions, and experiences can be studied and these can be used for further development of the metaverse. Lastly, there is a need for ongoing monitoring to analyze the hype and interests of users, researchers, and media to act accordingly. For this purpose, social media platforms can be used where most people express their interest in specific technology.

From the results of this research, it can be predicted that in near future Metaverse will emerge in many fields including Education, Health care, and Entertainment. It can always be improved and considering a lot of research in this field, it can be predicted that in near future it will be used by e-commerce, networking, sales, marketing, crypto business, etc. In the future, this work can be improved by adding more recent data sources including social media and news data. The research results can help to our understanding of how this evolving technology phenomenon might impact the social fabric, economy, and daily lives of individuals in the years to come by analyzing the various paths of metaverse development and acceptance.

References

1. Stephenson, N.: Snow Crash: Roman, 2nd edn. FISCHER Tor (2021)
2. Wadzpay: Ready player one? The metaverse went from fiction to reality. Wadzpay, http://wadzpay.com (2025) Accessed 4 Jun 2025
3. Lee, L.-H., et al.: All one needs to know about metaverse: A complete survey on technological singularity, virtual ecosystem, and research agenda. *arXiv* 2110.05352 (2021)
4. Han, B.: Mobile immersive computing: Research challenges and the road ahead. IEEE Commun. Mag. **57**(10), 112–118 (2019)
5. Zuckerberg, M.: Founder's letter, 2021. Meta, https://about.fb.com/news/2021/10/founders-letter/ (2025). Accessed 4 Jun 2025
6. Fischer, S.: Metaverse bull market. Axios, https://www.axios.com/metaverse-wall-streets-favorite-buzzword-eeb58edb-2c30-490a-bd91-47116195a48f.html (2025). Accessed 4 Jun 2025
7. Farough, A.: The metaverse will feel alive once 'storytelling' becomes 'storyliving'. VentureBeat, https://venturebeat.com/2021/01/28/the-metaverse-will-feel-alive-once-storytelling-becomes-storyliving/ (2025). Accessed 4 Jun 2025
8. Nagwani, N.K.: Summarizing large text collection using topic modeling and clustering based on MapReduce framework. J. Big Data. **2**(1), 6 (2015)
9. Alang, N.: Facebook wants to move to 'the metaverse'—here's what that is, and why you should be worried. Toronto Star, https://shorturl.at/frC26 (2025). Accessed 4 Jun 2025
10. Yoo, C.-M.: Seoul to offer new concept administrative services via metaverse platform. YONHAP News Agency, https://en.yna.co.kr/view/AEN20211103002700315 (2025). Accessed 4 Jun 2025
11. Narin, N.G.: A content analysis of the metaverse articles. J. Metaverse. **1**(1), 17–24 (2021)
12. Díaz, J., Saldaña, C., Avila, C.: Virtual world as a resource for hybrid education. Int. J. Emerg. Technol. Learn. **15**(15), 94–109 (2020)
13. Cabero-Almenara, J., Barroso-Osuna, J., Martinez-Roig, R.: Mixed, augmented and virtual, reality applied to the teaching of mathematics for architects. Appl. Sci. **11**(15), 7125 (2021)
14. Lee, J.Y.: A study on metaverse hype for sustainable growth. Int. J. Adv. Smart Converg. **10**(3), 72–80 (2021)
15. García-Avilés, J.A.: Diffusion of innovation. In: The International Encyclopedia of Media Psychology, pp. 1–8 (2020)
16. Moore, G.A.: Crossing the Chasm. HarperBusiness, New York (1999)
17. Linden, A., Fenn, J.: Understanding Gartner's Hype Cycles. Strategic Analysis Report N° R-20-1971. Gartner, Inc. (2003)
18. Zhang, G., Cao, J., Liu, D., Qi, J.: Popularity of the metaverse: Embodied social presence theory perspective. Front. Psychol. **13**, 997751 (2022)

Sustainable Development of Rural and Remote Areas: A Case Study from Saudi Arabia

Salem Ahmed Alabdali[1,2](✉), Salvatore F. Pileggi[2], and Dilek Cetindamar[2]

[1] Department of Management Information Systems, Jazan University, Jazan 45142, Saudi Arabia
salem.a.alabdali@student.uts.edu.au

[2] School of Computer Science, University of Technology Sydney, Ultimo, NSW 2007, Australia

Abstract. Rural and remote areas have recently gained relevance in sustainable development efforts. Due to the lack of studies in these contexts, many researchers emphasized the need to find long-term solutions to increase living standards. In fact, despite many challenges in these settings, the sustainable development approach is needed to satisfy pressing needs and address inequality gaps. A quantitative study was conducted through an online survey. The results show that multiple factors are perceived to be significant and a determinant for sustainable development in remote and rural regions. However, the results did not support the influence of socio-cultural aspects in these contexts. Technology is expected to play a significant role in promoting sustainable practices, and thus it might be perceived as challenging for certain local people, especially those who have limited exposure to technology. The study offers insight into these settings, from a socio-technical and economic perspective in line with the Saudi Vision 2030.

Keywords: Rural Areas · Sustainable Development · Innovative Solutions · Socio-technical Systems

1 Introduction

Rural and remote areas represent a significant part of many countries around the world; however, they often remain outside the mainstream of the innovative development process. In global discourse, the concept of sustainable development is often presented when looking at these areas as strategic priorities [1]. At the same time, they are often seen as untapped areas with the potential to implement in fact promising and innovative solutions to increase living standards. Therefore, many researchers emphasize the need to find long-term solutions to address the existing inequality gaps. This is critical, also because of the lack of specific research in the field, considering their unique characteristics and peculiarities [2], such as limited resources and digital divide [3, 4].

This survey-based study focuses on potential driving factors for a sustainable development in rural and remote areas by adapting and validating a previous conceptualization targeting rural settings [5]. The considered factors are Management Support (MS), Human Expertise & Skills (HES), Financial Resources (FR), Relative Advantage (RA),

D. Berkouk et al. (Eds.): CCTS 2024, PTT, pp. 304–312, 2026.
https://10.1007/978-981-95-7607-4_34

Perceived Usefulness/Ease of Use (U/EU), IT Infrastructure (ITI), Government Policies (GP), Socio-Economic Situation (SES), and Socio-Cultural Perspective (SCP). Based on the framework, a set of hypotheses has been developed accordingly as shown in Fig. 1 and stated in Table 1. Smart PLS has been adopted to analyze the collected data [6], which resulted in exploring and analyzing the correlations and hidden patterns [7]. The analysis does not involve indicators for the assessment of sustainable development, since they are not in the scope of this paper.

The results offer a deeper insight into rural and remote settings, especially from a socio-technical and economic perspective to understand the environment's complexity. Results, further, suggest possible ways to adopt innovative solutions in these areas as part of the broader strategy of development (Saudi Vision 2030) aimed at increasing the quality of life. That becomes crucial since the major spaces of the country's land are considered as remote and rural areas.

The following section focuses on research model, while the remaining parts of the paper present and discuss the main results.

2 Research Model

As previously mentioned, the conducted research is based on an integrated framework adapted for rural settings [5]. It considers the main characteristics and peculiarities of rural and remote areas. The adapted model was originally designed to implement innovative solutions in these areas. It combines three different theories used in the information systems domain. The Technology Acceptance Model (TAM), Diffusion of Innovation (DOI) theory, and the Technology-Organization-Environment (TOE) framework.

It results from a critical analysis that identifies priorities in context. These combined use of these theories addresses the sustainable development goals from a socio-technical perspective aiming specifically at rural and remote areas [2]. TAM addresses user perceptions in a socio-technical context. It contributes to understand the individual's intention to use technology and the users' preferences [8]. DOI explicitly incorporates relative advantage in complex systems. It explains how new technology can be spread within a social system [9]. TOE deals with the gap between technical capabilities and social needs. It deals with the pathway for innovation to be assimilated within organizations [10].

The framework takes advantage of a variety of theories, which are deemed well-suited to analyse a range of dimensions in remote and rural areas related to sustainable development. All theoretical frameworks presented in the study have been empirically tested and validated in many existing studies such as [11, 12]. Therefore, it can be used to identify open issues and suggest specific hypotheses. In addition, it allows a detailed explanation of the key factors influencing the sustainable development goals, through adopting innovative solutions in these areas.

3 Measurement Model Analysis

The data collected was analyzed using the Smart PLS software, which provides advanced statistical analysis capabilities. Structural Equation Modeling SEM and Partial Least Squares PLS have been applied as suggested by many researchers (e.g [13].). This

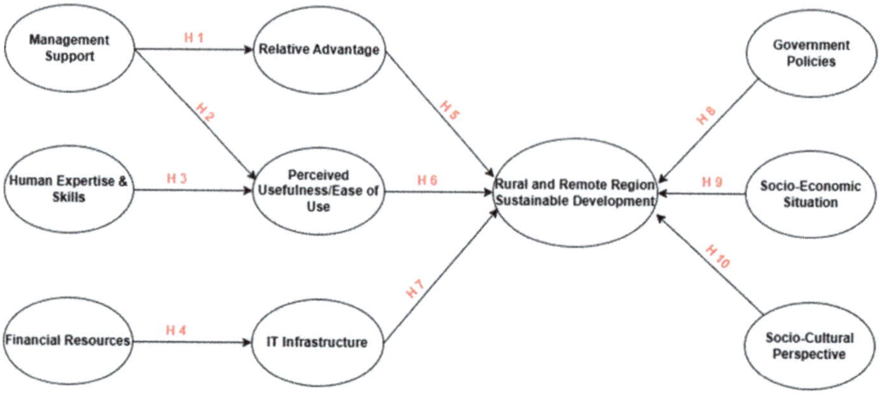

Fig. 1. Conceptual Framework.

section evaluates the validity and reliability of the model measurements. It is an important process for such a model that has many latent variables that need to be explored. The purpose is to provide diagnostic information for the phenomenon under investigation and to avoid incorrect estimates. The following subsections explain the measurement model and the discriminant validity to ensure the data sample is consistent with the proposed model.

3.1 Construct Reliability and Validity

Factor loadings is a statistical measure that indicates the variance in one variable, which can be explained by latent factors. In other words, it explains how well the item represents the underlying construct and generally ranges between -1.0 and $+\ 1.0$ [16]. Values are often assumed to be higher than 0.70, while a higher score is desirable to achieve reliability and validity. The results indicated that all items had a standardized factor loading of 0.70 or higher, meaning that they achieved an acceptable level of validity and reliability.

Cronbach's Alpha and Composite Reliability (CR) have been used to evaluate the reliability of internal consistency [14]. Reliability in this context refers to the degree to which a construct is measured by its relevant indicators. The Cronbach alpha and composite reliability values should be higher than 0.70. The results indicated that all the constructs had Cronbach's alpha values ranging between 0.783 and 0.913 and composite reliability values ranging between 0.795 and 0.918. Hence, the values of both measurements met the suggested criteria and achieved an acceptable level of reliability.

Average variance extracted (*AVE*) is commonly used to assess the validity of convergence, which refers to the extent to which measures of the same or similar constructs positively correspond to each other. According to the guidelines in [15], the AVE should be higher than 0.50. The results indicated that all constructs have AVE scores ranging between 0.590 and 0.825, which meet the suggested criteria. Hence, the convergent validity has been achieved.

3.2 Discriminant Validity (Fornell-Larcker and Collinearity Assessment)

Fornell-Larcker criterion is traditionally used to assess discriminant validity in which the AVE of each construct is distinct and larger than all the squared inter-construct correlations in the structural model. It refers to the extent to which a construct is empirically distinct from other constructs in the same model. This step is widely used in PLS to ensure the validity of the intended model [15, 16]. It simply provides a straightforward way to test the discriminant validity without the need for complex calculations. The results indicated that all the AVE scores that are represented on the diagonal of the matrix are higher than all the squared inter-construct correlations, represented below the diagonal of the matrix. Hence, the Fornell-Larcker criterion satisfies the discriminant validity.

Finally, *Collinearity Assessment* evaluates the high correlation between the variables and the Variance Inflation Factor (VIF). According to the guidelines, the VIF scores should be less than 5 to reflect no collinearity problem in the measurement [16]. It improves predictive power and statistical inference. In addition, it supports data interpretability and prevents bias estimation in the model. The results indicated that all VIF scores for the constructs are less than 5, indicating that there are no collinearity problems in the model.

3.3 Structural Model Assessment (Path Analysis PLS-SEM)

After assessing the validity of the measurement of the given constructs, a PLS-SEM analysis was performed to assess the causal relationship between exogenous variables (predictors) and endogenous variables (outcomes). The path coefficient uses the bootstrapping technique at a significance level (5%) to test the hypotheses (Fig. 2). Moreover, the significance of the path coefficients is typically evaluated using a nonparametric bootstrapping technique with 5000 subsamples [17].

Path Analysis – PLS-SEM was performed to assess the significance of the relationships between endogenous and exogenous variables by testing specific hypotheses that have been developed (Table 1). A path is considered significant if its P-value is less than 0.05 or the T- value statistically is higher than 1.96 of the constructs.

The results of the PLS-SEM analysis indicate that the GP and SEC factors have a significant and positive impact on sustainable development, while the SCP has a negative relationship due to some local cultural norms where the data come from (GP: $\beta = 0.369$, $t = 5.602$, $p < 0.001$; SEC: $\beta = 0.160$, $t = 3.110$, $p < 0.001$). Looking at the path coefficients β, it can be concluded that the impact of GP on sustainable development is stronger compared to SEC. Moreover, all three factors show a positive and significant impact on sustainable development, including RA, U/EU and ITI (RA: $\beta = 0.295$, $t = 4.195$, $p < 0.001$; U/EU: $\beta = 0.172$, $t = 2.799$, $p < 0.001$; ITI: $\beta = 0.114$, $t = 2.554$, $p < 0.001$). Looking at the path coefficients β, it can be concluded that the impact of RA on sustainable development is the strongest, followed by U/EU and ITI. The study has also hypothesized that [RA, U/EU, ITI] mediate the relationship between [MS, FR, HES] and 'Sustainable Development'.

The mediation analysis was performed, and the results were retrieved from the specific indirect effects of PLS-SEM. The results indicated that the RA significantly mediates the relationship between MS and sustainable development ($\beta = 0.043$, $t = 2.364$,

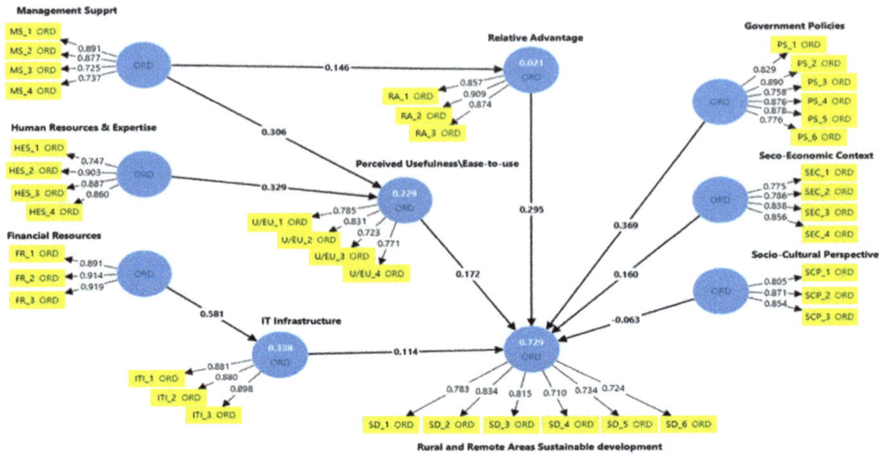

Fig. 2. Path Analysis.

Table 1. Results from PLS-SEM Analysis – Path Coefficients.

Hypothesis	Original sample	Sample mean	Standard deviation	T statistics	P values	Decision
MS → RA	0.146	0.153	0.059	**2.468**	**0.014**	Supported
MS → UEU	0.306	0.312	0.058	5.259	0.000	Supported
HES → UEU	0.329	0.335	0.063	5.253	0.000	Supported
FR → ITI	0.581	0.582	0.042	13.696	0.000	Supported
RA → SD	0.295	0.287	0.070	4.195	0.000	Supported
UEU → SD	0.172	0.177	0.062	2.799	0.005	Supported
ITI → SD	0.114	0.112	0.045	**2.554**	**0.011**	Supported
GP → SD	0.369	0.370	0.066	5.602	0.000	Supported
SEC → SD	0.160	0.162	0.051	3.110	0.002	Supported
SCP → SD	−0.063	−0.063	0.044	**1.455**	**0.146**	**Negative**
MS → RA → SD	0.043	0.043	0.018	**2.364**	0.018	Supported
HRE → UES → SD	0.057	0.060	0.026	**2.204**	0.028	Supported
FR → ITI → SD	0.066	0.065	0.027	**2.504**	0.012	Supported

$p < 0.018$). Moreover, U/EU significantly mediates the relationship between MS and sustainable development ($\beta = 0.053$, $t = 2.498$, $p < 0.013$). Lastly, the ITI significantly mediates the relationship between FR and sustainable development ($\beta = 0.066$, $t = 2.504$, $p < 0.012$).

4 Discussion

The interpretation of data used the Importance-Performance Map Analysis (IPMA) in the context of Structural Equation Modeling (SEM-PLS). The results reveal critical insights in light of priority and performance related to sustainable development in rural and remote areas. It became apparent that certain variables are important for achieving outcomes and where the performance can be improved to maximize the impact.

Allocated resources showed a moderate importance score (0.022) and strong performance (61.557). Similarly, the financial resources of organizations score slightly higher in importance (0.026) and have a performance score of 63.325. The quality of IT infrastructure, with an importance score of (0.025) and performance of (63.915), indicates a constant role in enabling innovation adoption. However, the technology relevance to professional/learning settings scores are lower for both importance (0.013) and performance (41.507), pointing to the need for more alignment of the technology people's needs. The relevance of technology skills to both study and job is important (0.018) but shows a relatively low-performance score (29.835), indicating a significant skill gap. Similarly, the need for additional skills has an importance of (0.018) and performance of (31.014), emphasizing the training needs. Technical support Availability scored (0.019) for importance and (35.849) for performance, pointing out a lack of enough technical support in these areas. From an infrastructure perspective, the resilience of IT systems and access to necessary equipment both show a high importance (0.043 and 0.042) and excellent performance (72.995). The ability of IT infrastructure to integrate new technology is also critical (0.046) with solid performance (72.528). This highlights that the infrastructure is developed well but still requires further improvements.

When it comes to the Government role, the effectiveness of the local Policies and strategies is important (0.072) and performs well (68.32). Similarly, alignment with rural needs has slightly higher importance (0.075) and performance (65.448). Other supportive policies, such as developing local infrastructure (0.063, 68.985) and fostering business investment (0.08, 68.524), are also critical. However, the policies for improvements of human capital (0.08, 66.981) and IT infrastructure improvement efforts (0.07, 63.443) could need to put further effort to enhance the outcomes. Respondents see the value of innovation adoption as highly important (0.1) but with moderate performance (50.472). The perception that better technology can improve practices scored even higher in importance (0.119) but still has a chance to grow in performance (51.297). Similarly, the belief that technology increases productivity is important (0.116) but moderately performing (50.825). In personal perception and community context, the relevance of technology in personal life is significant (0.049) with high performance (77.005). In addition, income-level influence on access to technology (0.048, 76.415) and the perception that education aligns with life goals (0.054, 74.882) indicate strong areas of performance. The view to technology, with the probability to enhance the community performance has a slightly lower importance score (0.046) but significant performance (83.844), making it a major area for sustaining success. On the downside, gender influence on access to technology is concerning, with a negative importance score (−0.02) and poor performance (38.679). Likewise, the belief that technology changes traditional aspects of life (−0.028, 42.925) and enhances the quality of life (−0.027, 43.278) scores negatively, showing societal resistance or dissatisfaction in these areas.

The study addresses a specific socio-cultural context, characterized by some gender imbalance due to cultural norms and social practices. Moreover, the collected data come from the most educated people, with access to digital resources and skills to use them [6]. While providing a consistent case study, such limitations do not allow a complete generalization of the results to other countries. Last but not least, the data was collected one time only, making it impossible to track development changes over time and related perceptions to explore causality.

5 Conclusions

This study investigated the key factors influencing sustainable development in rural and remote areas in Saudi Arabia. It adopts a quantitative approach, using an online survey. A total of 212 valid responses out of 374 were collected from Saudi Arabia in 2024 [6]. The tested factors resulted from a critical analysis to identify the priorities in these areas in light of the Vision 2030 efforts. The survey was distributed among different sectors of healthcare, business, and education to capture perceptions and different perspectives on sustainable and innovative solutions. These sectors may be considered among the main pillars of sustainability, shaping the economic growth, social inclusion, and life quality improvement to any community.

A set of hypotheses has been developed (Fig. 1) and tested (Table 2) by using statistical analysis software. In general terms, the study's results show a positive impact of all considered factors, except for the socio-cultural perspective. The analysis indicates some areas for potential improvement to strengthen the pathway towards a sustainable development. At an organizational level, leaders should prioritize skill training, technical assistance, and having a coherent plan for the incorporation of innovative solutions. These areas are overall growing, but the performance expectations are not currently fully met. Similar considerations apply for technology as a further investment in infrastructure and innovation is crucial. From a socio-cultural perspective, addressing the different social barriers is crucial to provide equal access and ensuring the maximum benefits of potential innovations.

This study contributes to the scholarly literature by offering a more comprehensive understanding of the key factors influencing the rural and remote areas' sustainability based on self-reported data from local communities. Additionally, it provides deep insight into the complexity of these areas, especially from a socio-technical and economic perspective. Although considering the previously discussed limitations, in practice it helps decision-makers to formulate appropriate strategies to minimize development gaps. Furthermore, it naturally defines future research steps, looking for a more integrated development approach to these communities. In the local context, it supports the Saudi Vision 2030 endeavors by shedding light on critical aspects that need to be considered when looking at innovative solutions in the context of current policies and practices.

References

1. Yin, X., Chen, J., Li, J.: Rural innovation system: Revitalize the countryside for a sustainable development. J. Rural. Stud. **93**, 471–478 (2022)

2. Alabdali, S.A., Pileggi, S.F., Cetindamar, D.: Influential factors, enablers, and barriers to adopting smart technology in rural regions: A literature review. Sustainability. **15(10):7908** (2023)
3. Statti, A., Torres, K.: The forgotten minority: Exploring deficiencies in access to education and technology in rural America. Peabody J. Educ. **95**(2), 173–182 (2020)
4. Mustafa, F., Nguyen, H.T.M., Gao, X.A.: The challenges and solutions of technology integration in rural schools: A systematic literature review. Int. J. Educ. Res. **126**, 102380 (2024)
5. Salem Ahmed Alabdali and Salvatore Flavio Pileggi. A research framework to identify determinants for smart technology adoption in rural regions. In *Proceedings of the 1st International Conference on Creativity, Technology, and Sustainability, CCTS* 2024, 15-16 May, Jeddah, Saudi Arabia, 2024.
6. Alabdali, S.A., Pileggi, S.F., Bharathy, G.: Self-reported data for sustainable development from people living in rural and remote areas. Data. **10**(1), 6 (2025)
7. Richter, N.F., Tudoran, A.A.: Elevating theoretical insight and predictive accuracy in business research: Combining PLS-SEM and selected machine learning algorithms. J. Bus. Res. **173**, 114453 (2024)
8. Maranguni'c, N., Grani'c, A.: Technology acceptance model: a literature review from 1986 to 2013. Univ. Access Inf. Soc. **14**, 81–95 (2015)
9. Meade, N., Islam, T.: Modelling and forecasting the diffusion of innovation–a 25-year review. Int. J. Forecast. **22**(3), 519–545 (2006)
10. Ramdani, B., Chevers, D., Williams, D.A.: SMEs' adoption of enterprise applications: A technology-organisation-environment model. J. Small Bus. Enterp. Dev. **20**(4), 735–753 (2013)
11. Baig, M.I., Shuib, L., Yadegaridehkordi, E.: A model for decision-makers' adoption of big data in the education sector. Sustainability. **13**(24), 13995 (2021)
12. Chatterjee, S., Rana, N.P., Dwivedi, Y.K., Baabdullah, A.M.: Understanding AI adoption in manufacturing and production firms using an integrated TAM-TOE model. Technol. Forecast. Soc. Chang. **170**, 120880 (2021)
13. Ringle, C., Da Silva, D., Bido, D.'o.: Structural equation modeling with the smartpls. Brazilian Journal of Marketing. **13**(2), 2015. Bido, D., da Silva, D., & Ringle, C. (2014)
14. Tavakol, M., Dennick, R.: Making sense of Cronbach's alpha. Int. J. Med. Educ. **2**, 53 (2011)
15. Hamid, M.R.A., Sami, W., Mohmad Sidek, M.H.: Discriminant validity assessment: Use of Fornell & Larcker criterion versus htmt criterion. In: Journal of physics: Conference series, vol. 890, p. 012163. IOP Publishing (2017)
16. Mooney, C.Z., Duval, R.D., Duvall, R.: *Bootstrapping: A nonparametric Approach to Statistical Inference*, vol. 95. Sage (1993)
17. Qahl, M., Sohaib, O.: Key factors for a creative environment in Saudi Arabian higher education institutions. *J. Inf. Technol. Educ. Innov. Pract.* (2023)

Shaping a Sustainable Future: The Role of Creativity in Societal and Economic Transformation

Effects of Public Spending on Economic Growth: Evidence from the GCC Countries

Salwa Abdulraquib[1](✉) and Rozina Shaheen[2]

[1] Dar Al-Hekma University, Jeddah, Saudi Arabia
sraquib@dah.edu.sa
[2] Effat University, Jeddah, Saudi Arabia
roshaheen@effatuniversity.edu.sa

Abstract. This research endeavor myopically examines the deep relationship between public spending and economic growth in five selected Gulf Cooperation Council (GCC) countries by employing panel data from 2003 to 2022. Multiple linear regression was operated by hypothesizing and theorizing five explanatory variables. The OLS specifications explicitly demonstrate that public spending including education expenditure, health spending, general government spending and foreign direct investment, have no significant impact on economic growth. It was also clearly revealed that the other relevant explanatory variables including income from natural resources and population growth fairly left some prominent impact on economic growth. Moreover, statistical analysis confirmed that income generated from natural resources has a positive association with the economic growth while population growth has an indirect relationship with the economic growth. A paradigm shift in fiscal policy is desired to offer more public spending on projects associated to optimal utilization of natural resources. Moreover, government spending on health and education services should be course-targeted and resource specific rather than of generic nature. In future, more rigorous data may be taken up to advance the empirical research in this fiscal domain so as to arrive at some holistic policy messages.

Keywords: Public spending · GDP · GCC countries

1 Introduction

This is an established fact that economic growth enhances national wealth and reduces poverty, unemployment, and social challenges. Therefore, it has been a key policy objective globally [1]. Public spending is a major driver of economic growth, as theorised in the Keynesian economics which argues that government intervention stimulates demand, particularly during economic downturns. In contrast, classical economics holds the narrative that markets are fairly self-correcting in the long run. Additionally, investments in human capital, such as on education and health, improve labor productivity and economic output [1]. Exceptions may be there but that needs to be researched how and under what situations conventional wisdom is disqualified.

D. Berkouk et al. (Eds.): CCTS 2024, PTT, pp. 315–323, 2026.
https://10.1007/978-981-95-7607-4_35

1.1 Economic Background in GCC Countries

It is projected that GCC economies may grow by 1% in 2023, 3.6% in 2024, and 3.7% in 2025 It has also been noted that the COVID-19 pandemic and oil price shocks significantly impacted the region and government interventions mitigated their effects [2]. During 2008 financial crisis, countries like Saudi Arabia, Kuwait, UAE, and Qatar have been experiencing sharp slowdowns in terms of GDP growth, while Oman and Bahrain faced moderate declines [3].

1.2 Research Focus

This research study examines the impact of public spending on economic growth in GCC countries from 2003 to 2022. This novel work is addressing the key question of how does public spending (general, government, education, and health) influence GDP per capita growth? A focus has been made on short run and long-term imperatives.

2 Literature Review

Paradoxically, the economic relationship between public spending and GDP growth has been widely debated in the literature. This research endeavor has largely focused on whether the composition of public expenditure influences long-term economic performance. Some studies argue that capital expenditures promote growth, whereas others caution that excessive investment in capital relative to current expenditures can lead to inefficiencies [4]. Misallocation of public spending, especially in developing economies remains a critical issue. This is often favoring capital investment at the expense of essential recurrent expenditures [5].

It has been observed that investment in human capital enhances economic productivity and long-term growth. With disparities in educational attainment is a key factor in economic performance gaps between developed and developing countries [6]. The effect of education spending on GDP per capita may be negative or insignificant in the short run [7]. Public expenditure in these sectors has a more significant impact on GDP growth [8]. Some studies argued that increased public spending on healthcare positively correlates with GDP growth [9]. The mediating role of other economic factors, such as oil revenues, may weaken the direct relationship between health expenditures and growth [10].

In this context, the relationship between public spending and economic growth is not universally consistent across economies. Variations in institutional quality, economic structure, and governance play crucial roles in determining the effectiveness of government expenditure [11]. This fact cannot be ignored that investments do not always translate into tangible economic benefits [12]. The long-term effects of public expenditure on GDP per capita in selected GCC countries have been examined. The most recent economic data of these countries was employed which fairly share similar economic structures and policy frameworks.

2.1 Hypotheses Development

Based on the theoretical framework, the following hypotheses have been tested at 1%, 5%, and 10% level of significance so as to examine the impact of seven explanatory variables on GDP per capita growth. The null hypothesis states that explanatory variables collectively have no significant effect:

$$\beta 1, \beta 2, \beta 3, \beta 4, \beta 5, \beta 6, \beta 7 = 0 \tag{1}$$

The alternative hypothesis suggests that at least one explanatory variable significantly influences GDP per capita growth:

$$\beta 1, \beta 2, \beta 3, \beta 4, \beta 5, \beta 6, \beta 7 \neq 0 \tag{2}$$

3 Data Specifications

The study initially considered all six GCC countries, but the United Arab Emirates was excluded due to incomplete and inconsistent data. The final sample includes Bahrain, Kuwait, Oman, Qatar, Saudi Arabia with 100 panel observations spanning over a period of 30 years from 2003 to 2022. Seven explanatory variables were carefully selected based on data availability from World Bank datasets [13, 14], except for income from natural resources, which serves as a proxy for oil returns and was gauged from global economy.com. Missing values (23 instances) were estimated using a three-year moving average. The study applies the specification of multiple linear regression (OLS method) to quantify the extent to which variations in explanatory variables influence GDP per capita growth. The general model structure is (Table 1):

$$\text{Yit} = \beta 0 + \beta 1 \text{ X it} + \beta 2 \text{ X it}, +.. \cdots + \beta \text{k X it} + \epsilon_{it} \tag{3}$$

i = 1,2,.....,n are the observations from the sample; t = time period;

Y_{it} = observation i of the dependent variable; $X_1, X_2, \ldots, X_k$ = independent variables;

β_0 = constant (free term of equation);$\beta_1, \ldots, \beta_k$ = coefficients of explanatory variables;

ϵ = error term of equation.

3.1 Descriptive Statistics and Correlations

This section presents the descriptive statistics of key variables and their correlation with GDP per capita growth. With 100 observations, trends in public spending, investment, and macroeconomic indicators was empirically captured (Table 2).

The descriptive statistics show that GDP per capita growth averages out at 0.5%, with a 0% median, thereby indicating symmetry. Standard deviation (4.5) suggests moderate variability and positive skewness (0.6) indicates a slight rightward distribution. GDP growth has a mean of 4.5%, median of 4.2%, and SD of 5.6 with a moderate kind of positive skewness (1.0).

Table 1. Summarizes variable definitions and measurement units.

The Variables	Unit	Definition
Dependent Variable		
GDP per capita growth	Percent	Annual percentage growth rate of GDP per capita based on constant local currency
Explanatory Variables		
Education expenditure as % of GNI	Percent	General government expenditure on education expressed as a percentage of total general government expenditure
General Government spending as percent of GDP	Percent	General government final consumption expenditure
Income from natural resources as percent of GDP	Percent	Total natural resources rents are the sum of oil rents, natural gas rents, coal rents (hard and soft), mineral rents, and forest rents.
Population growth	Percent	Annual population growth rate for year t
Health spending as precent of GDP	Percent	Level of current health expenditure expressed as a percentage of GDP
GDP growth (annual %)	Percent	Measured at market prices based on constant local currency
Foreign Direct Investment as percent of GPD	Percent	The net inflows of investment to acquire a lasting management interest

Government spending averages out at 19% of GDP, with minimal skewness (0.2) and a leptokurtic distribution. Education expenditure averages out at 13%, with an SD of 4.0, slight positive skewness (1.0) and a leptokurtic curve.

It was further testified in the shape of correlation analysis that DP growth (0.7) was the strongest positive determinant of GDP per capita growth. Education expenditure, foreign direct investment, and natural resource income have weak positive correlations (0.1–0.2), while government spending, health spending, and population growth show weak negative correlations (−0.1 to −0.2).

These results emphasize GDP growth as the key driver, while the influence of sectoral expenditures varies. Further econometric analysis has examined these relationships in detail.

3.2 Estimating the Multiple Regression Model and Selecting the Best Model

To examine the determinants of GDP per capita growth, a panel least squares regression model was estimated by using data from five GCC countries over a 20-year period (2003–2022). The estimated model is specified as follows (Table 3):

Table 2. Descriptive Statistics

	EDUC	GDP	GDP_GWTH	GOV	HEALTH	INVEST	NAT	POP
Mean	13.1	0.5	4.5	18.8	3.7	2.5	32.7	3.8
Median	12	0	4.2	18	4	1.43	33	3
Maximum	25	15	26.2	29	6	15.8	59	19
Minimum	5.9	−12	−8.9	10	2	−2.76	10	−3
Std. Dev.	4.2	4.5	5.6	4.4	1.2	3.01	12.5	3.8
Skewness	0.98	0.55	0.99	0.2	0.23	1.4	0.2	1.41
Kurtosis	3.1	4.2	5.5	2.06	2.2	5.9	2.043	6.5
Jarque-Bera	16.1	10.7	42.3	4.1	3.6	66.6	4.3	82.6
Probability	0.0003	0.005	0	0.13	0.17	0	0.12	0
Sum	1306.6	51	449.4	1882	374	248.7	3274	384
Sum Sq. Dev.	1770.3	2015	3074.1	1916.8	153.2	897.9	15433.2	1453.4
Observations	100	100	100	100	100	100	100	100

$$GDP\ Per\ Capita\ Growth_{it} = \beta_0 + \beta_1 EDUC_{it} + \beta_2 GOV_{it} + \beta_3 NAT_{it} + \beta_4 POP_{it} + \beta_5 HEALTH_{it} + \beta_6 GDP\ Growth_{it} + \beta_7 INVEST_{it} + \epsilon_{it} \quad (4)$$

Where:

i. i represents the cross-sectional unit (country), t represents the time period (year),
ii. ϵ_{it} is the error term.

The regression results identify GDP growth, natural resource income and population growth as key determinants of GDP per capita growth. GDP growth ($p < 0.01p < 0.01p < 0.01$) has the strongest positive effect, where a 1% increase leads to a 0.95% rise in per capita income. Natural resource income ($p < 0.01p < 0.01p < 0.01$) shows a moderate positive impact. With a 1% increase may result in a 0.015% rise in GDP per capita. In contrast, population growth ($p < 0.01p < 0.01p < 0.01$) negatively affects GDP per capita. It was noted that a 1% increase leads to a 1.03% decline, indicating potential economic challenges. Conversely, education expenditure, government spending, health expenditure, and investment are statistically insignificant ($p > 0.10p > 0.10p > 0.10$), suggesting no direct impact on GDP per capita growth in the short run. These sectors contribute to broader economic and social development. But their immediate effect on income growth is limited.

The model demonstrates a strong fit, with an R-squared value of 0.98, indicating that 98% of GDP per capita growth variation is reasonable explained by the included

Table 3. Panel Regression Results and Model Diagnostics

Variable	Coefficient	Std. Error	t-Statistic	Prob.
C (Intercept)	−0.0269	0.3559	−0.0755	0.9400
EDUC (Education Expenditure)	−0.0161	0.0147	−1.0948	0.2765
GDP_GWTH (GDP Growth)	0.9494	0.0114	83.6402	0.0000***
GOV (Government Spending)	−0.0049	0.0182	−0.2685	0.7889
HEALTH (Health Expenditure)	0.0059	0.0751	0.0783	0.9377
INVEST (Investment)	0.0081	0.0171	0.4714	0.6385
NAT (Natural Resource Income)	0.0147	0.0053	2.7752	0.0067***
POP (Population Growth)	−1.0285	0.0167	−61.4361	0.0000***
Model Diagnostics		Value		
R-squared		0.9896		
Adjusted R-squared		0.9888		
F-statistic		1251.091		
Prob (F-statistic)		0.0000***		
Durbin-Watson Statistic		1.8160		
Root MSE		0.4577		

variables. The F-statistic ($p < 0.01p < 0.01p < 0.01$) confirms the model's overall significance. Durbin-Watson statistic (1.8160) suggests no autocorrelation, thereby affirming result reliability. To refine the model, backward elimination was also applied, by systematically removing insignificant variables.

It was also analyzed that GDP growth, natural resource income, and population growth remained significant at 1% across all specifications. On the other hand, education expenditure, government spending, health expenditure, and investment were consistently eliminated. Introducing a COVID-19 dummy variable revealed a negative and significant impact (β = −0.37, p < 0.01\beta = −0.37, p < 0.01β = −0.37, p < 0.01). These results indicate that the pandemic contracted per capita growth of GDP. However, the 2008 financial crisis dummy was found to be insignificant. This suggests that there is no lasting effect on economic growth in the selected GCC economies.

3.3 Diagnostic Tests

To ensure the validity of the model, several diagnostic tests were conducted. The White test for heteroskedasticity ($\chi 2 = 0$, $p < 0.05$) confirms the presence of heteroscedasticity. This necessitates the use of robust standard errors. The Variance Inflation Factor (VIF) test indicates no severe multicollinearity, as all VIF values are below the conventional threshold of 10. Additionally, the Durbin-Watson statistic (1.72) and Breusch-Godfrey serial correlation LM test ($\chi 2 = 3.2$, $p = 0.2 > 0.05$) confirm the absence of autocorrelation in the residuals.

The Jarque-Bera normality test presents mixed results. These are suggesting that while the data meets normality assumptions at the 1% level, it is rejected at the 5% level, indicating the possibility of minor deviations from normality. However, given the large sample size (N = 100), the central limit theorem ensures that these deviations do not significantly impact the reliability of the statistical inferences.

3.4 Removing the Heteroskedasticity

Given the presence of heteroskedasticity, robust standard errors were employed to ensure reliable statistical inferences. The White test for heteroskedasticity was re-estimated after applying log-level transformation. This confirmed that the operated model is now homoscedastic ($\chi 2 = 0.33$, $p > 0.05$). These adjustments ensure that the statistical results remain unbiased and robust to heteroskedasticity concerns. All such diagnostics strengthen the empirical reliability of the findings.

4 Results and Discussions

In this research endeavor, it was testified that education, health, and government spending have no short-term impact on economic growth. However, their long-run effects may be explored in the future research with more advanced data and econometric techniques. The best-fit model includes GDP growth, natural resource income, and population growth all of which were significant at the 1% level though with low magnitude effects. GDP growth and natural resource income positively influence GDP per capita growth, whereas population growth has a negative impact. Diagnostic tests confirm that standard deviations are within an acceptable range. Jarque-Bera test supports normality at the 1% level. With no multicollinearity and some heteroskedasticity, the use of robust standard errors ensures result reliability. The Durbin-Watson and Breusch-Godfrey tests confirm no autocorrelation, while an R-squared value of 0.98 and a highly significant F-statistic indicate a strong predictive model. All such specifications clearly demonstrate the empirical justification of our findings.

5 Conclusion

This research effort has empirically employed multiple linear regression and revealed that there is no significant short-term impact of public spending on GDP growth. On the other hand, GDP growth, natural resource income, and population growth have been found as significant predictors. The alternative hypotheses for education expenditure, health spending, foreign direct investment, and government spending are rejected, while those for GDP growth, natural resource income, and population growth are accepted. Contrary to the conventional wisdom that there is direct association between population growth and economic expansion, this study finds otherwise, thereby revealing a potential economic strain. It was further testified that GDP growth and natural resource income remain positively correlated with GDP per capita growth. A paradigm shift in fiscal policy is highly desired to offer more public spending on public sector development projects associated to optimal utilization of natural resources. Moreover, government spending on health and education services should be targeted and resource specific rather than of generic nature following the traditions. The focus of future research should be on exploring the long-term effects of public spending by expanding the canvas of dataset across more regions and long time series and thus by applying advanced econometric techniques to reassess its significant impact of such spending on economic growth.

References

1. Ilkhom Shapirov: Contemporary economic growth models and theories: A literature review. Central and Eastern European Library (2015)
2. International Monetary Fund e-Library Homepage, https://www.elibrary.imf.org/view/journals/007/2022/052/article-A001-en.xml, last accessed in 2024/11/20.
3. Salah, I.: The Effect of the Financial Crisis on the Middle East, 1 Global Bus. L. Rev. **1**, 99–108 (2010)
4. Ada, A., Acaroglu: Human capital and economic growth: a panel data analysis with health and education for MENA region. Adv. Mgmt. & Applied Econ. **4**(4), 59–71 (2014)
5. Al-Yousif, Y.K.: Education expenditure and GDP per capita growth: Some empirical evidence from the GCC countries. J. Dev. Areas. **42**(1), 69–80 (2008)
6. Ficawoyi, D.-A., Pradhan, G., Basnet, H.C.: Human capital expenditure and income in developing countries. J. Dev. Areas. **55**(3), 113–128 (2021). https://doi.org/10.1353/jda.2021.0055
7. Kocevska, K.S.: Public expenditure on education and GDP per capita growth: Evidence from North Macedonia. J. Liberty & Int'l Aff. **9**(1), 22–34 (2023)
8. Mukhtarov, et al.: The relationship between government expenditures on education and GDP per capita growth: The case of Azerbaijan. Res. World Econ. **10**(2), 195–205 (2020)
9. Mladenović, I., et al.: Analyzing and management of health care expenditure and gross domestic product (GDP) growth rate by adaptive neuro-fuzzy technique. Comput. Hum. Behav. **64**, 524–530 (2016)
10. Hamdan, A.: Khamis, Reem: The mediating role of oil returns in relationship between investment in higher education and GDP per capita growth: Evidence from Saudi Arabia. Econ. Soc. **13**(1), 116–131 (2020). https://doi.org/10.14254/2071-789X.2020/13-1/8
11. Ramirez, M., Nazmi, N.: 2003, Public investment and economic growth in Latin America: an empirical test. Rev. Dev. Econ. **7**(1), 115–126 (2023)
12. Mercana, M., Sezerb, S.: The effect of education expenditure on GDP per capita growth: The case of Turkey. In: In: 2nd World Conference On Business, Economics and Management, Hungary (2013)
13. The World Bank Homepage, last accessed on 2025/02/25 https://datahelpdesk.worldbank.org/knowledgebase/articles/906519
14. The World Bank Homepage, last accessed on 2025/02/25 https://data.worldbank.org/indicator/NY.GDP.MKTP.KD.ZG?locations=1A

The Impact of ESG on Financial Performance and Market Value in Saudi Arabia

Roba Al Sharif and Karima Saci(✉)

Dar Al-Hekma University, Jeddah 222464872, Kingdom of Saudi Arabia
{rhalsharif,ksaci}@dah.edu.sa

Abstract. This paper examined the impact ESG has on financial performance and market value of businesses. Saudi Arabia is embracing ESG in alignment with its vision 2030. Additionally, ESG is beneficial in strengthening relationships with investors, prioritizing long-tern gains and encouraging transparency and accountability. These benefits increase sales and returns for the organization, resulting in economic growth and development. However, challenges such as corporate resistance, varying regulations, lack of standardized metrics and data reliability issues persist. These are found to be addressed through education, standardization, mandatory ESG practices and modern technologies.

Keywords: Environmental · Social · and Governance (ESG) · Saudi Vison 2023 · Financial Performance · Value of Businesses

1 Introduction

Environmental, Social, and Governance (ESG) principles assess the operations of an organization to ensure they observe governance, sustainability and ethical practices. The principles are important because they protect the long-term interests of all stakeholders, and promote risk management and transparency (Mrchkovska et al. 2023). The environmental aspect promotes the protection of the planet. This is done by reducing the carbon emissions, promoting resource conservation, proper management of waste and embracing energy efficiency (Baratta et al. 2023; Sun et al. 2024). Therefore, the application of environmental factors reduces the carbon footprint of the organization and promotes the fight against climate change. The social factors promote the wellbeing of all stakeholders. In this case, the organization must observe product safety, labor law, human rights and diversity and inclusion (Baratta et al. 2023; Sun et al. 2024). The Governance pillar involves adhering to ethical standards, promoting accountability and transparency and protecting the rights of all stakeholders (Baratta et al., 2024; Sun et al. 2024). In Saudi Arabia, ESG is increasingly being adopted because it matches with vision 2030. The Public Investment Fund (PIF) is heavily investing in industries like renewable energy, technology and infrastructure for Saudi Arabia to get to the leadership position when it comes to observing sustainability. Approximately 68% of businesses in the country have adopted ESG principles to promote environmental conservation and sustainability (Hussain et al. 2024). The adoption of ESG is increasing despite the market

D. Berkouk et al. (Eds.): CCTS 2024, PTT, pp. 324–332, 2026.
https://10.1007/978-981-95-7607-4_36

experiencing challenges like corporate resistance, difference in regulations, inconsistent and unreliable data and lack of standardized approaches to measure performance (Fometescu et al. 2024; Chopra et al. 2024). These challenges may be addressed through Saudi Arabia aligning ESG with vision 2030 more strategically and implementing the mandatory adoption of ESG principles, developing ESG regulations for every sector, promoting the growth of the green initiatives like green finance and the use of renewable energy, increasing governance, transparency and foreign investments, and integrating AI in ESG principles (Chopra et al. 2024). This paper analyzes the effect of ESG on financial performance and market value.

2 Literature Review

Many studies have supported the notion that ESG has an effect on financial performance and market value. Kim and Li (2021) explored the effect of ESG on financial risk and profitability. The results of the study showed that ESG increases profitability, especially for large corporations. These findings were further supported by Cheema-Fox et al. 2021who suggested that public and large corporations survived after the emergence of Covid-19 due to the integration of ESG into their activities and practices. Lee and Koh (2024) also found that ESG reduces risks for financial firms. Kim and Li (2021) went further to highlight that the most important factor is governance. With proper governance, the investors have confidence on the leadership of the organization since they focus on long-term gains. The social factor in ESG was considered the most important when it comes to credit rating. However, the environmental factor negatively affects credit rating. However, these findings were objected by Alshehri (2024) who found that financial performance determined by return on assets (ROA) and return on equity (ROE), were positively influenced by social and environmental factors. The study also found that governance negatively affected financial performance but the effect was insignificant.

Additionally, various studies discussed the factors that promoted the increased adoption of ESG in organizations. Fu and Li (2023) discussed that digital transformation encouraged the adoption of ESG, which in return promotes financial performance. This notion was supported by Sandberg et al. (2023) who investigated the relationship between ESG ratings and financial performance in the food industry of Europe. The study suggested that advancement of technology has made it easy to measure ESG ratings. This information is beneficial to investors and the stakeholders of the organization since it enables them to understand the performance of the business and make informed decisions.

ESG is beneficial to the organization in various ways. It cushions the organization from unpredictable events and crises like the 2008 financial crisis and Covid-19. Zhou et al. (2022) explained that ESG helps companies to prepare for such events, thus improving their financial performance and market share. The findings were supported by Cheema-Fox et al. (2021) and Lee and Koh (2024) who investigated ESG as a way of organizations responding and being resilient during crises like covid-19. During the period, large and public corporations were able to protect themselves from collapsing. These organizations were able to adjust to the situation. The ability to adjust to the situation strengthened the relationship between the organization and employees and reduced turnover rates, which increased market value. Another benefit of ESG is that it promotes transparency and accountability. Gao et al. (2022) found that ESG promotes transparency and accountability. This encourages investors to have confidence on the leadership of the organization and promotes customer trust.

Iazzolino et al. (2023) analyzed the influence ESG has on financial efficiency on firms in different sectors in the European market. The study suggested that ESG helps investors make informed decisions on where to invest their money. The results of the study showed that ESG promotes financial efficiency. However, the rate at which it promotes financial efficiency varies from one industry to another. It is more beneficial to risky businesses. Gao et al. (2022) claimed that ESG is beneficial for organizational financial performance because it focuses on long-term goals and the interest of all stakeholders. The implementation of ESG has a positive impact on the financial performance and also the reputation of the organization. Despite the benefits, Janah and Sassi (2022) suggested that ESG has its challenges, especially in developing countries. The authors held the opinion that ESG is only beneficial in developed economies. These findings were supported by Malekian and ALShakban (2024) who found that financial inflexibility negatively affects financial performance.

Overall, the current literature presents conflicting views on the relationship between ESG and financial performance. The claim that ESG has positive financial implications only in developed countries highlights the need for developing countries, such as Saudi Arabia, to thoroughly investigate the consequences of adopting and mandating ESG practices.

3 Discussion

3.1 Corporate Behavior

Integrating ESG provides competitive advantage that influence corporate behavior by reducing risk, driving growth, and enhancing resilience, increasing work productivity, and developing good reputation and stakeholder trust (Treepongkaruna and Suttipun 2024). Companies with strong ESG practices can mitigate risks related to regulations and public perception, as well as avoid legal liabilities and reputational damage, as shown by Volkswagen's emissions scandal (Hotten 2015; Yang et al. 2025). Moreover, since ESG roots from the purpose of sustainability, it uses a long-term strategy that promotes growth. Additionally, it fosters organizational resilience. Resilience is achieved by gathering knowledge, anticipation, and mastering adaptation. The corporation should learn how to adapt to shifting regulatory, technological, and market conditions to be able to go through any crisis (Leoni, 2024). Furthermore, workforce productivity creates positive work environments that attract and retain top talent, while boosting stakeholder trust through transparent and ethical practices, which, in turn, strengthens corporate reputation and builds investor confidence (Delevingne et al., 2020). Adopting ESG principles significantly enhances employee commitment and happiness, leading to increased productivity (Treepongkaruna and Suttipun, 2024). When employees see ethical behavior, they are more likely to stay motivated and perform better. However, when they encounter unethical practices or notice unfair treatment, it can lead to lawsuits, harm the company's reputation, and create internal dissatisfaction. Having a clear governance system in place, aligned with ESG principles, makes employees more ethical which enhances risk management (Kim and Li 2021; Lee and Koh 2024). It reduces risk of fraud which limits the potential for costly backlashes such as losing stakeholder trust or the license to operate.

3.2 Impact on Investor Decision-Making

ESG factors have become increasingly influential in shaping investor decision-making. Investors are no longer focused solely on financial performance; they are now looking at how companies handle sustainability, ethics, and corporate governance (Delevingne et al. 2020). Since ESG is growing in importance for investors, rating agencies like MSCI evaluate companies based on their ESG practices, assigning ratings from CCC (low) to AAA (high). These ratings help investors assess how well companies manage sustainability and long-term risks, making highly rated companies more appealing for long-term, responsible investment (MSCI 2024). Some investors increasingly favor sustainable funds during uncertain times due to their ability to manage risks and adapt to disruptions effectively (Delevingne et al. 2020; Landi et al. 2024). Another ESG pillar that influences the decisions made by investors is governance. Poor governance makes investors lose confidence in the leadership of the organization. For example, when the leadership of the organization decides to misreport its financials or embrace insider trading, it shows that all they care about are short term profits and not long-term or sustainable growth of the business (Park and Jang 2021). No investor would want to be associated with such a business. However, it is different for organizations that are accountable and transparent. Such businesses promote investors' confidence or trust, thus attracting and retaining them. It shows that the leadership is not only serving their interest but also those of other stakeholders (Delevingne et al. 2020).

3.3 Role of Regulations and Policy

Policies and regulations are important when it comes to the adoption of ESG. They help guide organization what is the best course of actions. In the beginning of ESG, its adoption was optional (Passas 2024). However, it has now became mandatory in many countries due to the benefits they present on the organizations and the economy as a whole. When it comes to organizations, ESG principles enable them to prepare and thrive in case of a crisis, which is beneficial to the growth of the economy (Zhou, Liu 2022; Luo 2022; Cheema-Fox et al., 2021; and Lee and Koh, 2024). These regulations encourage businesses to observe equality and fight fraud and climate change (Muigua, 2022). They enable organizations to not only focus on short-term benefits but also long-term ones. They also prevent organizations from greenwashing (Park and Jang, 2021). This means that companies cannot falsely claim that they are observing sustainable practices if they are not. The role of regulations and policy can be clearly understood by analyzing the EU and US markets. The EU leads the way with strong ESG regulations, mandating companies to integrate ESG into their operations. It is currently outpacing other regions like the United States where ESG regulations remain less standardized (LaForge, 2021). Europe is seen to remain the largest market with over $18 trillion in ESG assets by 2030 (Bloomberg, 2024). Environmental policies like the European Union's Green Deal and the Paris Agreement enforce strict climate goals, pushing companies to reduce carbon emissions and innovate in clean energy (European Commission, 2020). One key directive is the Corporate Sustainability Reporting Directive (CSRD). CSRD is an important EU regulation that aims to improve how companies report on their ESG impacts. It expands the scope of reporting to small and medium-sized enterprises. Under the CSRD, companies must disclose detailed information about how they manage sustainability risks and the measures they take to address climate change, employee welfare, and other social issues (European Commission, 2022). Meanwhile, in the U.S, during Ukraine war, the rising inflation led the Federal Reserve to increase interest rates to stabilize the U.S. economy. Higher interest rates made borrowing more expensive for companies, and since the U.S. lacked strong ESG policies, it reduced their ability to invest in sustainability and ESG initiatives (Kulkarni, 2024; OECD, 2022). Overall, having policies such as CSRD create legal and financial consequences for non-compliance, incentivizing businesses to integrate ESG into their operations.

3.4 ESG Challenges and Opportunities

Although companies acknowledge the potential financial risks of not incorporating ESG, some hesitate to fully embrace it due to concerns about short-term financial performance (Abdi, Li, and Càmara-Turull, 2022). Many businesses still treat ESG initiatives as supplementary rather than integrating them into their core strategies. Fully embedding ESG into operations, decision-making, and company culture requires a significant internal shift that often faces resistance, particularly from leadership or departments focused on short-term profitability (Deloitte 2024). Prioritizing ESG can be daunting, as it may conflict with immediate cost-cutting measures, and the benefits are often realized in the long term, requiring companies to endure short-term costs like cleaner technologies, employee welfare programs, or improved governance structures (Treepongkaruna and Suttipun 2024). This prolonged wait for positive outcomes can lead to resistance from shareholders focused on immediate returns. Additionally, the lack of standardized metrics for measuring ESG performance, due to varying methodologies used by organizations, results in inconsistent evaluations (MSCI 2023). This leads to the provision inconsistent, inaccurate, and unreliable results. Lack of standardized measures also discourages transparency and accountability. It is difficult to collect accurate and reliable information about ESG, especially companies that are not transparent and accountable and regions that have weak standards when it comes to reporting (Vermeulen 2021). ESG regulations are also different in different nations and regions. This makes it difficult for multinational corporations to adhere to the set regulations since it is expensive and takes a lot of time (MSCI 2023). The inability to observe the regulations may lead to financial losses due to expensive lawsuits being brought against the organization and a bad reputation, which promote distrust. When it comes to opportunities, technological advancement is promoting accurate valuation. This will enabled investors and other stakeholders to make informed decisions when deciding on where to invest. It also promotes transparency and accountability that result in the provision of consistent, accurate and reliable results (Vermeulen 2021). Another opportunity is that some countries are offering incentives and support to promote ESG. Getting support and incentives from the government means that the ESG market is expected to continue growing (PwC 2024).

They are also implementing strict measures on carbon emissions. The fight against climate change is an opportunity for the growth of ESG since the principles promote sustainability (Baratta et al. 2023). Social issues like Covid-19 increased awareness about ESG. The consumer demand shifting towards sustainability is increasing ESG adoption (Anandy 2024).

4 Recommendations

- Saudi companies should gradually implement ESG since it is a long-term gain and this allows for better capital management (Holland 2022).
- The regulatory bodies in Saudi Arabia should provide ESG guidelines for effectiveness and efficiency to be achieved (MSCI 2023).
- Investors, regulators, and company leaders should be trained on ESG principles for effective implementation and reduction of resistance (Delevingne et al. 2020).
- The ESG initiative should align with vision 2030, which focuses on diversification, social development, and sustainability (Coffin, M. (2022).
- Standardized reporting approaches should be adopted for acquisition of quality, comparable and consistent ESG data (MSCI 2023; Delevingne et al. 2020).
- Incentives should be offered for improved ESG performance for provision of reliable data and promotion of transparency and accountability (Grand View Research 2023).
- Companies should proactively engage all stakeholders to address their concerns. This can be achieved through effective communication strategies and the use of diverse communication channels (Delevingne et al. 2020).
- Technology should be embraced for provision of accurate, consistent and efficient data.
- Saudi Arabia should learn from EU, which has extensive experience on ESG (Hussain et al. 2024).

5 Conclusion

ESG is advantageous because it increases financial performance and market value of business. It focuses on long-term benefits. Additionally, ESG promotes transparency and accountability. This promotes trust and an increase in sales and profits. With increased profits, the market value of the business increases. ESG ratings help investors make the right decision when it comes to investments. Therefore, ESG increases financial performance and market value of organizations and promote the sustainable growth and development of the whole economy.

References

Abdi, Y., Li, X., Càmara-Turull, X.: Exploring the impact of sustainability (ESG) disclosure on firm value and financial performance (FP) in airline industry: The moderating role of size and age. Environ. Dev. Sustain. **24**(4), 5052–5079 (2022)

Alshehri, S. A. S. (2024). The impact of Environmental, Social and Governance (ESG) disclosures on financial performance of listed companies: A study of United Arab Emirates (UAE).

Educational Administration: Theory and Practice, 30(6), 2481–2490.
Anandy. 2024. The Future of ESG: Emerging Trends and Opportunities https://medium.com/%40anandv_62921/the-future-of-esg-emerging-trends-and-opportunities-fa5cc98d6e67
Baratta, A., Cimino, A., Longo, F., Solina, V., Verteramo, S.: The impact of ESG practices in industry with a focus on carbon emissions: Insights and future perspectives. Sustainability. **15**(8), 6685 (2023)
Bloomberg Intelligence. (2024). Global ESG assets predicted to hit $40 trillion by 2030, despite challenging environment, forecasts Bloomberg Intelligence. https://www.bloomberg.com/company/press/global-esg-assets-predicted-to-hit-40-trillionby-2030-despite-challenging-environment-forecasts-bloomberg intelligence/#:~:text=London%2C%208%20January%202024%20%E2%80%93%20Global, from%20Bloomberg%20Intelligence%20(BI).
Cheema-Fox, A., LaPerla, B.R., Wang, H., Serafeim, G.: Corporate resilience and response to COVID-19. J. Appl. Corp. Financ. **33**(2), 24–40 (2021). https://doi.org/10.1111/jacf.12457
Chopra, S.S., et al.: Navigating the challenges of Environmental, Social, and Governance (ESG) reporting: The path to broader sustainable development. Sustainability. **16**(2), 606 (2024)
Coffin, M. (2022). ExxonMobil is planning on climate failure despite 'Advancing Climate Solutions'. Carbon Tracker Initiative. https://carbontracker.org/exxonmobil-is-planning-on-climate-failure-despite-advancing-climate-solutions/#:~:text=ExxonMobil%20is%20plan ning%20on%20climate%20failure%20despite%20'Advancing%20Climate%20Solutions',-Oil%20%26%20Gas&text=Late%20last%20month%20ExxonMobil%20 launched,new%20%E2%80%9Cnet%20zero%20ambition%E2%80%9D.
Delevingne, L., Gründler, A., Kane, S., Koller, T.: The ESG premium: New perspectives on value and performance. Insights on Sustainability. (2020)
Deloitte: In: Deloitte (ed.) Sustainability Action Report: Enhancing business value through ESG integration (2024) https://www2.deloitte.com/content/dam/Deloitte/us/Documents/audit/2024-sustainability-action-report.pdf
European Commission. (2020). Corporate sustainability reporting directive (CSRD). European Commission. https://www.ec.europa.eu
European Commission. (2022). Corporate Sustainability Reporting Directive (CSRD). European Commission. https://www.ec.europa.eu
Fometescu, A., Hațegan, C.D., Cuc, L.D.: Challenges of measuring and reporting ESG performance. In: Exploring ESG Challenges and Opportunities: Navigating Towards a Better Future, pp. 61–82. Emerald Publishing Limited (2024)
Fu, T., Li, J.: An empirical analysis of the impact of ESG on financial performance: The moderating role of digital transformation. Front. Environ. Sci. **11**, 1256052 (2023). https://doi.org/10.3389/fenvs.2023.1256052
Gao, W., Li, M., Zou, C.: [Retracted] Analysis of the impact of ESG on corporate financial performance under the epidemic based on static and dynamic panel data. Wirel. Commun. Mob. Comput. **2022**(1), 6851518 (2022). https://doi.org/10.1155/2022/6851518
Grand View Research. (2023). ESG market analysis by application (Public equity, fixed income), by end use (Banks, asset managers, insurance companies), and segment forecasts, 2023-2030. Grand View Research. https://www.grandviewresearch.com/industry-analysis/esg-investment-market
Holland, B. (2022). Exxon joins peers in setting net-zero goals, but critics find them lacking. S&P Global Market Intelligence https://www.spglobal.com/market-intelligence/en/news-insights/articles/2022/1/exxon-joins-peers-in-setting-net-zero-goals-but-critics-find-them-lacking-68490050
Hotten, R. (2015). Volkswagen: The scandal explained. BBC News. https://www.bbc.com/news/business-34324772

Hussain, M.A., Alsayegh, M.F., Boshnak, H.A.: The Impact of environmental, social, and governance disclosure on the performance of Saudi Arabian companies: Evidence from the Top 100 non-financial companies listed on Tadawul. Sustainability. **16**(17), 7660 (2024)

Iazzolino, G., Bruni, M.E., Veltri, S., Morea, D., Baldissarro, G.: The impact of ESG factors on financial efficiency: An empirical analysis for the selection of sustainable firm portfolios. Corp. Soc. Responsib. Environ. Manag. **30**(4), 1917–1927 (2023). https://doi.org/10.1002/csr.2463

Janah, O.O., Sassi, H.: The role of ESG factors in mitigating crisis effects on Moroccan public companies' financial performance. Afr. Sci. J. **3**(15), 256–256 (2022)

Joudah, G.: Botulism outbreak traced to mayonnaise at Riyadh restaurant. Arab News. (2024) https://www.arabnews.com/node/2508531/saudi-arabia

Kim, S., Li, Z.: Understanding the impact of ESG practices in corporate finance. Sustainability. **13**(7), 3746 (2021). https://doi.org/10.3390/su13073746

Kulkarni, S. (2024). Russia-Ukraine military conflict: Key takeaways from our articles. https://www.spglobal.com/ratings/en/research/articles/220304-russia-ukraine-military-conflict-key-takeaways-from-our-articles-12300240

LaForge, J. (2021). ESG in the US vs Europe: The United States falls behind the curve?
https://www.trepp.com/trepptalk/esg-us-vs-europe-curve

Landi, G.C., Turriziani, L., Muto, V., Ricciardi, I.: Exploring the nexus between ESG risk variations and investment preferences: Insights from sustainable ETFs during the COVID-19 era. Socio-Econ. Plan. Sci. **95**, 102039 (2024)

Lee, J., Koh, K.: ESG performance and firm risk in the US financial firms. Rev. Financ. Econ. (2024). https://doi.org/10.1002/rfe.1208

Li, Y.: The case analysis of the scandal of Enron. Int'l J. Bus. & Mgmt. **5**(10), 37 (2024) https://www.researchgate.net/publication/46302792_The_Case_Analysis_of_the_Scandal_of_Enron

Malekian, E., ALShakban, A.: The role of corporate inflexibility moderator on the relationship between environmental, social, and corporate governance (ESG) and the cost of equity in listed companies in Iraq. Al-Qadisiyah Journal for Administrative and Economic Sciences. **26**(2) (2024)

MSCI. (2023). MSCI ESG Ratings and Cost of Capital. MSCI. https://www.msci.com/documents/10199/6c2115f3-5fef-b883-278d-853de6ffc019

MSCI. (2024). Sustainable Investing: ESG Ratings. MSCI. https://www.msci.com/our-solutions/esg-investing/esg-ratings

Mrchkovska, N., Dolšak, N., Prakash, A.: Does ESG privilege climate action over social and governance issues? A content analysis of BlackRock CEO Larry Fink's annual letters. PLOS Sustainability and Transformation. **2**(12), e0000090 (2023)

Muigua, P.D., 2022. Embracing Environmental, Social and Governance (ESG) Principles for Sustainable Development in Kenya. Social and Governance (ESG) Principles for Sustainable Development in Kenya (December 27, 2022).

OECD: Impacts of the Russian Invasion of Ukraine on Financial Market Conditions and Resilience: Assessment of Global Financial Markets. OECD Publishing, Paris. (2022). https://doi.org/10.1787/879c9322-en

Park, S.R., Jang, J.Y.: The impact of ESG management on investment decision: Institutional investors' perceptions of country-specific ESG criteria. International Journal of Financial Studies. **9**(3), 48 (2021)

Passas, I.: The Evolution of ESG: From CSR to ESG 2.0. Encyclopedia. **4**(4), 1711–1720 (2024)

PwC. 2024. Leveraging VAT to support ESG goals for government and business. https://www.pwc.com/m1/en/media-centre/articles/leveraging-vat-to-support-esg-goals-for-governments-and-businesses.html

Sandberg, H., Alnoor, A., Tiberius, V.: Environmental, social, and governance ratings and financial performance: Evidence from the European food industry. Bus. Strateg. Environ. **32**(4), 2471–2489 (2023). https://doi.org/10.1002/bse.3259

Sun, Y., Rahman, M.M., Xinyan, X., Siddik, A.B., Islam, M.E.: Unlocking environmental, social, and governance (ESG) performance through energy efficiency and green tax: SEM-ANN approach. Energ. Strat. Rev. **53**, 101408 (2024)

Treepongkaruna, S., Suttipun, M.: The impact of environmental, social and governance (ESG) reporting on corporate profitability: evidence from Thailand. Journal of Financial Reporting and Accounting. (2024)

Vermeulen, P.: Short-term steps for long-term value: The impact of ESG on business strategy and performance. EY Global.. https://www.ey.com/en_gl/insights/financial-services/emeia/short-term-steps-for-long-term-value-the-impact-of-esg-ratings-on-tsr#:~:text=There%20is%20growing%20evidence%20that, ESG%20risk%20management%20boosts%20returns (2021)

Yang, P., Ji, C., Prentice, C., Sthapit, E., Peng, Z.: Scale development and validation of corporate digital responsibility—A consumer perspective. Int. J. Consum. Stud. **49**(1), e70023 (2025)

Zhou, G., Liu, L., Luo, S.: Sustainable development, ESG performance and company market value: Mediating effect of financial performance. Bus. Strateg. Environ. **31**(7), 3371–3387 (2022). https://doi.org/10.1002/bse.3089

Zumente, I., Bistrova, J.: ESG importance for long-term shareholder value creation: Literature vs. practice. J. Open Innov.: Technol. Mark. Complex. **7**(2), 127 (2021)

Renewable Energy Investments and their Impact on Stabilizing Inflation

Karima Saci(✉) and Saida Khalifa

Dar Al-Hekma University, Jeddah 22246-4872, Kingdom of Saudi Arabia
{ksaci,skhalifa}@dah.edu.sa

Abstract. This study investigated how investments in renewable energy stabilize energy prices, lower the risk of inflation and improve the environmental effects of energy production and management, all of which help to lessen market volatility globally. The paper acknowledged both the opportunities and the challenges of transitioning from fossil fuels to renewable energy, emphasizing the need for strategic planning and implementation to maximize long-term economic and environmental benefits. Recommendations were formulated to smooth the path of transitioning by highlighting the importance of government policies, research driven technological advancement and stakeholders' collective actions.

Keywords: Renewable energy · Inflation risk · Sustainability · Fossil fuels

1 Introduction

The increasing importance of renewable energy investments and its relationship with inflation risk derives its importance from the profound consequences associated with various global economic and environmental crises (Yadav 2024). The financial issues posed to countries transitioning from fossil fuels is becoming more prominent. As a major challenge influencing consumers, investors and policymakers, inflation risk depends on several underlying factors, mainly related to energy prices and transportation cost fluctuations within the global supply chains (Lu et al. 2010). Renewable energy investments represent a growing strategy to achieve various goals, one of which is fighting inflation. Through more stable energy costs and decreased dependence on fossil fuel markets known for high price volatility, renewable energy helps maintain long-term financial stability, energy security and environmental sustainability (Adebayo et al. 2024).

Since the transition to renewable energy has become a global priority, sustainable energy must play a central role in inflation risk management. Beside the associated environmental benefits, the transition to renewables offers a fundamental economic benefit by reducing inflation and building financial and economic stability. This study examines how investments in renewable energy affect inflation risk, based on both theoretical insights and research findings. It explores the role of renewable energy in controlling energy costs which in turn engenders more stable financial markets and improve the economic stability as a whole (Yadav 2024 and Ma et al. 2024). The paper also addressed the

D. Berkouk et al. (Eds.): CCTS 2024, PTT, pp. 333–340, 2026.
https://10.1007/978-981-95-7607-4_37

main challenges associated with the initial stage of conversion as well as the long-term benefits once the conversion to renewable energy is fully achieved.

The various ways in which these investments impact inflation, from stabilizing energy costs to influencing financial markets and strengthening economic resilience (Lu et al. 2023). The paper also addresses key challenges tied to the transition to renewable energy, including the high upfront costs and infrastructure limitations that could drive inflationary pressures in the short term (Vyrostková et al. 2024). Through an in-depth review of existing research and trends, this study aims to shed light on the broader implications of renewable energy investments for the satiability of financial stability and long-term economic growth.

2 Literature Review

2.1 Trends in the Global Energy Market

Events such as COVID-19 and the Russian-Ukrainian war have negatively impacted the global economy. In 2021, as the economy adapted to COVID-19, there was an increase in overall demand and a recovery in energy prices (Bednárˇ *et al.*, 2022). Furthermore, Russia, the world's largest exporter of fossil fuels, faced sanctions due to the war, leading to a global surge in energy prices—including oil, gas, and electricity (Markowski and Kotli'nski 2023).

The rise in energy prices general leads to higher core inflation expectations both at the short-term and long-term horizon (Bednárˇ et al. 2022). This is primarily because energy costs significantly contribute to the Consumer Price Index (CPI), including electricity, heating and transportation expenses. Additionally, products in the agricultural and industrial sectors, which are part of the CPI, use energy as a production input. As a result, an increase in energy prices raises the costs of these consumer goods, ultimately causing a spike in core inflation (Markowski and Kotli'nski, 2023).

2.2 Renewable Energy Sources and Their Advantages

Renewable energy is defined as an energy resource that is durable and naturally renewable at a rate equal to or faster than its rate of its consumption. Sources of renewable energy are abundant and include wind, solar, hydropower, geothermal energy (Earth's heat), ocean energy-which further consists of wave, tidal and sea current energy (Maradin 2021). Chubraeva and Sergey (2018) report that global renewable energy consumption has quadrupled over the past decade.

There are many advantages of renewable energy. First, it produces zero greenhouse gas emissions, making it essential for environmental protection (United Nations Development Programme, 2000). Omri and Nguyen (2014), who analyzed 64 countries from 1990 to 2011, concluded that carbon dioxide emissions are a major driver of renewable energy adoption. Renewable energy reduces global warming and fosters a carbon-free environment. Second, renewable energy reduces the dependency on energy imports, especially for electricity generation (Maradin, 2021). The EU provides a strong example for the adoption of renewable energy, particularly after March 2022. Wind and solar energy now account for 24% of the EU's electricity. Their increased capacity avoided the import of 70 bcm of gas-saving €99 billion-and boosted electricity generation by 13% year-over-year (Pous *et al.* 2022).

Third, renewable energy promotes economic development and technological innovation, increasing demand for skilled workers and improving employment rates (Maradin 2021). Pohl et al. (2011), who investigated the factors that enhance renewable energy technology development in developing countries from 1980 to 2008, found that trade openness and economic development promote renewable energy technology adoption. Similarly, Armeanu et al. (2017), who assessed whether renewable energy contributes to the economic growth of EU-28 counties from 2003 to 2024, found that a 1% increase in renewable energy production leads to a 0.05–006% increase in GDP per capita.

3 Relationship Between Renewable Energy Consumption and Inflation Risk

3.1 Theoretical Review

The relationship between renewable energy consumption and inflation can be explored through numerous economic and environmental theories (Jin et al., 2024). Ngepah and Udeagha (2018) provide that the energy substitution theory suggests that shifts in energy prices driven by inflation can influence the selection of energy sources. For example, when the fossil fuel prices rise, consumers and industries may change their energy choices for more affordable or sustainable alternatives, thereby impacting environmental outcomes. Further, the pollution haven hypothesis suggests that rising inflation in certain countries can change production costs, potentially triggering shifts in industrial activity and energy usage. These inflation-driven changes may encourage pollution-intensive industries to relocate to countries with lower operational costs, leading to varying environmental benefits. Additionally, Udeagha and Breitenbach (2023) provide that the resource dependency theory explores how reliance on certain resources such as fossil fuels shapes environmental outcomes. When inflation impacts the prices of these resources, it can change consumption behaviors, thereby influencing environmental quality.

3.2 Empirical Review

During inflationary spikes—particularly due to the crowding-out effect—industries face rising prices for non-renewable energy inputs. As a result, they tend to switch to more cost-effective alternatives, such as renewable energy (Jin et al. 2024). Wang et al. (2019) examined the effects of energy prices, urbanization and GDP on per capita energy consumption while considering the income gaps between countries in a study covering 186 countries between 1980 and 2015. They discovered that higher energy prices drive the industrial sectors to shift their energy consumption from costly energy sources, i.e., fossil fuels to cost effective energy sources i.e., renewable energy. However, Vyrostková et al. (2024) argues that renewable energy sources have not yet reached their desired potential of cost efficiency. This is because it is only the developed countries that will be able to significantly utilize renewable energy whereas the developing countries still lack the capital for adequate investment in renewable energy technologies.

Furthermore, Ahmad et al. (2021), who assessed the relationship between inflation instability and pollution emissions in 40 Asian economies from 1990 to 2018, concluded that higher inflation instability is associated with lower carbon dioxide emissions. They claim that inflation instability promotes sustainability. This is because economic uncertainty reduces industrial investments and consumption which indirectly leads to environmental benefits through reduced carbon emissions.

On the other hand, Markowski and Kotli'nski (2023) assessed the relationship between the share of renewable energy production, the harmonized Index of Consumer prices (HICP) level and core inflation in the EU countries in a cross-sectional study during 2022 crisis. They argued that a higher share of renewable energy is associated with lower HICP and core inflation. Countries with similar energy mixes also tend to have similar inflation rates. Alternatively, Bilal et al. (2022) investigated the relationship between renewable energy, government expenditures, economic growth and environmental quality in Germany from 1971 to 2016. They found a long-run negative relationship between renewable energy sources, government expenditures and inflation and a positive relationship between economic growth and environmental quality.

Moreover, Deka et al. (2022) explored the link between exchange rate, inflation and renewable energy in Brazil from 1990 to 2019. Their study found that renewable energy consumption causes currency appreciation, and that inflation and exchange rates positively influence the use of renewable energy sources. Thus, renewable energy consumption not only promotes environmental quality but also results in improvements of the currency exchange rate values. Vyrostková et al. (2024) examined the relationship between the share of renewable energy and macroeconomic indicators such as Gross Domestic Product (GDP) per capita, unemployment, Financial Development Index (FDI), inflation, government efficiency and corruption control in the EU countries from 2006 to 2020. They found a positive correlation with GDP per capita and negative correlations with unemployment, inflation, and inefficiency.

Furthermore, Adi et al. (2024) examined the impact of Fossil Fuel prices on fossil and renewable electricity consumption, GDP, inflation and greenflation in Asia Pacific countries from 2016 to 2021. They found that there is a significant negative relationship between fossil fuel prices and fossil electricity consumption and the same for renewable electricity consumption. They also observed that while there is a significant negative relationship between fossil electricity consumption and GDP, the relationship is insignificantly negative with Inflation. Further, they affirmed that although renewable energy positively influenced GDP, its impact on greenflation was insignificantly negative.

Moreover, Lu et al. (2023) conducted a study in the MENA region from 2000 to 2016 examining links between economic growth, capital, labor, trade, inflation, and renewable energy consumption. They found a uni-directional causalities between inflation and renewable energy and renewable energy and economic growth. They also emphasized the importance of trade openness and institutional innovation for the development of sustainable energy development.

Rising global energy prices triggered by crises like COVID-19 and the Russian-Ud cost-effective renewable energy sources has become increasingly necessary. Numerous studies (e.g., Wang et al. 2019; Ahmed et al. 2020; Deka et al., 2022) suggest that inflation can promote sustainable energy transitions. Others (e.g., Markowski and Kotli'nski, 2023; Vyrostková et al., 2024; Bilal et al. 2022) indicate a negative correlation between renewable energy consumption and inflation. Overall, renewable energy appears to contribute not only to inflation stability but also to environmental preservation, economic growth, improved employment, and stronger exchange rates (Bilal et al., 2022; Deka et al., 2022; Vyrostková et al., 2024).

4 Discussion

The current literature demonstrated the importance of the interlink between the investment in transitioning to renewable energy and the satiability of inflation. This transition allows for less reliance on foreign sources of energy and the optimization of the usage to the national natural resources making more countries autonomous and less venerable in terms to energy centralization in certain geographical markets and the impact of possible political and economic conflicts (Ma et al., 2024). Beyond its benefits to national economies, this energy transition presents a potential for mitigating inflationary pressures experienced by consumers. Specifically, the reduction in energy related input costs for corporations may lessen the extent to which these costs are passed on to consumers in the form of elevated prices for general goods and services, particularly during periods of peak seasonal energy demand (Algarvio, 2023). The environmental benefits of transitioning to renewable energy, such as reduced CO2 emissions during production, allow corporations to enhance sustainability practices and improve ESG reporting, which can mitigate environmental penalties. This, coupled with decreased production costs, can exert downward pressure on product pricing, hence contributing to inflation stabilization (Radulescu et al., 2023; Alvarez & Griffoli, 2024).

While the above benefits are substantial, it is vital to acknowledge the associated challenges inherent in this transition. The required advanced technology required to smooth out and streamline the transition to renewable energy will undoubtedly cause a substantial increase in the cost of the initial investment causing prices to spike and inflation to increase (Descioli, 2024; Ma et al., 2024). Also, investors in renewable energy sectors may experience a low return given the level of risk they are exposed to in the initial phase of the transition which will further be adversely impacted by the associated increase in inflation. Investors may shay away from these sectors and look for better investment opportunities which may hinder the capital flows moving into energy related projects (Taghizadeh-Hesary & Yoshino, 2020).

5 Recommendations

This study emphasizes the distinction between the short-term inflationary challenges and long-term economic opportunities inherent in the transition from fossil fuels to renewable energy, with a specific focus on its impact on national sustainability. To mitigate the economic impact of these challenges on consumers and investors, this study recommends the implementation of the following measures:

- The government must introduce policies, grants, and tax incentives to promote investment in renewable energy sectors, ensuring a stable transition and mitigating inflationary pressures
- Prioritizing and diversifying renewable energy sources lead to enhanced energy security and reduced vulnerability to price volatility
- Incased funding of technology-based research to ensure the optimization of technology usage without excessive financial expenditure
- Policy makers, investors and consumers need to be more educated about the long-term impact of transitioning from fossil fuels to renewable energy and the future benefits associated with this transition
- Smart solutions should be put in place to mitigate the phases of this transition, ensuring financial investments are made accordingly to avoid the high risk of capital misallocation and projects failure

By implementing these recommendations, the economic and financial benefits of renewable energy investments can be maximized while mitigating potential inflationary risks. A well-structured transition has the potential to stabilize inflation risk, promote sustainable economic growth and enhance global financial stability.

6 Conclusion

This study critically reviewed the existing literature related to renewable energy investments and their impact on inflation stability, highlighting short-term challenges and long-term opportunities associated with the transition from fossil fuels. It was found that to realize the future financial, economic, and environmental benefits of this transition, all stakeholders must collaboratively manage the initial stage of this transition through strategic planning and coordinated actions. These benefits expand to the global level, allowing countries to experience improved energy independence, lessened exposure to volatile fossil fuel prices and a significant acceleration of sustainability initiatives.

References

Adebayo, Y.A., Ikevuje, A.H., Kwakye, J.M., Esiri, A.E.: A model for assessing the economic impact of renewable energy adoption in traditional oil and gas companies. GSC Advanced Research and Reviews. **20**(3), 298–315 (2024). https://doi.org/10.30574/gscarr.2024.20.3.0355

Adi, T.W., Susanto, E., Caswito, A., Yuwono, R.S., Warsokusumo, T., Nugroho, A.Y.A.: Influence of fossil fuel prices on fossil and renewable electricity consumptions, GDP, inflation and greenflation: A case study in the Asia Pacific Countries. Int. J. Energy Econ. Policy. **14**(4), 48–56 (2024). https://doi.org/10.32479/ijeep.15966

Ahmad, W., Ullah, S., Ozturk, I., Majeed, M.T.: Does inflation instability affect environmental pollution? Fresh evidence from Asian economies. Energy Environ. **32**(7), 1275–1291 (2021). https://doi.org/10.1177/0958305X20971804

Armeanu, D.Ş., Vintilă, G., Gherghina, Ş.C.: Does renewable energy drive sustainable economic growth? Multivariate panel data evidence for EU-28 countries. Energies. **10**(3) (2017). https://doi.org/10.3390/en10030381

Algarvio, H.: The economic sustainability of variable renewable energy considering the negotiation of different support schemes. Sustainability. **15**(5), 4471 (2023). https://doi.org/10.3390/su15054471

Alvarez, J., Mancini-Griffoli, T.: Do renewables shield inflation from fossil fuel-price fluctuations? IMF Working Papers. **2024**(112) (2024)

Bednář, O., Čečrdlová, A., Kadeřábková, B., Řežábek, P.: Energy prices impact on inflationary spiral. Energies. **15**(9) (2022). https://doi.org/10.3390/en15093443

Bilal, K., I., Tan, D., Azam, W., & Tauseef Hassan, S.: Alternate energy sources and environmental quality: The impact of inflation dynamics. Gondwana Res. **106**, 51–63 (2022). https://doi.org/10.1016/j.gr.2021.12.011

Chubraeva, L., Sergey, T.: Project of autonomous power plant with high-temperature superconductive devices. In: 2018 International Multi-Conference on Industrial Engineering and Modern Technologies (FarEastCon), pp. 1–5. Institute of Electrical and Electronics Engineers (2018). https://doi.org/10.1109/FarEastCon.2018.8602671

De Pous, P., Patuleia, A., Brown, S., & Rosslowe, C. (n.d.). More Renewables., Lesinflation Restoring Eu Economic Stability Through Investment In Renewables.

Deka, A., Cavusoglu, B., Dube, S.: Does renewable energy use enhance exchange rate appreciation and stable rate of inflation? (n.d.). https://doi.org/10.1007/s11356-021-16758-2/Published

Descioli, P.: Uncovering the Impact of Price Inflation on Sustainable Supply Chains in Renewable Energy. (2024). https://doi.org/10.20944/preprints202405.1283.v1

Jin, X., Xu, R., Farooq, U., Tabash, M.I., Aljughaiman, A.A.: How do inflation rate and energy alternatives influence environmental quality? A Brics analysis. Energy Strategy Reviews. **53** (2024). https://doi.org/10.1016/j.esr.2024.101369

Lu, W., Liu, T., Tseng, C.: Volatility transmissions between shocks to the oil price and inflation: evidence from a bivariate GARCH approach. J. Inf. Optim. Sci. **31**(4), 927–939 (2010). https://doi.org/10.1080/02522667.2010.10700003

Lu, X., Farhani, S., Soliman, A.M., Zhou, C., Su, K.: Renewable energy consumption, trade and inflation in MENA countries with augmented production function: Implications for the COP26. Technol. Forecast. Soc. Chang. **194** (2023). https://doi.org/10.1016/j.techfore.2023.122712

Ma, W., Wu, T., Stan, S.-E., Gao, B.: Ensuring environment sustainability through natural resources, renewable energy consumption, and inflation dynamics. Resources Policy. (2024). https://doi.org/10.1016/j.resourpol.2024.104676

Maradin, D. (2021). Advantages and disadvantages of renewable energy sources utilization. Int. J. Energy Econ. Policy, 11(3), 176–183. dooi:https://doi.org/10.32479/ijeep.11027

Markowski, Ł., Kotliński, K.: Renewables, energy mix and inflation in the European Union Countries. Energies. **16**(23) (2023). https://doi.org/10.3390/en16237808

Ngepah, N., Udeagha, M.: African regional trade agreements and Intra-african trade. The Journal of Economic Integration. **33**(1), 1176–1199 (2018). https://doi.org/10.11130/jei.2018.33.1.1176

Omri, A., Nguyen, D.K., Rault, C.: Causal interactions between CO2 emissions, FDI, and economic growth: Evidence from dynamic simultaneous-equation models. Econ. Model. **42**, 382–389 (2014). https://doi.org/10.1016/J.ECONMOD.2014.07.026

Pohl, B., Jakob, M., & Schlömer, S. (2011). Determinants of Renewable Energy Technology Adoption – Evidence for Developing Countries 1980–2008. African Economic Conference (AEC), African Development Bank.

Radulescu, R.; Ionescu, R.; Radulescu, D.; Pelinescu, E.(2023). Renewables, Energy Mix and Inflation in the European Union Countries. Energies 2023, 16(23), 7808.

Taghizadeh-Hesary, F., Yoshino, N.: Sustainable Solutions for Green Financing and Investment in Renewable Energy Projects. Energies. **13**(4), 788 (2020). https://doi.org/10.3390/EN13040788

Vyrostková, L., Lumnitzer, E., Yehorova, A.: Renewable Energy in the Eurozone: Exploring Macroeconomic Impacts via FMOLS. Energies. **17**(5) (2024). https://doi.org/10.3390/en17051159

Udeagha, M., Breitenbach, M.: Revisiting the nexus between fiscal decentralization and CO2 emissions in South Africa: fresh policy insights. Financial. Innovation. **50**(9) (2023). https://doi.org/10.1186/s40854-023-00453-x

United Nations Development Programme: World Energy Assessment: Energy and the Challenge of Sustainability. United Nations Development Programme, United Nations (2000)

Wang, Q., Su, M., Li, R., Ponce, P.: The effects of energy prices, urbanization and economic growth on energy consumption per capita in 186 countries. J. Clean. Prod. **225**, 1017–1032 (2019). https://doi.org/10.1016/J.JCLEPRO.2019.04.008

Yadav, A.: Promoting economic stability: The role of renewable energy transition in mitigating global volatility. International Journal of Energy Sector Management. (2024). https://doi.org/10.1108/ijesm-06-2024-0032

Mawhiba Creative Programs for Young Talented Minds and their Sustainable Impact

Fahad Alfaiz[1](✉), Saad Bakry[1], and Nada Alqahtani[2]

[1] King Saud University, Riyadh, Saudi Arabia
falfaiz@ksu.edu.sa
[2] Mawhiba, Riyadh, Saudi Arabia

Abstract. King Abdulaziz and his Companions Foundation for Giftedness and Creativity, known as "Mawhiba", directs its effort toward achieving its vision of being "a leading global reference in empowering the brightest young talented minds to create sustainable impact". Following a general introduction on Mawhiba's contribution to the Kingdom of Saudi Arabia (KSA) Vision 2030, this paper describes Mawhiba's methodology used for the development of sound creative education programs for empowering gifted K-12 students. Toward delivering these programs, the methodology attempts to achieve four main integrated purposes. The first purpose is concerned with identifying the targeted gifted students. The second is related to providing those students with the knowledge and skills that support their capabilities, while responding to the continuous progress of the world's innovation. The third is associated with exploring the thinking skills and creativity of the students. Finally, the fourth is related to putting the students into real-life creative competitions, both at the national level and at the international level. The paper then moves toward providing illustrative examples on Mawhiba' students in its various programs, and their increasing numbers and achievements

Keywords: Gifted Students · Education Programs · Sustainable Development

1 Introduction

1.1 What is Giftedness

People differ in their various abilities, and those gifted with special cognitive abilities require special attention from an early age to promote their abilities and enhance their potential creativity and innovation toward sustainable development. While there is no international definition for giftedness [1], there are various views on the matter. For example, the long-standing Intelligence Quotient (IQ) tests people considering their collective ability in four generic fields that include: verbal; numeric; logic; and spatial [2]. In addition, Boeree provides a general definition to intelligence, or high ability, stating that it is "a person's capacity to acquire knowledge that is to learn and understand; apply knowledge to solve problems; and engage in abstract reasoning" [3].

D. Berkouk et al. (Eds.): CCTS 2024, PTT, pp. 341–348, 2026.
https://10.1007/978-981-95-7607-4_38

Gardner produced the theory of multiple intelligence that considers nine abilities a gifted person may enjoy. These are verbal; logical-mathematical; visual-spatial; naturalistic; interpersonal; intrapersonal; musical; bodily-kinesthetic; and existential [4]. The USA National Association for Gifted Children (NAGC) considers various giftedness abilities that are recognized by schools in the different states of the country. These abilities include intellectual ability; academic ability; creativity; cultural diversity; thinking productivity; high performance; visual arts; music and theater; psychomotor; in addition to leadership [5].

Recognizing the importance role for nurturing gifted students in shaping future development, various countries all over the world have organizations concerned with nurturing these gifted individuals, such as NAGC in the USA, and Mawhiba in Saudi Arabia. In addition, international organizations nurturing the gifted also exist; an example here is the World Council for Gifted and Talented Children (WCGTC) [6].

1.2 Gifted Education Programs

The literature of gifted education emphasized the importance of providing various choices for gifted students to improve their academic potential and skills [7]. These include but are not limited to self-contained classes, summer enrichment programs, mentorships, etc. [8]. Researchers have concluded that giving students a range of options to participate in gifted education programs would enhance their outcomes and benefit them from their participations [9].

1.3 Giftedness in Saudi Arabia

In Saudi Arabia, there are two main institutions for nurturing gifted students, one of which is King Abdulaziz and His Companions Foundation for Giftedness and Creativity, known as Mawhiba. Mawhiba is a non-profit organization, established in 1999 and concerned with giftedness associated with science, technology, engineering, and mathematics (STEM). These fields are of special value to sustainable development [10].

At present, Mawhiba is enjoying the development environment of KSA Vision 2030 [11]. The vision targets sustainable development, and in this respect, it is based on three main pillars that view the country as an ambitious nation, a vibrant society, and a thriving economy. In addition, the vision has various implementation programs, including the Human Capability Development Program (HCDP). This program supports Mawhiba's role in nurturing and promoting giftedness, as one of its strategic objectives hopes to "develop our brightest minds in priority fields" [12].

Mawhiba's methodology used for the development and implementation of creative education programs for gifted students is based on the following four main principles [7]: (a) establishing an environment of excellence in both developing and implementing the targeted programs. This involves working with partners, including national and international experts and specialized organizations, in addition to Saudi K-12 schools, while, in the meantime, encouraging passion for acquiring knowledge and for seeking creativity and innovation; (b) identifying K-12 interested students who enjoy high ability in multiple capabilities that indicate their high potential in STEM subjects, using a sound, distinguished, and highly recognized identification approach; and (c) designing and implementing education programs that integrate acquiring suitable knowledge and skills with supporting research and innovation, in addition to exploring national and international competitions. This would promote gifted students' potential toward building up excellence supported by competitive abilities that can contribute to sustainable development, and (d) using modern technology to support the implementation of the programs. Such technology would involve what is needed for creativity and innovation in STEM topics on the one hand, in addition to the digital technology, including Artificial Intelligence (AI), that supports the wide spreading of the programs and their benefits on the other.

1.4 Purpose of the Study

The purpose of this research is to discuss Mawhiba's methodology in developing integrated programs for identifying gifted students, supporting their capabilities, and enabling and testing their creativity toward sustainable development.

1.5 Research Question

Q1: What programs did Mawhiba offer for gifted students in Saudi Arabia to promote their giftedness and creativity?

Q2: What impact did these programs have on increasing the number of students benefiting from Mawhiba's programs during 2022 and 2023?

2 Method

A mixed-method design was used in the current study. This approach is suitable for analyzing quantitative and qualitative data [13]. All gifted students who participated in Mawhiba Programs in 2022 and 2023. Two resources of data were used to answer the research questions. They included documents and raw data from Mawhiba. The authors communicated with staff in Mawhiba to collect the data.

3 Results

3.1 Mawhiba Programs for Gifted Students in Saudi Arabia

Mawhiba offers a variety of programs for gifted students in Saudi Arabia which include identification, supporting knowledge and skills, research, creativity and innovation, and competition. These programs are discussed as follows:

Identification. For identifying gifted students who enjoy high ability, Mawhiba uses its Multiple Cognitive Aptitude Test, which was developed with its partners, and which involves the following four main sections: (a) creativity or mental flexibility, (b) linguistic reasoning and reading comprehension, (c) mathematical and spatial reasoning, and (d) scientific and mechanical reasoning [14]. The application of the test has three levels. Level 1 is for students ages 9 to 11, who are in grades 3 to 5. Level 2 is for students ages 12 to 14, who are in grades 6 to 8. Level 3 is for students ages 15 and 16, who are in grades 9 and 10.

Programs Concerned with Supporting Knowledge and Skills. These programs have the following objectives. (a) Enriching students' understanding with new, in-depth knowledge. (b) Developing students' important skills, such as, thinking reasoning and communication skills. (c) Developing students' attitudes toward cooperation, openness, research, innovation, and others.

The programs have three main groups. The first group has two programs designed and implemented in collaboration with schools. The first program is Mawhiba Classes, which covers various scientific subjects; and it is usually introduced during normal school days. The second program is Advanced Program in Science and Mathematics, which is introduced through online technology.

The second group is concerned with knowledge enrichment and includes the following programs. (a) Academic Enrichment Program, which is concerned with essential scientific subjects; it is introduced both physically and online. (b) Mawhiba Ambassadors' Program, which is also concerned with essential scientific subjects, but introduced in the USA, UK, and Ireland to explore Mawhiba's international dimension. (c) Other knowledge enrichment programs. It should be noted that although the above programs are identified as knowledge enrichment, they are also partly skills enrichment for the benefit of students.

The third group is particularly concerned with skills enrichment and includes the following programs. (a) Basic Skills Enrichment Program, which is concerned with public speaking, creative writing, and innovation. (b) English Language Program, which is concerned with promoting English language skills, and it is introduced online. (c) Other skills enrichment programs including those preparing Mawhiba students to pass tests concerned with admission to highly recognized universities.

Programs Concerned with Research, Creativity & Innovation. Mawhiba programs concerned with research, creativity and innovation include the following. (a) Basics of Scientific Research Program, which is introduced online, and which is concerned the conceptual aspects of research and their increasing importance, while simultaneously encouraging research in various scientific fields. (b) Research Enrichment Program, which is concerned with promoting research skills, and with research collaboration in various scientific fields, and with different specialized research organizations. The program is introduced both physically and online. (c) Research Mentorship Program, which involves research in 22 scientific fields, and which is introduced physically and online. (d) Other research collaborations concerned with specific projects take place upon agreements with various specialized organizations.

Programs Concerned with Competition. Mawhiba has programs concerned with the participation of its students in various national and international competitions; these programs include the following. (a) The National Olympiad (Ibdaa), which is an annual competition in science among students of intermediate and secondary education schools. (b) Kangaroo-Mawhiba Math Competition, which is also an annual competition. It emphasizes math and it is introduced online. (c) Bebras-Mawhiba Informatics Competition, which is an annual competition, and which is also introduced online. (d) International Olympiads, which include six annual Olympiads in various countries considering different subjects. (e) National Artificial Intelligence (AI) and Programming Olympiad, which was established in 2023 in repose to AI development. It is a two-phase competition, with the first being online, and the second being by attendance.

The above four types of programs are of integrated nature, from identification of the gifted to their knowledge and skills acquisition, to research and innovation enablement, and finally to national and international competitions. This integration explores the special abilities of the gifted toward future distinguished scientific achievements.

Based on the above four methodological principles, Fig. 1 illustrates the integrated development stages of the targeted programs.

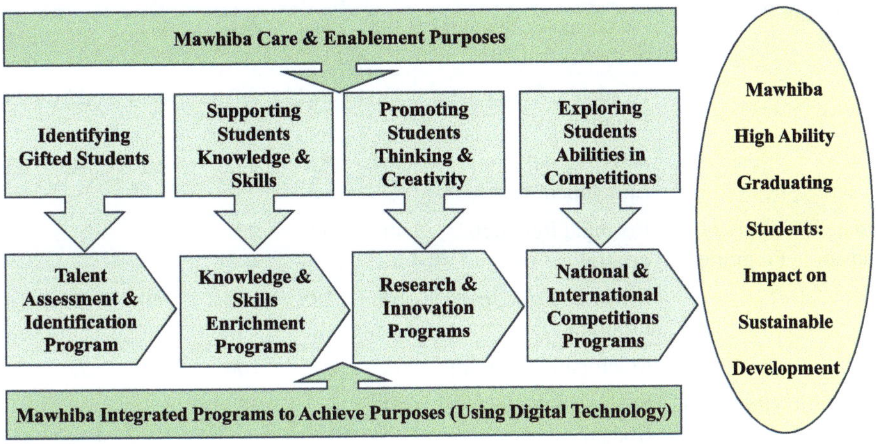

Fig. 1. Mawhiba methodology of designing and implementing its integrated education programs

3.2 Number of Gifted Students Benefiting from Mawhiba's Programs During 2022 and 2023

The number of gifted students who benefited from Mawhiba's programs has been increased sharply from 2022 to 2023. For example, the percentage of gifted who were identified as gifted was increased sharply in 2023 by 47% for male students and by 55% for females. Another example was the number of gifted students who enrolled in Advanced Program for Science and Math in 2023. We can notice that the percentage was increased by 73% for male and by 72% for female students. Table 1 showed the number of students and their percentage in Mawhiba programs.

Table 1. Number of Gifted students in Mawhiba' programs

Group	Program	Students					
		2022		2023		Inc. (%)	
		M	F	M	F	M	F
Assessment of giftedness & identification program: The outcome (gifted students)		7860	9143	11577	14146	47	55
Knowledge & Slills Enrichment Programs	Mawhiba classes	2847	2179	3806	2759	34	27
	Advanced Program for Science and Math	384	402	664	691	73	72
	Academic Enrichment Program	3264	3823	3985	4914	22	29
	After School Enrichment Program	304	326	432	402	42	23
Research, Creativity & Innovation Programs.	Essential Research Program	40	50	88	108	120	116
	Research Enrichment Program	100	225	150	260	50	16
	Mentorship Program	63	215	109	264	73	23
Competition Programs (Winners)	National Olympiad (Ibdaa)	40		40		0	
	Kangaroo Math Competition	1079	441	1508	489	40	11
	International Olympiad	40		41		5.2	
	National AI Olympiad (Established 2023)			70			

4 Discussion

In implementing and continuously running its programs, Mawhiba has successfully followed its path toward achieving its vision of empowering the brightest young talented minds to create sustainable impact. The given numbers in Table 1 illustrate the expected sustainability of Mawhiba programs through showing the percentage increase of these numbers in two years that are between the years 2022 and 2023 for both male and female gifted students.

The study results reveal the significant progress Mawhiba has made in identifying and nurturing gifted students through its diverse programs. Programs such as the "Assessment of Giftedness and Identification Program" have seen a substantial increase in participation between 2022 and 2023, particularly among females, indicating successful outreach and testing methodologies. Similarly, enrichment programs like the "Mawhiba Classes" and the "Advanced Program for Science and Math" show steady growth, reflecting their effectiveness in enhancing students' academic and STEM-related skills. Research and innovation programs, including the "Essential Research Program" and the "Mentorship Program," have also experienced notable increases in participation, highlighting Mawhiba's emphasis on fostering creativity and innovation. Furthermore, competition programs like the "Kangaroo Math Competition" and the newly introduced "National AI Olympiad" demonstrate Mawhiba's commitment to preparing students for future technological challenges and global competitiveness.

Overall, the results reveal Mawhiba's success in creating a sustainable framework for nurturing gifted students and contributing to their holistic development. The rise in female participation across most programs underscores efforts to bridge the gender gap, particularly in STEM fields. Additionally, the steady growth in STEM-related programs and competitions exemplifies Mawhiba's focus on equipping students with the skills required for future advancements. Moving forward, Mawhiba could further expand its outreach efforts, focus on increasing participation in international competitions, and continue to monitor gender representation to ensure balanced growth. These efforts will help solidify Mawhiba's role in fostering gifted students and enabling their contributions to sustainable development.

Considering the above, four main issues need to be emphasized. The first issue is concerned with the fact that Mawhiba's exploration of the potential of gifted K-12 students is an important issue toward seeking sustainable development. This agrees with the HCDP of the KSA Vision 2030, which benefits from Mawhiba's contribution to development. The second issue to emphasize is linked to another fact that there are many more young, gifted students than Mawhiba can accommodate in its various programs. Therefore, further support for Mawhiba capabilities is needed to promote its sustainable impact on development. The third important issue to emphasize is supporting the widespread benefits of Mawhiba programs through the widespread use of the online services capabilities. Although Mawhiba is already using online services in its programs, further use needs to be considered to widen Mawhiba's impact on sustainability. Finally, the fourth important issue to emphasize in support of Mawhiba's impact is the use of the increasing capabilities of generative AI in its various programs. This would support human intelligence (HI) in enhancing creativity and promoting impact on development [15].

5 Conclusion

Mawhiba' vision is concerned with "empowering the brightest young talented minds to create sustainable impact". At the national Saudi level, this is part of the HCDP of the KSA Vision 2030. Mawhiba programs, toward achieving its vision, are experiencing an increasing number of students who benefit from their integrated deliverables and consequently enhancing impact on sustainable development. Cooperation with different supportive sources concerned with sustainable development is highly recommended for Mawhiba toward widening and promoting its programs and their sustainable impact. In addition, by increasing its use of online services, Mawhiba would increase the widespread of its benefits. Furthermore, its future use of the increasing capabilities of generative AI would promote its impact on its targeted creativity and innovation needed for its future impact.

References

1. Rutigliano, A., Quarshie, N.: Policy approaches and initiatives for the inclusion of gifted students in OECD countries. OECD Education Working Papers. **262** (2025)
2. Eysenck, H.J.: Know Your own IQ. Penguin Books Ltd., Auckland, New Zealand (1981)
3. Boeree, C. G.: Intelligence and IQ, Shippensburg University, http://webspace.ship.edu/cgboer/intelligence.html. (Accessed July 2022).
4. Gardner, H.: Multiple Intelligences: New Horizons in Theory and Practice. Basic Books, New York, USA (2006)
5. NAGC.: Policy and Practice Data, NAGC & The Council of State Directors of Programs for the Gifted (2015).
6. WCGTC.: World Council for Gifted and Talented Children, https://world-gifted.org/ (Accessed: July 2022).
7. Toth, Nancy W.: Gifted education: A critical discussion. (1999).
8. Borders, C., Woodley, S., Moore, E.: Inclusion and giftedness. In: Gifted education: Current perspectives and issues, pp. 127–146. Emerald Group Publishing Limited (2014)
9. Kim, M.: A meta-analysis of the effects of enrichment programs on gifted students. Gifted Child Quarterly. **60**(2), 102–116 (2016)
10. Mawhiba: King Abdulaziz and his Companions Foundation for Giftedness and Creativity, Homepage: https://www.mawhiba.org/En/, last accessed 08/01/2024.
11. KSA Vision 2030, Homepage: https://www.vision2030.gov.sa/media/rc0b5oy1/saudi_vision203.pdf, last accessed 08/01/2024.
12. HCDP: Human Capability Development Program (2021), Homepage: https://na.vision2030.gov.sa/en/explore/programs/human-capability-development-program, last accessed 08/01/2024.
13. Creswell, W.J.: Research design, qualitative, quantitative and mixed methods approaches. SAGE (2014)
14. Mawhiba Multiple Cognitive Aptitude Test. Homepage: https://former.mawhiba.org/en/Initiatives/Identification/Pages/Tests.aspx, last accessed 10/01/2024.
15. Routray, S.K., Jha, M.K., Sharmila, K.P., Javali, A., Pap, M., Singh, M.: Generative Artificial Intelligence: Principles, Potentials and Challenges, 2024 2nd International Conference on Self Sustainable Artificial Intelligence Systems (ICSSAS), pp. 211–216. Erode, India (2024)

Enhancing Innovation Through Interdisciplinary Collaboration: A Study on Interior Design and Construction Management Students in Augmented Reality Learning Environments

Amani Khan[1], Nathaniel B. Walker IV[2], and Henry Wanakuta[2](✉)

[1] Dar Al-Hekma University, Jeddah, Kingdom of Saudi Arabia
akhan@dah.edu.sa

[2] Stephen F. Austin State University, Nacogdoches, USA
{walkernb,henry.wanakuta}@sfasu.edu

Abstract. In higher education, we often see a gap between practical interdisciplinary collaboration and what happens in the classroom. This study sets out to close that gap by examining how collaborative projects involving interior design and construction management students, enhanced by augmented reality (AR) technology, can influence learning outcomes and promote innovation. The research study involved a mixed-methods approach. Seventy-two undergraduate students from both programs worked together on a semester-long project in an AR-supported learning environment. The study combined quantitative surveys with a qualitative focus to get a well-rounded view of the collaboration's effect. The data collection process included: a before-and-after survey instrument (which measured students' teamwork, communication, and creative problem-solving abilities). At the end of the project, the researchers conducted focus group interviews. The results showed notable improvements in interdisciplinary skills and innovative output. On average, students rated their collaboration skills nearly one point higher on a five-point scale from the pre-survey to the post-survey, which is a significant increase ($p < 0.001$). The interviews revealed themes of enhanced creativity, improved communication across both disciplines, and a better grasp of each industry's processes. Both the interior design and construction management students attributed the use of AR technology to their ability to establish common project goals that led to integrative and creative design and construction solutions. The findings show that interdisciplinary learning experiences and immersive technology can significantly enhance students' innovation and readiness for collaborative work in their chosen careers.

Keywords: Interdisciplinary collaboration · Augmented reality · Interior design education · Construction management · Innovative learning · Mixed-methods study

D. Berkouk et al. (Eds.): CCTS 2024, PTT, pp. 349–356, 2026.
https://10.1007/978-981-95-7607-4_39

1 Introduction

In the real world, interior designers and construction managers need to collaborate closely to make projects successful, but often, traditional academic programs train these professionals separately. Closing this gap in education is crucial to meet the industry's need for collaborative problem-solvers. New teaching methods are looking into how technology can enhance teamwork to encourage this collaboration. One exciting tool is augmented reality (AR), which has shown great potential in creating immersive learning experiences that unite various disciplines. AR can bring complex design and construction scenarios to life in real time, giving students a shared space to interact and play around with ideas. Research indicates that these immersive settings can lead to more effective interdisciplinary learning compared to traditional approaches [2].

Through enabling students to visualize and tweak shared 3D models, AR helps break down communication barriers and aligns perspectives between design and construction [5]. Even though the advantages of interdisciplinary education are recognized, there are still limited structured opportunities for interior design and construction management students to collaborate. Previous studies have shown that integrated project-based learning can enhance outcomes throughout all stages of design and construction education. The goal of the research is to assess how this interdisciplinary, tech-enhanced approach impacts student learning outcomes, especially in areas like creativity, communication, and problem-solving skills [4].

2 Literature Review

Higher education is undergoing a transformation as more educators embrace interdisciplinary learning, realizing that tackling complex real-world issues often calls for teamwork across various fields. When students collaborate in academic environments, it not only sparks creativity and innovation but also enhances critical thinking by introducing them to a variety of viewpoints. Take the design and construction sectors, for instance; when interior designers and construction managers join forces, the results can be much more comprehensive and effective [3]. A great example is the interdisciplinary architecture-engineering course highlighted [7], which thrive by promoting mutual respect, open communication, and a clear understanding of each student's role, regardless of their discipline. These case studies illustrate that with well-defined collaborative frameworks, students can successfully navigate disciplinary boundaries and come up with groundbreaking solutions [11].

However, the literature also points out some hurdles in interdisciplinary education. Academic "silos" can leave students out of the loop when it comes to the terminology or methods used in other fields, which can create initial roadblocks in teamwork due to misunderstandings and different approaches. This highlights the importance of structured activities and guidance [10]. Theoretical frameworks like constructivist and experiential learning back the notion that students learn most effectively when they engage actively in realistic projects - something that interdisciplinary collaborations excel at. By tackling real-world inspired challenges together, students build knowledge collectively, which resonates with Kolb's experiential learning cycle of hands-on experience and reflective observation [5].

Technological advancements are making it easier for different fields to collaborate. One standout innovation is augmented reality (AR), which has proven to be a fantastic tool for connecting theoretical knowledge with hands-on learning. Immersive AR environments in construction and design education have allowed students to dive into the material, leading to deeper collaboration and comprehension [9]. AR creates a visual and interactive space where concepts from interior design and construction can come together, helping students visualize how their ideas overlap. This approach tends to level the playing field, as both groups can literally see the effects of design changes or construction limitations in a shared AR model [2].

3 Methodology

3.1 Participants and Context

The research took place at a mid-sized university and included interior design and construction management students. A total of 72 undergraduate students participated, with a fairly even split of 52% female and 48% male, with 36 students from each program. Most of these students were in their third or fourth year, averaging about 21 years old. Their participation was part of a required course for each program, which helped ensure a comprehensive sample from the class cohorts. While all participants had some basic knowledge of design or construction principles, this was the first time they had worked together on a project. Before starting, students gave their informed consent for their data to be included in the study, and the university's review board granted ethical approval.

3.2 Study Design and Procedure

The research utilized a mixed-methods approach. The project spanned an entire semester and involved students from both interior design and construction management, who were grouped into teams of 4 to 5, ensuring a balanced mix from each discipline. Each team was given the exciting challenge of creating a realistic design-build proposal for a small commercial space, like a community youth center, which required insights from both the interior design and construction management perspectives. To enhance teamwork, students made use of an AR application during their collaborative studio sessions. This AR tool, available on tablets and headsets, enabled teams to visualize their interior layouts, structural components, and materials in a dynamic 3D overlay of their project. This shared AR space served as a collaborative platform where design concepts and construction strategies could be viewed and adjusted by all team members at once. Over the course of 12 weeks, teams moved through various stages of concept development, design detailing, and construction planning, and regular meetings in a shared studio lab. Instructors from both fields were there to supervise and steered discussions to keep everything on track, and students were encouraged to take charge of the decision-making process.

3.3 Data Collection Instruments

The study gathered quantitative data through surveys that were given out both before (pre-survey) and after (post-survey) the interdisciplinary creative project. The survey was designed to assess important aspects of collaboration and learning outcomes, using established educational scales as a foundation. It featured Likert-scale questions ranging from 1 (strongly disagree) to 5 (strongly agree) on various topics, including teamwork skills, communication effectiveness, confidence in problem-solving, and attitudes towards interdisciplinary work. For instance, some of the statements included were "I feel confident in my ability to work well in a team with students from different disciplines" and "Using technology like AR can enhance collaboration on design and construction projects." The pre-project survey helped us establish a baseline for each student's self-evaluated skills and attitudes, while the post-project survey allowed us to see any changes that occurred after the collaborative AR experience. Additionally, we collected demographic information and any prior experience with AR or interdisciplinary projects. To ensure the survey's reliability, we found that the internal consistency of the collaboration scale was quite high, with a Cronbach's α of 0.88 on the pre-survey.

Qualitative data was gathered through semi-structured focus group interviews after completing the project. We held two focus group sessions, each with 6 students totaling 12 participants. They were chosen to represent both the interior design and construction management groups, ensuring a mix of genders. Each session lasted around 60 min. The discussion prompts were designed to help students reflect on their experiences, the challenges they faced, the benefits they gained, and their suggestions for improvement. For instance, we asked questions like, "How did collaborating with students from the other program shape your approach to the project?" and "How did the AR technology influence your teamwork and creativity?" The focus group environment encouraged participants to build on each other's responses, often highlighting shared views or contrasting opinions between the two fields.

3.4 Data Analysis

The researchers analyzed the quantitative survey data using statistical software, starting with descriptive statistics to highlight overall trends. To assess the impact of the interdisciplinary AR project, we ran paired sample t-tests to compare pre- and post-survey scores on key outcome measures like teamwork, communication, and problem-solving skills. We set a significant threshold of $p < 0.05$ for our hypothesis testing. Additionally, we examined differences between subgroups using independent t-tests and chi-square tests. For example, we compared responses from interior design and construction management students on specific items to see if one group gained more or if both experienced similar benefits. We also employed a chi-square test to explore any links between students' disciplines and their overall satisfaction with the project, such as the percentage of students who would recommend these types of collaborations to others.

On the qualitative side, we analyzed focus group data through thematic analysis. Two researchers independently read and coded the transcripts to pinpoint recurring concepts. We started with initial codes like "learning from other disciplines," "communication challenges," "creative inspiration," and "technology usability," which we then discussed and grouped into broader themes. The study involved an iterative process of refining these themes to ensure they truly reflected the range of student experiences. From the data, three main themes emerged, which we validated by cross-referencing with the raw transcripts for representative quotes.

4 Results

4.1 Qualitative Findings

The focus group discussions revealed rich insights into the students' experiences. Three key themes emerged:

Enhanced Cross-Disciplinary Understanding Students reported a significant expansion in their understanding of the other discipline's perspective. Interior design students gained appreciation for construction logistics and constraints, while construction management students developed sensitivity to design considerations. *"I learned how important clear communication is when working with someone from another field,"* noted one construction management student, highlighting how the project taught them to explain technical details in accessible terms.

Creative Synergy and Innovation Working in interdisciplinary teams sparked creative thinking that exceeded what either group could have achieved alone. Students felt that combining design creativity with construction practicalities led to more innovative solutions. An interior design student commented that *"Brainstorming with the construction team members sparked ideas I wouldn't have thought of on my own."* This sentiment was common: exposure to different problem-solving approaches pushed students to think outside their traditional box. The AR environment further stimulated creativity by allowing rapid prototyping of ideas, teams would try out a design tweak in AR, see its feasibility, and iterate on the spot.

Challenges and Growth in Collaboration Students did acknowledge challenges, especially early in the project. Differences in terminology and working style led to some friction in the initial stages. For instance, construction students initially focused on structural feasibility, whereas design students were talking in terms of aesthetics and user experience. One theme that emerged was the importance of guidance participants noted that instructor-facilitated sessions early on helped set common goals and clarify roles, preventing misunderstandings. Technical challenges with the AR tool also arose; a few students mentioned a learning curve in using the devices and occasional software glitches. However, rather than impeding progress, these challenges became part of the learning experience. Students collaborated to troubleshoot issues (whether communication gaps or AR hiccups), and as a result, students-built resilience and adaptability.

4.2 Quantitative Results

The quantitative analysis of the survey data corroborated the positive outcomes described by students. Table 1 summarizes a selection of survey statements and the percentage of students who agreed or strongly agreed with each statement after completing the interdisciplinary project. The results indicate overwhelmingly positive perceptions of the collaborative experience. For example, 92% of participants agreed that *"The AR environment enhanced our collaboration effectiveness,"* and 88% agreed that *"I developed a better understanding of the other discipline through this project."* Moreover, nearly all students (95%) stated that they would recommend similar interdisciplinary projects to their peers, reflecting a high level of satisfaction with the experience (only a very small number were neutral or unsure, and none disagreed).

Table 1. Post-project survey responses (n = 72)

Survey statement	Agree/Strongly agree (%)
The AR environment enhanced our collaboration effectiveness	92%
I developed a better understanding of the other discipline	88%
My creative problem-solving skills improved through this project	90%
I would recommend interdisciplinary projects to other students	95%

In addition to these post-survey perceptions, the pre- vs. post-surveys provide objective evidence of skill development. We observed significant increases in self-rate competencies on all three key skill scales (measured on a 5-point Likert scale). Figure 1 shows the average pre-project and post-project ratings for students' teamwork, communication, and creative problem-solving skills. Prior to the project, students' confidence in these areas was moderate (mean scores around 3.0–3.5). After the interdisciplinary AR experience, mean scores in all categories rose to approximately 4.0 or higher. For instance, the mean teamwork skill rating improved from 3.4 (pre) to 4.5 (post). A paired-samples t-test confirmed that this improvement in teamwork confidence is statistically significant, $t(71) = 10.6$, $p < 0.001$. Similarly, communication skills rose from a mean of 3.2 to 4.3, and creative problem-solving from 3.0 to 4.1, both showing significant gains ($p < 0.001$). The effect sizes for these changes were substantial (Cohen's d ranging from ~0.8 to 1.0), indicating that the project had a strong educational impact on the participants.

5 Discussion

This study aimed to explore whether using augmented reality to enhance collaboration between interior design and construction management students could boost their educational experience and spark innovation. The results from both qualitative and quantitative analyzes provide strong evidence that it indeed did. Students not only saw improvements in measurable skills like teamwork, communication, and problem-solving, but they also underwent significant shifts in perspective that, while harder to quantify, are just as crucial. By tackling real-world problems together, these students learned to appreciate each other's expertise and approach challenges in a more integrated way. These findings resonate with trends highlighted in existing literature, reinforcing key points while also offering fresh insights.

To start, the notable improvements in teamwork and communication skills we observed aligned with results from previous interdisciplinary educational studies. Our data backs up the often-discussed idea that bringing together diverse students on problem-based projects enhances their collaborative abilities. This aligns with [7] observation that effective communication and mutual respect are vital for the success of interdisciplinary projects. In our study, students faced initial language barriers between the design and construction fields; exactly the type of challenge often mentioned in broader discussions about interdisciplinary teaching. The fact that both groups achieved similar positive outcomes suggests that our structured approach; mixing teams evenly, providing early facilitation, and using AR as a shared tool was effective in leveling the playing field and engaging everyone. Secondly, the use of AR technology was validated as a powerful enhancer of interdisciplinary learning.

While the results were largely positive, the experience also revealed some challenges and lessons worth discussing. One issue was the initial hiccup in team communication. challenge was the technical side of AR - there were a few glitches that reminded us of the importance of having contingency plans (and technical support) ready when using advanced tools in the process.

6 Conclusion

This research shows that boosting innovation through interdisciplinary collaboration is not just feasible but also incredibly effective in an academic context. By bringing together students from interior design and construction management in an augmented reality learning environment, we saw significant improvements in their educational outcomes. The students walked away from the project with enhanced teamwork and communication skills, increased creative confidence, and a better understanding of how interconnected real-world projects can be. This interdisciplinary experience effectively bridged the gap between creative design and technical construction knowledge, demonstrates how blending these fields can lead to better solutions for complex challenges. Key findings reveal that students can quickly learn to navigate disciplinary boundaries when provided with the right support and tools. The augmented reality aspect played a crucial role in this success, offering a shared platform for exchanging ideas and keeping students actively engaged in their learning. Given these results, we suggest that academic programs in design, construction, and similar fields should integrate structured collaborative projects into their curricula.

References

1. Bhattacharjee, S. (2014). Effectiveness of role-playing as a pedagogical approach in construction education. In 50th ASC Annual International Conference Proceedings (Vol. 1, 1, pp. 81–105).
2. Tsolakis, M.Z.: College, Career, and Postsecondary Success: An Analysis of Building Trade Academies in California High Schools. Occidental College, Los Angeles (2008)
3. Messadi, T., Newman, W.E., Braham, A., Nutter, D.: Immersive learning for sustainable building design and construction practices. J. Civ. Eng. Archit. **11**, 841–852 (2017)

4. Güney, K.: Contemporary practices in pre-service professional development of candidate teachers in Türkiye and the world. In: Bayar, A., Kayalar, F. (eds.) Innovative Research In Educational Sciences, pp. 21–43. Duvar Publishing, İzmir, Türkiye (2023)
5. Sleipness, O., Ryan, K., Krikac, R., Gomez, S.: Rural interdisciplinary service-learning projects: Frameworks for engagement within regional rural development centers. Landsc. Res. Rec. **5**(1), 181–195 (2016)
6. Chou, S., Gunasagaran, S.: Sustainable Interior Design Management Education: Challenges and Opportunities. School of Engineering, Taylor's University, Subang Jaya, Malaysia (2021)
7. Łątka, J.F., Michałek, J.: Interdisciplinary methods in architectural education. World Trans. Eng. Technol. Educ. **19**(1), 102–107 (2021)
8. Bhattacharjee, S. (2014 March). Effectiveness of role-playing as a pedagogical approach in construction education. In 50th ASC Annual International Conference Proceedings (Vol. 1 1, pp. 81–105).
9. Figgess, G.W., Vogt, R.G.: Building career-ready students through multidisciplinary project-based learning opportunities-a case study. In: 2017 ASEE Annual Conference & Exposition (2017, June)
10. Tadesse, N.S.: Multidisciplinary Collaboration in the Current Architecture, Engineering, and Construction Education of Ethiopia. (Master's thesis. Izmir Institute of Technology (Turkey) (2021)
11. Debs, L., Hubbard, B.: Integrating energy simulation in design and construction studios: Experiences from two undergraduate courses. In: Proceedings of the Conference CIB W78, vol. 2021, pp. 11–15 (2021, October)

Fog Computing-Powered Driving Assistant System Under Fog Weather to Support Sustainable Tourism in Rural Area

Tami Alwajeeh(✉)

Department of Computer Science, College of Computing and Information, Al-Baha University, Alaqiq, Saudi Arabia
Talwajeeh@bu.edu.sa

Abstract. Foggy weather is one of the main causes of road traffic accidents. The existing driving assistant systems (DAS) are based on cloud computing environment. In today's world, where Internet of Thing (IoT) and smart cities are commonplace, the need to process and make decisions in real-time has never been more important. To solve this problem, this research proposes a fog computing-powered DAS under fog weather. The proposed solution utilizes fog computing, a local and real-time information processing model. Kalman filter algorithm is used to predict the position of vehicles in a traffic scenario. The system processes data locally, avoiding the need to send large amounts of data to remote cloud servers that consume a lot of energy and contribute to environmental degradation. An einxperiment was conducted to generate dataset and simulate the proposed DAS. Simulation results showed a promising performance in terms of prediction accuracy, with Root Mean Square Error (RSME) value of 0.45 and average of computational time of 53.71%.

Keywords: Fog Computing · Driving Assistant System · Machine Learning · Kalman Filter · Sustainable Tourism

1 Introduction

Albaha City is one of the most popular tourist destinations in Saudi Arabia. However, it comes with its own unique set of challenges. Due to the presence of heavy fog, the natural beauty and charm of the city is often overshadowed by the danger of driving in heavy fog. This atmospheric phenomenon is enchanting, but it also poses a serious risk to road safety for both tourists and residents. Foggy weather is one of the main causes of road traffic accidents. Fog reduces visibility, making it more likely for accidents to occur. This atmospheric phenomenon, while enchanting, poses a significant risk to road safety [1–3]. According to the Ministry of Health of Saudi Arabia [4], road traffic accidents caused 5,754 deaths and 32,910 injuries in 2019. Foggy weather is a major contributing factor to these accidents, as it reduces visibility and increases the likelihood of collisions and run-off accidents [5].

D. Berkouk et al. (Eds.): CCTS 2024, PTT, pp. 357–365, 2026.
https://10.1007/978-981-95-7607-4_40

With smart cities and the Internet of things (IoT), the amount of data being generated by different devices, including vehicles has increased exponentially. The DASs, which rely mainly on cloud computing models face significant challenges in processing this large amount of data in real time.

To solve this problem, this research aims to create a fog computing-powered driving assistant system under fog weather. This solution utilizes fog computing, a local and real-time information processing model, to help drivers navigate under foggy weather. Unlike traditional cloud computing systems that experience data transmission delays and network connectivity issues, fog computing provides prompt and efficient assistance to drivers even in the hilly terrain of Albaha. By addressing the environmental parameters contributing to fog-related accidents, the proposed system aims to enhance road safety metrics within the Al Bahah region, while simultaneously preserving public accessibility to its unique environmental and cultural attributes.

2 Related Works

Fog computing involves deploying computing, networking, storage and acceleration elements at the edge and at the network layer in a multi-level, distributed and possibly cooperative way [6]. These elements are virtualized computing functions that are deployed at edge devices or at network elements as needed to realize the "computing everywhere" concept [7]. A Fog Computing-based framework to support autonomous driving was proposed in a study by Maheswaran et al. [8]. The framework leverages overhead views from cameras as well as data streams from the vehicle sensors to create an edge twin on fog machines. This edge twin is constantly updated with the location of both the autonomous and man-piloted vehicle on the road segments, and the vehicle locations are collected from overhead cameras and the location feeds from the vehicle itself. In another study [9], researchers proposed a technique that included the development of a vision enhancement system known as a Perceptive Driving Assistant system (PDAS) to improve real-time video visibility of the road ahead of the vehicle for heavy earth-moving operators operating in opencast mines during extreme weather conditions. The PDAS system is composed of high-resolution Internet Protocol (IP) cameras and thermal cameras to process real-time image data and other clearly defined devices. Vehicular Fog Computing (VFC) architecture was proposed to implement cooperative sensing among multiple adjacent vehicles driving in the form of a platoon [10]. Based on the VFC architecture, greedy and Support Vector Machine (SVM) algorithms were adopted respectively to enhance the sensing coverage and accuracy in the platoon. Using VFC in a vehicle platoon creates a fast, local processing network for immediate coordination. A greedy algorithm optimizes resource distribution, such as sensor range and processing power, among vehicles. An SVM then refines the data, boosting object recognition and overall awareness, which ultimately improves the safety and efficiency of autonomous driving.

Multi-sensor fusion is one methodology used to achieve vehicle surrounding perception. Such sensors include radar, camera, lidar, and ultrasonic wave. Multi-sensor fusion is a highly desired strategy because different sensors have their own capabilities [11]. Several reviews have been published recently on the topic of multi-sensor fusion, some of them describing the architectural structure and sensor technologies in autonomous vehicles [12, 13], or focusing on the processing stages like sensor calibration, state estimation, object and tracking [14, 15]. Fayyad et al. [16] has reviewed research works on the use of deep learning sensor fusion for autonomous vehicle perception and localization.

Many research works have also been carried on the DAS under fog using conventional or intelligent methods such as works in [17–19]. Maheswaran et al. [20] present the comprehensive fog computing framework for the ADS called edge twin for auto-drive assist and evaluates key elements of the proposed framework through simulations based on a realistic dataset of vehicle position trace data on a highway.

3 Method

The methodology for developing a fog computing-based driving assistant system under fog weather consists of components shown in Fig. 1. The components are described in the following sections.

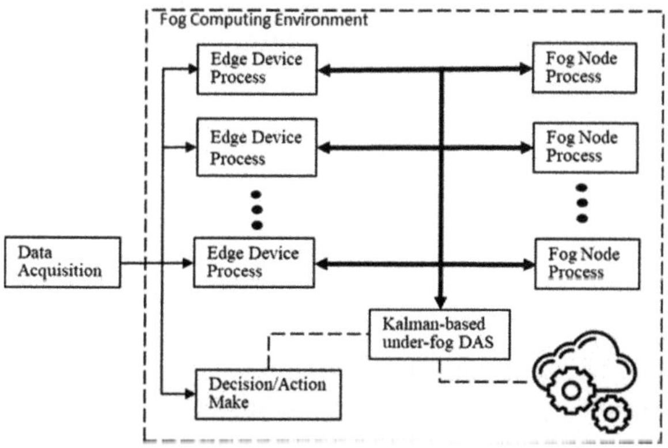

Fig. 1. The proposed fog computing-powered DAS under fog weather

Data Acquisition: Data will be collected during the testing phase for analysis, as the vehicle navigates through the city, it continuously collects data from its sensors, including: distance to nearby vehicles, pedestrian detection, traffic sign, recognition, weather conditions (e.g., fog density).

Local Processing at Fog Nodes: The fog nodes receive data from multiple vehicles within their vicinity and perform data fusion and local analysis.

Decision and Action Making: The fog node processes the information and produce real-time alerts to nearby vehicles and updates its driving assistance system such as automatically adjust its speed based on the alert.

Kalman-based under-fog DAS: The Kalman filter is a way to combine noisy and uncertain information to get the best possible estimate of a system's state. The Kalman filter is employed in driving assistance systems to filter noisy sensor data, such as GPS, radar, and Lidar inputs, ensuring accurate tracking of the vehicle's state even in challenging conditions. This is critical for applications like lane keeping, adaptive cruise control, and obstacle detection.

In this proposed model, a cloud service as a function for logging all interactions and alerts are used to improve machine learning models for better future predictions and enhance the overall driving assistance algorithms based on historical data and patterns.

3.1 Step-by-step Kalman Filter Algorithm on the Three Servers

The Kalman filter for the proposed DAS is implemented as follows.

Assumptions:

- Sensors: accelerometers sending data to the edge servers.
- State Estimation: The Kalman Filter estimates the state of a system based on noisy measurements.
- Edge Servers: three edge servers in the environment. Each server has its own role and performs part of the Kalman filtering algorithm. Server 1 is for sensor data preprocessing and measurement update; Server 2 is for prediction and process update; Server 3 is for fusion and state estimation aggregation.

1. Sensor Data Preprocessing (Server 1):

- Objective: Preprocess sensor data (i.e., filtering, normalization, or noise reduction).
- Data from sensors (e.g., accelerometer, GPS) are first received by Server 1.
- Server 1 filters out outliers or irrelevant data, removing noise and ensuring data quality.
- The data (measurements) are sent to Server 2 for the prediction step.

2. Prediction and Process Update (Server 2):

- Objective: Predict the next state of the system.
- Server 2 receives the preprocessed data from Server 1.
- The Kalman Filter prediction equations are executed on this server using (1) and (2) [18].

State prediction:

$$\hat{x}_{k|k-1} = A\hat{x}_{k-1|k-1} + BU_K \tag{1}$$

Covariance prediction:

$$P_{k|k-1} = AP_{k|k-1}A^T + Q \tag{2}$$

$\hat{x}_{k|k-1}$ is the predicted state estimate, and $P_{k\ |\ k\ -\ 1}$ is the predicted error covariance.

- The predicted state is sent to Server 3 for the measurement update step.

3. Measurement Update (Server 1):

- Objective: Update the state estimate based on new measurements.
- Server 1 also applies the Kalman Filter measurement update equations when it receives new sensor measurements using (3)–(5) [18].

The Kalman gain:

$$K_k = P_{k|k-1}H^T\left(HP_{k|k-1}H^T + R\right)^{-1} \tag{3}$$

Update the state estimate:

$$\hat{x}_{k|k-1} = \hat{x}_{k|k-1} + K_K\left(z_k - Hx_{k|k-1}\right) \tag{4}$$

Update the covariance:

$$P_{k|k} = (I - PK_kH) + P_{K|K-1} \tag{5}$$

The proposed Fog computing environment is shown in Fig. 2. The topology consists of network peripheral devices such as access point (AP) routers, switches; sensors and cameras; and three (3) high-end servers. The proposed DAS will be implemented as a mobile app. Users communicate with servers through WIFI or 4G access provided by telecommunication companies. Communication between sensors/cameras uses WIFI. However, for the experimental purpose, this research collects the GPS data from the users' cars.

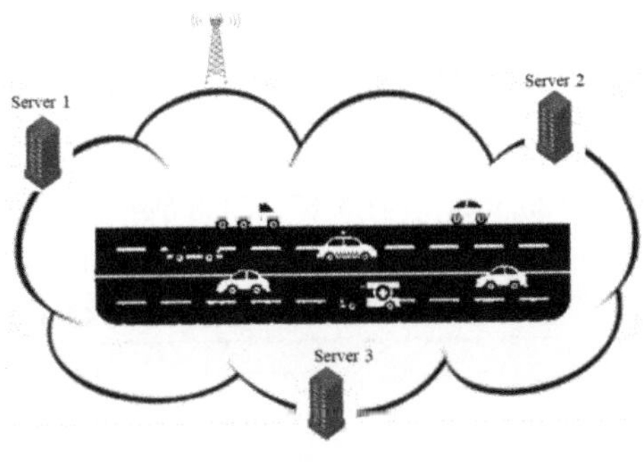

Fig. 2. Topology of fog computing environment

3.2 Dataset

The dataset for the experiment was taken from 4 cars running 18 h per- week for 8 weeks duration. The GPS data and velocity of each car are captured using the cars' GPS system. The data are captured for the interval of 5 min. Thus, 1312 data are recorded.

4 Preliminary Result

Initial experiment has been conducted that took place in the area close to Albaha University campus. Four cars are driven with different route and the proposed DAS is run. At the same time, a mobile phone that running simple App developed using MathWorks is used to capture the true value of the position of the car every 5 min. Cars' positions and velocity are captured from the GPS of the mobile phone. The experiment was run for 8 weeks, 1 h per day. Table 1 shows sample of data taken from the first car during the experiment for the first one hour. Then, manually, the errors are calculated. Overall, the estimated positions are relatively closed to the manual measurements, with RMSE value of 0.45.

Table 1. Sample of the data

	True value		Measurements		Estimation		Error	
T = 5	Lat	Long	Lat	Long	Lat	Long	Lat	Long
0	0	0			23.36627	42.5478	3.91623	1.2550
5	20.1857	41.2975	23.35521	41.569	23.36613	42.5429	3.18043	1.2454
10	20.69954	41.2971	23.35844	41.5615	23.36892	42.5369	2.66938	1.2398
15	20.69861	41.292	23.36087	41.5681	23.36974	42.5498	2.67113	1.2578
20	20.93994	41.3416	23.36186	41.5623	23.36994	42.5453	2.43	1.2037
25	20.93922	41.3499	23.36085	41.5691	23.37	42.5345	2.43078	1.1846
30	20.94499	41.3466	23.35705	41.5573	23.36786	42.562	2.42287	1.2154
35	22.93895	43.492	23.36109	41.554	23.37301	42.5464	0.43406	−0.9456
40	22.94469	43.4896	23.36208	41.5603	23.37242	42.5435	0.42773	−0.9461
45	23.35548	42.5557	23.35831	41.5695	23.36874	42.5595	0.01326	0.0038
50	23.35066	42.5662	23.36512	41.5387	23.36574	42.5528	0.01508	−0.0134
55	23.35925	42.5559	23.36382	41.5693	23.37406	42.5509	0.01481	−0.005
60	20.6964	41.2975	23.36627	41.5478	23.3658	42.5525	2.6694	1.255

4.1 Processing Time Comparison

Table 2 shows the computation time of the Kalman Filter computed centrally at cloud server compared to the results using Fog computing environment. From Table 2, it is observed that the average computational time of the proposed Fog-based DAS provides improvement of 55.47%, 51.81%, 50.972%, and 56.58% compared to standard DAS. As expected, the error percentage of the Fog-based DAS provides more accurate prediction for the position.

4.2 Simple Statistical Analysis

A t-test is a statistical test used to compare the means of two groups and determines if the differences between them are statistically significant. The hypotheses are defined as follows.

H_0: There is no significant difference between the two means.

Table 2. Processing time comparison fog-computing DAS vs. centralized DAS (Second)

	Car 1		Car 2		Car 3		Car 4	
t	KF	KF-Fog	KF	KF-Fog	KF	KF-Fog	KF	KF-Fog
0	5	3.2	6	4.4	4.5	2.1	5.1	4.2
5	5	2.6	5.5	3.8	4.5	2.1	5	3.2
10	4.3	1.9	5	2	4.6	2.4	5.1	4.1
15	3.9	3	4.9	2.1	4	1.9	4.4	3
20	3.2	2.4	4.2	1.9	3.8	1.8	4.5	3
25	2.5	0.9	4.2	1.8	3.8	2.1	4.3	1.9
30	2.5	0.7	3.6	2.3	3.6	2	4.3	1.8
35	2.6	0.5	3.5	2.4	3.5	1.9	4.3	1.8
40	3	2.1	3.7	1.9	3.5	2	3	1
45	2.9	1.8	3	1.3	3.8	1.9	3	1.1
50	2.4	1.1	2.9	1	2.3	1.2	2.7	0.9
55	2.4	1.1	2.9	1.1	2.2	1.2	2.6	0.8
60	2.3	2	2.9	1.1	2.2	1	2.6	2
Average:	3.2308	1.79231	4.0231	2.0846	3.5615	1.81538	3.91538	2.2154

H_1: There is a significant difference between the two means.

t value is calculated using (6).

$$t = \frac{\overline{x}_1 - \overline{x}_2}{\sqrt{\frac{S_1^2}{n_1} + \frac{S_2^2}{n_2}}} \tag{6}$$

$\overline{x}_1$ and S_1^2 is the mean and standard deviation of computational time on the cloud, while $\overline{x}_2$ and S_{21}^2 is the mean and standard deviation of computational time on the fog environment, respectively. n is the sample size. Table 3 shows the t-test results for the 4 cars with 95% confidence level.

Table 3. t-test results

Car	t-value	Inference
Car 1	−0.437	Reject H_0
Car 2	0.006	Reject H_0
Car 3	−0.752	Reject H_0
Car 4	0.019	Reject H_0

Statistically, it is shown that the computational time on cloud is significantly different from the computational time on Fog environment for every car. Thus, Fog computing environment provides promising platform for process time efficiency.

5 Conclusion

This paper has introduced Fog computing-powered driving assistant system that presents a promising solution for enhancing safety and promoting sustainable tourism in rural areas. By leveraging fog nodes for real-time processing and sensor fusion, the system enhances safety and navigational support in challenging weather scenarios. The effectiveness of Kalman filtering for GPS location prediction is heavily dependent on the accuracy of the system model and the characteristics of the noise. Environmental factors that affect GPS signal quality, also heavily impact the accuracy of the Kalman filter. One of mitigate solution to these limitations is to deploy sensor fusion techniques to integrate data from multiple sensors, such as accelerometers, gyroscopes, and magnetometers. In the near future, by leveraging machine learning models such as: Artificial Neural Networks and Support Vector Machines, and focusing on energy efficiency such as: Optimization of Driving Behavior and the use of sensor fusion, the proposed DAS can significantly improve the driving experience in adverse weather conditions.

References

1. Traffic Accidents Injuries Death - وزارة الصحة السعودية. https://www.moh.gov.sa/en/Ministry/Statistics/book/Documents/Book-Statistics-2019.pdf. (2025). Accessed 3 Jan 2025
2. Elshambaty, Y., et al.: An unusual cause of road traffic accident with an unusual outcome in Saudi Arabia. Int. J. Med. Dev. Ctries. **3**(1), 1–6 (2023)
3. Gulf News Report, "Video: Rare snow and hail hit Al Baha region of Saudi Arabia, disrupting traffic, Gulf News, Apr. 2023. https://gulfnews.com/world/gulf/saudi/video-rare-snow-and-hail-hit-al-baha-region-of-saudi-arabia-disrupting-traffic-1.95057167. (2025). Accessed 3 Jan 2025
4. Al Bahah, Al-Bahah, Saudi Arabia Weather Forecast|AccuWeather. https://www.accuweather.com/en/sa/al-bahah/317773/weather-forecast/317773
5. Albaha, Al-Baha, Saudi Arabia Weather-The Weather Channel. https://weather.com/en-MT/weather/tenday/l/Albaha+Al+Baha+Saudi+Arabia?canonicalCityId=2167aa5917b42bfb7c3469afed2d1183d994b3f28b66f060eae8780243320808. (2025). Accessed 3 Jan 2025
6. Bonomi, F., et al.: Fog computing and its role in the internet of things. In: Proceedings of the First Edition of the MCC Workshop on Mobile Cloud Computing, pp. 13–16 (2012)
7. Chiang, M., Zhang, T.: Fog and IoT: An overview of research opportunities. IEEE Internet Things J. **3**(6), 854–864 (2016)
8. Singh, J., Singh, P., Singh-Gill, S.: Fog computing: A taxonomy, systematic review, current trends and research challenges. J Parallel Distrib Comput. **157**, 56–85 (2021). https://doi.org/10.1016/j.jpdc.2021.06.005
9. Choudhary, M., et al.: Perceptive driving assistant system for opencast mines during foggy weather. Min. Metall. Explor. **39**, 2431–2447 (2022)
10. Du, L., et al.: A new vehicular fog computing architecture for cooperative sensing of autonomous driving. IEEE Access. **8**, 10997–11006 (2020)

11. Kim, J., Han, D.S., Senouci, B.: Radar and vision sensor fusion for object detection in autonomous vehicle surroundings. In: 2018 Tenth International Conference on Ubiquitous and Future Networks (ICUFN), pp. 76–78 (2018). https://doi.org/10.1109/ICUFN.2018.8436959
12. Velasco-Hernandez, G., Yeong, D.J., Barry, J., Walsh, J.: Autonomous driving architectures, perception and data fusion: A review. In: Proceedings of the 2020 IEEE 16th International Conference on Intelligent Computer Communication and Processing (ICCP 2020), Cluj-Napoca, Romania, 3–5 September 2020 (2020)
13. Campbell, S., et al.: Sensor Technology in Autonomous Vehicles: A review. In: Proceedings of the 2018 29th Irish Signals and Systems Conference (ISSC), Belfast, UK, 21–22 June 2018 (2018)
14. Wang, Z., Wu, Y., Niu, Q.: Multi-sensor fusion in automated driving: A survey. IEEE Access. **8**, 2847–2868 (2019)
15. Yeong, D.J., Barry, J., Walsh, J.: Review of multi-sensor fusion system for large heavy vehicles off road in industrial environments. In: Proceedings of the 2020 31st Irish Signals and Systems Conference (ISSC), Letterkenny, Ireland, 11–12 June 2020 (2020)
16. Fayyad, J., Jaradat, M.A., Gruyer, D., Najjaran, H.: Deep learning sensor fusion for autonomous vehicle perception and localization: A review. Sensors. **20**, 4220 (2020)
17. Fahad, A.: Infrastructureless Kalman filter based driving assistance system under fog condition. J. Basic Appl. Sci. **3**, 1–6 (2019)
18. Ross, R., Hoque, R.: Augmenting GPS with geolocated fiducials to improve accuracy for mobile robot applications. Appl. Sci. **10**, 146 (2020)
19. Kerim, B.: Improving the accuracy of under-fog driving assistance system. J. Signal Inf. Process. **11**, 23–33 (2020). https://doi.org/10.4236/jsip.2020.112002
20. Maheswaran, M., Yang, T., Memon, S.: A Fog Computing Framework for Autonomous Driving Assist: Architecture, Experiments, and Challenges. https://arxiv.org/pdf/1907.09454.pdf. Accessed 23 Sep 2023

Adapting New Technologies Within Cultural Heritage in Islamic Studio Ceramics

Zelal Basodan(✉)

University of Jeddah, Jeddah, Saudi Arabia
bazelal@uj.edu.sa

Abstract. Islamic art had and continues to be created, and ceramics production played a significant role in it. Recently, there has been a growing interest in replicating traditional Islamic patterns in both the East and West, particularly geometric patterns. However, heritage is a dynamic and ongoing cultural process, not just a reflection of the past. Furthermore, generally with the shift in interests from material to digital based applications. Therefore, this project is practice-based research employing a multi-method approach with the goal of producing hybrid contemporary Islamic ceramics using both traditional and new technology approaches. Thus, the research's creative practice also considered natural inquiry and serendipity as methods to inform its creative exploration. This resulted in a body of hybrid creative work, and a novel discovery regarding the application of lasers on fired ceramic pieces. Revealing a unique technique that practitioners can utilise and potentially benefiting the creative and industrial sectors.

Keywords: laser engraving · 3D printing · ceramics · Alhambra · Nasrid polygons · serendipity

1 Introduction

Clay is an ancient material that has been utilised since prehistoric times, which has historical and cultural significance. Ceramics are regarded as one of the most important artefacts, and civilisations learnt about one another through artefact exchanges. Since, heritage is a continues and ongoing cultural process, and not solely a reflection of the past [1, 2]. Therefore, the elements selected in this investigation was inspired by Alhambra Palace's cultural heritage, seeking to revisit and preserve these cultural ties within the outcome of contemporary ceramic creative practice.

Due to the advent of new technologies applications as laser cutting, 3D scanning, and 3D printing, many users' interests shifted from material-based disciplines to digital-based disciplines. Three main factors seem to be responsible for the growing decrease in the utilization of tactile sensation with real materials during the creation process: (a) the innovative digital tools used in contemporary design; (b) the vast amount of material knowledge that has been accumulated; and (c) the accessibility or reasonable cost [3]. Since studio ceramics is a craft area that its practitioners primarily relay on hand skill techniques, this led the field to be one of the most impacted by new technologies.

D. Berkouk et al. (Eds.): CCTS 2024, PTT, pp. 366–374, 2026.
https://10.1007/978-981-95-7607-4_41

Having said that it is important to remember that the history of ceramics has been marked by similar challenges and rejections, beginning with coiling and progressing to throwing, moulding, and now 3D printing and other modern digital technologies [4, 5]. Furthermore, Laura Grey [6] points out that the closure of ceramics programs in specialised universities that could improve knowledge and discourses in the field is a significant barrier facing the ceramics discipline in the UK. Thus, it is crucial to consider a dual perspective of contemporary studio ceramics that contemplates both traditional and recent technological advancements.

Therefore, this project aimed to investigate the feasibility of creating hybrid contemporary Islamic ceramics, incorporating hand skills to maintain the tactile sensation. As well as digital applications to potentially attract digital oriented users or younger users. Which raised the question of how can a combination of hand skills and new technologies be developed as a hybrid approach to Nasrid ceramics, in order to make an innovative contribution to contemporary Islamic Studio Ceramics? This paper discusses and explores the notion serendipity and how it can on some occasions inform the artistic practice, demonstrating a novel discovery that was achieved during the course of this research[1].

2 Methodology

The Constructive view as a Naturalistic inquiry were used as the foundation for this study's paradigm [7]. Accordingly, this study's overall design is qualitative [8]. Practice-based research can be under the overarching methodologies of applied research, as it allows practitioners to consider and analyse their own creative output. More precisely, the research's creative practice component employs a circular systematic approach of making, reflecting, evaluating, and remaking [9, 10]. This study focusses on studio ceramics practice and tracks new developments in making and thinking.

Moreover, this study is also employing a multi-method approach, which commonly used in design and art research. This approach is different from what is so-called mixed-methods that combines qualitative and quantitative data. Which is also referred to as 'triangulation' multi-methods, that involves the use of two or more techniques and tools in gathering, investigating, and analysing data [7]. While creative researchers must be adaptable, they can use two or more approaches in their research, depending on the needs and nature of the investigation. As it assists researchers to gather data from a variety of perspectives, this strategy helps to trustworthy the data collected [7].

The potential use of a hybrid approach was examined in this study, including the potential integration of new technology with conventional ceramic techniques within the framework of cultural heritage. However, as this study also refers to a historical setting, it also aligns with the idea that creative hybridity's status should be primarily historical in order to comprehend its presence [11]. The fact that the researcher is a Saudi ceramic artist female conducted the research in the UK to create work that is influenced by Islamic cultural heritage, created an additional layer of complex hybridity.

[1] This paper is based on parts from research conducted for a PhD thesis at the School of Design and Creative Arts, Loughborough University, UK, titled "Reimaging the Nasrid Palace and Alhambra Vases in Emerging Islamic Studio Ceramics – A Multi-method Approach"[17].

Both technical drawings and hand sketches were used during the design process and exploration, and some of the explorations also made use of digital programs such as Illustrator and TinkerCAD. Similarly, both hand crafting methods and digital technology such as 3D printing and laser cutting were used in the making process. Self-reflective analysis, observation and testing were the methods used throughout. Certainly, the creative process in ceramics relies on serendipity or accidents, in contrast to other fields of study.

2.1 Hybrid Creative Ceramic Expression Utilising Digital Technology Alongside Conventual Methods

This section highlights the research process of exploring the application digital technology and how it can be used in a simplified manner in combination with conventional ceramic techniques. According to Cathy Treadaway, as quoted in Shillito [12] "The complex interface of many graphics and computer-aided designing (CAD) packages is an obstacle to designer makers being heavy on features".

Correspondently, the aim was to employ accessible and simple CAD tools such as Illustrator and Tinkercad. This is through exploring the use of new technology that can be added as a new method to the toolbox and replace some time-consuming or challenging traditional hand methods, which can enhance the ceramic creative practice. The reasons behind using a software as Tinkercad are its accessibility for public and being user-friendly, as well as its possibility to achieve desired outcomes. In addition to the recommendations made by Jonathan Keep [13] and Bri Murphy [14], which also served as the foundation for this choice. Laser cutting and 3D printing are the two primary digital techniques examined in the creative practice.

Bri Murphy [15] notes that in her ten years of working with ceramics and in particular 3D printing, she has witnessed a great deal of concern about the effects of integrating digital applications into such traditional art practice. Having said that, it is key to remember that the role of hand-building or throwing, will not be replaced by 3D printing, as each tool has its own limitations and advantages. As 3D printing does not make ceramics better, yet differently with numerous of ways to explore its usage [13, 15]. The main two digital applications explored in this project are 3D printing and laser cutting, which are discussed in the following sections.

Exploring 3D Printing Polygons with Clay

To produce a printable clay, multiple consistencies of two types of clay pastes were prepared. In accordance with Jonathan Keep's guidelines and recommendations [13], water was added, the paste was mixed with a fork, and then it was pressed through a strainer. The exploration initially started with printing with clay using the Hyrel 3D printer, which prints through a syringe (about 30 ml) attached to the printer and can be filled with any kind of paste such as clay. 3D-printing Arabic texts was tested on this printer and achieved promising results. Printing Arabic text was also tested on the FormLab 3D printer by using ceramic resin instead of pure clay, the print had high resolution, yet beside the martial is not being organic the firing process needs to be set on a specific programming to fire out the resin first.

The most recent investigation was conducted on the Tronxy Moore 1 3D printer, an affordable and user-friendly Chinese printer. Due to the small size of the 3D polygons especially the angled shapes it can take up to 3–4 h for each piece using hand-building technique. While by using this printer to 3D print these 5 cm size polygons, it took 8 min to be printed per form. When multiple pieces are required, this method is advised because it saves a great deal of time in comparison to the hand-building method (Fig. 1). Nevertheless, the Tronxy Moore 1 3D printer does not allow for the addition of the grid that was used in the handcrafted item to the interior.

Fig. 1. Hand-built Alhambra-like vase, hand-built vase, the Nasrid birds wing, 3D printed polygons, underglaze colours, 65 × 50 cm, 2023, photo by Zelal Basodan.

Exploring Laser on Clay and Ceramics

Regarding the application of laser cutting on ceramics, the exploration started on various types of clay in different clay stages greenware (dry), bisque fired, and glazed ceramic tiles. Using a carbon dioxide flat-bed laser machine (Fb1500- 40 W). Testing the laser on greenware it did not show any promising results, while for fired clay there were potential successful results. The exploration with the laser cutter was taken further on a very thin unfired paper porcelain, which resulted in a new discovery by chance or serendipity.

The unexpected result was when glazed lines occurred, after using the laser cutter to engrave the circular grid of the pattern on an unglazed hand-made paper porcelain piece fired up to (1260 °C) which is the highest temperature (Fig. 2-a). The setting selected in the laser cutter machine was 'balsa wood 6m.m', yet when setting the machine on 'acrylic' the ceramic piece was burned by the heat of laser. Using the same setting this result was taken further by testing different types of unglazed fired clays, which are white earthenware, stoneware, porcelain, and terracotta, resulting in varied sheen effects. For porcelain and paper porcelain it resulted in clear glaze, yet the results slightly varied from one another. While the result on stoneware and earthenware showed clear glaze with some metallic effect. Terracotta pieces showed very different results with black marks with slightly glaze-like effect. Additional examinations were applied on fired ceramics using the laser cutter to explore the possibility of engraving of one line then two parallel lines to change thickness, then tested engraving Arabic text, which all showed successful results.

It is worth noting that, Chrisoula Konstantakou is also exploring the use of laser on terracotta and had obtained these black marks on terracotta. Konstantakou is specifically interested in terracotta as it is an ancient material with the application of laser as digital methods to decorate her hand made ceramics. However, she believes that the only type of clay that laser can be applied on is terracotta as it contains iron which reveals numerous tones of blacks, purples, and reds, which cannot be obtained on white clays [16].

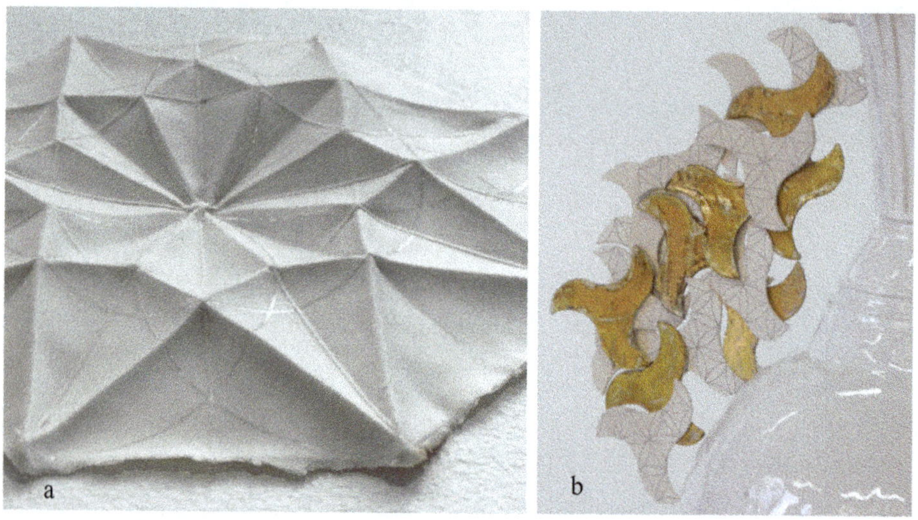

Fig. 2. (a) Circle grid using laser cutter, showing glazed sheen, 2022. (b) Details - The hybrid Nasrid birds wing, hand-cut polygons, reduction lustre and laser engraved grid, 2024, photos by Zelal Basodan.

However, it is worth noting that from the glaze affect the laser cutter left on the fired white clays, which no similar results are found to date, demonstrates the possibility of applying laser on white clays. For the current practice known in the field of ceramics the application of laser engraving is commonly on glazed pieces, which results in black marks on a white glaze. Another common practice is engraving with laser after covering the piece with glaze, but before firing the glaze that also results in black marks, wither it is white or red clay.

This novel discovery is not only useful for the current creative practice part of this research, but also provides a new technique that can be used by practitioners the field of ceramic and other industries. This is due to the permanent marks this method makes on unglazed fired ceramics, contrasting the applications of paints or pigments that can fade.

The Application of 3D Scanning in Ceramics and Future Opportunities

The method of 3D scanning of existing artefacts in museums is already established, which assist in preserving, repairing, or studying objects for education or research or purposes. The process in the current project involved making a hand-built vase, then 3D scanning it, and then 3D printing in filament, by the 3D scanned file is in printable format allowing more copies to be printed.

There is threefold rationale behind this approach: First, it allows ceramic practitioners to create a mould of a ceramic object without and saving the original piece from damage. Second, if new ideas arise after glazing a piece, 3D scanned file can print a duplicate to create a mould instead of rebuilding the form. Third, from a design perspective, this method can be more sustainable as after 3D scanning a piece, a digital copy of the design will be available to provide a variety of designs and scales. Fourth, the availability of 3D scanned allows and printable files having access to 3D scanning and printing data allows practitioners and researchers to conduct multiple experimentations with clay and any other different materials. Finally, from a practitioner viewpoint, 3D scanning became more accessible to implement, suggesting democratising access to digital applications. By demonstrating this process of designing and making that incorporate new technologies, wider audience can be engaged and encourage interest in the field of ceramics.

In terms of the creative practice of the project further designs and ideas have been developed, which include both an exploration of contemporary Alhambra vase wings and the utilisation of symbolism and concepts inspired from the Alhambra palace. Two of these designs were developed further in a collaboration with a Chinese 3D designer to create 3D digital models that are printable or partially printable. The designs used an arch as an architectural feature from the palace to highlight the intertwined relationship between Alhambra vases and the palace. These two designs are the most recent of what has been researched so far, and represent the beginning of the creative inquiry, leading to the next phase.

3 Results and Discussion

The previous sections demonstrated how research was investigating the possibility of achieving a hybrid creative body of work, by exploring digital technologies such as laser cutting, 3D scanning, and 3D printing to add new methods to the toolbox of making ceramics. Additionally, user-friendly CAD software such as Tinkercad and Illustrator were applied for the aim of accessibility for ceramic practitioners. This benefited the creative making by replacing the conventional making process that is time-consuming or challenging to achieve by hand, which can also be combined or attached to hand-made pieces using traditional methods.

An unexpected innovation as discovered during this project, achieved by engraving unglazed fired ceramic objects with the laser cutter. The novel results were glazed, or metallic permanent marks appear on the ceramic pieces. This innovative finding proposes a new technique that can be utilised by ceramic practitioners or designers and can influence other creative industries.

Furthermore, part of the research was aiming to combine the digital techniques with the significant ancient 9th century reduction lustre technique, this was through the application of digital decals and 3D printed pieces, some progressive results were archived, but will be carried further. The latest practice was a continuous exploration of Alhambra vases' wings by combining the novel discovery of laser with the ancient reduction lustre technique, resulting in a hybrid piece that contains hand-cut polygons with the application on both ancient and new technology methods (Fig. 2-b).

4 Conclusion

Due to the cultural heritage significance of Alhambra Palace and its aesthetic excellence it was chosen as source of inspiration in this project. Additionally, ceramic was selected as medium to adapt traditional patterns, symbolism, concepts, and traditional methods alongside digital methods. This is an example to encourage the need of further revisiting of other Islamic cultural heritage examples to be explored by contemporary ceramic artists. It can be concluded that this study should thus be of value to creative practitioners, not only in the field of Islamic art, but also in the wider ceramics field, and that may extend it impact to fields beyond these two fields.

In terms of contemporary creative practice related to Alhambra Palace and Alhambra vases, this project aimed to further re-interpret element from the palace and the vases as the wings of superb artefacts, which inspired explorations of new experiments and ideas. Aimed to create contemporary hybrid artefacts inspired by historical artefacts and following conventional methods and techniques while also integrating digital applications. Digital technology applications such as laser cutting, 3D scanning, 3D printing, were also adapted to assist the creative practice. This innovative finding therefore has great opportunity to influence diverse disciplines due to its accessibility and adaptably in designing and producing both small- and large-scale ceramic work.

Acknowledgments. Sincere thanks to my supervisor at Loughborough University, Dr Roberta Bernabei, and Prof Phillip Lindley for his advice. Special thanks to Ben Brierley and Andrew Chong for their technical support. Would also appreciatively acknowledge the funding provided by the Saudi Arabian Cultural Bureau and University of Jeddah.

References

1. Harvey, D.C.: Heritage pasts and heritage presents: temporality, meaning and the scope of heritage studies. In: A Museum Studies Approach to Heritage, vol. 7258, pp. 14–28 (2018). https://doi.org/10.1080/13581650120105534
2. Harrison, R.: Heritage Critical Approaches, 1st edn. Abingdon, Routledge (2013)
3. Leader. E.: Materializing craft: evaluating the effects of experiencing actual materials during the design process. Des. Principles Pract. Int. J. **4** (2010)
4. Keep, J.: The Fourth Way. Ceramic Review, 32–37 November/December (2014)
5. Keep, J.: Potting in a digital age. Studio Potter **47**, 16–21 (2019)
6. Gray, L.: Quiet revolution: contemporary curatorial approaches to ceramics in the White Cube. In: Myzelev, A. (ed.) Exhibiting Craft and Design: Transgressing the White Cube Paradigm, 1930-Present, pp 89–106. Routledge, New York (2017)
7. Gray, C., Malins, J.: Visualizing Research. Ashgate Publishing Limited, Aldershot (2004)
8. Creswell, J.W., Creswell, J.D.: Research Design; Qualitative, Quantitative & Mixed Methods Approaches, 5th edn. SAGE Publications, Inc., Los Angeles (2018)
9. Schön, D.A.: The Reflective Practitioner: How Professionals Think In Action. Basic Books, New York (2008)
10. Candy, L.: The Creative Reflective Practitioner, 1st edn. Routledge, Oxon, New York (2020)
11. Levinson, J.: Hybrid art forms. J. Aesthetic Educ. **18**, 5–13 (1984)
12. Shillito, A.M.: Digital Craft,1st edn. Bloomsbury Publishing, London (2013)
13. Keep, J.: A Guide to 3D Printing (2020)
14. Mulvin, L.S.: A pioneer of "Moresque" revival in Continental Europe: James Cavanah Murphy (1760–1814). Art Translation **11**, 148–180 (2019). https://doi.org/10.1080/17561310.2019.1658989
15. Murphy, B.: Communities of Altruism: ceramics and maker culture. NCECA J. **42**, 82–83 (2022)
16. Konstantakou, C.: Let there be light in the fire. Studio Potter (2021)
17. Basodan, Z.: Reimaging the Nasrid palace and Alhambra Vases in emerging Islamic studio ceramics – a multi-method approach. Ph.D., Loughborough University (2024)

Identifying and Reducing Waste in Parkinson's Disease Diagnosis: A Lean Approach to Sustainable Healthcare

Dannah AlSafran[1](✉), Mostafa Fawzy[1], Raneem AlQahtani[2], Sarah AlSafran[3], and Maha AlOtaibi[4]

[1] Dar Al-Hekma University, Jeddah, Kingdom of Saudi Arabia
dsafran@dah.edu.sa
[2] King Fahd University of Petroleum and Minerals, Dhahran, Kingdom of Saudi Arabia
[3] Imam Abdulrahman Bin Faisal University, Dammam, Kingdom of Saudi Arabia
[4] Najran Education Administration, Ministry of Education, Najran, Kingdom of Saudi Arabia

Abstract. Current Parkinson's Disease (PD) diagnostic methods, such as Magnetic Resonance Imaging (MRI), Positron emission tomography (PET), and Single Photon Emission Computed Tomography (SPECT) scans, require extensive resources, generate significant waste, and rely on energy-intensive procedures, contributing to high operational costs and environmental impact. These current PD diagnostic methods produce various types of lean waste, including excess inventory of disposable materials, overproduction of redundant tests, defects due to misdiagnoses requiring repeat scans, motion waste from patient movement between departments, and long waiting times for scheduling and processing. This research examines the inefficiencies inherent in current PD diagnostic methods and evaluates the potential sustainability benefits of a novel voice recognition-based device as an alternative. The findings show that the novel method significantly reduces various types of waste, such as excess inventory, overproduction, and over-processing … etc. It offers a non-invasive, cost-effective alternative with minimal infrastructure needs, making it particularly beneficial in resource-limited settings by improving diagnostic accessibility and speeding up patient care. The study underscores the potential of integrating advanced technology using lean principles to promote environmentally and socioeconomic responsible healthcare innovations.

Keywords: Parkinson's Disease · Lean Principles · Waste · Voice Recognition · Healthcare Sustainability

1 Introduction

The Lean Principles (LP), originating in manufacturing, focus on waste reduction and value generation through process optimization. They aim to improve workflows, reduce resource consumption, and eliminate inefficiencies while maintaining high-quality outputs. In various industries, LP help optimize time, energy, and resources by addressing non-valuc-adding operations [1].

D. Berkouk et al. (Eds.): CCTS 2024, PTT, pp. 375–383, 2026.
https://10.1007/978-981-95-7607-4_42

In healthcare, LP offers benefits like enhanced operational efficiency and sustainability. Diagnostic processes, which often require significant resources, present opportunities to integrate LP for reducing waste, energy use, and improving patient care. Sustainable healthcare innovation promotes accessible, eco-friendly solutions with long-term positive impacts [2].

Building on previous research on machine learning-based early PD diagnosis using voice recognition and research on other common diagnostic approaches [3], this study aims to compare all these techniques from the sustainability perspective using LP. It pinpoint waste and inefficiencies in conventional diagnostic methods, with an emphasis on how the recently developed novel method can maximize PD diagnosis while upholding sustainability using LP [4]. Identifying waste in PD diagnosis involves spotting inefficiencies or non-value-adding activities in the diagnostic process, such as repeating diagnostic tests unnecessarily. If patients undergo the same imaging studies (like MRI or PET scans) multiple times without clear clinical reasons or a structured review of prior results, this is considered overproduction waste. Another example is the long waiting times for specialist consultations or test results. Patients may need to wait weeks or even months for a neurologist's appointment, or delays in receiving test results (e.g., blood tests, neuroimaging), constitutes waiting waste.

2 Literature Review

2.1 Categories of Lean Waste

LP are methodologies aimed at improving operational efficiency, reducing waste, and delivering value to customers. Initially developed in manufacturing (Toyota Production System), Lean has since been adapted to many industries, including healthcare and service sectors. The central idea of Lean is to identify and eliminate non-value-adding activities, thereby improving process efficiency and overall outcomes [5–7].

One of the key components of Lean is identifying and eliminating waste (Muda). Waste in Lean is identified as anything that does not add value to the customer. Below are the seven categories of waste that Lean aims to reduce [1, 8, 9]: (1) Unneeded inventory by preventing excess stock while ensuring resources meet demand. (2) Overproduction by producing only what is needed, following the Just-in-Time (JIT) approach. (3) Defects by ensuring products meet customer standards to avoid rework and waste. (4) Unnecessary motion by reducing needless movements to save time and energy. (5) Over-Processing by eliminating non-value-adding steps to enhance efficiency. (6) Waiting by minimizing downtime to ensure smooth workflows. (7) Transportation by decreasing unnecessary travels to cut costs and resource waste. These waste categories based on LP prioritize efficiency and sustainability, making them relevant across diverse industries, including healthcare [9, 10].

2.2 Applications of LP in Healthcare

LP enhance efficiency and sustainability in healthcare by optimizing patient care, resource use, and workflow. Their application is particularly valuable in diagnostic processes and medical device evaluation, where structured, waste-reducing methodologies can significantly improve diagnostic accuracy and minimize costs associated with redundant tests and procedures [7]. Hospitals have successfully implemented LP to reduce wait times, improve service delivery, and minimize energy consumption. However, challenges such as leadership support, employee resistance, and cost concerns remain barriers to full adoption [11, 12]. Despite these obstacles, LP provides a structured framework for improving healthcare sustainability and efficiency [12]. For instance, Lakeview Healthcare improved patient flow and reduced wait times through Lean-based bed management and medical record optimization [13]. Similarly, an oncology hospital enhanced patient outcomes by streamlining procedures and clearly defining roles, demonstrating how Lean methodologies contribute to better healthcare delivery [14]. Such implementations show how LP can directly influence the effectiveness of diagnostic technologies and approaches, including innovations like the AI-based method for Parkinson's disease detection discussed in this study. These approaches also help improve patient satisfaction and reduce the overall cost of care [7]. Continued research on Lean-based innovations, such as the novel PD diagnostic method discussed in this study, is essential to further improving healthcare efficiency and sustainability.

3 Research Approach

The goal of this research is to pinpoint waste and inefficiencies in conventional diagnostic methods, with an emphasis on how the recently developed novel method can maximize PD diagnosis while upholding sustainability using LP. This research employs a mixed-methods approach, combining qualitative analysis with LP and quantitative evaluation using the Pugh Method, also known as the Pugh Decision Matrix.

The Pugh Method is a tool used for decision-making, especially in situations where there are multiple options and criteria to evaluate. It is a structured technique that helps teams or individuals compare a set of alternatives against a baseline option and systematically select the best one based on defined criteria. The method was introduced by Stuart Pugh in the 1980s, and it is widely used in product development, engineering design, project management, and business decision-making [15, 16]. For this research, the Pugh Method provides a structured quantitative comparison, with the Voice-Based Diagnostic Method (VBDM) as the baseline. The alternative diagnostic methods are evaluated across the seven Lean waste categories. Scoring within the matrix was derived from the researchers' analysis and assessment of the literature review findings.

4 Waste Assessment of PD Diagnostic Methods

This section examines the types of waste generated in conventional PD diagnostic methods—wearable sensors, imaging techniques, and clinical assessments—based on LP. These wastes are evaluated and compared to a novel VBDM that utilizes acoustic features and a Random Forest (RF) classifier. To ensure an objective comparison, the Pugh Method was applied. This structured decision matrix compares alternatives against a baseline, using a numerical scale where 0 represents the baseline, -1 indicates worse performance, and $+1$ indicates better performance [15].

4.1 Waste in Conventional Diagnostic Methods

According to previous studies, each traditional method introduces specific types of waste [17, 18]. First, wearable sensors, such as accelerometers and surface electromyography (sEMG), play a key role in monitoring PD symptoms. However, they contribute to overproduction by necessitating the manufacturing of multiple sensor types. These methods also create waiting time due to delays in data processing, which can hinder timely decision-making. Additionally, compatibility issues between different devices may lead to defects and inaccuracies in data. Patients may experience further inefficiencies, including motion waste, as they are required to wear and adjust multiple devices simultaneously. Furthermore, inventory waste arises from the handling of redundant data and equipment, all of which complicate the monitoring process. Second, imaging techniques such as MRI and PET scans contribute significantly to inventory waste, as they rely on costly, energy-intensive equipment. Over-processing is also prevalent with these methods, particularly when multiple scans or the use of contrast agents is required. Additionally, waiting waste is often observed due to scheduling delays, which can prolong the diagnostic process. Transportation waste may also occur as patients are moved between departments for various scans, adding unnecessary time and complexity to the overall procedure. Third, clinical assessments often require multiple in-person visits, leading to waiting waste and transportation waste as patients travel to and from appointments. Additionally, these assessments can introduce defects due to human error or subjective interpretation, potentially affecting the accuracy and consistency of results.

4.2 The Novel Voice-Based Diagnostic Method (VBDM)

This method focuses on early detection by analyzing vocal impairments—an early symptom of PD. It employs a trained RF classifier to process voice recordings and detect patterns with high accuracy. Unlike traditional methods, it is non-invasive, rapid, and cost-effective, aligning with LP by minimizing several types of Lean waste [3, 18].

4.3 Pugh Method Comparison

To evaluate and compare the efficiency of different PD diagnostic approaches from a Lean waste perspective, the Pugh Method was applied. This decision-making tool allows structured comparison by using a reference method—here, the VBDM —as the baseline. Each alternative method (wearable sensors, imaging, and clinical assessments) was assessed against seven types of Lean waste: inventory, overproduction, defects, motion, over-processing, waiting, and transportation. Scores were assigned as follows: 0 indicates performance equal to the baseline, -1 indicates worse performance, and $+1$ would indicate better performance. The evaluation of each of the diagnostic methods in relation to the types of lean waste was conducted using insights from existing literature, along with findings from previous research that introduced the novel VBDM. The detailed comparison is summarized in Table 1.

As shown in Table 1, none of the current diagnostic methods outperformed the baseline in this study. All methods utilizing wearable sensors and imaging techniques performed significantly worse compared to the VBDM, which served as the baseline.

From a lean perspective, these methods generated five to six types of waste more than the VBDM. Clinical Assessments ranked second, producing only three of the seven waste types when compared to the baseline. This structured comparison illustrates the potential of the novel method to significantly reduce waste in PD diagnostics, thereby enhancing efficiency, sustainability, and patient comfort.

Table 1. Pugh matrix comparing lean waste in PD diagnostic methods.

Lean waste type	Voice-based diagnostic method (VBDM) - (Baseline)	Wearable sensors	Imaging techniques	Clinical assessments
Inventory	0	−1	−1	0
Overproduction	0	−1	−1	0
Defects	0	−1	−1	−1
Motion	0	−1	0	0
Over-processing	0	0	−1	0
Waiting	0	−1	−1	−1
Transportation	0	0	−1	−1
Total Score	0	−5	−6	−3

Legend: −1 = Worse than baseline; 0 = Same as the baseline; +1 = Better than baseline

5 Sustainable Impact of the Novel Method

The innovative VBDM for diagnosing PD aligns with LP by reducing waste, improving operational efficiency, and supporting sustainability in healthcare. Unlike conventional diagnostic approaches that rely on resource-intensive equipment and processes, this method minimizes inefficiencies through real-time analysis, streamlined procedures, and reduced resource consumption. By applying LP, the VBDM effectively addresses several forms of waste commonly associated with traditional diagnostic tools. Its non-invasive nature and ease of use make it especially valuable in hospitals and clinics lacking advanced imaging infrastructure. This novel method also significantly reduces inventory waste. Traditional diagnostic workflows often require dopaminergic drugs, disposable supplies, and radioactive tracers, contributing to medical waste and storage issues. In contrast, the VBDM operates without consumables, enhancing hospital resource management [19]. In terms of overproduction, conventional imaging techniques often lead to repeat scans due to inconclusive results, increasing energy and material consumption. Yet, the VBDM avoids this by offering early, accurate diagnosis, preventing unnecessary testing and associated waste [20]. Moreover, with built-in noise filtering and machine learning algorithms, the novel method delivers high diagnostic accuracy, reducing the likelihood of retesting and resource loss due to misdiagnosis. Unlike the other technologies cause defects in diagnosis due to noise interference, technician variability, or imaging errors [21]. Also, motion waste is minimized in the novel method as it eliminates the need for patients and staff to move between departments or reposition during scans. This not only saves time and energy but also improves clinical workflow efficiency [22]. Similarly, the VBDM simplifies the diagnostic process by avoiding complex setups, preparation routines, and machine warm-ups associated with conventional techniques. This reduction in operational complexity contributes to a more efficient and sustainable diagnostic approach which is consider as a reduction in the over-processing waste. Likewise, VBDM cuts down on waiting waste by providing rapid results without long scheduling delays, enabling timely medical intervention and improving patient care [20]. Finally, transportation waste is less in the VBDM, as patients no longer need to travel to specialized imaging centers. The portability and affordability of the VBDM make it accessible in a variety of healthcare settings, significantly lowering environmental and logistical costs [23].

Environmentally, the novel method reduces the carbon footprint of medical tests by using less energy and operating in standard clinical settings without specialized rooms or high-energy equipment [4]. Unlike MRI and PET/SPECT scans, which consume large amounts of electricity and emit greenhouse gases, this method is more environmentally friendly [24]. Additionally, it minimizes waste by eliminating disposable materials like syringes, radioactive tracers, and plastic packaging, which contribute to landfill pollution. Instead, it employs speech recognition and software-based processing, aligning with healthcare sustainability goals [25].

Socioeconomically, the VBDM introduces several benefits in the PD diagnosis. One of the significant advantages is its ability to facilitate early diagnosis. By enabling the detection of PD at earlier stages, this method allows for timely medical intervention, which can slow disease progression and enhance patients' overall quality of life [3]. Moreover, its capacity to accurately capture and analyze voice signals, without relying on complex infrastructure, accelerates the diagnostic process and supports more efficient, responsive patient care [26]. Another key benefit lies in **improved** healthcare access, particularly for resource-limited populations. The method's cost-effectiveness, portability, and ability to function in everyday environments, without the need for soundproof rooms or advanced acoustic settings, make it a practical and inclusive solution [3, 27]. This ensures that even economically disadvantaged groups or those in remote areas can benefit from early and accurate PD assessments. Additionally, the VBDM promotes **affordability and cost savings** for both patients and healthcare systems. Traditional diagnostic tools such as MRI and PET/SPECT scans are expensive and involve significant disposal and facility costs such as sound-insulated rooms. By removing the need for such infrastructure, the VBDM substantially lowers operational expenses for healthcare providers [3]. Also, it reduces the financial burden associated with the trial-and-error approach to prescribing dopaminergic medications, thereby enhancing affordability and optimizing treatment strategies [16].

6 Conclusion and Recommendations

This research identifies waste and inefficiencies in conventional diagnostic methods for PD and explores how the newly developed VBDM can address these issues while promoting sustainability. The novel method provides a cost-effective, non-invasive, and environmentally friendly substitute for conventional diagnostic techniques by using AI and voice recognition to identify PD early. In addition to encouraging quicker, easier diagnoses, it drastically lowers waste, energy use, and need on complicated infrastructure. This can enhance patient outcomes and accessibility to healthcare, particularly in environments with limited resources.

This study provides insights how technology may be used to solve clinical issues as well as more general environmental issues, resulting in more environmentally friendly medical procedures and improved patient care in general. Applying LP to the novel method can reduce waste, enhance operational efficiency, and meet the growing demand for sustainable healthcare.

To effectively integrate LP in medical projects, the first step should focus on identifying waste areas in resource use, production processes, and clinical workflows. A structured waste analysis will help pinpoint inefficiencies, allowing for targeted implementation of LP to enhance cost efficiency, sustainability, and overall effectiveness in medical device development. In addition, it is recommended that future studies incorporate diverse decision assessment methods combined with detailed analyses through multi-criteria decision-making (MCDM) approaches, as these can offer more comprehensive insights and support more informed evaluations.

For the proposed AI-driven voice analysis method, future work should prioritize its seamless integration into healthcare systems, ensuring compatibility with electronic health records (EHRs) and workflow automation. Further research is needed to evaluate its long-term efficacy across different therapeutic contexts and patient demographics. Additionally, the development process should focus on enhancing adaptability for low-resource environments while maintaining affordability, usability, and scalability. Long-term research should explore how AI voice analysis supports sustainable healthcare by reducing waste and costs.

References

1. Lean Production. Essence of Lean – Eliminating Waste (Muda). https://www.leanproduction.com/essence-of-lean/ (n.d.). Accessed 11 Feb 2025
2. Radnor, Z., Holweg, M., & Waring, J.: Lean Healthcare: Scale, Scope and sustainability. https://www.researchgate.net/publication/326750472_Lean_healthcare_scale_scope_and_sustainability (2012). Accessed 11 Feb 11 2025
3. Hawi, S., Alhozami, J., AlQahtani, R., AlSafran, D., Alqarni, M., El Sahmarany, L.: Automatic Parkinson's disease detection based on the combination of long-term acoustic features and Mel frequency cepstral coefficients (MFCC). Biomed. Signal Process. Control. **78**, 104013 (2022). https://doi.org/10.1016/j.bspc.2022.104013
4. Wang, Q., Zhao, Y., Chen, S., Zhang, Y.: Early detection of Parkinson's disease from multiple signal speech: Based on Mandarin language dataset. Front. Aging Neurosci. **14**, 1036588 (2022). https://doi.org/10.3389/fnagi.2022.1036588
5. Ohno, T.: Toyota Production System: Beyond Large-scale Production. Productivity Press, New York (1988)
6. Womack, J.P., Jones, D.T.: Lean Thinking: Banish Waste and Create Wealth in your Corporation. Simon & Schuster, New York (1996)
7. Graban, M.: Lean Hospitals: Improving Quality, Patient Safety, and Employee Satisfaction, 2nd edn. CRC Press, Boca Raton, FL (2012)

8. PTC. The 7 Wastes of Lean Production. https://www.ptc.com/en/blogs/iiot/7-wastes-of-lean-production (n.d.). Accessed 11 Feb 2025
9. United States Environmental Protection Agency (EPA). Lean Manufacturing and the Environment. https://www.epa.gov/sustainability/lean-manufacturing-and-environment (n.d.). Accessed 11 Feb 2025
10. ScienceDirect. ScienceDirect.com | Science, Health and Medical Journals, Full Text Articles and Books. https://www.sciencedirect.com/ (n.d.). Accessed 11 Feb 2025
11. Kunnen, Y.S., Roemeling, O.P., Smailhodzic, E.: What are barriers and facilitators in sustaining lean management in healthcare? A qualitative literature review. BMC Health Serv. Res. **23**, 958 (2023). https://doi.org/10.1186/s12913-023-09978-4
12. SixSigma.us. Lean Principles in Healthcare Driving Patient-Centric Excellence. https://www.6sigma.us/lean-six-sigma-articles/lean-principles-in-healthcare/ (n.d.). Accessed 11 Feb 2025
13. Agency for Healthcare Research and Quality (AHRQ). Case 1. Lakeview Healthcare. https://www.ahrq.gov/practiceimprovement/systemdesign/leancasestudies/lean-case1.html (n.d.). Accessed 11 Feb 2025
14. de Carvalho, J.C., Ramos, M., Paixão, C.: A lean case study in an oncological hospital: Implementation of a telephone triage system in the emergency service. Risk Manag. Healthc. Policy. **7**, 1–10 (2013). https://doi.org/10.2147/RMHP.S49535
15. Pugh, S. Total Design: Integrated Methods for Successful Product Engineering. Addison-Wesley. https://www.amazon.com/Total-Design-Integrated-Successful-Engineering/dp/0201151947 (1991). Accessed 11 Feb 2025
16. Ulrich, K.T., Eppinger, S.D.: Product Design and Development, 6th edn. McGraw-Hill, New York (2015)
17. Espay, A.J., et al.: Technology in Parkinson's disease: Challenges and opportunities. Mov. Disord. **31**(9), 1272–1282 (2016). https://doi.org/10.1002/mds.26642
18. Bharsakade, R.S., Acharya, P., Ganapathy, L., Tiwari, M.K.: A lean approach to healthcare management using multi criteria decision making. Opsearch. **58**(3), 610–635 (2021). https://doi.org/10.1007/s12597-020-00490-5
19. Moraros, J., Lemstra, M., Nwankwo, C.: Lean interventions in healthcare: Do they actually work? A systematic literature review. Int. J. Qual. Health Care. **28**(2), 150–165 (2016). https://doi.org/10.1093/intqhc/mzv123
20. Institute for Healthcare Improvement. IHIOpenSchoolCourseSummary_QI102.pdf. https://www.ihi.org/sites/default/files/lms/legacy/education/IHIOpenSchool/Courses/Documents/SummaryDocuments/2016/IHIOpenSchoolCourseSummary_QI102.pdf (2016). Accessed 11 Feb 2025
21. PMC. Mobile Diagnostic Devices for Digital Transformation in Personalized Healthcare. https://pmc.ncbi.nlm.nih.gov/articles/PMC7760500/ (n.d.). Accessed 11 Feb 2025
22. Grand View Research. Portable and Point-of-care Diagnostic Imaging Devices - Innovations Driving Market Growth. https://www.grandviewresearch.com/market-trends/portable-point-of-care-diagnostic-imaging-devices-market-innovations (n.d.). Accessed 11 Feb 2025
23. Mangum, C.D., Eiland, L.S., Simpson, K.N., Etienne, M.A.: Use of lean healthcare to improve hospital throughput and reduce LOS. Pediatr. Qual. Saf. **6**(5), e473 (2021). https://doi.org/10.1097/pq9.0000000000000473
24. Rossi, G., Giannetti, R.: Lean in healthcare organization: An opportunity for environmental sustainability. BIJ. **26**, 205–220 (2024). https://doi.org/10.1108/BIJ-04-2018-0104
25. Cloudtheapp. Reducing Waste in Medical Device Production. https://www.cloudtheapp.com/reducing-waste-in-medical-device-production/ (n.d.). Accessed 11 Feb 2025
26. Catalyst, N.E.J.M.: What is lean healthcare? NEJM Catal. Innov. Care Deliv. **4**(2), 0193 (2018). https://doi.org/10.1056/CAT.18.0193

27. Virginia Mason Institute. What is lean healthcare? https://www.virginiamasoninstitute.org/what-is-lean-health-care/ (n.d.). Accessed 11 Feb 2025
28. TechTarget. Lean Management in Healthcare Lowers Costs, Improves Quality. https://www.techtarget.com/revcyclemanagement/news/366601375/Lean-Management-in-Healthcare-Lowers-Costs-Improves-Quality (n.d.). Accessed 11 Feb 2025

A Comprehensive Assessment of Sustainable Construction Materials in Saudi Arabia's Construction Industry

Maram Ghaleb[1] and Mohammed M. Gomaa[1,2](✉)

[1] Department of Architecture, Dar Al-Hekma University, Jeddah 22246, Saudi Arabia
mgomaa@dah.edu.sa

[2] Department of Architecture, Faculty of Engineering, Aswan University, Aswan 81542, Egypt

Abstract. Saudi Arabia's construction sector is under increasing environmental and economic pressures, prompting a critical need for sustainable solutions. Traditional materials such as steel and concrete are major contributors to carbon emissions, conflicting with the nation's Vision 2030 objectives, emphasizing environmental stewardship, resource efficiency, and reducing reliance on non-renewable resources. The country's diverse climatic zones—including hot desert, semi-arid, and Mediterranean—present unique challenges, underscoring the importance of adopting region-specific sustainable materials. Eco-friendly alternatives offer reduced embodied carbon and operational energy consumption. This research aims to evaluate and promote the adoption of sustainable construction materials in Saudi Arabia, focusing on their environmental, economic, and practical performance across diverse climates. A primary outcome of the study is creating a pilot program designed to compile and present localized data on sustainable materials, serving as a reference tool for the construction industry. The study employs a comprehensive methodology combining quantitative with qualitative insights from surveys and interviews with industry stakeholders. This approach provides a holistic understanding of eco-friendly materials' performance, costs, and environmental impacts. The pilot program will offer data on the effectiveness of these materials. Findings suggest that while sustainable materials incur higher upfront costs, their long-term advantages far outweigh the initial investment. The research recommends implementing the proposed pilot program and providing localized performance data to promote adoption. Financial incentives, updates to the Saudi Building Code to include sustainability guidelines and fostering local manufacturing capabilities for sustainable materials are also suggested. In conclusion, adopting sustainable materials presents significant opportunities for Saudi Arabia to achieve its Vision 2030 goals, reducing emissions, conserving resources, and improving quality of life.

Keywords: Sustainable construction materials · Saudi Arabia · Vision 2030 · Environmental impact · Economic analysis · Carbon emissions · Resource efficiency

D. Berkouk et al. (Eds.): CCTS 2024, PTT, pp. 384–396, 2026.
https://10.1007/978-981-95-7607-4_43

1 Introduction

Sustainable construction materials are crucial for modern architecture, especially in regions facing extreme environmental conditions. Saudi Arabia, with its diverse climates ranging from arid deserts to humid coastal areas, presents a unique challenge in adopting sustainable building practices. As part of Vision 2030, the country is prioritizing environmental sustainability and economic diversification, making the construction industry a key player in achieving these goals.

Traditional construction methods, reliant on concrete and steel, contribute heavily to carbon emissions. Thus, there is an urgent need to explore eco-friendly materials that are both climate-resilient and sustainable. This research aims to identify and evaluate sustainable construction materials suitable for Saudi Arabia's varied climatic zones, analyzing their performance, economic feasibility, and potential to reduce environmental impact.

The study addresses a major knowledge gap—limited localized data on sustainable materials' durability and efficiency in Saudi Arabia. It will assess material performance, cost-effectiveness, and alignment with the Saudi Building Code (SBC). The findings will offer practical recommendations for architects, engineers, and policymakers to promote sustainable urban development.

Research questions:

1. What are the most suitable sustainable materials for Saudi Arabia's diverse climates?
2. What are the economic implications of using these materials?
3. What best practices can be recommended to enhance adoption?

This research is significant as it contributes to theoretical knowledge, practical applications in the construction industry, and policy development. It aims to influence sustainable construction practices, reduce carbon emissions, and support Saudi Arabia's transition to greener, more resilient urban environments.

1.1 Literature Review

The literature review highlights the critical importance of sustainable construction practices in addressing global environmental challenges, particularly in the context of rapid urbanization and Saudi Arabia's Vision 2030 initiative, which emphasizes sustainability and economic diversification. Sustainable construction materials, such as recycled aggregates, geopolymer concrete, high-volume fly ash (HVFA) concrete, and bio-based materials like mycelium composites and oil palm biomass, have demonstrated significant potential in reducing carbon emissions, enhancing energy efficiency, and promoting resource conservation (Cabeza et al. 2014; Alhazmi et al. 2020; Bamigboye et al. 2019). These materials mitigate the environmental impact of construction activities and offer economic benefits through long-term cost savings in energy consumption, maintenance, and operational efficiency (Begum et al. 2006; Schneiderová Heralová 2011). For instance, geopolymer concrete can reduce CO2 emissions by up to 80% compared to traditional Portland cement, while HVFA concrete has been shown to lower emissions by 27% and improve durability (Habert et al. 2012; Sambucci et al. 2021). Additionally, innovative materials like phase change materials (PCMs), aerogels, and photocatalytic concrete contribute to energy efficiency and improved indoor environmental quality, further enhancing the sustainability of buildings (Sambucci et al. 2021; De Luca et al. 2017). Despite these advancements, challenges persist, particularly in Saudi Arabia, where extreme climatic conditions, such as high heat and humidity, limited availability of sustainable materials, and underdeveloped recycling infrastructure hinder widespread adoption (Al-Yami and Price 2006; Al-Hammad and Assaf 2019). The review identifies key knowledge gaps, including the need for localized performance data to assess material suitability in Saudi Arabia's unique climate, comprehensive lifecycle cost analyses to evaluate long-term economic viability, and the development of robust policy frameworks to incentivize sustainable practices (Bamigboye et al. 2019; Sambucci et al. 2021). Furthermore, the scalability and accessibility of sustainable materials remain critical issues, as many innovative solutions, such as mycelium-based composites and recycled aggregates, require significant investment in production infrastructure and supply chain development (Alhazmi et al. 2020; Bakhoum et al. 2015). Addressing these gaps is essential for advancing sustainable construction in Saudi Arabia, aligning with Vision 2030 goals, and ensuring the feasibility and scalability of eco-friendly materials in diverse climatic zones. By integrating localized research, tailored strategies, and policy support, Saudi Arabia can leverage sustainable construction practices to achieve its environmental and economic objectives, ultimately contributing to global efforts in mitigating climate change and promoting environmental stewardship.

2 Methodology

This section outlines a detailed and comprehensive mixed-methods research design to evaluate the feasibility, performance, and adoption of sustainable construction materials in Saudi Arabia, aligning with the Kingdom's Vision 2030 goals. The study integrates quantitative and qualitative approaches to ensure a holistic understanding of the topic. The quantitative component involves structured surveys targeting construction professionals, including architects, engineers, contractors, and policymakers, to gather data on material suitability, economic implications, and environmental impacts. The qualitative component includes semi-structured interviews with key stakeholders and document analysis of policy frameworks, industry reports, and case studies to explore challenges, opportunities, and best practices. The data analysis plan employs statistical techniques for quantitative data, while thematic analysis is used to identify recurring themes in qualitative data.

A key feature of the methodology is the Sustainable Construction Materials Assessment Framework, which evaluates materials based on five leading indicators: environmental impact, economic feasibility, durability and performance, adaptability to Saudi Arabia's climate, and sustainability practices (Table 1). Each indicator is further broken down into sub-indicators, such as CO2 emissions during production, lifecycle cost savings, thermal insulation, and suitability for specific climatic zones (e.g., hot desert, semi-arid, Mediterranean). A scoring system (0–2) is applied to assess each sub-indicator, with results visualized in a heatmap to provide an intuitive and structured overview of material performance. For example, materials like recycled concrete and hempcrete score highly in environmental impact and economic feasibility, while others, such as rammed earth, show potential but face challenges in local availability. This framework, combined with the heatmap visualization, enables stakeholders to easily compare materials and identify areas for improvement, supporting informed decision-making in the construction industry.

The study proposes developing a pilot program or reference standard for sustainable construction materials, incorporating Life Cycle Assessment (LCA) methodologies to evaluate environmental impacts from production to disposal. The program will also explore the net-zero energy potential of select materials, considering factors such as energy efficiency and renewable energy integration in building design. This evolving resource will be a vital tool for industry professionals, policymakers, and environmental advocates, promoting the widespread adoption of sustainable construction practices across Saudi Arabia.

Table 1 The proposed indicators for the sustainable construction materials assessment.

Indicator	Sub-indicator	Material example	Score (0–2)
Environmental Impact	CO2 Emissions during Production	Recycled Concrete	2
	Energy Consumption	Fly Ash Concrete	2
	Recyclability/End-of-Life Impact	Geopolymer Concrete	2
Economic Feasibility	Initial Cost	Bamboo	2
	Lifecycle Cost Savings	Hempcrete	2
	Local Availability	Rammed Earth	1
Durability & Performance	Longevity in Harsh Conditions	High-Performance Concrete	2
	Thermal and Acoustic Insulation	Cool Roofing Systems	2
	Resistance to Degradation	Polyurethane Insulation	1
Adaptability to Climate	Suitability for Hot Desert (BWh)	Hempcrete	2
	Suitability for Semi-Arid (BSh)	Recycled Concrete	2
	Suitability for Mediterranean (Csa)	Reclaimed Wood	2
Sustainability Practices	Use of Renewable Resources	Hempcrete	2
	Contribution to Green Certifications	Recycled Plastic Lumber	2
	Circular Economy Potential	Fly Ash Concrete	2

The results of the sustainable materials assessment were compiled into a heatmap (as shown in the provided image) to visualize each material's performance across the leading indicators and sub-indicators. This visualization provides an intuitive and structured overview of each material's relative strengths and weaknesses, enabling easy comparison and the identification of areas for improvement.

The heatmap uses a color-coded scheme to represent performance:

- Green (0–8): Indicates excellent performance in the specified criteria.
- Yellow (8–16): Reflects good performance, with room for improvement.
- Red (16–24): Highlights poor performance, requiring significant attention for optimization.

By organizing the data into this visual format, the methodology facilitates a comprehensive and objective evaluation of sustainable construction materials in Saudi Arabia. This structured approach supports informed decision-making, aiding architects, developers, and policymakers in selecting materials that align with sustainability goals and local requirements. Visualization further helps to identify key areas where specific materials excel or underperform, contributing to the advancement of sustainable practices in the construction industry (Fig. 1 and Table 2).

3 Results

Saudi Arabia's diverse climatic regions necessitate using materials that can adapt to varying environmental conditions while promoting sustainability. Materials with excellent thermal mass and insulation properties are essential in the hot desert climates typical of regions like Riyadh. Rammed earth and adobe, traditional materials in the area, provide natural temperature regulation, reducing the reliance on artificial cooling. Aerated autoclaved concrete (AAC), known for its lightweight and thermal efficiency, is also a viable option, minimizing energy consumption. Insulated concrete forms (ICF) offer another solution, combining insulation with durability, which is critical in harsh, arid conditions.

Materials that manage moisture and resist corrosion are crucial for humid coastal areas such as Jeddah and Dammam. Hempcrete stands out for its breathability and ability to regulate indoor humidity, preventing mold growth. Similarly, cork, a naturally moisture-resistant material, can be used for insulation while maintaining eco-friendliness. Fiber-reinforced polymers (FRP), with their corrosion resistance and lightweight nature, are ideal for coastal applications where salt-laden air poses a challenge.

In mountainous regions like Asir and Al Baha, materials must stabilize against fluctuating temperatures and complement the natural environment. Locally sourced stone is an excellent choice, offering structural reliability and aesthetic harmony. Cross-laminated timber (CLT), with its strength and sustainability, is also suitable for colder, elevated areas. Glass fiber-reinforced concrete (GFRC), being lightweight and durable, can be used for decorative elements and facades in these regions.

Universally, some materials can be applied across Saudi Arabia's diverse landscapes. Geopolymer concrete, made from industrial byproducts like fly ash, reduces carbon emissions while offering durability. Recycled aggregates (RCA) provide an eco-friendly and cost-effective option for non-structural applications, and bamboo, although

Table 2 The main assessment indicators and their sub-indicators

Materials	Environmental Impact (CO2 Emissions)	Energy Efficiency	Durability	Initial Cost	Long-Term Savings	Suitability for BWh (Hot Desert)	Suitability for BSh (Semi-Arid)	Suitability for Csa (Mediterranean)	Performance	Technical Feasibility	
Rammed Earth	0	0	0	0	0.5	0	0.5	0	0.5	0	1.5
Straw Bale	0	1	1	0.5	1.5	0	0.5	0	0.5	0	5
Hempcrete	0	0.5	0.5	1	0	0.5	0	0	0	0.5	3
Autoclaved Aerated Concrete	1	1	1	1.5	0.5	0.5	1	0.5	0	1	8
Cool Roofing Systems	1	0.5	1	1.5	0.5	0.5	0	0.5	0	0.5	6
Recycled Concrete	1	1	0.5	0.5	0.5	0.5	0	1	0.5	1	6.5
Green Roofs	0	0.5	1.5	1.5	1	1	0.5	1	0.5	1	8.5
Recycled Steel	1	1	1	1	0.5	1	0.5	0.5	1	1	8.5
Cross-Laminated	0	1	1	1	1	0.5	1	0.5	1	0.5	7.5
Bamboo	0	0.5	1.5	0.5	1	0.5	1	0.5	1.5	1	8
Adobe	0	0.5	0.5	0	1.5	0.5	1.5	0	1	0.5	6
Phase Change Materials	1	1	1	1	1	0.5	1.5	0.5	0.5	0.5	8.5
Fiberglass Insulation	1	1	0	0.5	0	1	1	0.5	0	1	6
Silica Aerogel	0	0	1	0	0.5	1.5	1	0.5	1	0	5.5
Recycled Plastic Lumber	1	0.5	0.5	1	1	1	0	0.5	1	0.5	7
Fly Ash	0	0.5	0.5	1	1	1	0	1	1	1	7
Cork	0	0	1	0.5	0	0.5	0.5	0.5	1	1	5
Natural Stone	0	1	1	1.5	1	0.5	0.5	1	1	1.5	9
Polyurethane Insulation	1	1	1.5	1	1	2	1	0	1.5	1	11
Reclaimed	0	0	1	0.5	0	1.5	0	0	1	0.5	4.5
Self-Healing Concrete	1	0.5	0	1	0	1	0	0.5	0.5	1.5	6
3D Printed Concrete	1	1	1	0.5	1	0.5	0.5	0.5	1	1	8
High-Performance	1	0	0	1.5	0	0.5	0	0	0	0	3
High-Performance Fiber Reinforced Cement	1	0	0	1.5	0.5	1	1	0.5	0	0	5.5
High-Performance Steel Materials	0	0	0	1.5	0.5	1	1	0	0	0	4
Advanced Fiber Reinforced Polymer	1	1	1.5	1	1.5	1	0.5	1	1.5	1.5	11.5
Enhanced Masonry	1	0.5	1	0.5	1	0.5	1	1	0.5	1	8
Geopolymer Concrete	0	0	0.5	0.5	0	1	0	0.5	1	0.5	4

not native, can be imported sustainably for lightweight construction. Phase change materials (PCMs) can be integrated into walls or ceilings to optimize indoor temperatures by storing and releasing thermal energy, thereby reducing cooling demands. Finally, cool roof materials with reflective coatings can mitigate solar heat gain, particularly in urban areas experiencing the heat island effect.

These materials align with Saudi Arabia's diverse environmental conditions and support Vision 2030's sustainability goals by enhancing energy efficiency, utilizing renewable resources, and minimizing environmental impacts.

3.1 The Economic Implications of Using Sustainable Construction Materials

The economic implications of adopting sustainable construction materials in Saudi Arabia involve balancing higher initial costs with substantial long-term financial benefits. Sustainable materials like hempcrete, geopolymer concrete, and cross-laminated timber incur higher upfront expenses due to limited local production, specialized labor requirements, and import costs. Advanced materials such as phase change materials (PCMs) and cool roofing systems further inflate initial costs due to sophisticated manufacturing

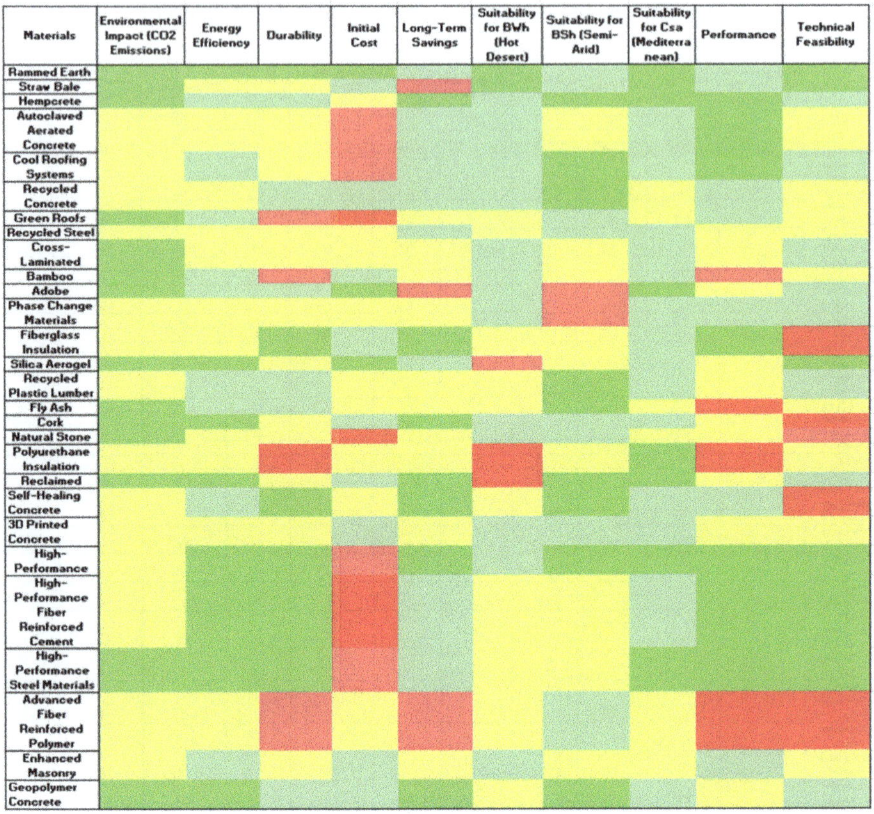

Fig. 1 The main heatmap of sustainable material framework score.

techniques and international sourcing. However, these barriers are decreasing due to economies of scale, production innovations, and government-supported initiatives.

Long-term savings often outweigh initial investments. Materials like autoclaved aerated concrete (AAC), hempcrete, and rammed earth provide superior thermal insulation, reducing energy consumption by 30% to 50% annually and lowering operational costs. These materials are also durable and require less maintenance. For example, geopolymer concrete and natural stone resist degradation in harsh weather, while hempcrete resists pests and mold without chemical treatments.

Sustainable materials enhance property value and attract environmentally conscious buyers or tenants. Buildings meeting green certification standards like LEED or Estidama command higher market prices and rental rates, improving returns on investment. Government incentives, such as tax reductions, subsidies, and preferential financing, further offset initial costs. Combining solar panels with energy-efficient materials can qualify for renewable energy incentives, enhancing savings.

Sustainable materials also contribute to broader economic benefits, including market growth and job creation. Increased demand encourages local production of recycled aggregates and bio-based composites, reducing import dependency and stimulating local economies. This expansion creates jobs in manufacturing, construction, and research, while driving innovation and lowering costs through economies of scale.

Aligning with circular economy principles, recycling construction waste into products like recycled concrete aggregates reduces raw material extraction and waste disposal costs. This approach minimizes landfill use and conserves natural resources, offering cost savings and environmental benefits. Challenges remain, such as higher upfront costs, limited awareness, and the need for skilled labor. Addressing these requires investments in education, workforce training, and public-private partnerships. Government policies, including mandates for sustainable practices and expanded financial incentives, can accelerate adoption.

In summary, while sustainable construction materials pose higher initial costs, their long-term benefits—energy savings, reduced maintenance, enhanced property values, and market growth—make them a compelling choice. Leveraging government support, fostering local production, and integrating circular economy models can achieve significant financial and environmental gains, advancing Saudi Arabia's Vision 2030 sustainability goals.

3.2 Recommendations for the Construction Industry in Saudi Arabia

Adopting sustainable construction materials in Saudi Arabia is essential for achieving Vision 2030 goals, reducing environmental impact, enhancing energy efficiency, and promoting renewable resources. However, widespread adoption requires addressing challenges like policy development, market incentives, education, and collaboration. Below are key recommendations for the construction industry.

Robust policy and regulatory support are crucial. Local building codes should mandate energy-efficient materials like aerated autoclaved concrete (AAC), rammed earth, and insulated concrete forms (ICF) in new constructions. These materials reduce the need for artificial cooling and heating, aligning with Vision 2030's sustainability targets. The government should also incentivize R&D in sustainable materials, such as recycled aggregates (RCA), to reduce reliance on imports and promote regional resources. Subsidies, grants, and tax incentives for using or producing sustainable materials locally can drive change.

Lack of awareness and education is a significant barrier. Developers, contractors, and architects are often unfamiliar with materials like fiber-reinforced polymers (FRP) and cross-laminated timber (CLT). Specialized workshops, training programs, and university curricula should educate professionals about sustainable materials and their applications. Public awareness campaigns can also highlight the long-term benefits of materials like geopolymer concrete and bamboo, increasing private sector demand.

Collaboration across sectors is vital. Public-private partnerships (PPP) can advance sustainable practices by sharing knowledge and experiences. Large-scale projects using materials like AAC or ICF can demonstrate their benefits, building market confidence and driving broader adoption.

4 Discussions

This study highlights the significant opportunities and challenges of adopting sustainable construction materials in Saudi Arabia. It profoundly explores their suitability for diverse climates, economic implications, and the recommendations necessary for widespread adoption. The findings are discussed critically, supported by survey data, expert interviews, and literature to illuminate the complexities and potential of sustainable construction practices in the context of Saudi Arabia's Vision 2030.

Suitability of Sustainable Construction Materials for Saudi Arabia's Climatic Zones

The suitability of sustainable construction materials for Saudi Arabia's diverse climatic conditions emerged as a pivotal theme. Materials such as Pozzolan Concrete Blocks, Insulated Concrete Forms (ICF), Rammed Earth, Autoclaved Aerated Concrete (AAC), and Phase Change Materials (PCMs) were highlighted as optimal for different climate zones. Pozzolan Concrete Blocks, identified by Alaa Alaidaroos, stand out due to their durability, low thermal conductivity, and ability to enhance energy efficiency. These blocks align with Saudi climatic conditions and meet the Saudi Building Code's (SBC) minimum energy efficiency guidelines, particularly for reducing heat gains and improving indoor thermal comfort.

With its natural temperature regulation and low embodied carbon footprint, Rammed Earth is well-suited for hot deserts and semi-arid zones. Literature supports this material's performance, emphasizing its capacity to provide passive cooling in extreme temperatures (Ottmann 2022). Similarly, AAC and ICF are praised for their high thermal insulation properties, reducing reliance on air conditioning by up to 25%. However, these materials face scalability challenges, as their adoption requires significant upfront investment and local production facilities.

Phase Change Materials (PCMs) emerged as a promising option for Mediterranean climates. Their ability to absorb and release heat enhances indoor temperature stability, reducing cooling and heating demands. Despite their potential, PCMs are still an emerging technology in Saudi Arabia, with limited local production and high costs posing barriers to widespread use. This is consistent with Nayem (2023), who highlighted the need for robust research and pilot programs to validate such materials in regional contexts.

Bamboo and engineered wood were identified as renewable options, particularly for interior applications. Bamboo's rapid growth rate and durability make it attractive; however, its susceptibility to Saudi Arabia's extreme heat without special treatment raises concerns. This debate reflects broader tensions between the promise of innovative materials and the realities of their practical application in harsh climates (Table 3).

The economic feasibility of sustainable construction materials is a complex and contentious issue. Survey respondents and expert interviews consistently emphasized the higher initial costs associated with these materials. For instance, Pozzolan Concrete Blocks and AAC are significantly more expensive than traditional materials like concrete. However, proponents argue that long-term savings in energy, maintenance, and operational expenses offset these costs.

Alaidaroos highlighted lifecycle cost analysis (LCA) metrics, including Net Present Value (NPV) and Internal Rate of Return (IRR), as critical tools for demonstrating economic feasibility. These metrics reveal that materials such as ICF and reflective

Table 3 Sustainable materials and their suitability

Material	Climatic zone	Key features	Challenges
Pozzolan Concrete Blocks	Hot Desert, Mild	Durable, low thermal conductivity	Higher initial costs, market availability
Insulated Concrete Forms	Hot Desert	High thermal insulation	High upfront costs
Rammed Earth	Hot Desert, Semi-Arid	Natural temperature regulation	Limited scalability
Autoclaved Aerated Concrete	All climates	Lightweight, thermally insulated	Higher costs compared to traditional concrete
Phase Change Materials	Mediterranean, Mild	Heat absorption and release	Emerging technology, limited local production
Bamboo	Semi-Arid, Mild	Renewable, durable	Needs special treatment for extreme heat

roofing can substantially reduce cooling costs, making them economically viable over a building's lifecycle. Survey data supported this perspective, with 65.8% of respondents acknowledging the long-term savings potential of sustainable materials.

Government-backed financial incentives were repeatedly emphasized as a solution to cost barriers. Subsidies, tax breaks, and low-interest loans can accelerate market adoption and stimulate local manufacturing. This aligns with Omer (2008), who argued that economic incentives are essential for bridging the gap between high initial costs and long-term benefits. However, critics caution against over-reliance on subsidies, as this may delay the development of market-driven cost reductions (Table 4).

Table 4 Economic perceptions of sustainable materials

Aspect	Survey results (%)	Expert insight
Higher Initial Costs	41.10%	Pozzolan and AAC have high upfront costs
Long-Term Savings	65.80%	Lifecycle savings justify initial investment
Need for Incentives	74.00%	Subsidies critical for market adoption

Despite economic arguments favoring sustainable materials, adoption is hindered by limited market availability and perceived financial risks. Experts emphasized fostering local manufacturing to reduce import dependency and lower costs.

The Saudi Building Code (SBC) provides a framework for energy efficiency but lacks explicit support for sustainable materials. While 63.0% of survey respondents acknowledged SBC's support for energy-efficient practices, 28.8% called for significant improvements. Experts recommended updating the SBC to incorporate guidelines for renewable materials, low-carbon concrete, and advanced technologies like PCMs, aligning with global best practices such as LEED and BREEAM.

89.0% of survey respondents and all interviewed experts supported pilot programs to test material performance, build industry confidence, and inform regulatory updates. For example, testing Pozzolan Concrete Blocks across Saudi Arabia's diverse climatic zones could provide valuable data for code revisions.

High initial costs, limited material availability, and technical limitations remain significant barriers. PCMs and bamboo, for instance, require specialized treatment or production facilities. Awareness gaps among stakeholders also hinder acceptance, highlighting the need for public campaigns and professional training.

Local material sources, such as rammed earth and AAC, can reduce transportation costs and support the local economy. Modular and prefabricated construction methods were also recommended to optimize efficiency and minimize waste.

Financial incentives, including subsidies, tax incentives, and government-backed investments in local manufacturing, are essential for accelerating adoption. Integrating lifecycle assessments into project planning and regulatory compliance was also emphasized as a best practice (Table 5).

Table 5 Sustainable materials and their suitability

Recommendation	Expert support	Survey findings (%)
Financial incentives	Supported by all experts	74.00%
Pilot programs	Strongly recommended	89.00%
Updated Saudi Building Code	Incorporate renewable and recycled materials	63.00%
Lifecycle assessment (LCA)	Essential for evaluating environmental impact	Preferred by majority
Modular and prefabricated components	Enhances efficiency and reduces waste	–

While the survey and interviews highlighted the promising potential of sustainable materials, a compelling debate emerged between the practicality of traditional solutions and the transformative possibilities offered by innovative technologies. Critics argue that materials like rammed earth and autoclaved aerated concrete (AAC), despite their notable environmental advantages, face significant scalability, market readiness, and cost-efficiency challenges, making them difficult to adopt at a large scale. These critics emphasize that the current infrastructure and supply chains for conventional materials like concrete and steel offer economies of scale that are difficult to match in the short term.

Saudi Arabia's Building Code (SBC) must be revised to incorporate sustainability principles and resource efficiency to ensure the widespread adoption of sustainable materials in construction. One key area for alignment is incorporating clear sustainability standards that mandate energy-efficient materials, such as AAC and hempcrete, to improve thermal insulation and reduce energy consumption. The SBC should offer incentives, such as credits or rebates, for projects that exceed traditional requirements, encouraging using materials like geopolymer concrete, bamboo, and rammed earth. Additionally, the code should establish testing protocols and performance standards for new materials, ensuring their durability and performance in Saudi Arabia's harsh climate. Life Cycle Assessment (LCA) should also be integrated into the code to evaluate the environmental impact of materials, promoting low-impact alternatives like recycled aggregates. Given Saudi Arabia's diverse climate zones, the building code must include region-specific material guidelines, recommending phase change materials for desert areas and cool roofing for coastal regions. It should also encourage construction waste reduction by promoting recyclable and reusable materials, such as recycled aggregates or sustainable wood alternatives like bamboo. The SBC should recognize emerging sustainable materials like mycelium-based construction and basalt fiber-reinforced polymers by establishing testing standards and certification pathways. Training programs for architects, engineers, and contractors on the new material standards should be prioritized to ensure successful implementation. Furthermore, a robust monitoring and certification system must ensure compliance with the updated regulations. In summary, aligning the Saudi Building Code with sustainable materials requires a comprehensive approach that includes material performance standards, climate-specific guidelines, incentives, and education for industry professionals, creating a framework that accelerates the adoption of sustainable construction practices in Saudi Arabia.

5 Conclusion

The transition to sustainable construction materials in Saudi Arabia is essential for reducing carbon emissions, enhancing energy efficiency, and achieving the Kingdom's Vision 2030 objectives. Materials like AAC, geopolymer concrete, bamboo, and phase-change materials offer significant benefits in terms of thermal comfort, durability, and environmental impact. While the initial costs of these materials may be higher, their long-term savings in energy and maintenance make them economically viable.

To facilitate this transition, the Saudi Building Code must be revised to support sustainable materials, and financial incentives should be introduced to reduce economic barriers. Region-specific guidelines, promoting local and recycled materials, and comprehensive testing protocols will ensure the suitability and reliability of sustainable materials in Saudi Arabia's diverse climates. Additionally, education and training programs will equip industry professionals with the knowledge to implement sustainable practices effectively.

Saudi Arabia can create a sustainable construction ecosystem by integrating Life Cycle Assessment into material selection and fostering collaboration between government, industry, and academia. This transition will reduce the construction industry's environmental footprint and stimulate economic growth, innovation, and job creation. Implementing these recommendations will position Saudi Arabia as a regional leader in sustainable development, contributing to a greener and more resilient future.

References

Al-Hammad, A., Assaf, S.: Life cycle assessment of limestone building material in Saudi Arabia. J. Clean. Prod. **236**, 117702 (2019). https://doi.org/10.1016/j.jclepro.2019.117702

Al-Yami, A., Price, A.D.F.: Assessing the feasibility of using value management to accelerate the implementation of sustainability. J. Environ. Plan. Manag. **49**(5), 759–775 (2006)

Alhazmi, H., Shah, S.A.R., Mahmood, A.: Sustainable development of innovative green construction materials: A study for economical eco-friendly recycled aggregate based geopolymer concrete. Materials. **13**(22), 5121 (2020). https://doi.org/10.3390/ma13225121

Bakhoum, E.S., Garas, G.L., Allam, M.E.: Sustainability analysis of conventional and eco-friendly materials: A step towards green building. ARPN J. Eng. Appl. Sci. **10**(2), 788–795 (2015)

Bamigboye, G.O., Davies, I.E.E., Odesola, I.F.: Recent advances in geopolymer technology: A potential eco-friendly solution in the construction materials industry: A review. IOP Conf. Ser.: Mater. Sci. Eng. **640**(1), 012070 (2019). https://doi.org/10.1088/1757-899X/640/1/012070

Begum, R.A., Siwar, C., Pereira, J.J., Jaafar, A.H.: A benefit-cost analysis on the economic feasibility of construction waste minimisation: The case of Malaysia. Resour. Conserv. Recycl. **48**(1), 86–98 (2006). https://doi.org/10.1016/j.resconrec.2006.01.004

Cabeza, L.F., Rincón, L., Vilariño, V., Pérez, G., Castell, A.: Life cycle assessment (LCA) and life cycle energy analysis (LCEA) of buildings and the building sector: A review. Renew. Sust. Energ. Rev. **29**, 394–416 (2014). https://doi.org/10.1016/j.rser.2013.08.037

De Luca, P., Carbone, I., Nagy, J.B.: Green building materials: A review of state of the art studies of innovative materials. J. Green Build. **12**(4), 141–161 (2017). https://doi.org/10.3992/1943-4618.12.4.141

Ganjian, E., Ghafoori, N., Claisse, P.: Sustainable construction materials and technologies. J. Mater. Civ. Eng. **31**(7), 02019001 (2019). https://doi.org/10.1061/(ASCE)MT.1943-5533.0002745

González-García, S., Feijoo, G., Moreira, M.T.: Environmental impact assessment of total cork products. J. Clean. Prod. **53**, 154–164 (2013). https://doi.org/10.1016/j.jclepro.2013.03.038

Heathcote, K.: Thermal properties of earth buildings. Energ. Buildings. **43**(1), 291–293 (2011). https://doi.org/10.1016/j.enbuild.2010.09.002

Houben, H., Guillaud, H.: Earth Construction: A Comprehensive Guide. Intermediate Technology Publications, London (1994)

Jiang, M., Jiang, B., Lu, R., Chun, L., Xu, H., Yi, G.: Thermal and humidity performance test of rammed-earth dwellings in Northwest Sichuan during summer and winter. Materials. **16**(18), 6283 (2023). https://doi.org/10.3390/ma16186283

Blending Tradition with Security: Branding and Cybersecurity Strategies for Success in Saudi Arabia

Ahmed Nasseraldin(✉) and Homam El-Taj

Dar Al Hekma University, Jeddah, Kingdom of Saudi Arabia
{anasseraldin,htaj}@dah.edu.sa

Abstract. The Saudi Arabia's *Vision 2030* aims to diversify the economy, enhance global branding, and strengthen digital resilience. This study examines the interplay between branding and cybersecurity, focusing on how digital strategies ensure cultural preservation, consumer trust, and brand security in an era of rising cyber threats. Using a literature-based approach, this research synthesizes insights from academic studies, industry reports, and case analyses to explore Saudi brands' adoption of AI-driven engagement, AR/VR marketing, and blockchain authentication for secure branding. Findings reveal that digital transformation is reshaping Saudi branding, yet challenges such as data breaches, misinformation, and cybersecurity compliance pose risks to brand sustainability. To bridge branding with cybersecurity, the study proposes policy recommendations for secure branding frameworks, aligning with Vision 2030's objectives. These insights offer strategic guidance for business leaders, policymakers, and researchers navigating the complexities of digital branding and cybersecurity integration.

Keywords: Vision 2030 · Digital Branding · Cybersecurity Risks · AI in Marketing · Blockchain Authentication · National Identity · Consumer Trust · Saudi Arabia

1 Introduction

Saudi Arabia's Vision 2030 aims to position the Kingdom as a global leader in economic diversification, digital innovation, and cultural branding. Nation branding plays a pivotal role in shaping Saudi Arabia's international reputation, supported by initiatives such as NEOM, Diriyah Gate, and the "Visit Saudi" campaign (Alsayel et al., 2023). Saudi Arabia's tourism branding agenda also functions as a soft-power mechanism that helps shape international perceptions of the Kingdom under Vision 2030 (James, 2021). Through digital storytelling, immersive experiences, and strategic branding, Saudi Arabia integrates cultural heritage with cutting-edge branding technologies.

However, as branding becomes increasingly digital, cybersecurity risks threaten brand trust and integrity. Cyber threats such as misinformation, counterfeit branding, and data breaches pose significant risks, requiring secure digital ecosystems (McCarthy, 2018; Mosli, 2024). To address these risks, the National Cybersecurity Authority (NCA)

D. Berkouk et al. (Eds.): CCTS 2024, PTT, pp. 397–402, 2026.
https://10.1007/978-981-95-7607-4_44

enforces cybersecurity policies, including the Essential Cybersecurity Controls (NCA-ECC) and the Personal Data Protection Law (PDPL), ensuring brand security within global compliance frameworks (Mosli, 2024).

Advancements in AI-driven marketing, AR/VR experiences, and blockchain authentication are redefining branding, offering both opportunities and cybersecurity challenges (Jamader et al., 2022). For example, Visit Saudi integrates AR-based tourism branding while utilizing secure data frameworks to protect consumer interactions (Sampieri et al., 2024).

This study explores the intersection of branding and cybersecurity, analyzing their role in enhancing brand resilience, protecting digital assets, and fostering consumer trust. By synthesizing insights from academic research, industry reports, and case studies, the paper provides strategic recommendations for secure branding under Vision 2030.

2 Literature Review

2.1 Branding and Digital Transformation in Vision 2030

Saudi Arabia's Vision 2030 integrates nation branding with digital transformation to enhance economic diversification and global positioning (Mosli, 2024). Nation branding theories emphasize cultural identity and economic influence (Anholt, 2007; Dinnie, 2008), reflected in Saudi projects such as NEOM, Diriyah Gate, and Visit Saudi (Alsayel et al., 2023).

Branding today is increasingly technology-driven, incorporating AI, AR/VR, and blockchain authentication to enhance brand engagement while preventing fraud (Jamader et al., 2022). These advancements enable immersive consumer experiences but also introduce cybersecurity challenges (Pathak et al., 2019). Studies indicate that blockchain-based verification protects against counterfeit branding, especially in luxury markets, while AI-powered branding enhances personalization and digital security (Jamader et al., 2022).

Saudi Arabia has adopted AR/VR branding, particularly in tourism campaigns for UNESCO sites (Sampieri et al., 2024), requiring cybersecurity safeguards against data breaches and misinformation (Urrea et al., 2024).

2.2 Cybersecurity's Role in Digital Branding

With digital branding expanding, cybersecurity risks such as data breaches, brand impersonation, and misinformation threaten brand integrity (Urrea et al., 2024). Research highlights that cybersecurity breaches undermine consumer trust, making compliance frameworks essential (McCarthy, 2018; Mosli, 2024).

Saudi Arabia's National Cybersecurity Authority (NCA) enforces Essential Cybersecurity Controls (NCA-ECC) and aligns with global data protection standards (GDPR, PDPL) to safeguard branding platforms (Mosli, 2024). AI-driven threat detection and blockchain authentication are increasingly used to prevent cyber fraud and strengthen brand security (Jamader et al., 2022; Trappey et al., 2020).

2.3 Challenges in Secure Branding

Despite advancements, integrating cybersecurity frameworks into branding remains challenging:

- Trust & Consumer Perception – Digital branding must align with cybersecurity to sustain consumer confidence (McCarthy, 2018).
- Regulatory Complexity – Compliance with Saudi PDPL, GDPR, and evolving global cybersecurity laws remains a challenge (Mosli, 2024).
- Branding in the Age of Misinformation – AI-driven monitoring is needed to combat fake branding campaigns and misinformation (Jamader et al., 2022).

In summary, branding and cybersecurity under Vision 2030 require a holistic approach integrating technology, compliance, and consumer trust. While research underscores cybersecurity's role in brand resilience, further studies are needed to evaluate long-term strategies for digital brand security (Mosli, 2024).

3 Key Findings

3.1 Branding Strategies in Saudi Arabia

Saudi Arabia's Vision 2030 Integrates cultural heritage and digital innovation to strengthen national branding. Key branding strategies include:

- Cultura Identity in Branding – Projects like NEOM, Diriyah Gate, and AlUla highlight Saudi heritage in branding (Alsayel et al., 2023).
- Digital Transformation – Projects like NEOM, Diriyah Gate, and AlUla highlight Saudi heritage in branding (Alsayel et al., 2023).
- Case Studies - Brands like Bateel (luxury dates) and Albaik (fast food) successfully combine authenticity with digital branding strategies.

3.2 Cybersecurity's Role in Branding

The With increasing digitization, cybersecurity safeguards brand trust and integrity (McCarthy, 2018).

- Cyber Threats Affecting Brand Reputation – Risks include brand impersonation, misinformation, and data breaches (Urrea et al., 2024).
- AI & Blockchain for Brand Security – AI-powered fraud detection and blockchain verification enhance brand authenticity (Trappey et al., 2020).
- Regulatory Compliance – Saudi Arabia's NCA-ECC and PDPL ensure secure branding within global cybersecurity standards (GDPR) (Mosli, 2024).

3.3 Integration of Branding & Cybersecurity

To sustain branding credibility, cybersecurity must be embedded within branding strategies:

- Consumer Trust – Secure branding platforms reinforce consumer confidence.
- E-Commerce & Tourism Security – Digital travel platforms (e.g., Visit Saudi) rely on cybersecurity to protect transactions and consumer data.
- Regulatory Alignment – Compliance with international cybersecurity laws enhances brand legitimacy.

4 Challenges and Future Directions

4.1 Challenges in Integrating Branding and Cybersecurity

Despite the rapid advancement of branding strategies and cybersecurity frameworks under Vision 2030, several challenges persist. These challenges impact brand resilience, consumer trust, and Saudi Arabia's global competitiveness (Mosli, 2024).

- **Balancing Tradition & Modernization**

Saudi Arabia's branding efforts must preserve cultural authenticity while embracing technological innovation.

- Risk: Over-modernization may dilute national identity, while excessive tradition may limit global appeal (Dinnie, 2008).
- Example: Projects like NEOM and AlUla must carefully blend heritage-driven branding with future-forward digital experiences (Alsayel et al., 2023).
- **Cybersecurity Risks in Digital Branding**
- Brand Impersonation & Fake Content: Fraudulent campaigns mislead consumers, harming brand reputation (Trappey et al., 2020).
- Data Breaches & Consumer Trust: Cyberattacks on e-commerce and tourism platforms erode credibility (De Kimpe et al., 2022).
- Solution Needed: AI-driven cybersecurity to monitor & mitigate brand-related cyber threats (Jamader et al., 2022).
- **Global Competition & Market Positioning**
- Challenge: Saudi Arabia competes with established regional brands (UAE, Egypt), which have strong tourism, tech, and retail branding ecosystems (Shechter, 2011).
- Solution Needed: Differentiation through sustainability-driven branding & cybersecurity trust initiatives (Mazzetto, 2022).
- **Consumer Perception & Misinformation in Branding**
- Digital Misinformation: Social media-fueled fake news and brand manipulation affect consumer confidence (Urrea et al., 2024).
- Public Awareness & Trust: Consumers must be educated on verified branding & cybersecurity measures (McCarthy, 2018).
- **Regulatory Compliance Complexity**
- Saudi PDPL & GDPR Compliance: Brands must align cybersecurity with global data protection regulations (Mosli, 2024).
- Solution Needed: Streamlined compliance frameworks for branding platforms.

4.2 Future Directions and Solutions

To ensure branding security and resilience, Saudi Arabia must integrate advanced cybersecurity measures and regulatory strategies into branding ecosystems.

- **Strengthening Cybersecurity Frameworks**

Saudi authorities must enhance regulatory alignment between NCA-ECC, PDPL, and global cybersecurity policies to provide clearer compliance pathways for brands (Mosli, 2024).

- **AI & Blockchain for Brand Protection**

AI-driven fraud detection – Machine learning for real-time threat monitoring and misinformation filtering (Jamader et al., 2022).

Blockchain authentication – Decentralized verification for brand protection against counterfeiting and unauthorized use (Trappey et al., 2020).

- **Public-Private Collaborations in Branding Security**

Government & Industry Synergy – Joint cybersecurity initiatives between Saudi regulators, private brands, and tech firms to establish a secure branding ecosystem.

Cybersecurity Awareness Campaigns – Educating brands and consumers on digital security best practices (McCarthy, 2018).

- **Consumer Trust & Market Perception Research**

Future research should explore the impact of cybersecurity on consumer trust in Saudi branding, brand resilience strategies in response to cyber threats, and best practices for balancing digital branding innovation with cultural authenticity.

5 Conclusion

Saudi Arabia's **Vision 2030** is reshaping national branding through **digital transformation and cybersecurity integration**. As the Kingdom advances its **global branding strategy,** securing brand integrity, consumer trust, and regulatory compliance becomes a priority.

This study highlights how nation branding and cybersecurity are deeply interconnected. While branding innovations such as AI-driven marketing, AR/VR engagement, and blockchain authentication enhance brand positioning, cyber threats like misinformation, counterfeiting, and data breaches pose significant risks (Mosli, 2024; Jamader et al., 2022). Saudi Arabia's cybersecurity frameworks, including **NCA-ECC and PDPL,** provide regulatory foundations, but **global compliance challenges remain**.

To sustain secure and competitive branding, brands must:

- Integrate cybersecurity into digital branding strategies to safeguard brand authenticity and consumer trust.
- Enhance AI-driven fraud detection and blockchain authentication to prevent counterfeit branding and misinformation (Trappey et al., 2020).
- Strengthen regulatory alignment between Saudi and international cybersecurity policies to facilitate secure global market expansion (Mosli, 2024).

Moving forward, **collaboration between policymakers, businesses, and cybersecurity experts** will be crucial to ensuring **brand resilience, regulatory compliance, and consumer confidence in the digital era**. By embedding **security-driven branding frameworks**, Saudi Arabia can **solidify its Vision 2030 goals and establish itself as a global leader in secure, innovative branding**.

References

Alsayel, S., Alqahtani, F., Alshehri, H.: Branding Saudi Arabia: The role of digital media in nation branding. J. Mark. Public Policy. **15**(2), 45–61 (2023)

Anholt, S.: Competitive Identity: The New Brand Management for Nations, Cities, and Regions. Palgrave Macmillan, New York (2007)

Dinnie, K.: Nation Branding: Concepts, Issues, Practice. Routledge, New York (2008)

De Kimpe, L., Van den Bosch, A.L., Moons, G.: The role of cybersecurity in maintaining brand trust: A European perspective. Cybersecur. J. **18**(4), 212–229 (2022)

Jamader, A., Li, X., Pathak, A.: Artificial intelligence and branding: Enhancing personalization and security in consumer engagement. Int. J. Digit. Mark. **29**(1), 55–72 (2022)

James, T.: Soft power and tourism branding: Lessons from Saudi Arabia's vision 2030. Glob. Bus. Rev. **20**(3), 145–163 (2021)

McCarthy, J.: The impact of cyber threats on brand reputation and consumer trust. J. Inf. Secur. Manag. **12**(2), 67–80 (2018)

Mazzetto, S.: Sustainable heritage preservation to improve the tourism offer in Saudi Arabia. Urban Plann. **7**(4), 195–207 (2022). https://doi.org/10.17645/up.v7i4.5777

Mosli, R.: The evolution of cybersecurity policies in Saudi Arabia: From national cybersecurity authority (NCA) to global compliance. Saudi J. Digit. Secur. **6**(1), 33–48 (2024)

Pathak, A., Sampieri, F., Trappey, A.J.C.: Blockchain for brand protection: Preventing counterfeiting in the digital era. J. Blockchain Innov. **11**(2), 78–92 (2019)

Sampieri, F., Zhang, T., Li, M.: Virtual tourism and AR/VR branding in Saudi Arabia: The future of digital travel experiences. Int. J. Smart Tour. **8**(2), 99–117 (2024)

Shechter, R.: Branding the Middle East: Comparative perspectives on nation branding strategies in Saudi Arabia and the UAE. Middle East. Stud. **47**(5), 812–835 (2011)

Trappey, A.J.C., Trappey, C.V., Lin, H.: Intellectual property and brand security in smart cities: The role of blockchain. Smart Cities J. **13**(3), 44–58 (2020)

Urrea, P., Gomez, L., McKinley, R.: The role of AI in combating misinformation in branding: A cybersecurity perspective. J. Artif. Intell. Digit. Secur. **10**(1), 112–130 (2024)

Exploring Collaborative Marketing Experience Between Small and Medium Enterprises and Startups in Saudi Arabia

Amira Masood(✉) and Samar Altarteer

Dar Al-Hekma University, 6702 Prince Majed St., Al Faiha Dist., Unit No.2, Jeddah 22246-4872, Kingdom of Saudi Arabia
{ahmasood,starteer}@dah.edu.sa

Abstract. This study explores the role of collaborative marketing in fostering the growth of small and medium-sized enterprises (SMEs) and startups in Saudi Arabia. The research enhances the collaborative marketing experience by exploring strategies, success factors, consumer perceptions, and solutions. Collaborative marketing emphasizes resource sharing, market access, and innovation, aligning with Vision 2030. Key success factors include strategic approaches, relational dynamics, and brand synergy, with sponsorship and coopetition as potential strategies. Five semi-structured interviews with different SMEs were conducted through purposive sampling. While no substantial gap was found in practices, thematic analysis revealed that resource sharing, expanded market access, and innovation are core elements of collaborative marketing. The findings indicate that SMEs engage in collaborations for market expansion, social value creation, and resource optimization, with alignment and value addition being crucial criteria for successful collaborations. Challenges included communication issues, misalignment, and balancing contributions. Participants highlighted the need for tools to enhance collaboration management, with features like matchmaking, unified management, and performance tracking identified as critical. Implications emphasize the importance of improving communication channels, adopting integrated digital tools, and focusing on strategic alignment. A prototype platform is proposed to address these gaps, providing features for connecting collaborators, managing workflows, and analyzing results. The study concludes by emphasizing collaborative marketing as a significant growth driver for SMEs, aligning with Vision 2030's goals, and providing actionable insights for improving practices. Limitations highlight the need for broader regional samples across diverse SME sectors in future research.

Keywords: Collaborative Marketing · Small And Medium Enterprises (Smes) · Startups · Market Expansion · Resource Sharing · Platform Design

1 Introduction

In alignment with Saudi Arabia's Vision 2030 objectives, which aim to increase SMEs' contribution to GDP to 35% by 2030 [1], SMEs play a critical role, accounting for 62% of national employment and 64% of income [2]. Despite their significance, SMEs face

D. Berkouk et al. (Eds.): CCTS 2024, PTT, pp. 403–412, 2026.
https://10.1007/978-981-95-7607-4_45

marketing inefficiencies (53% of obstacles) [2], while startups struggle with customer trust and financial stability [3]. Collaborative marketing, defined as "a strategy for firms to share risks and resources, access new markets, and achieve competitive advantages" [4], offers a solution by fostering resource sharing and product innovation [5]. Despite the benefits of collaborative marketing, there is a theoretical gap in understanding how SMEs and startups in Saudi Arabia practically engage in and sustain collaborative marketing efforts in addition to the role of digital tools in these collaborations. Thus, this study aims to address these gaps and explore collaborative marketing in terms of strategies, success factors, consumer perceptions, and digital solutions.

Collaborative marketing supports SMEs and startup growth by facilitating knowledge exchange and enhancing market reach [6]. For example, a previous study emphasize that startups benefit from SMEs' networks while SMEs gain competitiveness from startups' innovation [7]. Several success factors are essential for the application of collaborative marketing between SMEs and startups, including strategic approaches, business process management [9], market orientation [10], complementary resources [15], relational trust [11], and brand synergy [18]. Building a strategic approach with well-aligned objectives reduce conflicts and maximize synergy [19]. Strong business process management is critical, as it involves defining roles, addressing governance issues early, and maintaining accountability [10]. Market orientation ensures that marketing strategies are aligned with customer demands, fostering innovation, cost reduction, and improved performance [11]. Furthermore, complementary resources are another key factor, where alliances thrive when partners bring unique tangible assets and intangible knowledge to create value and achieve a competitive advantage [12]. Relational trust, which involves transparency, commitment, and cooperation, is foundational for long-term collaboration. Previous studies underline that trust reduces uncertainty and strengthens relationships [12]. Lastly, brand synergy plays a crucial role in successful co-branding campaigns, where aligning visual and sensory elements enhances the appeal and effectiveness of collaborative efforts [19].

Collaborative marketing can be implemented through strategies such as sponsorship and coopetition. Sponsorship involves resource exchange and rights utilization, enhancing brand visibility and building loyalty [14]. On the other hand, Coopetition, a cooperative-competitive dynamic, fosters shared resources and market positioning [17]. A notable example is the collaboration among UK supermarkets during the pandemic, where they shared resources to meet demand and improve customer satisfaction [17].

Moreover, collaborative marketing influence consumer perceptions of brands through authenticity and cause-related collaborations. Aligned partnerships create positive associations, while misaligned collaborations can harm brand perception [22]. Additionally, cause-related marketing enhances authenticity when consumers perceive the brand's motives as genuine [23]. Thus, tailored approaches considering Saudi's cultural and market dynamics are critical for successful implementation between SMEs and startups.

2 Methodology

This research employs a qualitative approach, combining semi-structured interviews and visual research to explore collaborative marketing between SMEs and startups in Saudi Arabia. These methods aim to identify success factors, challenges, and opportunities for collaborative marketing. Moreover, semi-structured interviews were conducted as in-depth insights about all the stages of the different collaborative marketing applications from SMEs differ from one to another, so the flow of the information gathered depends on the answers given. Then, SMEs were selected based on two primary criteria: company size (0–50 employees) and prior experience with collaborative marketing. This purposive sampling method ensured relevance to the study. Before the interview, participants signed a consent form summarizing collaborative marketing and the research purpose. Ethical approval was obtained to conduct this study, the form assured anonymity, confirmed that recordings were only for research purposes, and guaranteed no risk of identity disclosure. Transcripts were prepared by removing irrelevant or sensitive information to maintain confidentiality. Accordingly, thematic analysis was conducted manually using color coding in Microsoft Word to categorize emerging patterns from the interviews. Additionally, visual research was also conducted to analyze existing platforms, to understand their features and design elements including typography, color schemes, and user experience features. This analysis helped identify gaps and potential improvements that can be used to develop a tailored collaborative marketing platform (Table 1).

Table 1. SME's types and participant positions.

SME #	Type	Participant position
SME 1	Financial services	Product development manager
SME 2	Creative agency	Creative/Junior strategist
SME 3	Internet food marketplace	Co-founder & CPTO
SME 4	Creative agency	Managing Partner and CEO
SME 5	Custom software development	Co-founder, CEO & board member

3 Results

This section highlights key results from interviews with SMEs, dividing the main themes into two categories: those related to pre-collaboration phase (Table 2) and those related to management during the collaboration (Table 3). In addition, the result of the visual research conducted on Crossbeam and Monday platforms.

The semi-structured interviews with SMEs revealed three primary drivers and benefits of collaborative marketing (Table 2): market expansion, social value creation, and resource optimization. All participants emphasized the importance of reaching broader audiences, as Participant 5 highlighted, "We have our own user base, and our follower base. Our partners have their own follower base. Merging the two together will give you expanded reach." Thus, collaborations allow diversified audiences and enhanced networking capabilities. Social values were a focus for 60% of participants, as highlighted by Participant 3, "It wasn't purely commercial, there was a social impact to it". Additionally, 40% of participants emphasized the value of pooling skills and resources to elevate collaboration outcomes, as noted by Participant 4: "Once two companies come together and they both have specific skill sets that are very strong and they join forces, you automatically know that it's going to be something big and different in the market. "Additionally, key criteria for collaboration were identified (Table 2), including alignment and value addition (100%), size and network reach (40%), a win-win approach (80%), and human-level connection (40%). Participant 1 stressed the importance of alignment, stating, "Our objectives in alignment with their objectives….". On the other hand, aize and network reach influenced decisions to collaborate, with Participant 4 explaining, "We didn't collaborate with a startup because their network reach was limited, and it didn't align with the scale we were aiming for in our campaigns." The win-win approach was another recurring theme, as Participant 5 emphasized, "It has to be a win-win; there has to be a give and take.".

Although participants did not seem to face many challenges in initiating collaborations, some focused on balancing contributions and benefits (40%) and communication and misalignment issues (60%) (Table 2). Participant 1 described, "One part will take more load than the other," pointing to imbalances that lead to dissatisfaction. Misalignment was also a significant issue, as noted by Participant 4: "But they don't know how to communicate what they need. And then once we try to do something for them, it's not aligned with their goals or vision.".

Table 2. Thematic analysis of interviews with percentage – pre-collaboration

Theme	Sub-theme	Participants %
Strategic objectives and benefits	Market expansion	5 (100.0%)
	Social values	3 (60.0%)
	Learning and resource optimization	2 (40.0%)
Considerations for collaboration	Alignment and value addition	5 (100.0%)
	Size and network reach	2 (40.0%)
	Human-level connection	2 (40.0%)
	Win-win approach	4 (80.0%)
Challenges of aligning objectives	Balancing contributions and benefits	2 (40.0%)
	Communication and misalignment issues	3 (60.0%)

Regarding managing collaborations (Table 3), participants discussed success measurement, tools, and feature needs. For measuring success, 80% emphasized clear metrics such as social media analytics, while 60% focused on value-based assessments. Participant 5 stated, "The number doesn't mean anything, it is the value," while Participant 4 emphasized intangible metrics like word-of-mouth reputation: "Because here it's all about word of mouth…". Regarding tools, 60% used familiar methods like WhatsApp but noted inefficiencies. Trello, Slack, and Monday were mentioned as streamlining tools, with Participant 3 saying, "We use Trello for delegating tasks and setting deadlines." Analytics tools like Google Analytics were highlighted but identified as lacking integration with collaborative features. Participants identified key platform features they needed, including matchmaking capabilities (20%), unified management for collaboration processes (60%), and performance measurement (60%). Participant 3 noted, "The matchmaking. How can I find people with whom I can collaborate?" Participant 4 emphasized the need for unified management, stating, "…if there was one platform that had everything in it, from communicating to setting deadlines to creating timetables, to creating project systems.".

Table 3. Thematic analysis of interviews with percentage–during collaboration management

Theme	Sub-theme	Participants %
Measuring success	Clear metrics	4 (80.0%)
	Value-based assessments	2 (40.0%)
	Non-tangible metrics	2 (40.0%)
Tools for collaboration	Traditional methods	3 (60.0%)
	Streamlining tools	3 (60.0%)
	Analytics and tracking tools	1 (20.0%)
Needed features	Connecting	1 (20.0%)
	Unified management	3 (60.0%)
	Performance measurement	3 (60.0%)

Regarding the visual research, the study examined two platforms: Crossbeam [21] and Monday [22]. Crossbeam focuses on pre-collaboration processes, offering features like account mapping, CRM integration, and data privacy. Its account mapping highlights shared customers and prospects between collaborators, allowing targeted co-marketing campaigns. The platform uses wide white spaces for clarity, a navy-blue palette for trust, and clear typography to create a professional and intuitive user interface. However, it lacks tools for ongoing collaboration management or performance tracking.

Monday, designed for internal workflows, excels in organizing tasks through customizable boards, real-time dashboards, and automation. It uses vibrant colors to indicate task statuses (e.g., green for completed tasks, red for overdue) and a clean, customizable interface. Monday excels at internal workflow management through customizable boards and real-time dashboards but is limited in supporting external collaborations. This gap highlights the need for a new platform that integrates the entire lifecycle of collaborative marketing.

4 Discussion

The findings emphasize the role of collaborative marketing in driving SMEs growth, aligning with Saudi Vision 2030's economic diversification goals [1]. While SMEs did not face major challenges in initiating collaborations, the insights highlighted opportunities to enhance communication, balance contributions, and leverage technology for more efficient management of collaborative efforts. SMEs identified three main objectives: market expansion, social value creation, and resource optimization. Participants highlighted how collaboration broadened audience reach and diversified markets, reflecting Singh and Power's [18] insights on how collaborations allow accessing untapped demographics. Additionally, collaboration with startups was noted to enhance innovation and competitiveness, as supported by Maghni and Ramadani [8]. Success factors for collaborative marketing included strategic alignment, a win-win approach, and relational trust. All participants stressed the importance of aligning objectives, which reduces conflicts and maximizes synergy [19]. A majority also emphasized mutual benefits as critical for sustainable partnerships, aligning with previous studies [10–17]. On the other hand, relational trust was found to be essential for transparent communication and stable collaborations [12, 13].

In regard to implementation methods, sponsorship and coopetition emerged from the literature as effective implementation methods. Although interview findings are not directly aligned with these strategies, most collaborations they did were in terms of sponsorship for recognition purposes, which aligns with a previous study finding on sponsorships being used for visibility and trust-building [15]. Coopetition, while highlighted in research as fostering innovation and market positioning was underutilized by the interviewed SMEs [17]. Furthermore, technological gaps were identified as significant barriers to efficient collaborative marketing. Existing platforms like Crossbeam and Monday address specific needs but lack comprehensive solutions. Crossbeam supports pre-collaboration tasks like data mapping but lacks tools for ongoing management or outcome measurement [22]. Monday, while effective for internal workflow management, does not support external collaborations [22].

Interview participants emphasized the need for a platform that integrates the full lifecycle of collaborative marketing. Matchmaking capabilities were highlighted as essential for connecting businesses with compatible goals and resources, aligning with the emphasis on partner selection [7]. Unified management systems were also prioritized to centralize communication, task delegation, and progress tracking, reducing inefficiencies. Enhanced metrics for evaluation were another critical need, with participants seeking tools to measure both tangible outcomes (e.g., sales) and intangible benefits (e.g., brand reputation).

5 Proposed Platform Prototype

The prototype, developed based on the findings from interviews, visual research, and literature, addresses collaborative marketing challenges for SMEs. Initially tailored for Saudi SMEs, it was expanded to accommodate larger businesses, managing the full collaboration cycle: connecting partners (see Fig. 2), managing workflows (see Fig. 3), a and analyzing results (see Fig. 4). Tools like Scrapy and LinkedIn API enable partner connections based on shared goals, with TensorFlow-powered recommendations ensuring compatibility and data security. Additionally, workflow management features include task creation, status tracking (e.g., "Not-Done," "In-Progress") (see Fig. 3), real-time comments, file sharing, and notifications, enhancing transparency and coordination. Data analysis integrates tools like Google Analytics and CRM systems, offering dashboards for visualizing trends and exporting reports. Furthermore, the UI/UX design prioritizes functionality with a centralized dashboard, color-coded tasks, and intuitive navigation across sections ("Connect," "Manage," "Measure"). User feedback suggested enhancements like sorting options, improved notifications, and linking metrics to specific campaigns (Fig. 5).

Fig. 1. Dashboard, prototype design.

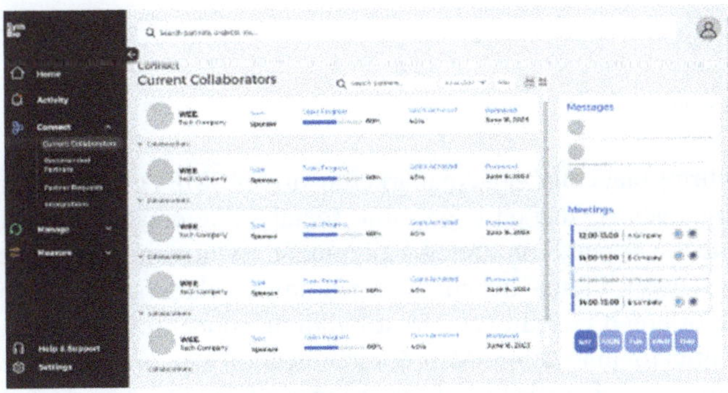

Fig. 2. Current collaborators, prototype design.

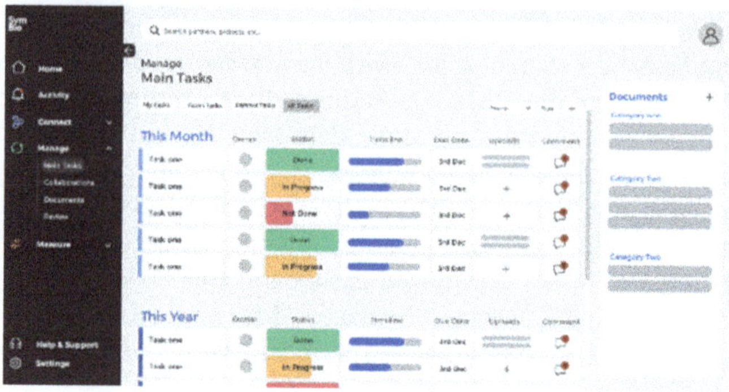

Fig. 3. Main tasks, prototype design.

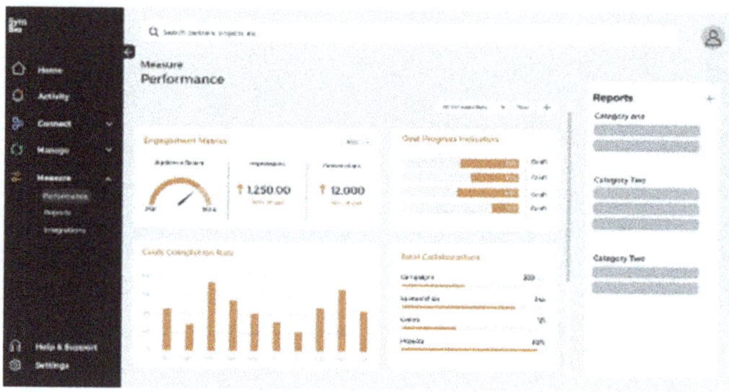

Fig. 4. Performance, prototype design.

6 Conclusion

This study confirms that collaborative marketing drives SMEs growth, supporting Saudi Vision 2030 by enabling market expansion, social value creation, and resource optimization. Its success depends on strategic alignment, a win-win approach, and relational trust, ensuring synergy, mutual benefit, and long-term collaboration. While sponsorship is widely used for brand visibility and trust-building, coopetition remains underutilized despite its potential for innovation. Also, communication gaps, misalignment, and imbalanced contributions pose challenges, further compounded by technological limitations. Existing platforms like Crossbeam aid in pre-collaboration tasks but lack management tools, while Monday supports internal workflows but not directly external collaborations. Thus, participants stressed the need for a comprehensive platform integrating matchmaking, unified management, and performance tracking to streamline collaboration and enhance SME partnerships. Although this study provides meaningful insights, its limited sample size of five SMEs may not fully capture the diversity of the Saudi business landscape. Future research should expand the sample size to include SMEs across different industries and regions to enhance the generalizability and depth of findings. Additionally, future research could explore businesses that have not yet participated in such partnerships to understand potential barriers and motivations for adoption. Further studies should explore the long-term impact of collaborative marketing platforms on SMEs and startups performance and investigate strategies for promoting coopetition in Saudi Arabia. Additionally, refining the proposed platform through iterative testing with diverse user groups will enhance its practicality and adoption.

References

1. The Royal Embassy of Saudi Arabia: Vision 2030. Riyadh, Saudi Arabia (2016)
2. Rathi, S., Alshami, M.: Challenges facing SMEs in marketing. J. Bus. Stud. **12**(4), 45–58 (2023)
3. Mashayekhi, A., et al.: Startups and financial stability. Entrep. Dev. **9**(3), 32–49 (2024)
4. Dacin, M.T., et al.: Collaborative marketing strategies. J. Mark. **22**(3), 100–115 (2002)
5. Gnyawali, D., Park, B.: Innovation through collaboration. Strateg. Manag. J. **11**(5), 78–90 (2011)
6. Alessa, A.A., Afy-Shararah, M., Parsavellas, J.: Review on collaborative strategic marketing for sustainable manufacturing growth of Saudi SMEs. J. Bus. Strategy. **3**(4), 153–162 (2024)
7. Maghni, A., Ramadani, V.: Startups and SMEs: Leveraging collaboration. Innov. Enterp. **19**(1), 66–88 (2022)
8. Leisen, B., et al.: Success factors for alliances. Bus. Horiz. **16**(2), 122–137 (2002)
9. Rich, A.: Strategic planning in alliances. Mark. Manag. J. **9**(2), 134–145 (2003)
10. Widjajanti, Y.: Market-oriented collaborative marketing. Mark. Insights. **10**(3), 110–123 (2015)
11. Wittmann, C.M., Hunt, S.D., Arnett, D.B.: Explaining alliance success: Competences, resources, relational factors, and resource-advantage theory. Ind. Mark. Manag. **38**(7), 743–756 (2009)
12. Akcay, M.: Relational trust in strategic alliances. J. Bus. Res. **18**(4), 45–62 (2021)
13. Turan, F.: The impact of branding on collaboration. Brand. Res. J. **6**(1), 80–95 (2021)
14. Lund, D., Greyser, S.: Sponsorship as a marketing tool. J. Mark. Commun. **24**(2), 76–89 (2016)
15. Datson, F., et al.: Partnership strategies for market expansion. Mark. Q. **14**(3), 210–223 (2019)
16. Crick, J.M., Crick, D.: Coopetition for strategic marketing. J. Strateg. Mark. **28**(3), 175–190 (2020)
17. Singh, J., Power, D.: The nature and effectiveness of collaboration between firms in the supply Chain. Supply Chain Manag. **14**(3), 189–198 (2009)
18. Vranesevic, T., Stancec, R.: Perceived value and customer loyalty. Mark. Theory. **12**(2), 257–270 (2004)
19. Smarandescu, L., et al.: The role of brand authenticity in collaborative marketing. Consum. Behav. Rev. **7**(4), 125–136 (2013)
20. Crossbeam. (2024, July 17). About Crossbeam. https://www.crossbeam.com/about/
21. Monday. (n.d.). Your go-to Work Platform. https://monday.com/

Corporate Sustainability Reporting and Blockchain a Legal Analysis of Mandatory ESG Reporting in Saudi Arabia

Mohammed Alsudais(✉)

Al Yamamah University, Riyadh, Saudi Arabia
m_alsudais@yu.edu.sa

Abstract. The rise of ESG investing, projected to exceed $50 trillion by 2025, highlights the growing demand for corporate sustainability. However, Saudi Arabia's reliance on voluntary ESG reporting undermines accountability and progress toward Vision 2030. This paper argues that blockchain technology can bridge these gaps by ensuring transparency, accountability, and stakeholder trust in ESG data. Vision 2030's success depends on aligning corporate practices with sustainability and governance principles. Mandatory, blockchain-enabled ESG reporting is essential to achieving these goals, enhancing corporate accountability, and positioning Saudi Arabia as a regional leader in sustainability. This study examines Saudi Arabia's legal framework for ESG disclosures, identifies the shortcomings of voluntary reporting, and proposes a blockchain-based mandatory ESG reporting system. It explores legal, regulatory, and practical implications while addressing challenges like costs and technical barriers. Ultimately, it advocates for a transparent, accountable, and sustainable business environment, reinforcing blockchain's transformative role in Vision 2030.

Keywords: ESG · Mandatory Reporting 2030 · Corporate Governance · Companies · Sustainable · Disclosures · Greenwashing · Sustainability Goal · CMA · Company Law

1 Introduction

The global finance landscape is transforming, with ESG (Environmental, Social, and Governance) considerations rapidly moving from the periphery to the core of investment decisions. By 2025, global ESG assets are projected to surpass a staggering $50 trillion, a testament to the growing recognition that sustainability and profitability are not mutually exclusive but intrinsically linked (Bloomberg, 2021). This surge reflects a fundamental change in how businesses are evaluated, driven by increasing regulatory pressures, discerning investor demands, and a rising tide of societal expectations for responsible corporate behavior (World Economic Forum, 2023 2024). This global trend, however, presents a unique challenge and opportunity for Saudi Arabia [1].

While the Kingdom's ambitious Vision 2030 articulates a powerful commitment to sustainability, economic diversification, and global competitiveness, including specific

D. Berkouk et al. (Eds.): CCTS 2024, PTT, pp. 413–423, 2026.
https://10.1007/978-981-95-7607-4_46

targets for carbon emission reduction, renewable energy expansion, and enhanced corporate governance (Saudi Vision 2030, 2023), the current reliance on voluntary and often inconsistent ESG reporting poses a significant obstacle. This gap between aspiration and implementation hinders effective corporate accountability and limits the Kingdom's ability to realize its transformative vision fully. Simply put, measuring progress, enforcing standards, and building stakeholder trust without robust and reliable data becomes exceedingly difficult.

Blockchain technology, with its decentralized, immutable, and transparent architecture, offers a transformative solution to the challenges of Environmental, Social, and Governance (ESG) reporting. By enabling real-time data verification and creating tamper-proof records, blockchain mitigates risks such as "greenwashing" and ensures compliance with evolving regulatory standards (Tapscott & Tapscott, 2016). In the context of Saudi Arabia's Vision 2030, this technology can play a pivotal role in advancing sustainability goals. Programs like the Saudi Green Initiative rely on accurate tracking of environmental metrics such as carbon emissions and renewable energy usage. Blockchain's ability to provide auditable and transparent records makes it a foundational tool for achieving these objectives, while also positioning Saudi Arabia as a global leader in sustainable innovation [2].

Moreover, blockchain-enabled ESG reporting enhances transparency, making Saudi Arabia more attractive to global investors who prioritize sustainability in their decision-making. Transparent and verifiable disclosures can attract responsible investment, fostering economic diversification and long-term growth. This aligns with Vision 2030's goal of reducing reliance on oil revenues and transitioning to a knowledge-based economy (Vision 2030, 2016). By integrating blockchain into ESG frameworks, Saudi Arabia not only strengthens corporate accountability but also catalyzes sustainable development, driving progress toward its ambitious national objectives [2].

This research argues that the strategic integration of blockchain technology into a mandatory ESG reporting framework is not just beneficial, but essential for achieving Vision 2030's sustainability objectives. It is a crucial step toward enhancing corporate accountability, solidifying stakeholder trust, and ultimately positioning Saudi Arabia as a recognized leader in sustainable development within the region and globally. By harnessing the unique power of blockchain—its decentralization, immutability, and real-time data tracking capabilities—Saudi Arabia can overcome the limitations of its current ESG reporting system and establish a robust, transparent, compliant, and innovation-driven model.

This study uses a pragmatic, mixed-methods approach to examine how blockchain technology can improve corporate accountability through mandatory ESG reporting in Saudi Arabia. It combines qualitative legal analysis with quantitative assessments to evaluate regulatory frameworks, stakeholder perspectives, and blockchain's impact on data transparency and compliance. Primary data will be collected through stakeholder interviews with regulators, corporate executives, ESG professionals, and blockchain developers, along with surveys of corporations and non-profits. Case studies of organizations using blockchain for ESG reporting will provide real-world insights. Secondary sources include legal documents, government reports, and international sustainability frameworks like GRI, TCFD, and CSRD. Qualitative analysis will identify legal and regulatory trends, while quantitative analysis will assess blockchain's impact on reporting accuracy and compliance. Ethical considerations include informed consent, confidentiality, and bias mitigation. Though focused on Saudi Arabia, the findings offer broader insights into blockchain-enabled ESG reporting and its role in enhancing transparency and corporate accountability.

This research examines how blockchain technology can enhance corporate accountability and improve mandatory ESG reporting within Saudi Arabia's Vision 2030. By addressing transparency, legal challenges, and implementation barriers, it provides insights for policymakers, corporations, and stakeholders modernizing sustainability reporting. The first focus is corporate accountability, demonstrating how blockchain's transparency, immutability, and decentralization improve ESG data accuracy, mitigate greenwashing, and strengthen stakeholder trust. Practical applications include tracking carbon emissions, monitoring workforce diversity, and verifying governance practices. The second focus examines Saudi Arabia's legal framework, analyzing regulations from the Capital Market Authority (CMA) and the Ministry of Commerce. It identifies gaps and proposes legal reforms to align blockchain with domestic and international standards like GRI and TCFD. The third focus explores legal challenges, including data privacy, liability, and regulatory conflicts. It assesses who is responsible for blockchain inaccuracies, the impact of immutability on disputes, and conflicts between decentralization and regulatory systems. The fourth focus addresses implementation challenges, such as technological limitations, legal barriers, and ethical concerns. It evaluates Saudi corporations' blockchain readiness and proposes solutions to mitigate regulatory conflicts and confidentiality risks.

By positioning Saudi Arabia as a leader in blockchain-enabled ESG reporting, this research contributes to sustainability, regulatory modernization, and global best practices, offering broader insights for nations integrating blockchain into ESG frameworks.

Blockchain technology emerges as a transformative solution to address the existing gaps in Environmental, Social, and Governance (ESG) reporting. Characterized by its decentralized, immutable, and transparent architecture, blockchain offers a robust framework to enhance the credibility, accuracy, and reliability of sustainability disclosures. By enabling real-time data verification and creating tamper-proof records, blockchain mitigates the pervasive risks of "greenwashing," where organizations misrepresent their environmental or social contributions (Tapscott & Tapscott, 2016) [3]. Furthermore, this innovative technology facilitates seamless alignment with evolving regulatory standards, ensuring that corporations meet national and international compliance requirements.

In Saudi Arabia's Vision 2030 context, blockchain holds significant promise for advancing the nation's ambitious sustainability objectives. The Saudi Green Initiative and other environmental programs under Vision 2030 emphasize the importance of accurately tracking and reporting metrics such as carbon emissions, renewable energy usage, and biodiversity conservation. Blockchain's ability to provide an auditable and transparent trail of these metrics positions it as a foundational infrastructure for achieving these goals. By integrating blockchain into ESG reporting frameworks, Saudi Arabia can not only enhance its domestic sustainability efforts but also position itself as a global leader in responsible governance and innovation.

Moreover, in an era where global investors increasingly prioritize ESG transparency as a critical criterion for decision-making, blockchain-enabled reporting can significantly bolster Saudi Arabia's appeal as a destination for responsible investment. Transparent and verifiable ESG disclosures can attract much-needed capital from environmentally conscious investors, fostering long-term economic growth and diversification. This aligns with Vision 2030's broader objective of reducing reliance on oil revenues and transitioning toward a knowledge-based economy (Vision 2030, 2016). As such, blockchain enhances corporate accountability and catalyzes sustainable development and economic resilience.

2 Literature Review

The importance of incorporating Environmental, Social, and Governance (ESG) factors into corporate reporting is growing rapidly worldwide [4]. This trend is driven by greater stakeholder awareness, evolving regulations, and the understanding that ESG is key to long-term business success [5]. This global trend is evident in the Middle East, where ESG integration into corporate governance and policies is increasing [5]. Investors, including those in the GCC, also prioritize ESG-compliant companies. Saudi Arabia's Vision 2030 emphasizes sustainable development, creating a policy foundation for advancing ESG practices [6]. To support this, initiatives like the Saudi Green Initiative aim to achieve sustainability targets, such as increasing renewable energy use. Blockchain technology can improve ESG reporting by enhancing data accuracy, transparency, and traceability. Its features, such as decentralization and immutability, can build trust in ESG reports [7]. ESG reporting in Saudi Arabia is mainly voluntary, with encouragement from bodies like the Saudi Exchange (Tadawul) [8]. While guidelines exist, there are concerns about the quality of ESG reports compared to international standards [9]. These guidelines often suggest aligning with global frameworks such as the GRI, SASB, and TCFD [9]. Saudi Arabia's move toward mandatory ESG reporting is driven by investor demand, a global shift toward mandatory reporting, stakeholder expectations, and alignment with Vision 2030 [10]. Mandatory reporting can enhance transparency, improve data quality, attract investment, and support sustainability goals. Integrating blockchain into ESG reporting involves legal and practical considerations [11]. These include the legal admissibility of blockchain data, the validity of smart contracts, data privacy by regulations like the Personal Data Protection Law (PDPL), and the need for a clear regulatory framework.

2.1 Existing Laws and Regulations: A Snapshot of Saudi Arabia's Evolving ESG Framework

Saudi Arabia has made significant strides in integrating Environmental, Social, and Governance (ESG) considerations into its legal and regulatory frameworks, reflecting a broader commitment to sustainability under Vision 2030. The New Companies Law (2023) marks a pivotal shift in corporate governance by embedding stakeholder-centric principles and sustainability risks into corporate operations. For instance, Article 23 expands fiduciary duties beyond shareholders to include employees, communities, and the environment, acknowledging the interconnectedness of business success with societal well-being (Royal Decree No. M/132, 2023) [12]. Similarly, Article 45 mandates robust risk management frameworks addressing sustainability risks, while Article 67 enhances transparency by requiring disclosure of material information that could impact stakeholders, including ESG factors (Ministry of Commerce, 2023) [13].

Tadawul's efforts have led to significant advancements in ESG reporting among listed companies. While full mandatory compliance is not yet enforced across all entities, the Exchange has adopted a 'comply or explain' approach to encourage adoption. The Tadawul ESG Framework emphasizes critical areas such as climate action, gender diversity, and governance transparency. Stricter eligibility criteria for inclusion in the Tadawul ESG Index now require verified carbon reduction targets and a minimum of 20% female representation on boards. Furthermore, Tadawul aims to achieve 70% compliance with its ESG reporting guidelines by 2025, underscoring its commitment to aligning the Saudi capital market with international sustainability standards (Saudi Exchange, 2024) [14].

2.2 Challenges in Voluntary ESG Reporting

Voluntary Environmental, Social, and Governance (ESG) reporting in Saudi Arabia faces significant challenges that undermine its effectiveness and credibility. These include inconsistent metrics, a lack of robust verification, limited stakeholder pressure, and data fragmentation (Deloitte, 2021; World Economic Forum, 2020). These issues hinder meaningful sustainability disclosures and impede progress toward Vision 2030's ambitious goals [15].

Blockchain technology offers a potential solution. Its decentralized, transparent, and immutable nature makes it well-suited to address these shortcomings. Blockchain can facilitate standardized ESG metrics through smart contracts, ensuring consistency and comparability (Saberi et al., 2019) [17]. Its immutability enhances verification and transparency, deterring greenwashing and bolstering stakeholder trust (Tapscott & Tapscott, 2016). Real-time monitoring via blockchain can increase accountability (IBM Institute for Business Value, 2020), while improved data accessibility and interoperability streamline analysis (Accenture, 2021). Finally, by enhancing data reliability and accessibility, blockchain can strengthen stakeholder engagement (PwC, 2022) [18].

Therefore, while voluntary ESG reporting in Saudi Arabia faces challenges, blockchain presents a transformative opportunity to overcome these obstacles. By ensuring data accuracy, enhancing transparency, and fostering stakeholder trust, blockchain can elevate the quality and impact of ESG disclosures, playing a crucial role in achieving Vision 2030's sustainability objectives [18].

2.3 Gaps in Saudi Arabia's ESG Reporting Framework and the Potential of Blockchain

Saudi Arabia's ESG reporting framework faces significant gaps that hinder its effectiveness and the Kingdom's ability to achieve Vision 2030's sustainability objectives. These weaknesses include a lack of mandatory requirements, weak enforcement mechanisms, and a unified reporting standard [18].

Lack of Mandatory Requirements: Unlike jurisdictions like the EU with mandatory ESG disclosures (e.g., CSRD), Saudi Arabia's voluntary approach results in fragmented reporting, primarily by large corporations, leaving out smaller firms and diminishing overall quality and coverage (Al-Mulhim & Al-Dosari, 2021; Khan et al., 2020). This also reduces the perceived importance of ESG considerations among corporate leaders (Deloitte, 2021) [19].

Weak Enforcement Mechanisms: The absence of a dedicated regulatory body and penalties for non-compliance or inaccurate reporting undermines the reliability of disclosures and creates opportunities for greenwashing (Saberi et al., 2019; World Economic Forum, 2020). This lack of accountability erodes stakeholder confidence (PwC, 2022) [20].

Lack of Unified Reporting Standard: The use of varied methodologies and frameworks makes it difficult to compare ESG performance across entities and complicates data aggregation for policymakers (Accenture, 2021; Tapscott & Tapscott, 2016). This also increases the administrative burden on companies (IBM Institute for Business Value, 2020) [21].

Blockchain as a Solution: Blockchain technology offers a promising solution to these challenges. Its decentralized, immutable, and transparent nature can ensure data integrity by providing a tamper-proof platform (Saberi et al., 2019). Smart contracts can automate compliance (Tapscott & Tapscott, 2016) [3], while blockchain can facilitate standardization and interoperability (Accenture, 2021). Furthermore, it can promote transparency and stakeholder trust by providing real-time data access (PwC, 2022). Finally, blockchain aligns with Vision 2030's ambition to leverage technology for sustainable development (World Economic Forum, 2020) [22].

Therefore, integrating blockchain into the legal framework can bridge these gaps by enhancing data integrity, compliance, and stakeholder trust. This can contribute to a robust ESG reporting system that supports Vision 2030 and positions Saudi Arabia as a leader in sustainable development.

3 Challenges and Opportunities in Blockchain-Enabled Corporate Sustainability Reporting

Integrating blockchain with corporate sustainability reporting presents complex challenges. Existing literature lacks a structured approach to blockchain's application within the evolving regulatory landscape. Interoperability with traditional systems poses technical barriers, while data privacy concerns and the risk of greenwashing complicate trust in sustainability claims.

3.1 Technical Barriers

Integrating blockchain faces technical barriers and infrastructure limitations. Organizations must address inadequate technological frameworks and a lack of skilled personnel. High upfront costs, especially for emerging markets, can detract from sustainability efforts. Cross-sector collaboration is crucial for establishing standardized data-sharing protocols [23].

3.2 Regulatory and Compliance Issues

Stringent regulations necessitate accurate and transparent reporting, creating challenges in data integrity and accessibility. Efficient IT solutions are crucial. Blockchain adoption must navigate diverse regulatory landscapes [23].

3.3 Resistance to Change Within Organizations

Resistance to change can hinder blockchain implementation. Organizations face challenges in data sourcing and management. Overcoming this resistance is crucial for leveraging blockchain in sustainability efforts [23].

4 Solutions and Future Directions

Blockchain's transparency, security, and immutability can enhance the credibility of ESG data. Implementation aligns with regulatory initiatives. Integrating digital tools with blockchain can optimize ESG index performance. Addressing technological complexity and lack of awareness is essential [24].

4.1 Best Practices for Successful Implementation

Successful implementation requires an understanding of the blockchain's capabilities. Training programs can reduce resistance to change, interdepartmental collaboration can streamline integration, and stakeholder engagement ensures alignment with expectations [24].

4.2 Collaborative Efforts Between Stakeholders

Collaboration between businesses, regulators, and technology providers is essential. Collaboration can improve ESG index performance. Collective involvement streamlines reporting and builds trust [24].

4.3 The Future of Blockchain in Corporate Sustainability Reporting

Blockchain has the potential to enhance ESG reporting credibility. However, challenges exist in implementation due to regulations and technological complexities. Further research into practical applications is needed. Blockchain's performance compared to traditional systems requires further investigation [24].

5 A Proposed Framework for Blockchain-Enabled Mandatory ESG Reporting

The increasing emphasis on ESG factors necessitates a robust reporting framework. Blockchain offers a solution for mandatory ESG reporting. This framework leverages blockchain's characteristics to ensure accurate, real-time reporting. It will explore key elements, stakeholder roles, capacity building, and alignment with relevant principles [24].

5.1 Blockchain Enhancing ESG: Empirical Evidence and Global Applications

Empirical research in Saudi Arabia indicates a positive relationship between ESG disclosure and financial performance [25]. Studies of Tadawul-listed companies (2017–2022) found that stronger ESG practices correlated with improved financial metrics [26]. This aligns with stakeholder and signaling theories. However, ESG reporting faces reliability, transparency, and standardization challenges, especially in emerging markets like Saudi Arabia [27]. Blockchain technology offers solutions, as shown by global examples. IBM and Maersk use blockchain for transparent supply chains, supporting ethical sourcing [28]. The World Bank has explored blockchain for emissions reduction reporting and carbon offset trading, enabling more transparent carbon markets [29]. Companies like Unilever and Starbucks use blockchain for supply chain traceability, providing consumers with product origin and sustainability information [30]. Almarai, a Saudi dairy company, uses blockchain to verify ethical sourcing [31]. Blockchain also helps address greenwashing—companies like Carbon Connect AG and DoxyChain issue green certifications. Organizations like OpenSC and WWF use blockchain to certify food sustainability and track seafood supply chains. Platforms like Data Gumbo and Topl facilitate the secure sharing of ESG data [32].

In Saudi Arabia, blockchain's ability to align ESG disclosures with Maqasid al-Shariah could strengthen the perceived link between ESG practices and financial performance. Projects like the BE Blockchain Demonstrator can streamline ESG reporting for regulatory compliance.

Integrating blockchain into ESG reporting can address data transparency, reliability, and standardization limitations. Global case studies illustrate blockchain's versatility in enhancing supply chain traceability, combating greenwashing, and supporting carbon markets. These applications are relevant to empirical findings on ESG disclosure in Saudi Arabia. As Saudi Arabia prioritizes sustainability under Vision 2030, blockchain offers a transformative opportunity to improve ESG practices and drive financial performance.

6 Conclusion

Integrating blockchain technology into ESG reporting is crucial to achieving Saudi Arabia's Vision 2030. By providing an immutable, transparent, decentralized platform, blockchain enhances corporate accountability, attracts green investments, and ensures compliance with global sustainability standards. This innovation fosters stakeholder trust while supporting economic diversification, environmental sustainability, and digital transformation.

Despite its potential, blockchain adoption in ESG reporting faces legal and regulatory challenges, particularly regarding data privacy and compliance. Addressing these issues requires amendments to Saudi Company Law, stronger Capital Market Authority (CMA) regulations, and the development of dispute resolution mechanisms that leverage blockchain's transparency. These reforms would create a robust legal framework, ensuring widespread adoption and alignment with Vision 2030's objectives.

Successful implementation depends on the commitment of policymakers, regulators, and businesses. Policymakers must align domestic regulations with global standards, regulators should develop centralized ESG data repositories, and businesses must invest in blockchain integration. By taking decisive action, Saudi Arabia can lead in transparency, accountability, and sustainability, accelerating progress toward a more innovative and responsible economic future.

References

1. McKinsey & Company. 2025. Does ESG Really Matter—and Why?. Accessed 15 Jan 2025. https://www.mckinsey.com/capabilities/sustainability/our-insights/does-esg-really-matter-and-why. Also World Economic Forum. 2023–2024. Annual Report 2023–2024. Accessed 15, Jan, 2025. https://www3.weforum.org/docs/WEF_Annual_Report_2023_2024.pdf
2. Tapscott, D., Tapscott, A.: Blockchain Revolution: How the Technology Behind Bitcoin Is Changing Money, Business, and the World. Penguin Random House. Also See, Vision 2030. (2016). Kingdom of Saudi Arabia Vision 2030 (2016). https://vision2030.gov.sa
3. Tapscott, D., Tapscott, A.: Blockchain Revolution: How the Technology Behind Bitcoin Is Changing Money, Business, and the World. Penguin Random House (2016)
4. The developing legal and regulatory regime for ESG in the Middle East and emerging D&O exposures | United States - Norton Rose Fulbright, accessed March 17, 2025, The developing legal and regulatory regime for ESG in the Middle East and emerging D&O exposures|United States | Global law firm | Norton Rose Fulbright
5. MIDDLE EAST: ESG Revolution - | Asian Legal Business. https://www.legalbusinessonline.com/features/middle-east-esg-revolution. Accessed 17 Mar 2025
6. ESG Reporting Requirements in 2025 Insights KSA. https://insightss.co/blogs/esg-reporting-requirements-in-2025/. Accessed 17 Mar 2025
7. The Influence of Fintech Innovations, ESG Reporting, and Blockchain Technology on Financial Transparency and Accountability – ResearchGate. https://www.researchgate.net/publication/382414437_The_Influence_of_Fintech_Innovtions_ESG_Reporting_and_Blockchain_Technology_on_Financial_Transparency_and_Accountability. Accessed 17 Mar 2025
8. ESG Guidelines – Tadawul. https://www.saudiexchange.sa/wps/portal/saudiexchange/listing/issuer-guides/esg-guidelines. Accessed 17 Mar 2025
9. The Effects of Mandatory ESG Disclosure Around the World - European Corporate Governance Institute (ECGI). https://www.ecgi.global/sites/default/files/working_papers/documents/kreugersautnertangzhongfinalfeb_0.pdf. Accessed 17 Mar 2025. Also See Government-Mandated ESG Disclosure: It's Not Easy Being Green - Houston Law Review, accessed March 17, 2025, https://houstonlawreview.org/article/117184-government-mandated-esg-disclosure-it-s-not-easybeing-green
10. ESG in Saudi Vision 2030|Sustainability Goals - Astra ESG Solutions, accessed March 17, 2025, https://astra.grandviewresearch.com/country-esg-outlook/saudi-arabia-vision Saudi

Arabia's Vision 2030: A New Era for Sustainable Tourism - Seneca ESG, accessed March 17, 2025, https://senecaesg.com/insights/saudi-arabias-vision-2030-a-new-era-for-sustainable-tourism/
11. The Influence of Fintech Innovations, ESG Reporting, and Blockchain Technology on Financial Transparency and Accountability – ResearchGate. https://www.researchgate.net/publication/382414437_The_Influence_of_Fintech_Innovations_ESG_Reporting_and_Blockchain_Technology_on_Financial_Transparency_and_Accountability. Accessed 17 Mar 2025
12. Royal Decree No. M/132. New Companies Law. Kingdom of Saudi Arabia (2023). https://laws.boe.gov.sa/BoeLaws/Laws/LawDetails/a8376aea-1bc3-49d4-9027-aed900b555af/1
13. Ministry of Commerce. Implementation Guidelines for Corporate Governance (2023). https://mc.gov.sa/en/Regulations/Pages/details.aspx?lawId=ea0fc797-4127-4667-962d-aec400ee9f72
14. Saudi Exchange. ESG Guidelines 2025. https://www.saudiexchange.sa/wps/portal/saudiexchange/listing/issuer-guides/esg-guidelines. Accessed 15 Jan 2025
15. Deloitte, A.: Year of Connection, Impact and Action: The Middle East Impact Report. (2021). https://www.deloitte.com/middle-east/en/about/governance/dme-impact-report-2021.html. Accessed 15 Jan 2025
16. World Economic Forum. 2023–2024. Annual Report 2023–2024. https://www3.weforum.org/docs/WEF_Annual_Report_2023_2024.pdf. Accessed 15 Jan 2025
17. Saberi, S., et al.: Blockchain for Good: Driving Sustainability Through Technology. Blockchain technology and its relationships to sustainable supply chain management. Int. J. Prod. Res. **57**(7), 2117–2135 (2019). Accenture. (2021). Retrieved from ibm.com. Title of Accenture Report. Accenture. Accessed 15, Jan 2025 https://www.accenture.com/content/dam/accenture/final/a-com-migration/pdf/pdf-165/accenture-fiscal-2021-annual-report.pdf
18. Tapscott, D., Tapscott, A.: Blockchain Revolution: How the Technology Behind Bitcoin Is Changing Money, Business, IBM Institute for Business Value (2020)
19. Al-Mulhim, F., Al-Dosari, A.: The role of ESG reporting in enhancing corporate governance in Saudi Arabia. Int. J. Bus. Manag. Stud. **10**(2), 45–62 (2021)
20. PwC. Family Businesses and ESG in Saudi Arabia (2023). https://www.pwc.com/m1/en/publications/family-business-survey-2023.16html. Accessed Jan 2025
21. IBM Institute for Business Value. Blockchain for Good: Driving Sustainability Through Technology (2020). Retrieved from ibm.com
22. World Economic Forum. 2023–2024. Annual Report 2023–2024. Accessed 15 Jan 2025
23. Chopra, S.S., et al.: Navigating the challenges of environmental, social, and governance (ESG) reporting: the path to broader sustainable development. Sustainability **16**(2), 606 (2024). https://doi.org/10.3390/su16020606
24. Kabir, M.S., Ko, I.: Collaboration with Digital Capabilities enhance firm\u27s ESG Management: A Systematic Literature Review AIS Electronic Library (AISeL) (2024). https://core.ac.uk/download/616565712.pd
25. Shalhoob, H.: ESG disclosure and financial performance: survey evidence from accounting and Islamic finance. Sustainability **17**, 1582 (2025). https://doi.org/10.3390/su17041582
26. The Impact of Environmental, Social, and Governance Disclosure on the Performance of Saudi Arabian Companies: Evidence from the Top 100 Non-Financial Companies Listed on Tadawul - ResearchGate. https://www.researchgate.net/publication/383650216_The_Impact_of_Environmental_Social_and_Governance_Disclosure_on_the_Performance_of_Saudi_Arabian_Companies_Evidence_from_the_Top_100_Non-Financial_Companies_Listed_on_Tadawul. Accessed 17 Mar 2025
27. ESG Reporting in the GCC: Trends and Challenges 2025. https://www.orennow.com/blog/esg-reporting-gcc-2025. Accessed 17 Mar 2025. Also see The impact of blockchain

technology on regulatory compliance. accessed March 17, 2025, https://community.trustcloud.ai/docs/grc-launchpad/grc-101/compliance/the-impact-of-blockchain-technology-on-regulatory-compliance-opportunities-and-challenges/
28. Blockchain for Good: Driving Sustainability Through Technology, Op. cit
29. Banking, Blockchain, and ESG (Chapter 11) - The Role of. https://www.cambridge.org/core/books/role-of-distributed-ledger-technology-in-banking/banking-blockchain-and-esg/05043C13ED2914B9A0976CAE261C062B. Accessed 17 Mar 2025
30. Blockchain in ESG | Real World Use Cases – Kaleido. https://www.kaleido.io/industries/esg. Accessed 17 Mar 2025
31. Driving ESG Transformation in the GCC: The Role of Technology, CX, and EX in Building a Sustainable Future - New Metrics. https://www.newmetrics.net/insights/driving-esg-transformation-in-the-gcc-the-role-of-technology-cx-and-ex-in-building-a-sustainable-future/. Accessed 17 Mar 2025
32. Blockchain in ESG|Real World Use Cases Op. cit

Impact of Knowledge Sharing, Information and Communication Technology Usage, and Green Market Orientation on Sustainable Performance

Jalal Rajeh Hanaysha(✉)

School of Business, Horizon University College, Ajman, United Arab Emirates
jalal.hanaysha@hu.ac.ae

Abstract. This study focuses on examining whether knowledge sharing, information and communication technology (ICT) usage, and green market orientation have any effect on sustainable performance. The required data was gathered based on a structured survey being administered to the managers/owners of small and medium enterprises (SMEs) in the United Arab Emirates (UAE). During the data gathering process, convenience sampling technique was used to ensure that the required sample size can be reached. In total, 206 usable questionnaires were received from the respondents. Furthermore, the SmartPLS software was utilized for analyzing the data and drawing conclusions. The findings verified that knowledge sharing and ICT usage positively affect sustainable performance. The analysis also confirmed that green market orientation is positively associated with sustainable performance. These outcomes contribute to the empirical literature on sustainable performance and provide valuable implication for policy makers in SMEs in order to maintain the long-term prosperity and growth of their businesses.

Keywords: Green market orientation · ICT usage · Knowledge sharing · Sustainable performance

1 Introduction

The evolving business landscape, characterized by new rules, policy alterations, consumer behaviour shifts, and technological advancements pose considerable expectations on enterprises to protect the ecosystem and pursue sustainable performance [1]. Sustainability is crucial for companies to adeptly address environmental issues. Companies are facing increasing pressure from environmentally concerned stakeholders to adopt green practices, such as information exchange to improve sustainable performance [2]. The dissemination of environmental knowledge is essential for a firm's sustainable performance. Numerous firms have acknowledged that information is essential and serves as a critical resource for attaining sustainable performance across all industries [3]. For several organizations, achieving a competitive advantage is contingent on their capabilities to generate and utilize intellectual expertise [4]. Moreover, A firm's network

D. Berkouk et al. (Eds.): CCTS 2024, PTT, pp. 424–434, 2026.
https://10.1007/978-981-95-7607-4_47

of knowledge-sharing enable it to seize new business opportunities and strengthen its commitment to environmentally friendly practices [5]. Therefore, firms that invest in the preservation of surrounding environment can get a first-mover advantages, facilitating unique positioning strategies, enhanced green image, and competitive benefits.

These days, technological advancements exert a profound influence on individuals, with information transmitted globally in efficient ways via ICT, benefiting both the society at large and businesses [6]. Within organizations, ICT is a powerful approach for transforming the market-oriented attributes of the offerings, as well as manufacturing processes, personnel workflows, and management approaches [7]. The Internet has fundamentally transformed business practices. The literature reported numerous beneficial consequences of ICT use in companies, including minimal efforts for performing tasks, enhanced communication among business stakeholders and shareholders, enhancing the efficiency and internal operations, maximizing profits and lowering costs [8]. The utilization of ICT resources also minimizes the efforts in different types of tasks, which enhances employee satisfaction and motivation [9]. However, there exist limited studies that examined the impact of ICT usage on the sustainability performance in SMEs' setting, particularly in developing and emerging economies [10].

Furthermore, as environmental issues intensify, it is imperative for companies to consider the natural environment when adopting green market orientation [11]. Wang [12] reported that companies should broaden their market orientation by addressing the requirements of environmental concerns. GMO denotes that an organization cultivates and enhances environmental competencies, services, or products to achieve exceptional sustainable business performance [13]. Organizations that follow a GMO approach recognize the importance of environmental management, track the activities of their rivals, and project an image regarding environmental responsibility to consumers [14]. Accordingly, such businesses are likely to establish environmental goals for their internal operations or production, work hard to achieve these goals, and invest more resources into promoting sustainable practices and performance. Nevertheless, while prior research emphasized the significance of the linkages between GMO and green performance, there is insufficient studies that have examined whether green innovation constitutes a missing link [12].

In accordance of the above gaps stated above, this research aims to explore the impact of knowledge sharing, ICT usage, and green market orientation on sustainable performance with novel findings from SMEs in the UAE. The aforementioned factors were examined separately in earlier research, and it is difficult to come across an empirical research paper that examined their impact on sustainable performance, particularly in the Middle East region.

2 Literature Review

2.1 Knowledge Sharing

Knowledge sharing can be conceptualized as the dissemination of important knowledge with all stakeholders of the firm [15]. The word "knowledge sharing" also refers to a firm's approaches and tactics for communicating and transmitting knowledge to its staff, enabling them to utilise and enhance it to generate new ideas [16]. Knowledge sharing plays a key part in the success of enterprises, as it facilitates the generation of new ideas, goods, or services [17]. The exchange of knowledge among organizational staff facilitates the dissemination of ideas, information, thoughts, and experiences, hence serving as a catalyst for creativity and innovation to ensure long-term prosperity [18]. Given the regular changes in business environment, it is widely acknowledged that organizations which are proficient in managing the knowledge rooted in their daily operations will excel over their rivals [19]. Furthermore, knowledge has transformed the traditional perception of competitiveness, especially in industrialized nations. There is a pressing need for businesses to find ways to get necessary knowledge and seek for the relevant ideas that will lead to ecologically sustainable outcomes and provide competitive strengths over rivals [4, 20]. Past studies confirmed that knowledge sharing positively affects sustainable performance [21]. Similarly, Shahzad et al. [22] reported that knowledge sharing reinforces the capabilities of an enterprise's green innovation and leads to sustainable performance. Accordingly, the subsequent hypothesis is advocated:

H1: Knowledge sharing positively affects sustainable performance.

2.2 ICT Usage

ICT capabilities entail a range of a firm's technological infrastructure, tools and individual technical know-how that enable firms to streamline operations and sustain themselves in digital environments [10, 23]. Aligned with the assumptions of the DC theory, studies reveal that ICT capabilities serve as the basis for enhanced firms' innovative practices, operational efficiency, and financial sustainability [24, 25]. The utilization of ICT can improve environmental growth and yield favourable sustainability results by creating real-time decision support systems that integrate government and an enterprise's ICT innovations, thereby empowering managers and decision makers to take appropriate approaches that align with environmental sustainability objectives [26]. Innovation capabilities and ICT breakthroughs play a significant role in advancing organizations and nations towards more ecologically sustainable results [27]. The utilization ICT enables organizations to substantially lower expenses and augment productivity, hence improving their performance [28]. Earlier researches confirmed that ICT usage positively affects sustainable performance [29]. In relation to the aforementioned arguments, the following hypothesis is proposed:

H2: ICT usage positively affects sustainable performance.

2.3 Green Market Orientation

Green market orientation promotes the establishment and maintenance of relationships with existing markets, positioning market intelligence as the focal point of corporate operations and actions [11]. GMO urges enterprises to detect innovative ideas by meticulously identifying and analyzing stakeholders' environmental needs, hence enhancing their proactivity in refining existing green products or launching new ones [30]. It also ensures that green businesses stay alert to changes in the market environment regarding sustainability issues and adapt accordingly [12]. A company with a robust GMO possesses a comprehensive awareness of consumers' needs and environmental requirements, enabling it to pursue ecological innovation and acquire further knowledge [31]. Furthermore, green market-oriented enterprises tend to acquire diverse knowledge to identify opportunities and respond to stakeholders' expectations [12]. Thus, enterprises that have green market orientation tend to implement strategies that promote the creation of products for meeting stakeholders' needs [32]. Moreover, certain researches indicated that enterprises adopting a market-oriented strategy often demonstrate greater social performance [31–33]. Hence, the subsequent hypothesis is proposed:

H3: GMO has a positive impact on sustainable performance.

3 Methodology

The primary aim of this research was to examine the impact of knowledge sharing, ICT usage, and GMO on sustainable performance of SMEs. In relation to that, the data was gathered from managers/owners of several SMEs operating in the UAE. These respondents are considered as the most suitable individuals who can provide accurate data to fulfil the aims of this research and then test the hypotheses of this research. Due to the high challenges in collecting the data from SMEs' owners/managers, convenience sampling method was used during the entire data collection process. In addition to that, both online and self-administered survey was employed to warrant that the required sample size can be obtained from them. In total, 630 questionnaires were distributed on the respondents, but only 214 were returned, accounting for 34% of response rate. Out of these questionnaires, eight were discarded due to incomplete answers. Hence, the usable sample which was run for data analysis included 206. Most of the participants are males, from service sector (57.8%), and had more than 10 years of experience.

The measurement items for the constructs used in this paper were adapted from prior studies. Particularly, the construct of knowledge sharing was measured using six validated measurement items adapted from Oliveira et al. [34]. To measure ICT usage, five items were further taken from León-Gómez et al. [7]. Furthermore, GMO was measured using six measurement items being adapted from Nigatu et al. [35]. Finally, sustainable performance was measured via a six items' scale taken from Akanmu et al. [36]. Each item of the constructs was assessed using a 5-point Likert scale that fall on the range of 1 (strongly disagree) to 5 (strongly agree).

4 Results

To deal with the potential problem of common method bias (CMB), the test of Harman's one-factor was utilized, as recommended by Podsakoff [37]. Such mechanism is built on the premise that CMB results in a singular component that accounts for the most of the covariance between the measurements, or that only one variable is extracted. The findings from Harman's [38] one-factor analysis demonstrated the absence of common method bias, since the primary factor explained 32.7% of the extracted variance. Additionally, a partial correlation analysis was conducted to determine whether any significant differences exist in the correlation values after controlling for marker factors.

The findings demonstrated that there was no indication of common method bias that could compromise the reliability of the data.

As suggested by Hair et al. [39], before the estimation of the main effects, it is necessary to assess the reliability and validity of the constructs. Hence, the current study performed confirmatory factor analysis (CFA) in SmartPLS. The results presented in Table 1 reveal that Cronbach's Alpha and Composite Reliability values for all constructs exceeded 0.7. Furthermore, the factor loadings for items were calculated, and those which have a minimum factor weight of 0.5 were maintained. An AVE test was also performed using the residual factor loadings of measurement items. When the AVE value is higher than 0.5, convergence validity and uni-dimensionality are considered satisfactory [40]. The research verified the existence of convergent validity across all variables, since AVE values are above 0.5.

Table 1 Results of CFA

Constructs	Items	Factor loadings	Cronbach's Alpha	CR	AVE
Knowledge Sharing	KS1	0.803	0.821	0.873	0.581
	KS2	0.740			
	KS3	0.724			
	KS4	0.735			
	KS5	0.803			
ICT Usage	ICTU1	0.741	0.836	0.885	0.609
	ICTU2	0.847			
	ICTU3	0.713			
	ICTU4	0.914			
	ICTU5	0.659			
Green Market Orientation	GMO1	0.678	0.818	0.870	0.533
	GMO2	0.822			
	GMO3	0.813			
	GMO4	0.585			
	GMO5	0.853			
	GMO6	0.578			
Sustainable Performance	SP1	0.825	0.853	0.890	0.576
	SP2	0.771			
	SP3	0.749			
	SP4	0.766			
	SP5	0.761			
	SP6	0.672			

This study also relied on Fornell and Larcker's [40] formula for calculating discriminant validity. Specifically, the test was run via the comparison of SQRT of AVE with the other correlation values among other variables. To establish discriminant validity, the SQRT of the AVE must exceed the corresponding inter-correlation coefficients [41]. Table 2 shows that all diagonal values exceed the corresponding inter-correlation coefficients, hence indicating that convergent validity has been attained. The coefficient values also were all below 0.8, signifying a lack of strong association between any pair of variables [42].

Table 2 Results of Discriminant Validity

Construct	1	2	3	4
1. Green Market Orientation	**0.730**			
2. ICT Usage	0.447	**0.781**		
3. Knowledge Sharing	0.321	0.455	**0.762**	
4. Sustainable Performance	0.329	0.245	0.375	**0.759**

After the estimation of measurement model, the next step was to estimate the structural model in order to examine the proposed hypotheses, as depicted in Table 3. A bootstrapping technique (5,000 resamples) was employed to assess the route estimates of the proposed hypotheses. Specifically, the study's first hypothesis (H1), which proposed a positive impact of knowledge sharing on a firm's sustainable performance was supported as the results revealed a significant positive effect ($\beta = 0.568$, t-value = 8.383, $p > .05$). Furthermore, the analysis verified the significant positive effect of ICT usage on sustainable performance ($\beta = 0.205$, t-value = 2.213, $p > .05$), thus supporting H2. Finally, the findings displayed that GMO has a positive as well as significant impact on sustainable performance ($\beta = 0.179$, t-value = 2.029, $p > .05$); therefore, H3 is accepted.

Table 3 Results of Hypotheses

Hypothesis			Beta	Std. Dev.	t-value	P
Knowledge Sharing	→	Sustainable Performance	0.568	0.068	8.383	0.000
ICT Usage	→	Sustainable Performance	0.205	0.093	2.213	0.027
Green Market Orientation	→	Sustainable Performance	0.179	0.088	2.029	0.043

5 Discussion and Conclusion

The main purpose of the present research was to investigate the effect of selected variables (knowledge sharing, ICT usage, and green market orientation) on sustainable performance in SMEs. By reviewing the literature on sustainable performance, it is hard to find an empirical study that collectively examined the impact of the aforementioned variables on it, particularly in SMEs' context. The findings displayed that knowledge sharing positively affects sustainable performance. Additional support was reported in past researches which concluded that there is a positive link amongst knowledge sharing and a firm's sustainable performance [20, 43]. Darroch and McNaughton [44] also stated that enterprises which use superior knowledge-sharing processes exhibit enhanced creativity and improved economic performance. The results indicate that acquisition of relevant and valuable knowledge from internal or external sources facilitates a more profound comprehension of information related to new goods/services, identification of potential suppliers, buyer preferences, and similar aspects. Thus, the acquired knowledge can be utilized to address complex business challenges, subsequently contributing to the improvement of organizations' reputation and positioning in the target markets.

The findings also verified that ICT usage positively influences sustainable performance. Further support was documented in prior studies which showed that ICT usage positively and significantly influence sustainable performance [29, 45]. Rutainurwa et al. [10] stated that ICT usage helps firms automate routine tasks, whilst bolstering sustainable performance through improved operational efficiency and minimizing transportation and inventory holding costs. Similarly, Tan et al. [46] asserted that ICT represents a cost-effective method for enabling SMEs to address the challenges in the competition with enterprises, utilizing the Internet as a novel approach to business operations. Integrating ICTs in the daily operations of SMEs will enable them to possess enhanced resources to undertake new endeavours [47]. ICT can enhance the ability of SMEs to create goods, processes, and services that are more appropriate to the sustainability requirements of their stakeholders [7]. Businesses can integrate their ICT innovations with environmental goals, enabling the use of ICTs to promote environmental protection through service and cost efficiency [48].

Finally, the outcomes displayed that GMO positively impact sustainable performance. More support can be noticed in prior researches which verified that positive association among GMO and sustainable performance [33]. Certain researchers also proposed that GMO enables enterprises to cultivate distinctive competencies for attaining exceptional performance in relation to environmental sustainability and business profits. GMO signifies a strategic stance and market-centric methodology that principally directs corporations to comprehend customers' environmental demands and monitor competitors' environmental strategies [11]. Green market orientation drives companies to improve their offerings by introducing ecologically friendly items for minimizing the detrimental effect of their daily operations on the environment and social wellbeing [49]. Thus, entrepreneurs and decision makers in SMEs should recognize that enhancing GMO will promote green innovations within the organization.

A few limitations were noticed in this research that should be carefully considered in future research endeavours. Primarily, this study relied on a cross-sectional survey for gathering the data; hence, upcoming studies may utilize longitudinal or experimental approaches to more effectively ascertain the causal impact of GMO, TCT usage, knowledge sharing on sustainable performance. Secondly, the findings of this study are derived from data collected in a specific nation (UAE); hence, generalizing the results to other countries or areas may be constrained due to cultural variations that may demonstrate differences in the results of hypotheses. Finally, this study focused on exploring the effect of three variables only on sustainable performance. Thus, future research can test other predictors, such as leadership and corporate social responsibility. Moreover, upcoming studies can extend the model by including mediators (such as sustainable leadership) for examining the fundamental mechanisms that elucidate how the selected variables affect sustainable performance in SMEs' setting.

References

1. Khan, R.U., Saqib, A., Abbasi, M.A., Mikhaylov, A., Pinter, G.: Green leadership, environmental knowledge sharing, and sustainable performance in manufacturing Industry: Application from upper echelon theory. Sustain Energy Technol Assess. **60**, 103540 (2023)
2. Riva, F., Magrizos, S., Rubel, M.R.B.: Investigating the link between managers' green knowledge and leadership style, and their firms' environmental performance: The mediation role of green creativity. Bus. Strateg. Environ. **30**(7), 3228–3240 (2021)
3. Al Koliby, I.S., Mohd Suki, N., Abdullah, H.H.: Linking knowledge acquisition, knowledge dissemination, and manufacturing SMEs' sustainable performance: The mediating role of knowledge application. Bottom Line. **35**(4), 185–202 (2022)
4. Alharbi, G.L., Aloud, M.E.: The effects of knowledge management processes on service sector performance: Evidence from Saudi Arabia. Humanit. Soc. Sci. Commun. **11**(1), 1–19 (2024)
5. Ma, L., Ali, A., Shahzad, M., Khan, A.: Factors of green innovation: The role of dynamic capabilities and knowledge sharing through green creativity. Kybernetes. **54**(1), 54–70 (2025)
6. Vermesan, O., Friess, P.: Internet of Things-Global Technological and Societal Trends from Smart Environments and Spaces to Green ICT. CRC Press, New York (2022)
7. León-Gómez, A., Santos-Jaén, J.M., Ruiz-Palomo, D., Palacios-Manzano, M.: Disentangling the impact of ICT adoption on SMEs performance: The mediating roles of corporate social responsibility and innovation. Oecon. Copernic. **13**(3), 831–866 (2022)
8. Mwantimwa, K.: ICT usage to enhance firms' business processes in Tanzania. J. Glob. Entrep. Res. **9**(1), 46 (2019)
9. Wang, B., Liu, Y., Parker, S.K.: How does the use of information communication technology affect individuals? A work design perspective. Acad. Manag. Ann. **14**(2), 695–725 (2020)
10. Rutainurwa, V., Mwenda, B., Israel, B.: ICT-enabled supply chain integration capabilities and firms' financial performance: An empirical study of Tanzania's manufacturing SMEs. Cogent Bus. Manag. **11**(1), 2393235 (2024)
11. Du, Y., Wang, H.: Green innovation sustainability: How green market orientation and absorptive capacity matter? Sustainability. **14**(13), 8192 (2022)
12. Wang, C.H.: An environmental perspective extends market orientation: Green innovation sustainability. Bus. Strateg. Environ. **29**(8), 3123–3134 (2020)
13. Aboalhool, T., Alzubi, A., Iyiola, K.: Humane entrepreneurship in the circular economy: The role of green market orientation and green technology turbulence for sustainable corporate performance. Sustainability. **16**(6), 2517 (2024)
14. Wong, S.K.-S.: Impacts of environmental turbulence on entrepreneurial orientation and new product success. Eur. J. Innov. Manag. **17**(2), 229–249 (2014)

15. Al-Busaidi, K.A., Olfman, L.: Knowledge sharing through inter-organizational knowledge sharing systems. VINE J. Inf. Knowl. Manag. Syst. **47**(1), 110–136 (2017)
16. Obeidat, A.M., Abualoush, S.H., Irtaimeh, H.J., Khaddam, A.A., Bataineh, K.A.: The role of organisational culture in enhancing the human capital applied study on the social security corporation. Int. J. Learn. Intellect. Cap. **15**(3), 258–276 (2018)
17. Azeem, M., Ahmed, M., Haider, S., Sajjad, M.: Expanding competitive advantage through organizational culture, knowledge sharing and organizational innovation. Technol. Soc. **66**, 101635 (2021)
18. Lin, Y.H., Chen, Y.S.: Determinants of green competitive advantage: The roles of green knowledge sharing, green dynamic capabilities, and green service innovation. Qual. Quant. **51**, 1663–1685 (2017)
19. Jordão, R.V.D., Novas, J.C.: Information and knowledge management, intellectual capital, and sustainable growth in networked small and medium enterprises. J. Knowl. Econ. **15**(1), 563–595 (2024)
20. Aamir, A., Jan, S.U., Qadus, A., Nassani, A.A., Haffar, M.: Impact of knowledge sharing on sustainable performance: Mediating role of employee's ambidexterity. Sustainability. **13**(22), 12788 (2021)
21. Abbas, J., Hussain, I., Hussain, S., Akram, S., Shaheen, I., Niu, B.: The impact of knowledge sharing and innovation on sustainable performance in Islamic banks: A mediation analysis through a SEM approach. Sustainability. **11**(15), 4049 (2019)
22. Shahzad, M., Qu, Y., Zafar, A.U., Rehman, S.U., Islam, T.: Exploring the influence of knowledge management process on corporate sustainable performance through green innovation. J. Knowl. Manag. **24**(9), 2079–2106 (2020)
23. Khin, S., Ho, T.C.: Digital technology, digital capability and organizational performance: A mediating role of digital innovation. Int. J. Innov. Sci. **11**(2), 177–195 (2019)
24. Cuevas-Vargas, H., Aguirre, J., Parga-Montoya, N.: Impact of ICT adoption on absorptive capacity and open innovation for greater firm performance. The mediating role of ACAP. J. Bus. Res. **140**, 11–24 (2022)
25. Al-Qubaisi, S.S., Ajmal, M.M., Khan, M.: Impact of knowledge management and ICT on operational efficiency: An empirical study. Int. J. Knowl. Based Dev. **9**(2), 174–202 (2018)
26. Krishnan, S., Teo, T., Nishant, R.: IT readiness, ICT usage, and national sustainability development: Testing the source-position-performance framework. In: Proceedings of Thirty Second International Conference on Information Systems. Shanghai, 2011
27. Melville, N.P.: Information systems innovation for environmental sustainability. MIS Q. **34**(1), 1–21 (2010)
28. Alam, S.S., Noor, M.K.M.: ICT adoption in small and medium enterprises: An empirical evidence of service sectors in Malaysia. Int. J. Bus. Manag. **4**(2), 112–125 (2009)
29. Jayaprakash, P., Pillai, R.R.: The role of ICT for sustainable development: A cross-country analysis. Eur. J. Dev. Res. **34**(2), 1–23 (2022)
30. Borah, P.S., Dogbe, C.S.K., Pomegbe, W.W.K., Bamfo, B.A., Hornuvo, L.K.: Green market orientation, green innovation capability, green knowledge acquisition and green brand positioning as determinants of new product success. Eur. J. Innov. Manag. **26**(2), 364–385 (2023)
31. Habib, M.A., Bao, Y., Ilmudeen, A.: The impact of green entrepreneurial orientation, market orientation and green supply chain management practices on sustainable firm performance. Cogent Bus. Manag. **7**(1), 1743616 (2020)
32. Afum, E., Agyabeng-Mensah, Y., Baah, C., Asamoah, G., Yaw Kusi, L.: Green market orientation, green value-based innovation, green reputation and enterprise social performance of Ghanaian SMEs: The role of lean management. J. Bus. Ind. Mark. **38**(10), 2151–2169 (2023)

33. Borazon, E.Q., Huang, Y.C., Liu, J.M.: Green market orientation and organizational performance in Taiwan's electric and electronic industry: The mediating role of green supply chain management capability. J. Bus. Ind. Mark. **37**(7), 1475–1496 (2022)
34. Oliveira, M., Curado, C., Balle, A.R., Kianto, A.: Knowledge sharing, intellectual capital and organizational results in SMES: Are they related? J. Intellect. Cap. **21**(6), 893–911 (2020)
35. Nigatu, T., Degoma, A., Tsegaye, A.: Green practices and economic performance: Mediating role of green innovation in Ethiopian leather, textile, and garment industries—An integrated PLS-SEM analysis. Heliyon. **10**(15), 1–18 (2024)
36. Akanmu, M.D., Hassan, M.G., Bahaudin, A.Y.B.: A preliminary analysis modeling of the relationship between quality management practices and sustainable performance. Qual. Manag. J. **27**(1), 37–61 (2020)
37. Podsakoff, P.M., MacKenzie, S.B., Lee, J.Y., Podsakoff, N.P.: Common method biases in behavioral research: A critical review of the literature and recommended remedies. J. Appl. Psychol. **88**(5), 879–903 (2003)
38. Harman, H.: Modern Factor Analysis. University of Chicago Press, Chicago, USA (1967)
39. Hair Jr., J.F., Hult, G.T.M., Ringle, C.M., Sarstedt, M.: A primer on partial least squares structural equation modeling (PLS-SEM), 2nd edn. Sage Publications, California, USA (2017)
40. Fornell, C., Larcker, D.F.: Evaluating structural equation models with unobservable variables and measurement error. J. Mark. Res. **18**(1), 39–50 (1981)
41. Bamfo, B.A., Dogbe, C.S.K., Mingle, H.: Abusive customer behaviour and frontline employee turnover intentions in the banking industry: The mediating role of employee satisfaction. Cogent Bus. Manag. **5**(1), 1522753 (2018)
42. Tabachnick, B.G., Fidell, L.S., Ullman, J.B.: Using Multivariate Statistics, 5th edn. Pearson, Boston, MA (2007)
43. Oxenswärdh, A.: Knowledge sharing and sustainable development. In: Leal Filho, W. (ed.) Encyclopedia of Sustainability in Higher Education, pp. 1057–1065. Springer International Publishing, Cham (2019)
44. Darroch, J., McNaughton, R.: Beyond market orientation: Knowledge management and the innovativeness of New Zealand firms. Eur. J. Mark. **37**(3/4), 572–593 (2003)
45. Acosta-Prado, J.C., Tafur-Mendoza, A.A.: Examining the mediating role of dynamic capabilities in the relationship between information and communication technologies and sustainable performance. VINE J. Inf. Knowl. Manag. Syst. **54**(3), 561–577 (2024)
46. Tan, K.S., Chong, S.C., Lin, B., Eze, U.C.: Internet-based ICT adoption: Evidence from Malaysian SMEs. Ind. Manag. Data Syst. **109**(2), 224–244 (2009)
47. Higón, D.A.: The impact of ICT on innovation activities: Evidence for UK SMEs. Int. Small Bus. J. **30**(6), 684–699 (2012)
48. Nica, E.: ICT innovation, internet sustainability, and economic development. J. Self-Gov. Manag. Econ. **3**(3), 24–29 (2015)
49. Tjahjadi, B., Soewarno, N., Hariyati, H., Nafidah, L.N., Kustiningsih, N., Nadyaningrum, V.: The role of green innovation between green market orientation and business performance: Its implication for open innovation. J. Open Innov.: Technol. Mark. Complex. **6**(4), 173 (2020)

The Influence of Latin Typography on Arabic-Speaking Female Consumers' Perception of Brand Identity

Abdelqader Kamal Alnobani(✉), Tahar Lazhar Ayed, Ahmed Khaled Nasseraldin, Lama Alkhail, Afnan Bahri, Farah Alfityani, and Joud Alhussan

Dar Al-Hekma University, Jeddah 22246, Kingdom of Saudi Arabia
anobani@dah.edu.sa

Abstract. Typography shapes consumer perceptions in packaging, influencing value and efficacy. Understanding Latin typography's impact can sustain brand equity and inform cross-cultural marketing. This study investigates the impact of typeface choice on Arabic-speaking female consumers' perception of hand cream products. A sample of 189 participants evaluated two identical product mockups, differing only in typeface—Bodoni (Serif) for Product 1 and Arial (Sans Serif) for Product 2. Consumer perceptions of price, hydration effectiveness, and healing effectiveness were assessed through survey responses. The findings indicate a statistically significant association between the perceived price of a product and its perceived effectiveness for hydration, with product 1 being seen as both more expensive and more effective for hydration. However, no significant relationship was found between price perception and healing effectiveness, suggesting that typography's influence varies depending on the specific attribute evaluated. These results highlight the role of typeface selection in shaping consumer expectations, particularly for price-related and hydration cues. The study underscores the importance of typography in brand communication and packaging design while acknowledging limitations related to sample specificity and product scope. Future research should explore typography's impact across different consumer demographics and product categories.

Keywords: Typography · Consumer Perception · Brand Identity · Packaging Design · Cross-Cultural Marketing

1 Introduction

Global brands increasingly rely on Latin typography in non-Latin markets to establish a consistent and recognizable brand identity, reinforcing business sustainability. Typography serves as a fundamental design element in marketing materials, playing a critical role in corporate communication. Scholars and industry professionals alike acknowledge that typeface design functions as a powerful visual tool in shaping consumer perception [1–4]. Among the various typeface categories, serif and sans-serif fonts stand out for

D. Berkouk et al. (Eds.): CCTS 2024, PTT, pp. 435–442, 2026.
https://10.1007/978-981-95-7607-4_48

their distinct impact on communicating product attributes. For instance, sans-serif typefaces are often associated with competence and authority [5], making them a strategic choice for brand messaging.

Despite the widespread use of Latin typography in hybrid linguistic markets, its influence on Arabic-speaking female consumers remains underexplored. Prior research has examined cross-cultural responses to typography, with Caldwell [6] demonstrating consistency in emotional reactions to typefaces across cultures. However, Peterson [7] identified cultural divergence in responses to stylized typefaces, while McMullen [8] highlighted persistent distinctions between Eastern and Western visual preferences in graphic design. These findings suggest that typography's impact on consumer perception is not universally uniform but rather influenced by cultural and contextual factors.

A key area where these cultural and functional nuances manifest is in the distinction between serif and sans-serif typefaces. Serif fonts, such as Bodoni, incorporate small decorative strokes that lend them a traditional, elegant aesthetic. Their design enhances readability in print media by guiding the reader's eye smoothly across text [9, 10]. Bodoni's sharp lines and high contrast make it a popular choice for formal, high-end designs [11, 12]. Conversely, sans-serif fonts like Arial offer a clean, modern appearance that improves clarity on digital screens, making them particularly effective for web design and user interfaces [13, 14]. Arial's simplicity and adaptability enhance readability across different sizes [15, 16], aligning with research that favors sans-serif fonts for digital consumption [17]. These functional advantages, combined with cultural connotations, suggest that typeface choices may influence consumer perceptions of product efficacy—such as hydration and healing properties—as well as their willingness to pay premium prices, both of which are central to brand equity.

Given these insights, this study explores how Arabic-speaking female consumers perceive Latin typography in brand identity, focusing specifically on product packaging for hand cream. By examining the effects of serif (Bodoni) versus sans-serif (Arial) typefaces on perceived price and product efficacy—including hydration and healing—this research aims to refine cross-cultural marketing strategies. The findings will contribute to a deeper understanding of how typography shapes brand perception, supporting brand sustainability in non-Latin markets.

2 Methodology

2.1 Participant Selection

The study involved 189 Arabic-speaking female consumers, selected through a combination of convenience and purposive sampling methods. This approach ensured that all participants were fluent in Arabic and had prior exposure to Latin-script branding. The sample was predominantly educated, with 82.0% holding a Bachelor's degree, and 58.2% of participants had a background in design, which could influence their perception of typography.

The study's focus on Arabic-speaking women provides a unique lens into how gender and cultural identity might mediate responses to branding in a globalized market, where Latin-script branding is increasingly prevalent. This demographic specificity, combined with the participants' educational and professional profiles, offers a rich foundation for exploring the interplay between cultural familiarity, design expertise, and consumer behavior in the context of typography [18–20].

2.2 Experimental Design

Participants were asked to evaluate two printed mockups of hand cream packages, which were identical except for the typeface used (see Fig. 1). Product 1 (P1) featured Bodoni (Serif), while Product 2 (P2) used Arial (Sans Serif). Both packages had a white background and minimal text, including the product name "Hand Cream" in large font, the phrase "For All Skin Types" in smaller font below, and the volume indication "50 ml" on the side. The primary independent variable in this study was the typeface (Serif vs. Sans Serif), allowing for an analysis of its influence on consumer perception.

Fig. 1. Bodoni (serif typeface) - product 1 (left) and Arial (Sans Serif) - product 2 (right).

2.3 Perception Measurements

To assess the impact of typography, participants physically interacted with the printed mockups and responded to survey questions. Their price perception was measured by asking, "Which product appears more expensive?", while efficacy perceptions related to hydration and healing were evaluated using the questions, "Which product is most effective for hydration?" and "Which product is most effective for healing?" These measures helped determine whether typography alone could shape consumer expectations regarding product quality and effectiveness.

3 Data Analysis

The collected data were analyzed using SPSS v.28, employing both descriptive and inferential statistical methods. Descriptive statistics, such as frequencies and percentages, were used to summarize participant demographics. The study sample consisted exclusively of female participants, predominantly aged 18–25 years (mean = 22), with

75% holding bachelor's degrees or higher. Regarding product perception, 65% viewed Product 1 as more expensive, while Product 2 was associated with superior hydration efficacy (50% vs. 35%). The dataset reflects a young, educated female cohort whose design literacy and age-specific preferences may systematically influence evaluations of branding and product attributes.

To test associations between typography and consumer perception, chi-square tests and Phi/Cramer's V were applied, with a p-value threshold of $< .05$ used to determine statistical significance.

The cross-tabulation analysis indicates a statistically significant relationship between the perceived price of a product and its effectiveness for hydration. The Pearson Chi-Square test ($\chi^2 = 21.579$, $p = .000$) and symmetric measures (Phi = 0.338, Cramer's V = 0.239) suggest a moderate association, implying that consumers are more likely to perceive P1 as more effective for hydration when it is seen as higher-priced.

Specifically, among the 146 cases where P1 was perceived as more expensive, 78 participants considered P1 the most effective for hydration, while 49 participants found P2 to be more effective. Conversely, only 17 out of 24 respondents who perceived P2 as more expensive also viewed it as the most effective for hydration (Table 1).

Table 1. Crosstabulation: "More expensive" X "More effective for hydrating".

		Most effective for hydrating			Total
		P1 most effective for hydrating	Equal	P2 most effective for hydrating	
More expensive	P1 more expensive	78	19	49	146
	Equal	3	6	10	19
	P2 more expensive	5	2	17	24
Total		86	27	76	189

Pearson Chi-Square 21.579, sig. = 0.000, Phi = 0.338, Cramer's V = 0.239.

These findings suggest that a significant portion of respondents (78 out of 189) associate a higher price for P1 with better hydration performance, whereas fewer respondents perceived P2 as both more expensive and more effective. However, the presence of small expected counts in certain categories (e.g., "Equal" and "P2 More Expensive") suggests that results should be interpreted with caution. Overall, the results emphasize the role of pricing strategies in influencing consumer perceptions of hydration efficacy.

The second cross-tabulation analysis reveals no statistically significant association between the perceived price of a product and its effectiveness for healing (Table 2). The Pearson Chi-Square test yields a p-value of 0.225, which is above the 0.05 threshold, indicating that any observed differences in consumer perceptions are likely due to random chance rather than a meaningful relationship. Similarly, the symmetric measures (Phi = 0.173, Cramer's V = 0.122) suggest a weak association, further supporting the lack of significance.

Among the 146 consumers who perceived P1 as more expensive, 68 found it to be the most effective for healing. However, the findings lack consistency to establish a definitive conclusion. Additionally, 56 out of 146 respondents who viewed P1 as more expensive believed that P2 was more effective for healing. This implies that factors other than price, such as brand reputation, ingredient quality, or personal experience, may play a more critical role in shaping perceptions of healing effectiveness. Marketers should consider emphasizing these alternative cues rather than relying solely on price to differentiate healing products (Table 2).

This methodology of analysis effectively revealed the role of typography in shaping perceptions among Arabic-speaking consumers. The research yielded significant findings regarding how typography affects consumer interpretation and perceived product value. The findings contribute to a broader understanding of design's impact on marketing and consumer behavior in regions where Latin typography is not the primary script.

Table 2. Crosstabulation: "More expensive" X "More effective for healing".

		Most effective for healing			Total
		P1 most effective for healing	Equal	P2 most effective for healing	
More expensive	P1 more expensive	68	22	56	146
	Equal	6	5	8	19
	P2 more expensive	6	6	12	24
Total		80	33	76	189

Pearson Chi-Square 5.670, sig. = 0.225, Phi = 0.173, Cramer's V = 0.122.

4 Discussion

The findings reveal a statistically significant association between the perceived price of a product and its perceived effectiveness for hydration. Specifically, participants who viewed Product 1 (P1) with the Serif typeface as more expensive also tended to perceive it as more effective for hydration. This aligns with existing literature suggesting that certain typefaces can evoke perceptions of luxury or high quality, thereby influencing consumer judgments about a product's value and efficacy. However, no significant association was found between perceived price and perceived effectiveness for healing, indicating that typography's influence may vary depending on the specific attribute being evaluated.

These results underscore the importance of typeface selection in packaging design, as it can subtly yet significantly shape consumer perceptions. Designers and marketers should consider how typographic choices align with the desired brand image and product attributes to effectively communicate value to the target audience.

4.1 Limitations

While this study provides valuable insights into the influence of typography on consumer perception, several limitations should be acknowledged. The sample was exclusively composed of Arabic-speaking female consumers, limiting the generalizability of the findings to other demographic groups. Male consumers or individuals from different cultural backgrounds may respond differently to typeface variations. Additionally, the study only examined two typefaces, Bodoni (Serif) and Arial (Sans Serif). While these are common categories, they do not capture the full range of typographic diversity. Future research could explore additional typefaces, such as script or decorative fonts, to assess their broader impact on consumer perception.

Another limitation lies in the study's focus on hand cream packaging, which may not be representative of all consumer goods. The influence of typography might vary across product categories, particularly those associated with luxury branding, medical efficacy, or technology. Furthermore, participants' backgrounds may have influenced their responses, as 58.2% had experience in design, making them potentially more attuned to typographic nuances than the general population. Additionally, this research was conducted as a cross-sectional study, capturing consumer perceptions at a single point in time. Since brand perception evolves with repeated exposure and market shifts, a longitudinal approach could provide deeper insights into the long-term impact of typography on consumer perception.

5 Conclusion

This study contributes to the understanding of how typography influences consumer perceptions among Arabic-speaking female consumers. The significant association between Serif typefaces and higher perceived price and hydration efficacy highlights the role of design elements in shaping consumer expectations. However, the lack of association with healing efficacy suggests that typography's impact may be context-dependent. Future research could explore these dynamics further by examining different product categories and diverse consumer demographics to develop more comprehensive insights into the role of typography in consumer behavior.

References

1. Childers, T.L., Jass, J.: All dressed up with something to say: Effects of typeface semantic associations on brand perceptions and consumer memory. J. Consum. Psychol. **12**(2), 93–106 (2002). https://doi.org/10.1207/S15327663JCP1202_03
2. McCarthy, M.S., Mothersbaugh, D.L.: Effects of typographic factors in advertising-based persuasion: A general model and initial empirical tests. Psychol. Mark. **19**(July–August), 663–691 (2002). https://doi.org/10.1002/mar.10032
3. Pan, Y., Schmitt, B.H.: Language and brand attitudes: Impact of script and sound matching in Chinese and English. J. Consum. Psychol. **5**(3), 263–277 (1996). https://doi.org/10.1207/s15327663jcp0503_03
4. Tantillo, J., DiLorenzo-Aiss, J., Mathisen, R.E.: Quantifying perceived differences in type styles: An exploratory study. Psychol. Mark. **12**(5), 447–457 (1995). https://doi.org/10.1002/mar.4220120507

5. Wang, Z., Shanat, M.B., Kanyan, L.R.: The effect of serif and sans serif typeface of luxury fashion logotype on Chinese consumers' brand perception. J. Bus. Adm. Res. **11**(2), 1–9 (2022)
6. Caldwell, J.: Cross-cultural emotion responses to Japanese typefaces. In: 2014 IEEE International Professional Communication Conference Proceedings, pp. 1–11 (2014). https://doi.org/10.1109/IPCC.2014.7020368
7. Peterson, M. E.: (2017). Rhetoric of Typography: Cross-Cultural Perceptions of Typefaces for Technical and Visual Communication. Master's thesis, Minnesota State University, Mankato. https://cornerstone.lib.mnsu.edu/etds/672/
8. McMullen, M.: The Eastern/Western cultural divide: A visual content analysis of printed graphic design. Int. J. Vis. Des. **14**(3), 1–14 (2020). https://doi.org/10.18848/2325-1581/CGP/v14i03/1-14
9. Richardson, J. T. E.: (2025). The Legibility of Serif and Sans Serif Typefaces. Springer, Cham. https://library.oapen.org/bitstream/id/165f35ac-584f-4a2e-ba84-6832faed7e69/978-3-030-90984-0.pdf
10. Akhmadeeva, L., Tukhvatullin, I., Veytsman, B.: Do serifs help in comprehension of printed text? An experiment with Cyrillic readers. Vis. Res. **65**, 21–24 (2012). https://doi.org/10.1016/j.visres.2012.05.013
11. Burt, C.: A psychological study of typography. Cambridge University Press, Cambridge (1959) https://www.worldcat.org/oclc/299574302 (OCLC permalink for library access)
12. Davis, R.C., Smith, H.J.: Determinants of feeling tone in type faces. J. Appl. Psychol. **17**(6), 742–764 (1933). https://doi.org/10.1037/h0074491
13. Ali, A.Z.M., Wahid, R., Samsudin, K., Idris, M.Z.: Reading on the computer screen: Does font type has [sic] effects on web text readability? Int. Educ. Stud. **6**(3), 26–35 (2013). https://doi.org/10.5539/ies.v6n3p26
14. Garcia, M.L., Caldera, C.I.: The effect of color and typeface on the readability of on-line text. Comput. Ind. Eng. **31**(1–2), 519–524 (1996). https://doi.org/10.1016/0360-8352(96)00189-1
15. Adams, M.J.: Models of word recognition. Cogn. Psychol. **11**(2), 133–176 (1979). https://doi.org/10.1016/0010-0285(79)90008-2
16. Adams, S., Rosemier, R., Sleeman, P.: Readable letter size and visibility for overhead projection transparencies. AV Commun. Rev. **13**(4), 412–417 (1965). https://doi.org/10.1007/BF02766846
17. Baron, N.S., Calixte, R.M., Havewala, M.: The persistence of print among university students: An exploratory study. Telematics Inform. **34**(5), 590–604 (2017). https://doi.org/10.1016/j.tele.2016.11.008
18. Zhang, Y., Khare, A.: The impact of cultural identity on consumer responses to global brands: A meta-analysis. Int. J. Res. Mark. **38**(2), 345–362 (2021). https://doi.org/10.1016/j.ijresmar.2020.12.004
19. Almousa, M., Alnawas, I.: The role of typography in shaping brand identity: A study of Arabic and Latin scripts. J. Brand Manag. **29**(4), 345–360 (2022). https://doi.org/10.1057/s41262-022-00287-5
20. Li, C., Li, D., Chiu, C.Y., Peng, S.: Strong brand from consumers' perspective: A cross-cultural study. J. Cross-Cult. Psychol. **50**(1), 116–129 (2019). https://doi.org/10.1177/0022022118799456

Sustainable AI Adoption: Legal Considerations for Cybersecurity in Saudi Arabia's Context

Homam El-Taj[1](✉) and Firas Najjar[2]

[1] Department of Cybersecurity, Dar Al-Hekma University, Jeddah, Saudi Arabia
htaj@dah.edu.sa

[2] Department of Information Security and Cybersecurity Department, Philadelphia University, Amman, Jordan
fnajjar@philadelphia.edu.jo

Abstract. The Kingdom of Saudi Arabia's ambitious Vision 2030 initiative aims to revolutionize its economy through technological innovation, with artificial intelligence (AI) playing a pivotal role. This transformation, while promising enhanced efficiency, sustainability, and economic diversification, also introduces significant cybersecurity risks. These include threats to data privacy, potential biases in AI algorithms, and vulnerabilities to cyberattacks. Addressing these challenges is critical for ensuring the secure and ethical deployment of AI technologies. This paper provides an in-depth analysis of Saudi Arabia's legal and regulatory frameworks governing AI adoption and cybersecurity. It explores the strategic roles of the Saudi Data and Artificial Intelligence Authority (SDAIA) and the National Cybersecurity Authority (NCA) in crafting policies that safeguard against cyber threats while fostering innovation. Additionally, the study examines data protection laws, encryption standards, incident response protocols, and liability frameworks that collectively support a resilient cyber environment. Through a comprehensive review, the paper highlights the importance of robust legal structures, cross-sector collaboration, and proactive cybersecurity measures in realizing the full potential of AI in alignment with Vision 2030's goals, thus offering insights into balancing technological advancement with security imperatives in a rapidly evolving digital landscape.

Keywords: Artificial intelligence (AI) · Sustainability · Cybersecurity · Legal considerations · Saudi Arabia · Vision 2030 · Regulatory framework · Cyber resilience · Data protection · Incident response

1 Introduction

In recent years, Saudi Arabia has embarked on a transformative journey outlined in its Vision 2030 initiative [1]. This ambitious roadmap seeks to diversify the country's economy and reduce its dependence on oil by fostering innovation and technological advancement across various sectors. One of the key areas of focus in this transformation is the integration of artificial intelligence (AI) into the fabric of Saudi society and economy.

D. Berkouk et al. (Eds.): CCTS 2024, PTT, pp. 443–451, 2026.
https://10.1007/978-981-95-7607-4_49

1.1 Background on Saudi Arabia's Vision 2030 and AI Integration

Vision 2030, inaugurated by Crown Prince Mohammed bin Salman in 2016, articulates an ambitious agenda targeting the comprehensive transformation of Saudi Arabia's economic and social spheres [1]. This strategic initiative underscores the pivotal role of advanced technologies, notably artificial intelligence (AI), in catalyzing innovation, enhancing productivity, and ameliorating the standard of living for the nation's populace. The Saudi government has demonstrated its commitment to this technological pivot by allocating significant investments towards AI research and development, thereby aspiring to establish the Kingdom as a preeminent hub for AI innovation on the global stage.

1.2 Significance of AI for Sustainability and Economic Growth

The integration of AI across various sectors of the Saudi economy is anticipated to usher in a new era of sustainability and robust economic growth. AI technologies are poised to revolutionize resource management, energy efficiency, and environmental stewardship across key industries, including but not limited to, energy, transportation, and agriculture [2]. By facilitating more informed decision-making and automating routine processes, AI is expected to deliver substantial productivity improvements and cost efficiencies, thereby contributing significantly to the overarching goals of economic advancement and development [18].

1.3 Emergence of Cybersecurity Challenges in AI Adoption

While the integration of artificial intelligence (AI) heralds considerable advantages, it concurrently ushers in a spectrum of cybersecurity challenges. AI systems, inherently susceptible to a myriad of cyber threats, encompass data breaches, malevolent attacks, and systemic vulnerabilities. The extensive integration of AI into vital infrastructural components and decision-making processes amplifies the potential repercussions of cyberattacks, elevating them to a matter of national security and economic stability concern [3, 4]. The profundity of such cyber threats accentuates the imperative need for robust cybersecurity measures to safeguard against the adverse impacts that could undermine the integrity of critical systems and destabilize socio-economic frameworks [19].

2 Legal Considerations for Cybersecurity in Saudi Arabia

2.1 Overview of the Regulatory Landscape

In the Kingdom of Saudi Arabia, a considerable endeavor has been made to formulate a rigorous regulatory structure that governs the application of artificial intelligence (AI) and addresses issues pertaining to cybersecurity [5]. Central to these efforts is the Saudi Data and Artificial Intelligence Authority (SDAIA) [21], which is instrumental in supervising AI-related policy formulations and regulatory enactments. In conjunction, the National Cybersecurity Authority (NCA) assumes the responsibility for the enactment of cybersecurity policies and the assurance of adherence to international standards

[6]. The collaborative efforts of these entities are directed towards establishing a holistic legal framework that not only facilitates the responsible utilization of AI but also ensures protective measures against cybersecurity threats (Fig. 1).

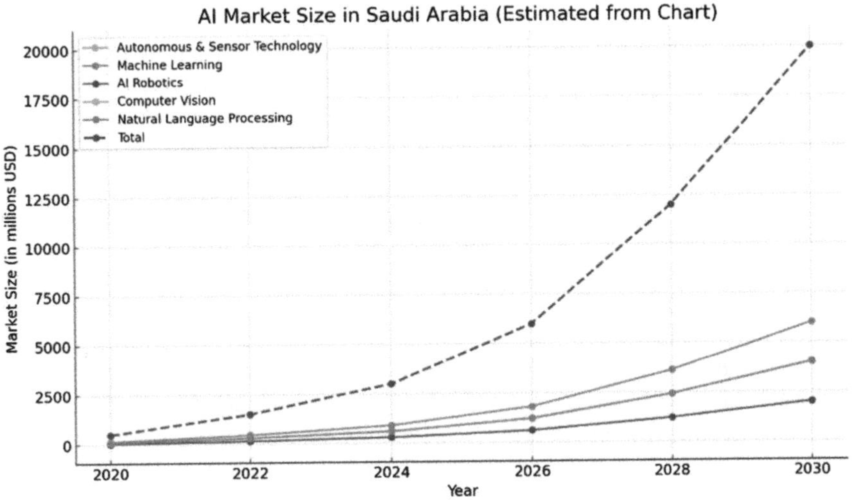

Fig. 1 Projected analysis of the artificial intelligence (AI) market size in Saudi Arabia from 2020 to 2030

The chart provided offers an analytical depiction of the anticipated growth within the artificial intelligence (AI) market in Saudi Arabia, charting its expansion from 2020 to 2030 across various technological domains pertinent to AI [15]. This analysis delineates a compound growth trajectory, with the market's valuation expected to surge from an initial estimation of approximately 500 million USD in 2020 to a forecasted 20 billion USD by the end of the projection period. The segmentation of the AI market into distinct domains—namely Autonomous & Sensor Technology, Machine Learning, AI Robotics, Computer Vision, and Natural Language Processing—reveals a consistent trend of growth across these sectors. Notably, Autonomous & Sensor Technology and Machine Learning are identified as the predominant contributors to the market size throughout the observed timeframe, underscoring their integral role in the advancement of the AI industry within the regional context. Despite their relatively smaller market size, the sectors of AI Robotics, Computer Vision, and Natural Language Processing also exhibit significant growth trajectories, reflecting the increasing applicability and integration of AI technologies across diverse fields [21]. The graphical representation employs distinct visual markers and lines to differentiate each AI segment, with a dashed line specifically demarcating the total market size, thereby facilitating a clear visual comparison between the growth of individual segments and the overall market expansion. This visual analytical tool underscores the robust growth prospects of Saudi Arabia's AI market, highlighting the nation's strategic initiatives and investments aimed at fostering the development of AI technology as a cornerstone of its broader economic diversification and modernization endeavors.

2.2 Analysis of Potential Risks in AI Adoption

The integration of AI technologies in the Saudi Arabian context raises several potential risks, particularly in the domain of cybersecurity. A significant concern is the susceptibility of AI systems to cyberattacks, wherein attackers could exploit vulnerabilities within AI algorithms or infrastructures to gain unauthorized access to sensitive information or to disrupt essential infrastructures [4]. Moreover, there exists the risk of AI systems inadvertently perpetuating biases or discrimination, thereby introducing ethical and legal complexities [2]. Mitigating these risks necessitates a proactive stance on cybersecurity and the development of stringent legal frameworks.

2.3 Importance of Robust Legal Frameworks for Cyber Resilience

The establishment of robust legal frameworks is imperative for ensuring cyber resilience amidst the continually evolving cyber threats. In pursuit of enhancing its legal infrastructure for cybersecurity, Saudi Arabia enacted the Cybersecurity Law in 2018 [7]. This legislation provides a comprehensive legal foundation for cybersecurity within the Kingdom, specifying the obligations of government agencies, the private sector, and individuals towards the maintenance of cybersecurity. Furthermore, the Kingdom has initiated measures such as the National Cybersecurity Strategy, aimed at augmenting cybersecurity capabilities and promoting a widespread culture of cybersecurity awareness [6]. These endeavors are indicative of the critical importance placed on legal frameworks in fortifying cyber resilience and safeguarding essential infrastructures against potential cyber threats.

3 Regulatory Framework for AI Adoption

3.1 Data Protection Regulations

In the contemporary landscape of technological advancement, Saudi Arabia has made significant strides in the domain of data protection, particularly concerning the application of artificial intelligence (AI). The enactment of the Personal Data Protection Law in 2019 exemplifies the Kingdom's commitment to safeguarding personal data within AI systems (Fig. 2). This legislation establishes comprehensive principles for the processing and security of personal data, emphasizing the necessity for obtaining explicit consent and ensuring robust data protection mechanisms [8, 16].

The presented line graph meticulously delineates three distinct scenarios for the total addressable market (TAM) of artificial intelligence (AI) in Saudi Arabia, covering the period from 2022 to 2030 [15]. These scenarios—Conservative, Moderate, and Optimistic—each propose a unique forecast for market growth, encapsulating varied assumptions about the socio-economic and technological milieu influencing AI development and adoption.

In the Conservative forecast, the market's progression is characterized by a linear growth pattern, escalating modestly from 100 million USD in 2022 to 390 million USD by 2030. This projection might signify a cautious approach to AI integration and investment, reflecting minimal risk-taking and innovation within the sector. The Moderate

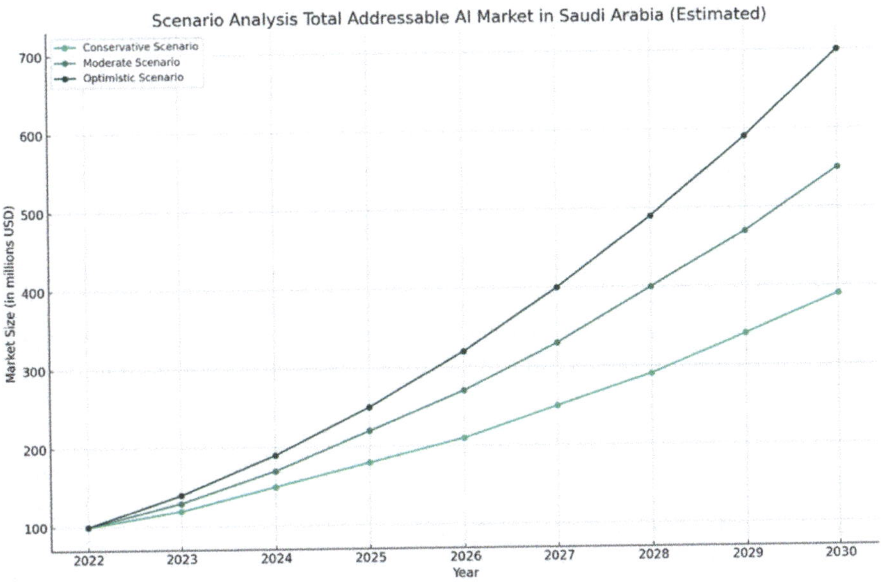

Fig. 2 Scenario analysis of the total addressable market (TAM) for artificial intelligence (AI) in Saudi Arabia.

scenario, with its projection extending from 100 million USD in 2022 to 550 million USD by 2030, suggests a more pronounced growth trajectory. It likely indicates a balanced confluence of factors—steady investment, consistent technological innovation, and a supportive economic and regulatory landscape.

The Optimistic forecast envisages the most substantial market expansion, with the market size burgeoning from 100 million USD in 2022 to an ambitious 700 million USD by 2030. This scenario presupposes an exceptionally favorable environment for AI, marked by significant technological breakthroughs, robust investment flows, conducive policy frameworks, and widespread market adoption. Graphical representations of each scenario utilize distinct markers and line styles, enabling an intuitive visual comparison of potential market sizes under varying conditions. This analytical tool is invaluable for stakeholders and policymakers, offering insight into the prospective magnitudes of AI market opportunities within Saudi Arabia, factoring in current exchange rates and the market impacts of the Russia-Ukraine conflict as of the latest update in August 2023.

3.2 Encryption Standards

Moreover, the adoption of international encryption standards underscores Saudi Arabia's dedication to the confidentiality and integrity of data in AI applications. By advocating for the implementation of advanced encryption methodologies, the Kingdom endeavors to shield sensitive information from unauthorized access, thereby enhancing data security within AI systems [9].

3.3 Incident Response Protocols

Furthermore, the development of incident response protocols marks a critical component of Saudi Arabia's cybersecurity strategy. These frameworks delineate systematic procedures for the detection, response, and recovery from cybersecurity incidents, facilitating organizations in mitigating the impacts of cyberattacks and safeguarding critical infrastructures [10].

3.4 Liability Frameworks

Lastly, the establishment of liability frameworks is pivotal in delineating accountability for damages or harm resulting from AI systems. These legal frameworks articulate the responsibilities of AI developers, users, and regulators, aiming to ensure that all parties are held accountable for the repercussions of AI utilization [20]. Collectively, these regulatory measures reflect Saudi Arabia's proactive approach in addressing the multifaceted challenges associated with AI and cybersecurity, laying the groundwork for a secure and responsible digital future [11].

4 Collaboration and Awareness

4.1 Government-Industry-Academia Collaboration

The imperative for collaboration across government, industry, and academic spheres is profoundly recognized in the context of enhancing cybersecurity amid the adoption of artificial intelligence (AI). The Kingdom of Saudi Arabia has been at the forefront of fostering such collaborative efforts, establishing partnerships that facilitate the exchange of knowledge, collaborative research, and technology transfer (Fig. 3). These endeavors aim to cultivate a dynamic ecosystem conducive to cybersecurity innovation, underpinned by the synergistic contributions of diverse stakeholders [12].

The line graph elucidates the anticipated contribution of the artificial intelligence (AI) market to the Gross Domestic Product (GDP) of Saudi Arabia from the years 2022 to 2030 [15], under three projected scenarios: Conservative, Moderate, and Optimistic. Each scenario posits a different contribution level to GDP, reflecting varying degrees of market performance and economic integration of AI technologies.

The Conservative scenario presents a steady, linear increase in AI's GDP contribution, incrementing from an estimated 10 billion USD in 2022 to 39 billion USD by 2030. This trajectory suggests cautious growth projections, potentially underpinned by a conservative uptake of AI applications and incremental advancements in technology. Under the Moderate scenario, the contribution exhibits a more assertive ascent, beginning at 10 billion USD and culminating at 55 billion USD by the end of the forecast period. This may imply a balanced approach to AI integration, reflecting a strong but realistic adoption curve and technological progression within the market.

The Optimistic scenario forecasts the most aggressive growth, with AI's GDP contribution projected to soar from 10 billion USD to 70 billion USD within the same timeframe. This optimistic outlook may be based on assumptions of rapid technology

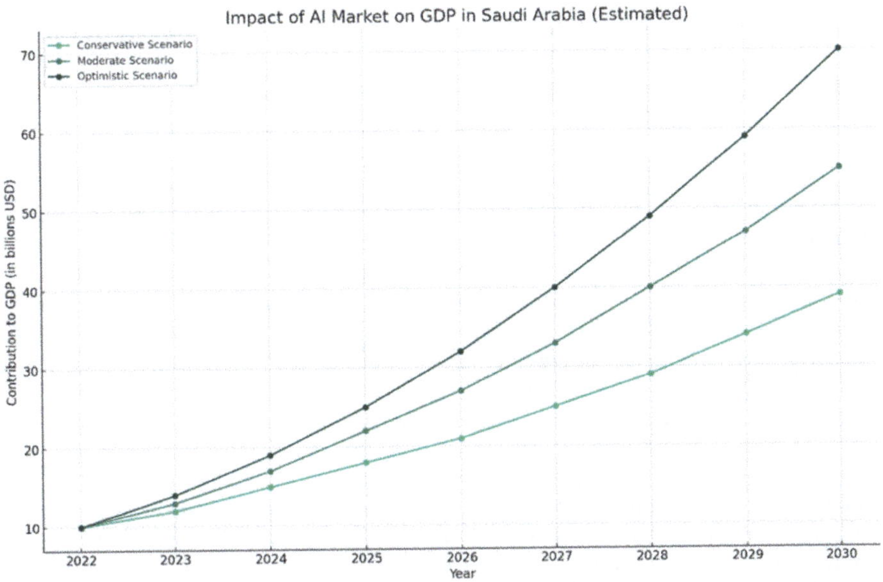

Fig. 3 Anticipated contribution of the artificial intelligence (AI) market to the Gross Domestic Product (GDP) of Saudi Arabia.

adoption, substantial investments in AI, and transformative market developments leading to significant economic impacts. The graph is marked by distinctive data points and trend lines for each scenario, facilitating a comparative visual analysis. It underscores the significant and increasing role that AI is expected to play in the Saudi Arabian economy, highlighting the strategic importance of AI as an economic driver in the nation's vision for the future. The note on the graph indicates that these projections take into account the current exchange rates and acknowledge the economic implications of the Russia-Ukraine conflict as of the most recent data update in August 2023.

4.2 Enhancing Cybersecurity Awareness and Research

Further augmenting this approach, Saudi Arabia has embarked on initiatives designed to elevate cybersecurity awareness among both individuals and organizations. The government's commitment to promoting cybersecurity education and training is evident in its efforts to cultivate a workforce proficient in navigating and mitigating cyber threats. Concurrently, research institutions within the Kingdom are actively engaged in studies aimed at advancing cybersecurity technologies and methodologies, thereby contributing to the broader knowledge base and practical capabilities in the field [13].

4.3 Promoting Innovation in Cybersecurity

In addition to these efforts, the establishment of innovation hubs and centers of excellence represents a strategic move by Saudi Arabia to stimulate innovation within the realm of cybersecurity. These platforms serve as incubators for startups and entrepreneurs dedicated to developing cybersecurity solutions [17], supported by government-provided funding and grants. This initiative underscores the Kingdom's dedication to fostering innovation and enhancing the cybersecurity landscape through a collaborative and supportive ecosystem [14].

References

1. Vision: The official Saudi vision 2030. www.vision2030.gov.sa/en (2030)
2. Alotaibi, N.S., Alshehri, A.H.: Prospers and obstacles in using artificial intelligence in Saudi Arabia higher education institutions—The potential of AI-based learning outcomes. Sustainability. **15**, 10723 (2023). https://doi.org/10.3390/su151310723
3. Alhalafi, N., Veeraraghavan, P.: Exploring the challenges and issues in adopting cybersecurity in Saudi smart cities: Conceptualization of the cybersecurity-based UTAUT model. Smart Cities. **6**, 1523–1544 (2023). https://doi.org/10.3390/smartcities6030072
4. El-Taj, H.: THE evolving threat landscape: An overview of attacks and response strategies against cybercrime in Saudi Arabia. Int. J. Netw. Secur. Appl. (IJNSA). **15**(3), 85–94 (2023). https://doi.org/10.5121/ijnsa.2023.1530685
5. Saudi Data and Artificial Intelligence Authority (SDAIA): About SDAIA. SDAIA. https://www.sdaia.gov.sa/en/about-sdaia (2024). Accessed 16 Feb 2024
6. National Cybersecurity Authority (NCA): About NCA. NCA. https://www.nca.gov.sa/en/about-nca (2024). Accessed 16 Feb 2024
7. WPS: Cybersecurity in the Kingdom, Cybersecurity Legislation and Regulations. https://www.my.gov.sa/wps/portal/snp/main (2024). Accessed 16 Feb 2024
8. Personal Data Protection Law of Saudi Arabia, Royal Decree No. M/22, 17/08/1440 H. (2019).
9. Saudi Standards, Metrology and Quality Organization (SASO): Encryption standards. SASO. https://www.saso.gov.sa/en (2024). Accessed 16 Feb 2024
10. National Cybersecurity Authority (NCA): Incident response framework. NCA. https://www.nca.gov.sa/en (2024). Accessed 16 Feb 2024
11. Ministry of Justice, Saudi Arabia: Liability frameworks for AI. Ministry of Justice. https://www.moj.gov.sa/en (2024). Accessed 16 Feb 2024
12. Saudi Data and Artificial Intelligence Authority (SDAIA): Collaboration initiatives. SDAIA. https://www.sdaia.gov.sa/en (2024). Accessed 16 Feb 2024
13. Ministry of Education, Saudi Arabia: Cybersecurity awareness programs. Ministry of Education. https://www.moe.gov.sa/en (2024). Accessed 16 Feb 2024
14. Saudi Federation for Cybersecurity, Programming, and Drones (SAFCSP): Innovation hubs. SAFCSP. https://www.safecsp.com/en (2024). Accessed 16 Feb 2024
15. Statista: Artificial intelligence-Saudi Arabia. https://www.statista.com/outlook/tmo/artificial-intelligence/saudi-arabia#market-size. (2023)
16. Weber, P.A., Zhang, N., Haiming, W.: A comparative analysis of personal data protection regulations between the EU and China. Electron. Commer. Res. **20**, 565–587 (2020) https://link.springer.com/article/10.1007/s10660-020-09422-3
17. Marican, M.N., Yusuff, S.A., Razak, A.S., Othman, S.H.: Cyber security maturity assessment framework for technology startups: A systematic literature review. IEEE Access. **11**, 5442–5452 (2022) https://ieeexplore.ieee.org/abstract/document/9989381
18. AlDaajeh, S., Saleous, H., Alrabaee, S., Barka, E., Breitinger, F., Choo, K.-K.R.: The role of national cybersecurity strategies on the improvement of cybersecurity education. Comput. Secur. **119**, 102754 (2022). https://doi.org/10.1016/j.cose.2022.102754

19. Quadri, A., Khan M. K.: Cybersecurity challenges of the kingdom of Saudi Arabia. https://www.researchgate.net/profile/Aman-Quadri/publication/331009167_CYBERSECURITY_CHALLENGES_OF_THE_KINGDOM_OF_SAUDI_ARABIA/links/5c6103f2299bf1d14cbb4e6c/CYBERSECURITY-CHALLENGES-OF-THE-KINGDOM-OF-SAUDI-ARABIA.pdf. (2019)
20. Hacker, P.: The European AI liability directives–critique of a half-hearted approach and lessons for the future. Comput. Law Secur. Rev. **51**, 105871 (2023). https://doi.org/10.1016/j.clsr.2023.105871
21. Saudi Data and Artificial Intelligence Authority: About SDAIA. https://sdaia.gov.sa/en/SDAIA/about/Files/GenAIGuidelinesForGovernmentENCompressed.pdf (2024). Accessed 8 Mar 2024

AI and Sustaining Legal Education for Creative Originality in the Gulf States: Saudi Arabia, Qatar, and United Arab Emirates

Samuel Samiái Andrews(✉)

Albany State University College of Professional Studies, Albany, Georgia 31705, USA
samuel.andrews@asurams.edu

Abstract. The digital era created disruptive technologies of which artificial intelligence (AI) constitutes part of that ubiquitous, and transformative system. The emergence of AI systems across various socioeconomic sectors has upended traditional orthodoxies and regimes concerning creativity, learning, personhood, and intellectual property. The Gulf states have invested humongous financial and human resources to accelerate the development of AI systems in their various countries. This paper explores how the Gulf states, particularly, the Kingdom of Saudi Arabia, Qatar, and the United Arab Emirates, could reconceptualize the colleges of law curriculum and pedagogical methods as an agency to achieving a sustainable development in legal education in the AI era. The AI systems with its generative capacities creates divergences within intellectual property jurisprudence and policies. An academic or law student who uses, for example, *ChatGPT* or *Grammarly* to write their academic work may infringe the plagiarism ethics, which is a long-standing feature of academic ordering and copyright laws. The traditional academic integrity ethical norms now fundamentally intersect with creative positive laws. The paper concludes with recommendations laying out a framework for a sustained policy and jurisprudential digital legal literacy for the region.

Keywords: Artificial Intelligence · legal education · Gulf States · Intellectual property · Legal writing · Curriculum development

1 Introduction

The digital era emphasizes the recognition of the interface of technology and every aspect of societal socioeconomic productions [1, 2]. At the beginning of this millennium legal regimes have adapted their objectives to achieving traditional effectiveness with resultant creations and innovations. Artificial intelligence (AI) has been one of such creations of the digital era [3]. Most sectors of the global economies particularly health, education, manufacturing, national security, and justice. and national multilateralism have interfaced with digital realism. Almost all countries of the Gulf States have the objective of attaining goals four and nine of the United Nations Sustainable Development

D. Berkouk et al. (Eds.): CCTS 2024, PTT, pp. 452–458, 2026.
https://10.1007/978-981-95-7607-4_50

Goals program (SDG), as part of their national policy initiative [4].[1] The Gulf states aspire to harness the evolving digital intellectual property jurisprudence, which enhances creativity of the fourth industrial revolution that is upon humanity. Professionally trained digitally efficient legal practitioners as subject matter experts form a significant chain of deliverables for achieving the sustained AI legal literacy.

The Gulf states' legal regimes, policies and cultural norms consciously set out proper, effective, and efficient balance in their socioeconomic spaces. Along this scope comes IP regimes as a controlling regime to impel creativity and also discourage the dilution of legal professionalism. Academic authorship and creative originality are areas where AI, especially GenAI output programs, have consequential impacts for the standards of learning in legal academia [5]. Introducing students to proper, legal, and ethical collaboration with AI systems in learning may be a starting point of sustaining developmental successes of society. The consequences of the disruptive nature of the digital era have given birth to new forms of laws to regulate and guide human activities. Creators and innovators are now exploring through lawsuits around the world the scope of these forms of legal protection available in IP laws [6–8]. The safe haven laws like fair use and fair dealing grants academics and students statutorily stipulated exemptions from legal liability.

2 Methodology and Literature Review

Current academic studies and literature on the impact of generative artificial intelligence (GenAI) input and output on the effectiveness, efficiency and professionalism of lawyers produced from law schools in the AI era, are sparse [1, 2]. In the Gulf states after extensive electronic searches this author could not find such work that focuses on the region. This article evolves into Three thematic sections. It has Section One, which includes examining background and analyzing existing literature of AI, and its global jurisprudence impacts. Section Two analyses the Gulf states' copyright laws as it relates to legal academic writings in the colleges of law, and how these laws impel sustainable techno logical development. Section Three sets a conceptualized legal and policy framework to achieve the goals of harnessing AI era's agency for efficiency, which the colleges of law forms part of. It concludes by proffering recommendations for the way forward.

The Gulf States' Sustainable Creativity for Development: The Gulf states have adopted the United nation's Sustainable Development Goals (SDG), particularly goals four and nine [9]. These regimes and policies properly couched to synchronize with the socioeconomic and political aspirations of the Gulf states is a panacea for success. The traditional and original objectives of intellectual property and their derivative rights were principally human centric and on a larger scale for societal good [10]. However, the current aspirational expansion of these objectives to include non-humans like AI, with its lack of transparency of creative systems goes contrary to societal innovative sustainability [11, 12]. The copyright laws of most Gulf States (Saudi Arabia, Qatar and

[1] Goals four: quality education, goal nine: promote inclusive and sustainable industrialization and foster innovation- build resilient infrastructure of the United Nations Sustainable Development Goals program.

UAE) conceptualize the originality of literary works from a strictly jurisprudential perspective. However, each Gulf state has slightly divergent copyrightability postulations in their laws. They have modicum-of-creativity standard, sweat of the brow, creative selection, and author's intellectual creation, which demands substantial human intervention in a creative process. Conceptually, the copyright laws of the Gulf states converge on fundamental themes because they frame their laws as members of the Gulf Cooperation Council. However, on their approaches to technology related creations like AI, Saudi Arabian law diverges distinctively from others because of its clear recognition of scientific creations.

These Gulf states have different approaches to copyright exemptions on the subject of scholarly research and literary works. Saudi Arabia's law exempt academic research but not commercial academic research while UAE and Qatar laws expressly does not exempt commercial academic research. However, the laws of these Gulf States have not clarified the proportionality of use of copyrighted work in academic research by AI generative systems which are generally commercial oriented [13–16]. Gulf states' courts have not yet evaluated the substantiality standards of human intervention in determining copyrightability of AI works. AI's impact on copyrightability standards is ongoing in the United States, China, and Europe [17]. The Gulf states could study which jurisprudence suits their legal system, socioeconomic and political goals for adoption. Each of the Gulf States referenced in this paper adopts a nuanced and particularly in conceptualizing authorship, ownership and delineating copyright exemptions- Saudi Arabia have a different approach from Qatar and UAE as it relates to scientific and technological creations, the scope of AI. The jurisprudential approaches by these States to the digital era realities is captured differently in their various laws. For example, answering whether IP laws of the Gulf States recognize GenAI outputs as literary works, including academic writing, research works, and educational teaching aids as copyrightable works? This paper acknowledges that copyright laws of the Gulf States support the position that these types of works attract least protection of copyright law. Copyright law in the Gulf States mostly protects creations made by humans. Therefore, academic writings done with a GenAI or machine learning tool may lose the protection and benefits accruable from copyright law.

Colleges of law have obligations to filter the issues of academic integrity further through the lenses of codes of professional conduct that regulate lawyers and legal professionals applicable to the Gulf states [18–21]. These rules of professional conduct demand the duties of candor, honesty, competence from lawyers and make them mandatory in every aspect. Therefore, legal scholars or law students using AI systems to create or write academic work must adhere to these rules. The duty of disclosure and attribution requires satisfying these questions- Who did the work? What was the amount of AI contribution? What was the amount of human contribution to creating the work?' Within the IP subjects allied regimes like neighboring rights laws, and moral rights laws, exist other forms of IP that interfaces with academic normative systems that promote integrity [22]. Moral rights regimes demand for attribution when including an author's work in a subsequent literary and non-derogation of an author's creative fidelities. Allied to this argument of derogatory imputation of an author's creation is the prohibition of disparagement of an author's work and in essence an author's privacy right of upholding her personhood, which aligns with sharia jurisprudence of the Gulf States [23].

3 An Experiential and Observatory Approach [24, 25][2]

My thesis is mostly derivable from experiential encounters and recollections garnered throughout a period of more than a decade in teaching law courses in colleges of law across four continents of the world, and targeted surveys conducted, including those in more than six colleges of law in the Gulf States, and Africa (Grammarly programs are AI enabled, therefore this faculty is incorrect. For strict privacy and confidentiality reasons identity of the school and faculty is hereby excluded.). For example, an instructor in a college of law gave this author this response- "…grammarly is not AI. I checked myself using the same program you used. My work was also flagged by the same program as being assisted by AI… you have something personal against the students…"-Such understanding from a law teacher indicates the need for expertise among law faculty on the intersection of AI and legal writing. Law students tend to depend solely on the instructor's PowerPoints, teaching notes and AI systems to prepare for examinations. Recently a survey concluded that in the United Kingdom, 'More than half of UK undergraduates say they use AI to help with essays.' [26]. Individualized proficiency in case briefing, extensive independent reading from recommended textbooks and preparing individual outlines to ease legal learning processes are lacking. The lack of self-engagement and analytical mindset about class tutorials is anathema to competency in legal practice. A Legal practitioner's capacity to acquire the necessary skill sets as an analytical person in the real world outside the theoretical cushions of the classrooms is intrinsically commensurate with the reality of contemporary digital legal academia. This is why most universities use AI checks programs to optimize their academic integrity. Colleges of law could conduct individualized and specialized studies to target how AI systems could produce a uniform manual for delivery of lectures to law students that supports academic excellence.

4 AI and Fundamentals of Legal Writing

In academia, the doctrine of attribution is set within the anti-plagiarism norms. Conceptualizing a doctrine of originality in academic research work may differ from how IP law regulates it [27]. Originality in academic research and writing is perceived and judged on the scale of plagiarism. Creative originality in academic work means as long as an author shows the source of her knowledge, and that work is not a mere reproduction of similar existing subject matter, then it could qualify. The study, learning, pedagogical or instructional methods of law subjects are distinctively different from those of other fields of academia. Legal writing is one of the core law courses that is fundamental to training

[2] Author has taught law subjects in law schools in Ethiopia, Saudi Arabia, Nigeria, United States and European. He conducted a semi structured survey and interviews with students and law faculties in these countries on the digital era curriculum part of the outcomes and conclusions has been published in [24].

a knowledgeable and excellent lawyer. The language of instruction and legal writing also impacts and imposes added layers of complexities on how learning. Instructors may teach law in Arabic, English, Amharic, or any other language. However, when English is the medium of instruction, there are inherent pedagogical distinctiveness for law subjects. Legal writing is a specialized professional art form and fundamentally driven by terms of in the field. It captures pure sciences, literature-in-English, social science and socio-legal academics in its pedagogy. Therefore, subject matter experts who are legal scholars and experienced authors should teach law courses in the Gulf states colleges of law particularly now that legal writing intersects with GenAI. Exceling in this art form would require human sentient and application of critical thinking.

Legal Academia and Learning in Second Language: Most received laws in this digital era come in English language or French or Latin. In the common law systems, English Language is mostly the language of the Statute. Without prejudice to the epistemological traits of the sources of these laws, which may be different from English language, the predominant language of these laws are different from those of the societies that received them into their legal system. Writing modern legal works in the language of the received laws comes with issues of legal authenticity and exactitude. GenAI programs have accelerated with ease, and at times inappropriately assisting students who use English as a secondary language (ESL) [28].

5 A Framework for Digital Legal Literacy

Using an AI system to discover novel technological devices or create literary work has become a controversial public interest issue globally. The scope of infringing acts for students or academics who use the AI systems are still muddy. Copyright law requires a modicum of creativity to avail a creator legal ownership and authorship. A plagiarized creation implicates copyright jurisprudential scope. Law students or academics who depends on GenAI to produce literary works may infringe ethics and IP laws. Law students and legal academics engage either deliberately or unintendedly in the “copy and paste” style because of GenAI’s application of minimal prompts. Without stalling this conduct, academia’s fundamental dependency on continuity in knowledge availability and accessibility risk jeopardy. Published legal scholarship that passes legal and ethical eligibility tests in affirmed peer reviewed scientific publications is one of the established sources of passing on generational knowledge. Legal writing created with GenAI would not get an audience in any reputable medium.

6 Conclusion

Artificial intelligence systems, being interruptive technology, could become an agency for legal technological excellence and developmental achievements in the Gulf states. However, a collaboration to achieve set goals may require active participation of the lawyer’s Associations in the Gulf States. Enforcing professional conducts and ethics particularly around the AI work products relates to candor, honesty, and competency should be maximized. The Gulf states colleges of law should emphasize teaching AI and Law experientially. Colleges of law should encourage only subject matter experts to teach specialized law courses, for example, legal writing and research. Legal academics with measurable experience, those who have authored and published peer reviewed academic papers in impactful publications are effectively suited for these tasks.

References

1. Perlman, A.: Generative AI and the future of legal scholarship. https://papers.ssrn.com/sol3/papers.cfm?abstract_id=5072765 (2025). Accessed 25 Jan 2025.
2. Noto, G., Koolen, C.: Generative AI. education, and copyright law: An empirical study of policymaking in UK universities. Eur. Intellect. Prop. Rev. **46**(1), 346 (2024)
3. Adams, R.: The evolution of intellectual property rights in the digital age. J. Mod. Law Policy. **3**(2), 52–63 (2023). https://doi.org/10.47941/jmlp.1554
4. D' Lazzaro, F.: Sustainable development goals and copyright: A proposal for reform in the field of education. Eur. Intellect. Prop. Rev. **47**(2), 63–64 (2025)
5. United States Copyright Office: Copyright and artificial intelligence-Part 2: Copyrightability https://www.copyright.gov/ai/Copyright-and-Artificial-Intelligence-Part-2-Copyrightability-Report.pdf (2025). Accessed 12 Feb 2025
6. Saudi Arabia Copyright Law: Royal Decree No. M/41 as amended by Decision of the Council of Ministers No. 536, 2018. Article 15
7. United Arab Emirates: Concerning Copyrights and Neighboring Rights Federal Law, No. (7), 2022. Section 22. (1)
8. Qatar Law: The Protection of Copyright and Neighboring Rights No. 7, 2002. Section 18 (2) & (3)
9. Webb, C.: National visions aligning Gulf states with UN SDG targets: GCC chief. Arab News. https://arab.news/6jvr5 (2025). Accessed 11 Jan 2025
10. Breyer, S.: The uneasy case for copyright: A study of copyright in the books, photocopies, and computer programs. Harv. Law Rev. **84**, 281–235 (1970)
11. Tschider, C.: Don't fear AI. Expose it. The Hill https://thehill.com/opinion/congress-blog/4100557-dont-fear-the-ai-expose-it/ (2025). Accessed 10 Feb 2025
12. Solove, D.J.: Artificial intelligence and privacy. Fla. Law Rev. **77**, 1–8 (2025)
13. Infopaq International A/S v Danske Dagblades Forening (2009) ECR 16569 (Infopaq).
14. Bezpečnostní softwarová asociace-Svaz softwarové ochrany v Ministerstvo Kultury (2010) ECR I-13971 (BSA)
15. Feist Publications Inc v Rural Telephone Service Co (feist) (1991) 499 US 340
16. Gervais, D.: Feist goes global: A comparative analysis of the notion of originality in copyright law. J. Copyr. Soc. USA. **49**, 949–951 (2009)
17. The United States Copyright Office: Copyright and artificial intelligence Part 2: Copyrightability. https://www.copyright.gov/ai/Copyright-and-Artificial-Intelligence-Part-2-Copyrightability-Report.pdf (2025). Accessed 11 Feb 2025
18. Qatar Lawyers Association: Code of ethic of the legal profession. https://qatarlawyersassociation.qa/code_of_conduct.html (2025). Accessed 25 Jan 2025
19. Al Meeza: The State of Qatar Ministry of Justice, "The Code of Law Practice" article 50 Al Meezan - Qatary Legal Portal|Legislations|Law (23) for year 2006 regarding Enacting the code of law practice. Accessed 26 Jan 2026
20. United Arab Emirates: Article 16 'Regulating the Legal Profession and Legal Consultation' Federal Decree Law No. (34), . https://uaelegislation.gov.ae/en/legislations/1566/download (2022). Accessed 26 Jan 2025
21. Saudi Arabia: Rules 3 & 5 'Rules of Professional Conduct for Lawyers'. https://www.moj.gov.sa/Documents/Regulations/pdf/En/Rules%20of%20Professional%20Conduct%20for%20Lawyers.pdf (2025). Accessed 26 Jan 2025

22. Ginsburg, J.: The most moral of rights: The right to be recognized as the author of one's work. Geo Mason J Intl Comm L. **8**, 44–86 (2016)
23. Suwondo, D.: The legal Protection of personal data in Islamic perspective. Int. J. law Reconstr. **7**, 221 (2023)
24. Andrews, S.: Developing a Copyright Curriculum for Nigerian Universities for the Creative Digital Space. GRUR J. Eur. Int. IP Law. **71**(4), 322–334 (2022)
25. Andrews, S.S.: Copyright originality in the digital space: The kingdom of Saudi Arabia's creatives. In: Gupta, I. (ed.) Handbook on Originality in Copyright, pp. 1–24. Springer, Singapore (2023)
26. Richard Adams: More than half of UK undergraduates say they use AI to help with essays. The Guardian. https://www.theguardian.com/technology/2024/feb/01/more-than-half-uk-undergraduates-ai-essays-artificial-intelligence (2025). Accessed 29 Feb 2025
27. Frye, B.: Should using an AI text generator to produce academic writing be plagiarism? Fordham IP Media Ent, Law J. **30**, 947 (2023)
28. Sabrina Clay and Danny Freeman: She graduated from high school with honors but can't read or write. Now she's suing. https://edition.cnn.com/2025/02/27/us/connecticut-aleysha-ortiz-illiterate-lawsuit-cec/index.html (2025). Accessed 28 Feb 2025

Towards Automation in OT Risk Assessment: A Review of Methods, Machine Learning Techniques, Challenges, and Gaps

Amal A. Alahmadi, Aryam A. Alnemari(✉), Deema A. Alfaleh(✉), Fatema A. Alqahtani(✉), Nouf M. Alabdallah(✉), Ghada M. Alrashedi(✉), and Abdulrahman A. Alharby

Imam Abdulrahman bin Faisal University, Dammam, Saudi Arabia
aryamalnamari@gmail.com

Abstract. The raised reliance on the OT infrastructure in vital sectors requires robust risk assessment methodologies to protect against evolving threats. Most of the current approaches and techniques followed for the risk assessment of OT infrastructure in critical sectors are largely static and usually performed with human intervention. This paper aims to fill this gap, proposing a Systematic Literature Review. A total of 19 studies met the inclusion criteria. Our analysis revealed that there is a lack in standardized risk assessment frameworks that constitute a holistic approach to risk monitoring and assessment across systems and devices via AI/ML algorithms for predictive analytics and decision enhancement. So Intelligent Automation will drive better processes of risk detection, proactive response towards vulnerabilities, and real-time insights. Finally, future research should place major emphasis on integrated assessment methodologies that have the potential for automation and through which manual efforts and operating costs can be reduced, thereby improving the overall performance of the risk assessment.

Keywords: Industrial control systems · Operational technology · Risk assessment · Machine learning

1 Introduction

OT environments were once secure due to air-gapped architectures, however with increasing connectivity now led to growing cyber risks. The number of interconnected devices in manufacturing alone is expected to exceed 30 billion by 2025 [1]. OT infrastructures are critical for the operation and safety of critical infrastructure in many industries. The need for high availability and reliability in OT infrastructure introduces a challenge in employing effective security measures. Cyberattacks, including malware, ransomware, and supply chain attacks, pose serious threats to infrastructure and human safety given these risks, automating OT security and risk assessment using machine learning is becoming a priority.

D. Berkouk et al. (Eds.): CCTS 2024, PTT, pp. 459–467, 2026.
https://10.1007/978-981-95-7607-4_51

2 Background and Definitions

2.1 Overview of OT Environments

Operational Technology (OT) systems enable and manage the automation of physical processes in several sectors, including manufacturing, energy, and transportation. The backbone of OT systems consists of Industrial Control Systems (ICS) that make use of Programmable Logic Controllers (PLC) in combination with Supervisory Control and Data Acquisition systems (SCADA). Furthermore, Human-Machine Interfaces (HMI) are used by engineers to interact with and manage the functions of PLCs.

2.2 Vulnerabilities Targeting OT Environments

OT systems are increasingly targeted by cybercriminals due to weak security configurations and outdated software. Groups like "CyberAv3ngers" and pro-Russian hackers exploit these vulnerabilities. Increased security measures are essential because high-profile attacks and hacktivists are increasingly targeting OT systems. Furthermore, OT systems are exposed to high potential risks due to the supply chain attacks. Strengthening security and risk management is crucial to counter these evolving threats. A summary of threats targeting OT environments is illustrated in Table 1.

Table 1 Threats Targeting OT Environments.

Threat category	Description
Malware	Malicious software targeting OT systems, including worms, trojans, and viruses.
Ransomware	Encrypts OT system data, demanding ransom for decryption.
Supply chain attacks	Compromises OT through third-party software, hardware, or vendors.
Insufficient Authentication	Weak or default credentials allowing unauthorized access.
Unpatched Vulnerabilities	Exploiting known OT software and hardware vulnerabilities.

2.3 Risk Assessment of OT Environments

In today's business environment, OT risk assessment is significant. An OT manual risk assessment is an examination of vulnerabilities in OT systems based on frameworks such as IEC 62443 or NIST SP 800-82. The importance of OT manual risk assessment is followed by hurdles. These include a lack of standardized methods for OT environments and continuous risk assessment costs. Recent research has explored how manual risk assessment functions in OT systems. Methods include a framework that provides insights into vulnerabilities by using online security catalogs. Additionally, conducting systematic mapping analysis related to ontologies relevant to OT systems risk management.

2.4 Machine Learning

Machine Learning (ML) is a branch of Artificial Intelligence where systems automatically learn from data and possess self-enhancement in performance over time without any direct programming. It's used in algorithms that assist in pattern recognition and decision-making optimization. It's a central technology for Industry 4.0 and acts to enhance automation, efficiency, and intelligent data-driven solutions across varied industries.

3 Research Methodology

The methodology used in this paper is a systematic literature review (SLR). The paper selection process followed Preferred Reporting Items for Systematic Reviews and Meta-Analyses (PRISMA) 2020. The identification phase, screening phase, and inclusion phase of the studies are demonstrated in Fig. 1.

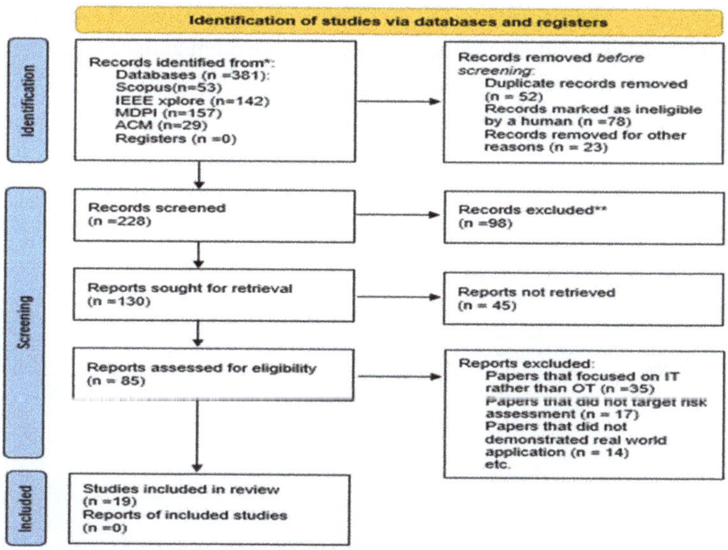

Fig. 1 PRISMA flow Diagram.

The key research questions (RQs) for this SLR included the following:

RQ1: Which parts of the risk assessment process of OT environments are mostly automated?

RQ2: What are the most followed frameworks for automating OT risk assessment?

RQ3: What are the most utilized machine learning algorithms in risk assessment?

RQ4: What are the primary challenges and gaps in current risk assessment methodologies of OT environments?

For this review, four databases were searched thoroughly, including Scoups, IEEE xplore, MDPI, and ACM. To find relevant studies and ensure consistency of the search results, the following search query was unified and used across all databases:

("OT" OR "operational technology") AND (" ML " OR " Machine Learning") AND ("Risk assessment") AND ("ICS" OR "Industrial Control Systems")

The extensive search generated 381 results; they were narrowed down to 19 based on the following inclusion criteria:

Relevant papers: only include papers that automate the process of risk assessment of OT environments. Open access paper. Recently published papers (from 2019–2024). And Contribution Papers: only studies that have a new methodology or approach to contribute to the field.

4 Results and Discussion

RQ1: Which parts of the risk assessment process of OT environments are mostly automated?

The increasing cyber threats to OT environments have driven the need to integrate automation into risk assessment processes. Even though manual risk assessment has been seen as ineffective for such an environment, different methodologies have integrated automation across different stages of risk assessment.

Table 2 An analysis of the automated parts in the reviewed studies.

Ref	Automated data analysis	Automated data gathering	Automated risk scoring	Report generation	Evaluation metrics
[2]	✓	✓	✓	✓	–
[3]	–	✓	–	–	–
[4]	✓	✓	✓	×	✓
[5]	✓	–	✓	✓	✓
[6]	✓	✓	✓	–	✓
[7]	✓	–	✓	✓	✓
[8]	✓	–	✓	✓	–
[9]	–	–	✓	–	✓
[10]	×	×	✓	×	✓
[11]	✓	✓	–	–	–
[12]	–	–	–	✓	–

Automated data analysis and risk scoring are key components of risk assessment, utilizing machine learning, structured frameworks, and mathematical models to improve accuracy and predictive capabilities. Several studies have incorporate both automated data analysis and risk scoring, such as Zahran et al. [2], who employ OCTAVE Allegro and used ISO/IEC 27030 to identify vulnerabilities, Kure et al. [4], who introduce i-CSRM, leveraging Cyber Threat Intelligence (CTI), Common Weakness Enumeration (CWE), fuzzy set theory, Naïve Bayes, and Neural Networks to automate both threat analysis and risk prediction. Phillips et al. [5] also contributes to data analysis and risk scoring automation by incorporating penetration testing tools. Similarly, Guo et al. [7] apply interval-valued intuitionistic fuzzy numbers (IVIFNs) to automate both attack path analysis and risk quantification. While Qin et al. [6] propose an association network-based methodology that automates historical attack data analysis and security incident probability calculation. Their approach is validated through a coupling tank system case study, confirming its efficiency and precision. Guarino et al. [8] further advance automation by integrating Bayesian Networks (BN) and Multi-Criteria Decision Making (MCDM) for both risk analysis and dynamic risk scoring. Other studies focus more on automated data analysis, and Alanen et al. [11], who introduce a RAMSS ontology to structure cybersecurity risk evaluation for improved traceability and decision-making. On the other hand, some methodologies prioritize automated risk scoring without a strong emphasis on data analysis, such as Fujimoto et al. [10] who dynamically adjust risk scores through real-time network traffic monitoring, Their evaluation shows a 67% risk level reduction after applying mitigation steps, confirmed through penetration testing. Tantawy et al. [9], who integrate hybrid automaton models within Cyber-Physical Systems (CPS) to assess risk in industrial environments, also validate their approach through a CSTR testbed, though they focus more on modeling relationships than providing numerical metrics.

Data gathering is another important component, with some methodologies automating this step in their risk assessment automation process. Zahran et al. [2] and Qin et al. [6] incorporate some level of automated data collection, while Georgiadou et al. [3] and Alanen et al. [11] rely more on structured frameworks rather than real-time extraction. Kure et al. [4] integrates Cyber Threat Intelligence (CTI) for enhanced data collection but still requires manual input.

Report generation remains one of the least automated aspects, though some methodologies incorporate partial automation. Phillips et al. [5], Guarino et al. [8], Wang et al. [12], Zahran et al. [2], and Guo et al. [7] integrate reporting functionalities to varying degrees, but manual synthesis is still required.

Table 2 shows that automated risk scoring and data analysis are the most automated components of the risk assessment process. Automated data gathering is less common, and report generation is the least automated. Table 3 provides automation methods used in reviewed papers.

Table 3 Automation methods for the reviewed papers.

Ref	Automation method
[2]	Uses ISO/IEC 27030 and OCTAVE Allegro with an agentless system to automate risk identification through structured data collection and prioritization.
[3]	Combines MITRE ATT&CK with a cybersecurity culture model to assess risks from both technical and human factors.

(*continued*)

Table 3 (*continued*)

Ref	Automation method
[4]	Employs CTI, CAPEC, CWE, and machine learning (Neural Networks) to automate risk prediction and assessment in critical infrastructure.
[5]	Uses the Spyderisk tool to simulate attack paths and cascading effects based on ISO 27005 and a knowledge-based threat model.
[6]	Applies a three-layer association network using historical data to automatically infer incident probabilities and assess risk.
[7]	Implements attack-defense trees with fuzzy logic (IVIFNs) to dynamically calculate and prioritize risk scores.
[8]	Integrates Bayesian Networks with Multi-Criteria Decision Making for continuous, automated risk scoring and system monitoring.
[9]	Uses a hybrid automaton model and CPS simulation to link physical and cyber threats in a model-based risk assessment.
[10]	Analyzes network traffic to detect attack paths and evaluate risk levels, validated through penetration testing on ICS testbeds.
[11]	Uses a hybrid RAMSS ontology and STA method to link threats, assets, and requirements for traceable, automated evaluations.
[12]	Applies fuzzy attack-defense trees and defuzzification to compute objective, scenario-specific risk levels in ICS.

RQ2: What are the most followed frameworks for automating OT risk assessment?

The most common standards used for risk assessment in Operational Technology (OT) infrastructure include IEC 62443, ISO/IEC 27000 series, NIST SP 800-82, OCTAVE, MITRE ATT&CK Framework, and ISA 99. IEC 62443 is used to secure industrial automation and control systems, while ISO/IEC 27000 series provide standardized methodologies for cybersecurity threat assessment and reduction. NIST SP 800-82 is mandatory for making industrial control systems immune to cyber-attacks. OCTAVE is used for identifying, prioritizing, and managing security. The MITRE ATT&CK Framework categorizes adversary tactics to improve detection and response against cyber-attacks. Fig. 2 below shows the usage frequency of the standards.

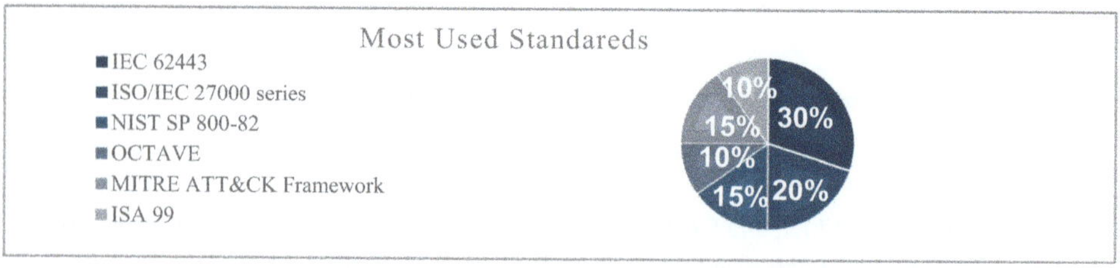

Fig. 2 The most used frameworks and standards in risk assessment methodologies.

RQ3: What are the most commonly utilized machine learning algorithms in risk assessment?

Machine learning algorithms are used in risk assessment across various sectors. Artificial neural networks (ANNs) [13], decision trees, and support vector machines (SVMs) [14] are commonly used for risk identification and analysis. Deep neural networks (DNNs) are effective in real-time data processing in the oil and gas industry. Yao and García de Soto [15] developed a framework for assessing cybersecurity risks using ML learning models. Jiang and Atif [16, 17] used ML for automating vulnerability assessments and inferring CVSS severity scores. Kozlov and Noga [18] introduced a method that combines fuzzy logic with regression analysis to predict security risks. Wang et al. [19] used a stacking ensemble approach to enhance risk assessment in power systems. Aprillia et al. [20] applied quantitative regression random forests and probability mapping to smart grid systems (Fig. 3).

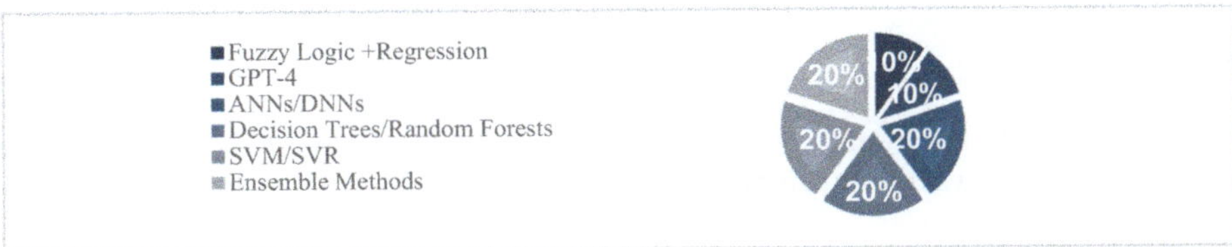

Fig. 3 ML algorithms applied in the literature.

RQ4: What are the primary challenges and gaps in current risk assessment methodologies of OT environments?

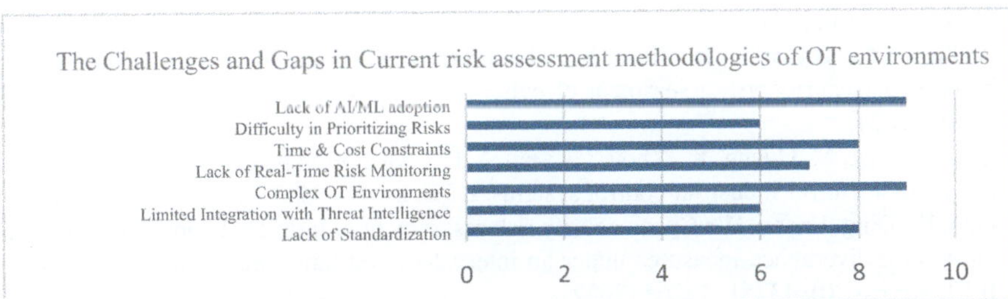

Fig. 4 The primary challenges and gaps in current risk assessment methodologies.

Figure 4 reveals the main challenges in current risk assessment methodologies, primarily due to the complexity of Operations Technology (OT) environments, which include physical security, software vulnerabilities, and human factors. The lack of AI/ML adoption reduces predictive capabilities. The main gap is the lack of standardization across risk assessment methodologies. Periodic risk evaluation approaches in security methodologies restrict their ability to detect newly emerging threats because they do not offer real-time risk monitoring capabilities (e.g., [4, 10]). Assigning appropriate risk levels to different threats is a significant challenge. One key reason for this difficulty is that some methodologies do not fully leverage Cyber Threat Intelligence (CTI) sources (e.g., [4, 7]).

5 Conclusion

This paper presents a systematic literature review (SLR) to analyze risk assessment methodologies in OT environments, highlighting the need for uniform frameworks and the potential of machine learning techniques. The research also reports a qualitative evaluation of reviewed papers. As OT environments become more complex, automated risk assessment methodologies, particularly those using machine learning, will be crucial for mitigating risks and securing critical infrastructure against evolving cyber threats. The study emphasizes the importance of standardized risk assessment frameworks in OT environments.

References

1. Rawal, D.: OT security for critical infrastructure. T-Systems. https://www.t-systems.com/de/en/insights/newsroom/expert-blogs/ot-security-for-critical-infrastructure-998620 (2024). Accessed 21 Sept 2024
2. Zahran, B., Hussaini, A., Ali-Gombe, A.: IIoT-ARAS: IIoT/ICS automated risk assessment system for prediction and prevention. In: Proceedings of the Eleventh ACM Conference on Data and Application Security and Privacy, New York, 2021
3. Georgiadou, A., Mouzakitis, S., Askounis, D.: Assessing MITRE ATT & CK risk using a cyber-security culture framework. Sensors. **21**(9), 3267 (2021)
4. Kure, H.I., Islam, S., Mouratidis, H.: An integrated cyber security risk management framework and risk prediction for the critical infrastructure protection. Neural Comput. Applic. **34**(18), 15241–15271 (2022)
5. Phillips, S.C., Taylor, S., Boniface, M., Modafferi, S., Surridge, M.: Automated knowledge-based cybersecurity risk assessment of cyber-physical systems. IEEE Access. **12**, 82482–82505 (2024)
6. Qin, Y., Peng, Y., Huang, K., Zhou, C., Tian, Y.T.: Association analysis-based cybersecurity risk assessment for industrial control systems. IEEE Syst. J. **15**(1), 1423–1432 (2021)
7. Guo, H., Ding, L., Xu, W.: Cybersecurity risk assessment of industrial control systems based on order-α divergence measures under an interval-valued intuitionistic fuzzy environment. IEEE Access. **10**, 43751–43765 (2022)
8. Guarino, S., Faramondi, L., Oliva, G., Del Prete, E., Setola, R.: Holistic risk assessment in industrial control systems: Combining multiple Bayesian networks with multi-criteria decision making. In: 2024 32nd Mediterranean Conference on Control and Automation (MED), Chania, 2024
9. Tantawy, A., Abdelwahed, S., Erradi, A., Shaban, K.: Model-based risk assessment for cyber physical systems security. Comput. Secur. **96**, 101864 (2020)
10. Fujimoto, M., Mitsunaga, T., Hash, Y., Matsuda, W.: Efficient industrial control systems risk assessment using the attack path to the critical device. In: Proceedings of the 2021 3rd International Conference on Management Science and Industrial Engineering, Osaka, 2021
11. Alanen, J. et al.: Hybrid ontology for safety, security, and dependability risk assessments and Security Threat Analysis (STA) method for industrial control systems. Reliab. Eng. Syst. Saf. **220**, 108270 (2024)

12. Wang, S., Ding, L., Sui, H., Gu, Z.: Cybersecurity risk assessment method of ICS based on attack-defense tree model. J. Intell. Fuzzy Syst. **40**(6), 10475–10488 (2021)
13. Hegde, J., Rokseth, B.: Applications of machine learning methods for engineering risk assessment—A review. Saf. Sci. **122**, 104492 (2020)
14. Paltrinieri, N., Comfort, L., Reniers, G.: Learning about risk: Machine learning for risk assessment. Saf. Sci. **118**, 475–486 (2019)
15. Yao, D., Garcia de Soto, B.: Cyber risk assessment framework for the construction industry using machine learning techniques. Buildings. **14**(6), 1561 (2024)
16. Jiang, Y., Atif, Y.: Towards automatic discovery and assessment of vulnerability severity in cyber–physical systems. Array. **15**, 100209 (2022)
17. Jiang, Y., Atif, Y.: An approach to discover and assess vulnerability severity automatically in cyber-physical systems. In: 13th International Conference on Security of Information and Networks, Merkez, 2021
18. Kozlov, A., Noga, N.: Applying the methods of regression analysis and fuzzy logic for assessing the information security risk of complex systems. In: 2021 14th International Conference on Management of Large-scale System Development (MLSD), Moscow, 2021
19. Wang, Y., Sun, Y., Dan, Y., Li, Y., Cao, J., Han, X.: Online load-loss risk assessment based on stacking ensemble learning for power systems. Front. Energy Res. **11**, 1281368 (2023)
20. Aprillia, H., Yang, H.-T., Huang, C.-M.: Statistical load forecasting using optimal quantile regression random forest and risk assessment index. IEEE Trans. Smart Grid. **12**(2), 1467–1480 (2021)

Cybersecurity for Sustainable Supply Chains in the Fourth Industrial Revolution

Fatima Al-ammari[1] and Imed Ben Dhaou[2](✉)

[1] Cybersecurity Department, Dar Al-Hekma University, Jeddah, Saudi Arabia
[2] Computer Science Department, Dar Al-Hekma University, Jeddah, Saudi Arabia
ibendhaou@dah.edu.sa

Abstract. This study explores the critical role of cybersecurity in fostering sustainable supply chain management. As digital transformation accelerates, supply chains face escalating cyber threats that compromise operational resilience, data security, and long-term sustainability. This paper examines key cybersecurity challenges and strategies, emphasizing the integration of cybersecurity frameworks, stakeholder engagement, and capacity-building initiatives to enhance supply chain security. Findings indicate that robust cybersecurity measures not only mitigate risks but also strengthen supply chain visibility, resilience, and regulatory compliance. Industry best practices underscore the importance of continuous innovation, collaboration, and policy development to create a secure and sustainable supply chain ecosystem.

Keywords: Cybersecurity · Sustainable supply chain · Sustainability

1 Introduction

In the era of globalization and the fast transition towards Industry 4.0, the supply chain is experiencing a leap transformation. In particular, the intensive integration of artificial intelligence and the Industrial Internet of Things (IIoT) has created new opportunities for business development, such as the platform economy. Figure 1 illustrates a typical supply chain of modern industry, which includes the supply of raw materials, manufacturing, distribution, and consumption. Also, shown in the figure is the increasing role of the AI and ICT (Information and Communication Technologies) in the supply chain management.

Today, industries face significant challenges in developing and maintaining sustainable supply chains that not only encompass supply and support sectors but also adapt to evolving customer needs and contribute to broader economic growth. These challenges have been further intensified by rising insecurity, affecting both physical logistics and the cyber infrastructure that underpins modern supply networks. In particular, small and medium enterprises (SMEs), which often serve as critical suppliers and partners within larger supply chains, are increasingly vulnerable to cyber threats. Recent studies, such [1, 2], highlight that SMEs and technology start-up companies frequently lack the resources

D. Berkouk et al. (Eds.): CCTS 2024, PTT, pp. 468–476, 2026.
https://10.1007/978-981-95-7607-4_52

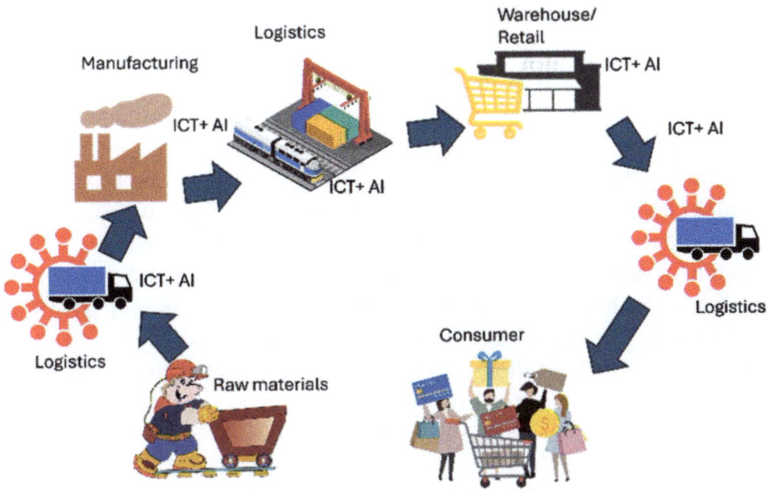

Fig. 1 Key Elements of the Supply Chain

and robust cybersecurity frameworks necessary to defend against sophisticated attacks, posing a risk not only to their operations but to the stability of entire supply ecosystems.

In the era of the fourth and fifth industrial revolutions, integrating cybersecurity into supply chains is crucial for enhancing the security and efficiency of supply chain networks. Supply Chain 4.0, which integrates advanced technologies like blockchain, AI, IoT, and cyber-physical systems, faces significant cybersecurity challenges due to poor interoperability and lack of security standards. These challenges are particularly pronounced in military supply chains, where cyber-attacks can have mission-critical consequences [3].

Cybersecurity threats in the supply chain can take many forms, including malware, phishing attacks, insider threats, Advanced Persistent Threats (APTs), Social engineering, Denial of Service (DoS) and Distributed Denial of Service (DDoS) attacks, Zero-day vulnerabilities, data breaches. According to the studies presented in [4], ransomware attacks are the most feared attacks by many CIOs in the US.

Numerous methods and techniques have been developed to enhance cybersecurity in supply chains. Among the most widely adopted are cybersecurity frameworks, which consist of structured guidelines, best practices, and controls designed to identify, assess, and mitigate cyber risks throughout the supply chain. Building on these existing approaches, this study aims to identify key cybersecurity risks, particularly those arising from the increasing involvement of less secure entities. It also evaluates the role of cybersecurity practices and frameworks in strengthening supply chain sustainability and resilience. Furthermore, the study proposes an integrated management framework that combines cybersecurity measures, sustainability initiatives, and collaborative strategies to mitigate risks and improve overall supply chain performance. By addressing these objectives, the study seeks to contribute to the development of a secure, sustainable, and collaborative supply chain capable of withstanding the complex challenges of the digital era.

2 Literature Review

The supply of services increases the number of challenges that are related to the supply chain's cyber-resilience, and this depends on the type of services the companies offer; an incident can affect the service delivery to receive, process, and respond to requests at the personal and system level; moreover, the cyber-resilience issue impacts the supply chain that provides services, including the operation cell center, website, and payment systems. The supply chain infrastructure is vulnerable to various types of attack. For example, denial of service (DoS), distributed denial of service (DDoS), routers, services, and switches can be hidden. Even though the concept of DOS and DDoS attacks is understood, applying the techniques to protect agents is not usually found in the supply chain's infrastructure; organizations that deal with services must invest in protecting their supply chain in all digital aspects. However, it is necessary to recognize the importance of threats and vulnerabilities in supply chain delivery.

Additive Manufacturing (AM), commonly known as 3D printing, is a transformative process that fabricates three-dimensional objects layer by layer from a digital design. AM has found applications across diverse industries, including healthcare, electronics, automotive, aerospace, and construction. As discussed in [5], there are three distinct models of the AM supply chain. The first model mirrors traditional manufacturing but integrates AM for producing specialized components—commonly adopted in the automotive industry. The second model treats the design file as the primary product, enabling end-users to print parts on-site, as demonstrated by the use of 3D printers aboard the International Space Station (ISS). The third model separates design and production responsibilities: customers obtain or develop design files and conduct virtual testing and validation, while contract-based print shops manage the physical infrastructure, including printers and raw materials. The authors further analyze cybersecurity risks in AM supply chains and propose three key mitigation strategies: establishing a robust information infrastructure, implementing anti-reverse engineering techniques, and securing and authenticating design files.

The author of [6] identified six types of cybersecurity threats: hacktivism, corporate espionage, government-driven, terrorism, criminal activities, and natural events. Cyberattacks exploit vulnerabilities arising from poor design or accidental factors, putting people, processes, physical assets, and technical systems at risk. The author emphasized a holistic approach to cybersecurity, integrating people, processes, physical, and technological aspects, and proposed a model based on the Parkerian hexad augmented with safety and resilience to address the cyber-resilience of supply chains.

Cyber-resilience failures can severely impact a product's lifecycle, as threats and vulnerabilities may compromise the integrity of critical information during the design phase. For example, unauthorized alterations to sensitive components in machinery, such as those used by manufacturers of high-availability pumps, can lead to product failures and an increase in warranty claims. Beyond manufacturing, the nature of the supply chain itself introduces additional cyber vulnerabilities across three key infrastructure elements: ports that handle goods and raw materials, delivery vehicle navigation systems, and global data processing, storage, network, and communication systems. A notable example of cybercriminal exploitation occurred in October 2013, when drug traffickers linked to organized crime breached the logistics system at the Port of Antwerp using spear-phishing emails. The attackers compromised the port's container management software, allowing them to manipulate container reference numbers and redirect shipments including those containing cocaine and heroin without detection [7].

During the past two decades, the globalization of integrated circuits (ICs) in the supply chain has been driven by cost reductions and the increasing complexity of semiconductor manufacturing. However, this globalization has introduced significant security risks associated with untrusted foundries and end-users. Untrusted foundries have full access to chip layouts, posing risks such as intellectual property theft and hardware backdoors, despite mitigation techniques like split manufacturing and logic obfuscation. Additionally, end-users can reverse engineer packaged ICs by repackaging, delayering, imaging, and processing them to extract circuit layouts. While reverse engineering is used for legitimate purposes, such as failure analysis and maintenance of legacy systems, it also presents serious threats, including intellectual property theft and malicious alterations.

To mitigate these risks, hardware security techniques such as logic locking, physically unclonable functions (PUFs), and encryption are employed to protect critical components from unauthorized reverse engineering and tampering [8]. In the IC supply ecosystem, Hardware Trojans (HT) can be inserted into NoC-based SoCs through malicious IP hardware, compromising both the reliability and security of integrated circuits. Mitigation techniques include the insertion of special circuitry that harnesses the potential of AI for the detection of HT or malicious IP [9]. However, PUFs, HT mitigation techniques, and other on-chip sensor-based solutions require sophisticated infrastructure and cannot address the traceability and visibility across the entire IC supply chain, which includes Integrated Device Manufacturers (IDMs), PCB assemblers, system integrators, and end-users. To address this gap, the authors propose an e-chain, a blockchain-based system for verifying the genuineness of ICs [10].

Within the Technology-Organization-Environment (TOE) framework, environmental factors such as government regulations significantly influence the adoption of cybersecurity measures in supply chains. While broader external risks like natural disasters and political tensions can affect supply chain sustainability, in the TOE context, regulatory policies are more central. Rising geopolitical tensions, such as the trade war between China and the United States, have heightened concerns over cyberattacks on Chinese supply chains. As discussed in [11], the Chinese government has established a legal framework to mitigate the growing cybersecurity threats. The three key laws are the Cybersecurity Law, the Data Security Law, and the Personal Information Protection Law. In addition, using the TOE model, the author identifies five main factors that influence the adoption of cybersecurity in the Chinese supply chain ecosystem. These factors are relative advantages, technological readiness, top management support, firm size, and government policy and regulation. Based on comprehensive review of cyberattacks on Chinese supply chains over the past decade, the author provided practical insights for managers and policymakers.

While extant literature extensively addresses the intersection of cybersecurity and supply chain management, certain gaps persist: Industry-Specific Frameworks: The majority of studies concentrate on overarching cybersecurity strategies, yet there is a paucity of research on bespoke frameworks for particular industries (e.g., healthcare, automotive, or agriculture supply chains). Economic Impact Analysis: Despite the financial ramifications of cybersecurity breaches, limited studies offer a cost-benefit analysis of cybersecurity investments in supply chain sustainability. Emerging Technologies: The potential role of blockchain, quantum cryptography, and AI-driven cybersecurity solutions in bolstering supply chain security remains inadequately explored. Regulatory Alignment: An expansion of research is requisite to evaluate the concordance of global cybersecurity regulations (e.g., GDPR, NIST, ISO 27001) with sustainable supply chain practices across diverse regions.

Table 1 Overview of existing Cybersecurity techniques in supply chain management

Techniques	Purpose	Identified gap
Cybersecurity frameworks	Provide guidelines for managing cyber risks in the supply chain	Lack of adaptation in a specific industry or sustainable integration
Hardware security measures	Protect integrated circuits and hardware components from reverse engineering	Focused mainly on the semiconductor industry, not generalized to full supply chain industry
Cyber-physical Infrastructure protection	Secure critical infrastructure	Limited focus on sustainability, lacks dynamic threat detection for distributed infrastructure
Resilience and recovery planning	Minimize the impact of cyber distribution and natural disasters	Does not integrate sustainability metrics
Regulatory compliance	Align supply chain practices with legal and cybersecurity requirements	Few studies connect regulatory compliance with sustainability frameworks

Many frameworks have advantages to provide a secure, sustainable supply chain. NIST Cybersecurity Framework (CSF) is a voluntary framework consisting of standards, guidelines, and best practices to manage cyber risks, but it lacks specific guidance on supply chain risks, with recent updates to address this gap. Another consideration is ISO/IEC, is an international standard for managing information security risk, it has a limited focus on supply chain-specific risks that emphasize internal controls over third-party risk. NIST Cybersecurity Supply Chain Risk Management(C-SCRM) guidance that manages cybersecurity risks in the supply chain, focusing on ongoing risk assessment. There is a challenge with this guidance, which is the complexity of implementing it, as it requires continuous monitoring across multiple supply chain tiers [12]. Table 1 summarizes the most current techniques focused on security with limited integration of sustainability goals.

3 Proposed Integrated Framework for Cybersecure and Sustainable Supply Chains

The literature review demonstrates the critical role of cybersecurity frameworks and collaborative strategies in ensuring improved supply chain sustainability. The results were summarized into three main themes: cybersecurity strategies, employee training, and collaborative benefits.

Of all the frameworks, NIST is the most efficient, with a success rate of 85% in risk evaluation and risk management. Likewise, threat detection through AI reduces false positives and attacks real-time menace with an astounding 90% accuracy, even if it calls for significant capital investment in the beginning.

Manufacturing companies should follow a standard framework to enhance cybersecurity strategies; hence, the framework and cybersecurity are important. However, when manufacturing communicates with third parties and external resources or vendors, it is critical to defend the data. For example, presenting access to special locations can be an additional security layer to save private data. Additionally, monitoring vulnerabilities is supportive of identifying weaknesses in systems. To protect data, monitoring should be combined with real-time reporting during the monitoring process. False positives can be an issue for detection, which leads to waste of resources; therefore, artificial intelligence can be useful for false positives as well as real-time cyber threats.

Categorizing tools and skills, determining concepts, and creating a cybersecurity framework in the manufacturing field can help achieve the goal of sustainable manufacturing [13]. The operation of hypothetical and conceptual work, the applicability of frameworks in the real world, and the management of cybersecurity risk are significant in the supply chain [14]. Collaboration can be an answer for many organizations or companies that have not found solutions to provide high-performance indicators. Implementing supply collaboration can be related to sustainability. The sustainability of collaboration also impacts supply chain partners as a result of organizations using collaborative relationships that rely on frameworks and tools to support them through the planning process to have a sustainable supply chain. Collaboration in the supply chain is required, as participants in Step Change Organic-Food Logistics Ecosystems (SCALE) admit that collaboration between partners in planning and designing supply chains can make the supply chain more sustainable. Collaboration can add a unique advantage: the active implementation of cleaner production, which positively influences sustainability indicators [15].

The influence of cybersecurity operations highlights the critical need for companies to create strategies to protect their supply chains. Cybersecurity throughout the supply chain identifies the importance of a vision arrangement. However, the different vendors within the organization must agree on common cybersecurity goals to achieve cybersecurity objectives, and integration between policies and the technical community is recommended to address the risks associated with critical infrastructure; Moreover, the co-production of guidance can be used to recognize the gaps that allow participation in cyber capabilities together with organizations, not only in terms of the supply chain, and collaborations can create frameworks that are so impotent as we mentioned before to improve cybersecurity [16].

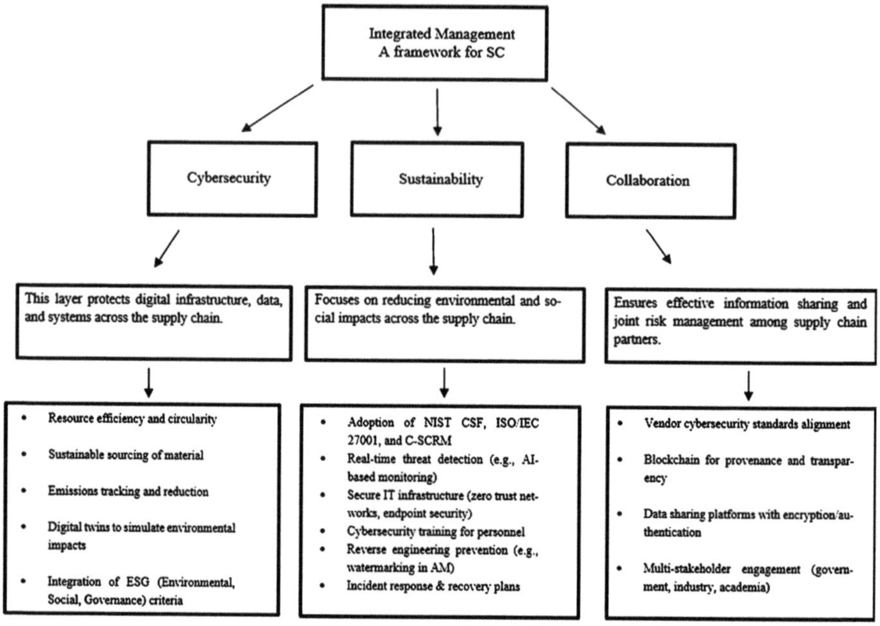

Fig. 2 Integrated management framework for supply chains.

However, training can be used to explain the threat analysis and scenario, including cybersecurity incidents and attacks that impact organizations. Training can consist of different methods related to knowledge, skills, and other concepts that are taught during training, which depends on companies or organizations, and the feedback of trainees that helps to learn the gaps in the training content to improve and evaluate cybersecurity practices in the supply chain [19]. Behavioral factors are considered to play a significant role in cybersecurity, as employees in the supply chain are a failure point. In the event of a lack of awareness and compliance, cybersecurity training and education will be required as the risk of the supply chain continues to increase [17]. Supply chain consists of emerging IT and IT security incidents that affect supply chain operations. Isolation of specific parts of IT threats can help to discover vulnerabilities to study risk mitigation strategies, which is important, as IT threats can come from various sources. The most common example of an IT threat is a malicious program developed by an employee in an organization. Malicious code can exploit the vulnerability of the supply chain; therefore, collaboration between IT and the supply chain must be highly demanded, as it can also affect the disruption flow of the supply chain, making it difficult to be sustainable. IT risk management must be implemented in terms of confidentiality, integrity, and availability in all supply chains [18].

As illustrated in Fig. 2, the integrated management framework for supply chains is built upon three interdependent pillars: cybersecurity, sustainability, and collaboration. Cybersecurity ensures the protection of digital infrastructure and systems through the implementation of standards (e.g., NIST CSF, ISO/IEC 27001), real-time threat detection, secure IT networks, and robust incident response strategies. Sustainability focuses on minimizing environmental and social impacts by advancing resource efficiency, sustainable sourcing, emissions reduction, digital simulation, and the integration of Environmental, Social, and Governance (ESG) criteria. Collaboration underscores the importance of shared risk management and secure information exchange, facilitated through vendor standard alignment, blockchain-enabled transparency, encrypted data sharing, and active engagement with stakeholders across government, industry, and academia. Collectively, these three pillars foster a supply chain ecosystem that is resilient, sustainable, and secure [4].

4 Conclusions

Cybersecurity practices must be implemented in supply chain management in the digital world. Many organizations are vulnerable to threats due to the lack of cybersecurity measures. These threats affect business operations, national security, and the economy. Implementing a cybersecurity framework and collaborating with internal and external partners can improve supply chain sustainability. Collaboration is vital to improve supply chains and reduce cyber threats. Companies can create secure and sustainable supply chain networks through collaboration between IT and sustainable teams and the development of cybersecurity strategies. By highlighting the importance of cybersecurity practices and promoting cooperation between supply chain organizations, stakeholders can manage cyber risks effectively and ensure a sustainable supply chain for the future.

References

1. Alahmari, A., Duncan, B.: Cybersecurity risk management in small and medium-sized enterprises: A systematic review of recent evidence. In: 2020 International Conference on Cyber Situational Awareness, Data Analytics and Assessment (CyberSA), pp. 1–5. IEEE, Dublin (2020) https://doi.org/10.1109/CyberSA49311.2020.9139638
2. Marican, M.N.Y., Razak, S.A., Selamat, A., Othman, S.H.: Cyber security maturity assessment framework for technology startups: A systematic literature review. IEEE Access. **11**, 5442–5452 (2023). https://doi.org/10.1109/ACCESS.2022.3229766
3. Sobb, T., Turnbull, B., Moustafa, N.: Supply chain 4.0: A survey of cyber security challenges, solutions and future directions. Electronics. **9**(11), 1864 (2020). https://doi.org/10.3390/electronics9111864
4. Berry, H. S.: The importance of cybersecurity in supply chain. In: 2023 11th International Symposium on Digital Forensics and Security (ISDFS), pp. 1–5. IEEE, Chattanooga, TN, USA (2023) https://doi.org/10.1109/ISDFS58141.2023.10131834
5. Gupta, N., Tiwari, A., Bukkapatnam, S.T., Karri, R.: Additive manufacturing cyber-physical system: Supply chain cybersecurity and risks. IEEE Access. **8**, 47322–47333 (2020)
6. Boyes, H.: Cybersecurity and cyber-resilient supply chains. Technol. Innov. Manag. Rev. **5**(4), 28 (2015)
7. Senarak, C.: Port cyberattacks from 2011 to 2023: A literature review and discussion of selected cases. Marit. Econ. Logist. **26**(1), 105–130 (2024). https://doi.org/10.1057/s41278-023-00276-8
8. Shamsi, K., Li, M., Plaks, K., Fazzari, S., Pan, D.Z., Jin, Y.: IP protection and supply chain security through logic obfuscation: A systematic overview. ACM Trans. Des. Autom. Electron. Syst. (TODAES). **24**(6), 1–36 (2019)

9. Hussain, M., et al.: Hardware trojan mitigation technique in Network-on-Chip (NoC). Micromachines. **14**(4), 828 (2023). https://doi.org/10.3390/mi14040828
10. Vashistha, N., Hossain, M.M., Shahriar, M.R., Farahmandi, F., Rahman, F., Tehranipoor, M.M.: eChain: A blockchain-enabled ecosystem for electronic device authenticity verification. IEEE Trans. Consum. Electron. **68**(1), 23–37 (2022). https://doi.org/10.1109/TCE.2021.3139090
11. Pérez-Morón, J.: Eleven years of cyberattacks on Chinese supply chains in an era of cyber warfare: A review and future research agenda. J. Asia Bus. Stud. **16**(2), 371–395 (2022). https://doi.org/10.1108/JABS-11-2020-0444
12. Kumar, S., Mallipeddi, R.R.: Impact of cybersecurity on operations and supply chain management: Emerging trends and future research directions. Prod. Oper. Manag. **31**(12), 4488–4500 (2022)
13. Alqudhaibi, A., Deshpande, S., Jagtap, S., Salonitis, K.: Towards a sustainable future: Developing a cybersecurity framework for manufacturing. Technol. Sustain. **2**(4), 372–387 (2023)
14. Lewis, R., Louvieris, P., Abbott, P., Clewley, N., Jones, K.: Cybersecurity information sharing: A framework for sustainable information security management in UK SME supply chains. Lewis, Riyana, Panos Louvieris (2014)
15. Allaoui, H., Guo, Y., Sarkis, J.: Decision support for collaboration planning in sustainable supply chains. J. Clean. Prod. **229**, 761–774 (2019)
16. Wallis, T., Dorey, P.: Implementing partnerships in energy supply chain cybersecurity resilience. Energies. **16**(4), 1868 (2023)
17. Chowdhury, N., Nystad, E., Reegård, K., Gkioulos, V.: Cybersecurity training in Norwegian critical infrastructure companies. Int. J. Saf. Secur. Eng. **12**(3), 299–310 (2022). https://doi.org/10.18280/ijsse.120304
18. Smith, G.E., Watson, K.J., Baker, W.H., Pokorski Ii, J.A.: A critical balance: Collaboration and security in the IT-enabled supply chain. Int. J. Prod. Res. **45**(11), 2595–2613 (2023)
19. Prümmer, J., van Steen, T., van den Berg, B.: A systematic review of current cybersecurity training methods. Comput. Secur. **136**, 103585 (2024)

The Effect of Vaccination and Investor Sentiment on the Crypto-Currency Market

Sonia Sayari[1,2](✉) and Faten Sameer Hanbali[1]

[1] Department of Business and Administration, College of Administration and Finance, Saudi Electronic University, Jeddah, Saudi Arabia
s.sayari@seu.edu.sa

[2] Higher Institute of Accounting and Business Administration, University of Manouba, Tunis 2010, Tunisia

Abstract. In this paper, we explore the impact of vaccination and investor sentiment on the cryptocurrency market, focusing on Bitcoin and Ethereum. We synthesize empirical studies on the influence of vaccination on financial markets and examine the unique relationship between vaccination efforts and the evolving dynamics of digital assets. Our study highlights the novelty of this subject by dedicating a section to the specific impact of vaccination on the cryptocurrency market. The findings reveal a positive influence of vaccination on global financial markets, contributing to increased stability and economic recovery. This positive effect extends to international markets, including cryptocurrencies, where vaccination plays a key role in mitigating risks and boosting investor interest in digital assets. The dual role of vaccination in stabilizing traditional financial markets and enhancing the appeal of cryptocurrencies underscores its broad and multifaceted impact on the global economy.

Keywords: Vaccination · Investor Sentiment · Social Media · Cointegration.First Section

1 Introduction

Over the past decade, cryptocurrency markets have experienced exponential growth in influence and relevance. Characterized by notable volatility and significant price fluctuations, these markets have attracted increasing interest from both investors seeking new financial opportunities and researchers aiming to understand the mechanisms behind these digital assets. Policymakers have also started to consider the potential impact of cryptocurrencies on the economy and financial systems.

The theory of behavioral finance, rooted in the work of psychologists and economists like Tversky & Kahneman (1973) [1], and Barberis and Thaler (2003) [2], has become crucial for understanding financial market anomalies. It shifts the focus from the traditional rational investor model to examine the psychological factors underlying financial decisions.

D. Berkouk et al. (Eds.): CCTS 2024, PTT, pp. 477–485, 2026.
https://10.1007/978-981-95-7607-4_53

In conventional financial markets, investors' responses to negative news, particularly fear and panic, are well-documented. Events like economic downturns, financial fraud, or recession warnings often lead to widespread asset sell-offs, resulting in steep price declines. At the start of the COVID-19 pandemic in March 2020, cryptocurrency prices plummeted as global markets reacted to widespread panic. Investors offloaded their crypto assets to secure liquidity, fearing a looming economic recession, which triggered a sharp drop in cryptocurrency valuations. This underscores how fear-driven behaviors affect these markets. On the other hand, excitement and optimism can have a profound impact on traditional financial markets

While extensive research exists on cryptocurrencies, certain areas remain relatively unexplored, such as how vaccination efforts impact these markets. Cryptocurrency markets have demonstrated a high level of sensitivity to news, global events, and investor sentiment, making it important to examine how vaccination, as a global response to a health crisis, might affect them.

Our paper is divided into two sections. The first part presents a literature review, analyzing previous studies that have investigated the impact of vaccination on financial markets, with a specific focus on the cryptocurrency market. In the second part, we gathered data from various sources. The study covers the period from the start of COVID-19 vaccinations, beginning December 4, 2020, through October 4, 2023, with data collected on a daily basis.

2 Literature Review

As the influence of vaccination on the cryptocurrency market is a relatively new area of study, there remains a limited body of research on this specific topic. However, our empirical analysis builds upon similar studies that have explored the impact of vaccination on traditional stock markets. While the effects of vaccination on financial markets, including cryptocurrencies, are still being explored, our approach leverages findings from earlier research on other market types. By doing so, we aim to address the gap in this emerging field, using a solid foundation of existing literature to guide our analysis. This approach enables us to gain insights into how vaccination may affect the cryptocurrency market, despite the scarcity of research directly focused on this subject.

Acharya et al. (2020) [3] assessed the value of a potential cure by analyzing the relationship between stock prices and vaccination progress during the COVID-19 pandemic. Their model demonstrated that the ability to end the pandemic carried substantial value, accounting for 5% to 15% of total wealth, with greater value observed in scenarios with higher uncertainty regarding the pandemic's duration. This study underscores the importance of understanding the factors influencing future pandemics and managing financial market crises effectively.

Rouatbi et al. (2021) [4] investigated how vaccination efforts have mitigated the negative effects of the COVID-19 pandemic and influenced the volatility of global stock markets. They found that the observed reduction in volatility cannot be solely explained by the pandemic or government interventions, as the negative correlation remains even when these factors are considered. Notably, the effects of vaccination were found to be more significant in developed markets compared to emerging ones.

Di and Xu (2022) [5] analyzed how the COVID-19 vaccine influenced the recovery of financial markets. Their study compared changes in Bitcoin's implied volatility with that of traditional financial markets, including the U.S. stock market, developed markets, emerging markets, gold, and crude oil. Al-Shboul et al. (2022) [6] examined the dynamic relationships between major cryptocurrencies under different market conditions, focusing on the effects of the COVID-19 pandemic and vaccination efforts. They also explored the uncertainty surrounding cryptocurrency regulations and the resulting price fluctuations.

In a related study, Havidz et al. (2023) [7] investigated the influence of vaccine confidence and blockchain characteristics on Bitcoin and Ethereum returns. Focusing on cryptocurrencies in Europe, the U.S., and Asia, the authors employed a panel model with autoregressive distributed lags and tested robustness using random effects and the generalized method of moments. Their results showed that greater vaccine confidence boosted economic recovery and increased demand for Bitcoin, whereas the opposite was observed for Ethereum. Additionally, blockchain characteristics were found to be significant for Bitcoin but not for Ethereum. The study also highlighted that, in the short term, network purchase rates and difficulty moved inversely during the pandemic, with governments playing a key role in shaping cryptocurrency policies during periods of uncertainty.

Abida and Mnif (2023) [8] investigated how vaccination and the spread of COVID-19 influenced investor interest in cryptocurrencies. They analyzed global and national reproduction rate data, geographic interest via Google Trends, and vaccination rates from January 2020 to 2022. Using Wavelet Mortelle and Continuous Wavelet models, the authors demonstrated that the spread of COVID-19 significantly increased cryptocurrency market volatility, while vaccination played a key role in mitigating risks and boosting investor interest. They concluded that forecasting Bitcoin market activity and price trends can create trading opportunities and provide essential insights for policymakers and investors, supporting the development of more informed investment strategies.

3 Empirical Validation

We focus on two prominent cryptocurrencies, Bitcoin and Ethereum, and gather daily global data on the number of vaccinated individuals and the vaccine dosage trend, which tracks the frequency of related search terms on Google Trends. Additionally, we utilize the cryptocurrency market sentiment index, a composite measure that factors in market volatility, social media activity, price fluctuations, and trading volumes. We also incorporate the investor sentiment index (ISV), which analyzes social media and news to gauge investor behaviors and market sentiment. Bitcoin and Ethereum prices are sourced from Coin. Desk and Yahoo Finance, while vaccination data comes from the World Health Organization (WHO), and vaccine dosage trends from Google Trends. Cryptocurrency market sentiment data is sourced from Kaggle.com, and investor sentiment from the Federal Reserve Bank of the United States. Our study spans from December 4, 2020, the first day of vaccinations, through October 4, 2023, with daily data collection.

3.1 Graphic Representation

We evaluate the non-stationarity and heteroscedasticity of Bitcoin, Ethereum, le nombre des gens vaccines (NGV), Trends of vaccine dosages (TVD), Sentiment of cryptocurrency markets, and Investor sentiment index through the Fig. 1.

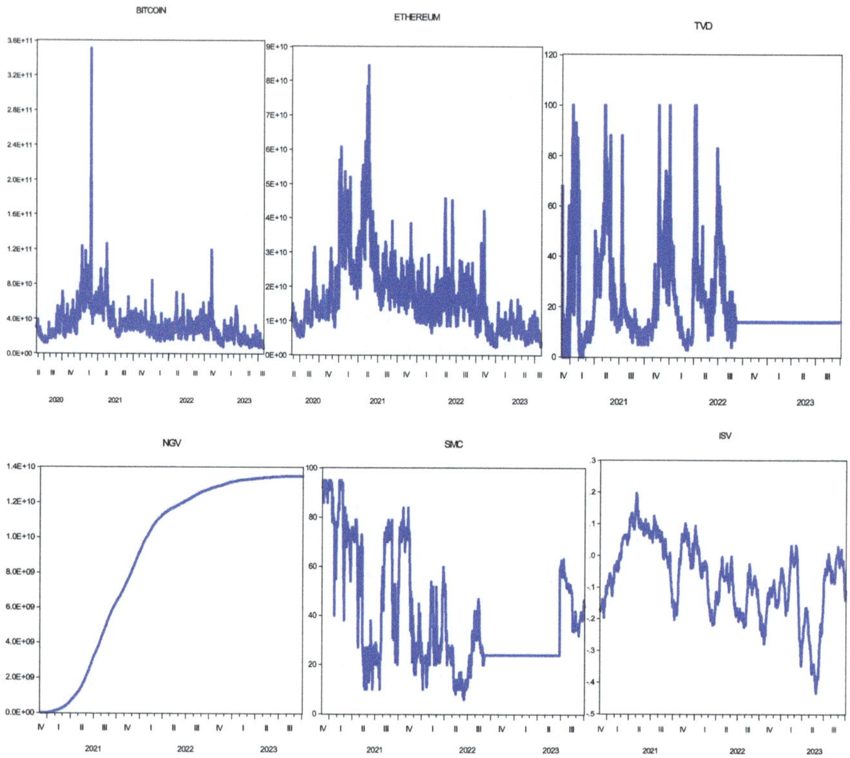

Fig. 1. The evolution of each variable during our study period

Notes: The prices of cryptocurrencies and the trend in vaccination doses have shown significant volatility. Additionally, both investor sentiment indices and cryptocurrencies have been highly vulnerable during the study period. Despite this, the number of vaccinated individuals has followed an upward trend throughout the period.

The graphs above indicate that both Bitcoin and Ethereum are stationary but exhibit significant volatility due to heterogeneous residual variance. Similarly, the investor sentiment index (ISV) is stationary but highly sensitive to fluctuations. In contrast, the number of vaccinated individuals is non-stationary and stable, while the vaccine dosage trend is stationary but prone to vulnerability.

3.2 Descriptive Statistics

We use three statistical indicators—position, dispersion, and shape—to evaluate the precision, linear fit to the mean, information distribution, flatness, and normality of each component shown above. Our resul show that transaction volumes are notably high for the two leading cryptocurrencies, as indicated by their inflated averages. However, the investor sentiment index shows a strong linear adjustment. The data is asymmetrically skewed to the right for cryptocurrencies, the cryptocurrency market sentiment, and the trend in vaccine doses, while it is skewed to the left for vaccinated individuals and the investor sentiment index.

3.3 Absolute and Relative Dependency Relationships

We examine the absolute and relative dependency relationships among the prices of Bitcoin and Ethereum, the two cryptocurrency market sentiment indices, investor sentiment, the number of vaccinated individuals, and the vaccine dosage trend. This analysis uses the Variance-Covariance matrix to assess the precision of each variable and the absolute interactions between them. The cryptocurrency market sentiment index shows a positive dependence on both cryptocurrencies and the investor sentiment index.

3.4 Non-Stationarity Tests

We assess non-stationarity at both the level and first difference using the Dickey-Fuller tests (1979–1981) for the two prominent cryptocurrencies, the investor sentiment index, the cryptocurrency market sentiment index, the vaccine dosage trend, and the number of vaccinated individuals.

3.5 Estimating and Adjusting Long-Term Relationships

We have referenced the Dickey-Fuller (1979–1981) [9] tests and confirmed first-order integration for the prices of Bitcoin and Ethereum, the two sentiment indices (investor and cryptocurrency market), the number of vaccinated individuals, and the vaccination dosage trend. We will estimate the long term relationship linking the natural logarithm of either Bitcoin or Ethereum to these logarithmic indices, the logarithmic number of vaccinated individuals, and the dosage trend. To prevent spurious regressions common in classical econometric methods, we will use co-integration theory. In the first step, we will estimate each long-term relationship using the Engle and Granger (1987) [10] procedure, which employs the Ordinary Least Squares (OLS) method for estimation. This procedure validates the relationship ex-post, requiring the residuals to be stationary at the level of each target in the long-term relationship. Each long-term relationship is represented by the model below.

$$\begin{gathered} Log(Ethereum_t) = \alpha_0 + \alpha_1 Log(NDV) + \\ \alpha_2 Log(ISV_t) + \alpha_3 Log(TVD_t) + \alpha_4 Log(SMC_t) + \varepsilon_t \end{gathered} \tag{1}$$

We examine the long-term adjustment of the relationship that links Bitcoin prices to the two sentiment indices, the vaccination trend, and the number of vaccinated individuals within a univariate Error Correction Model (ECM). The Table 1 presents the estimation of the ECM model using the Ordinary Least Squares (OLS) method.

Table 1. Estimation of the ECM model of Log (Bitcoin)

	Coefficient	Std Error	T-Statistic	Prob
Constant	0,0006	0,0273	0,0208	0,9834
ΔLog(NGV)	−0,0286	0,0979	−0,2926	0,7699
ΔLog(ISV)	−0,0314	0,1626	−0,1931	0,8469
ΔLog(TVD)	0,0011	0,0026	0,4313	0,6663
ΔLog(SMC)	−0,2536	2,0466	−0,1239	0,9014
$Residus_{t-1}$	−0,0187	0,006	-0,1223	0,0018

Note: The coefficient of the lagged residual must be negative and significant to confirm the presence of an adjustment mechanism.

Table 2. Johansen Co-integration Tests (1990)

	Test λ_{trace}			Test $\lambda_{\max}$		
(Log(Bitcoin); Log(NGV); Log(ISV); Log(TVD); Log(SMC))						
Null hypothesis	r = 0	r ≤1	r ≤2	r = 0	r = 1	r = 2
Alternative hypothesis	r ≥1	r ≥2	r ≥3	r = 1	r = 2	r = 3
LR- Statistics	**250,9133**	90,1087	39,5919	160,8046	50,5168	24,7241
Critical value at 5%	**76,9728**	94,0790	35,1928	34,8059	58,5881	32,2996
(Log(Ethereum); Log(NGV); Log(ISV); Log(TVD); Log(SMC))						
Null hypothesis	r = 0	r ≤1	r ≤2	r = 0	r = 1	r = 2
Alternative hypothesis	r ≥1	r ≥2	r ≥3	r = 1	r = 2	r = 3
LR- Statistics	**232,4801**	78,8102	33,4875	153,6699	45,3228	25,3878
Critical value at 5%	**69,8189**	87,8561	39,7971	33,8769	57,5843	31,1316

Note: Johansen's (1990) Co-integration tests are performed using the maximum likelihood procedure.

The estimation of each Ordinary Least Squares (OLS) model yields non-significant and weak estimators for the deterministic part, accompanied by a very weak negative adjustment speed for Bitcoin prices. As a result, there is no rectification mechanism to return the Bitcoin target to a relatively stable long-term state.

To investigate the presence or absence of a co-integration space, we apply Johansen's (1990) trace and maximum eigenvalue tests. We then estimate a Vector Error Correction Model (VECM) using the maximum likelihood method and assess long-term causality within the adjustment matrix. Table 2 presents the results of Johansen's (1990) [11] tests.

The Johansen (1990) [11] tests for trace and maximum eigenvalues indicate the presence of a single co-integrating vector associated with either Bitcoin or Ethereum prices, depending on the logarithmic investor sentiment, the logarithmic trend of vaccinations, the logarithmic crypto market sentiment, and the logarithmic number of vaccinated individuals. This unique co-integrating vector is confirmed by the significance of the likelihood ratio relative to the 5% critical values for both the trace test and the maximum eigenvalue, indicating a single co-integrating relationship. We estimate a Vector Error Correction Model (VECM) using the maximum likelihood method, with the results presented in the Table 3.

Table 3. Estimation of VECM using the maximum likelihood technique

The variables	Normalized Co-integrating Vectors (matrix β)		Error correction coefficients (matrix α)	
	Coefficients	Coefficients	Coefficients	Coefficients
Log(BITCOIN)	1.000000		−0.8125*	
Log(ETHEREUM)		1.000000		−0.5315*
Log(NGV)	1.8682*	1.6445*	−0. 2891*	−0.000395*
Log(ISV)	2.8146*	1.7244*	−0.6545*	−3.11E-05*
Log(TVD)	0.109544**	0.6984*	−0.1415*	−0.002074*
Log(SMC)	0.5878*	1.1537*	−0.2287*	−5.09E-07*

Note: The linear adjustment coefficient of each target variable is negative and significant, indicating the presence of a linear adjustment mechanism that helps return the target to a partially stable state.

The table above shows that Bitcoin exhibits a strong, positive, and significant elasticity in relation to the number of vaccinated individuals and the investor sentiment index. Additionally, Bitcoin demonstrates a weaker, yet still positive and significant relationship with the vaccine trend and the cryptocurrency market sentiment index. Likewise, Ethereum is positively and significantly sensitive to both the number of vaccinated individuals and investor sentiment, although its sensitivity to the vaccine trend is lower. Ethereum prices also show a less significant and positive response to cryptocurrency market sentiment. Adjustment speeds are negative and significant, indicating a long-term causal relationship among the various components in our database, as per the Granger representation theorem (1988). Specifically, 81.25% of the imbalance in Bitcoin prices is corrected by the vaccine trend and the number of vaccinated individuals, which have driven speculators to engage in the cryptocurrency market, bolstered by improved investor sentiment and the cryptocurrency market sentiment index. Similarly, Ethereum prices are corrected by 53.15% in response to the presence of COVID-19 vaccines, which have encouraged greater speculation among economic agents in the cryptocurrency market.

3.6 Dynamic Causality

In what follows, we examine Granger causality (1969) [12] among the natural logarithms of Bitcoin and Ethereum prices, the investor sentiment index, the vaccine trend, the number of vaccinated individuals, and the cryptocurrency market sentiment index. Our result confirms the presence of a Granger causality relationship between the first difference of the vaccine dosage trend and the first difference of Bitcoin prices. Additionally, there is a Granger causality relationship between the first difference of the vaccine dosage and the first difference of the cryptocurrency market sentiment index.

4 Conclusion

In this paper, we explored the significant impact of vaccination and investor sentiment on the cryptocurrency market, with a particular focus on Bitcoin and Ethereum. We synthesized key empirical studies that have examined the influence of vaccination on financial markets, while also dedicating a section to the specific relationship between vaccination and the cryptocurrency market, given the novelty of this topic. Our analysis revealed that vaccination positively affects financial markets, promoting stability and economic recovery globally, especially in international markets. Similarly, in the cryptocurrency market, vaccination has played a crucial role in reducing risks and stimulating investor interest in digital assets.

Our analysis used statistical indicators to evaluate the position, dispersion, and shape of each component in our dataset. We found that transaction volumes for Bitcoin and Ethereum are high; however, the adjustment for Bitcoin, Ethereum, the vaccine trend, the investor sentiment index in the cryptocurrency market, and the number of vaccinated individuals is poor due to significant positive outliers. In contrast, the investor sentiment index shows a good linear fit to the mean, with minimal outlier influence. The various components of our dataset exhibit asymmetric information, with some skewed to the left and others to the right. None of these components conform to a normal distribution, as the Jarque-Bera statistics exceed the critical value of the Chi-squared distribution with two degrees of freedom.

We assessed the non-stationarity of each component in our dataset at both the level and first difference. Using the Dickey-Fuller (1979–1981) [9] tests for unit roots without trend breaks, we found that Bitcoin is stationary but exhibits a trend break in December 2022, as confirmed by the Perron (1998) test. Thus, Bitcoin is integrated of order zero (I(0)). Similarly, the number of vaccinated individuals is also stationary at the level according to the Perron (1998) [13] test, indicating it is integrated of order one. The remaining components of our dataset are non-stationary at the level and show the presence of structural breaks; these components are integrated of order one (I(1)).

In this paper, we used cointegration theory to estimate each long-term relationship while avoiding spurious regressions, using the Engle and Granger (1987) [10] method. This approach examines the relationship between the prices of Bitcoin or Ethereum and the vaccine trend, the investor sentiment index, the number of vaccinated individuals, and the cryptocurrency market sentiment index. We accepted only the long-term relationship involving Bitcoin, as its target is stationary at the level. Our findings indicate that Bitcoin is more elastic with respect to the investor sentiment index but less responsive to the vaccine trend. Additionally, the number of vaccinated individuals negatively influences the prices of these cryptocurrencies. We confirmed that the adjustment of this target is weak within an error correction model. Furthermore, a single co-integrating vector was identified using Johansen (1990) tests, demonstrating long-term adjustments that correct deviations from the equilibrium of Bitcoin or Ethereum towards a relatively stable long-term state within the vector error correction models (ECM).

References

1. Tversky, A., Kahneman, D.: Availability: a heuristic for judging frequency and probability. Cogn. Psychol. **5**(2), 207–232 (1973)
2. Barberis, N., Thaler, R.: A survey of behavioral finance. In: Constantinides, G., Harris, M., Stulz, R. (eds.) Handbook of the Economics of Finance North-Holland. Amsterdam (2003)
3. Acharya, V. V., Johnson, T., Sundaresan, S., Zheng, S.: The value of a cure: an asset pricing perspective (w28127). National Bureau of Economic Research (2020)
4. Rouatbi, W., Demir, E., Kizys, R., Zaremba, A.: Immunizing markets against the pandemic: COVID-19 vaccinations and stock volatility around the world. Int. Rev. Financ. Anal. **77**, 101819 (2021)
5. Di, M., Xu, K.: COVID-19 vaccine and post-pandemic recovery: evidence from bitcoin cross-asset implied volatility spillover. Financ. Res. Lett. **50**, 103289 (2022)
6. Al-Shboul, M., Assaf, A., Mokni, K.: When bitcoin lost its position: cryptocurrency uncertainty and the dynamic spillover among cryptocurrencies before and during the COVID-19 pandemic. Int. Rev. Financ. Anal. **83**, 102309 (2022)
7. Havidz, S.A.H., Calvilus, I.C., Angelita, Z.: COVID-19 full vaccination and blockchain size: empirical evidence from the cryptocurrency market. EuroMed J. Bus. **18**(2), 229–247 (2023)
8. Abida, M., Mnif, E.: Investor attention in cryptocurrency markets: examining the effects of vaccination and COVID-19 spread through a wavelet approach. Int. J. Econ. Financ. Issues. **13**(5), 43 (2023)
9. Dickey, D.A., Fuller, W.A.: Distribution of the estimators for autoregressive time series with a unit root. J. Am. Stat. Assoc., 427–431 (1979)
10. Engle, R.F., Granger, C.W.: Co-integration and error correction: representation, estimation, and testing. Econometrica, 251–276 (1987)
11. Johansen, S., Juselius, K.: Maximum likelihood estimation and inference on cointegration—with appucations to the demand for money. Oxf. Bull. Econ. Stat. **52**(2), 169–210 (1990)
12. Granger, C.W.: Investigating causal relations by econometric models and cross-spectral methods. Econometrica, 424–438 (1969)
13. Perron, P.: The great crash, the oil price shock, and the unit root hypothesis. Econometrica, 1361–1401 (1989)

Author Index

A

Abas, Hassan A. 3, 21, 82
Abdalla, Tareg Abdalla 75
Abdulkhalek, Rasha 116
Abdullah, Saima 295
Abdulraquib, Salwa 315
Aboud, Abdulaleem 3
Ahmed, Ahmed 46, 107
Ahmed, Eman 107
Ahmed, Mobyen Uddin 125
Ahmed, Mostafa Khaled 29
Ahmed, Muna 37
Ajmire, Prafulla E. 206
Al Qurashi, Rawabi Saeed 56
Al Sharif, Roba 324
Alabdali, Salem Ahmed 304
Alabdallah, Nouf M. 459
Alahmadi, Amal A. 459
Al-ammari, Fatima 468
Alasseri, Areej 176
Alawad, Ethar 82
Albarak, Ali ben Saleh 256
Albeesh, Zahra A. 274
Albogami, Abdullah 75
Aldahwan, Nouf S. 282
Aldolaim, Rayhanah Jassim 231
Aldossary, Maryam M. 190, 274
Aldossary, Sara F. 274
Alfaiz, Fahad 341
Alfaleh, Deema A. 459
Alfityani, Farah 435
AlGhamdi, Abrar Saleh 231
AlGhamdi, Azza 98
Al-Ghamdi, Rahaf A. 274
Alghamdi, Rawan Hamed 231
Alghamdi, Teef Lahg 56
Al-Haddad, Mona 223
Alharby, Abdulrahman A. 459
Alharby, Abdulrahman M. 274
Alharthi, Alanoud Saad 56
Alharthi, Shahad Saed 56
Alhotailah, Solaf Mohammed 231
Alhothly, Jomana 145
Alhrbi, Braa 21
Alhussan, Joud 435
Alissa, Khalid A. 274
Aljindi, Alia 98
Aljohani, Yussra 21
AlKatheer, Aisha Saleh 231
Alkhail, Lama 435
Alkhaldi, Dalal Khalid 190
AlMahfoud, Reem Zaki 231
Almalki, Amjad Hasan 56
Almanjumi, Maram Meshal 56
Alnatah, Sara khaed 190
Alnemari, Aryam A. 459
Alnobani, Abdelqader Kamal 435
AlOtaibi, Maha 375
Alqahtani, Abdullah Ali M. 231
Alqahtani, Fatema A. 459
Alqahtani, Lujain Ali 190
Alqahtani, Nada 341
AlQahtani, Raneem 375
Alrashedi, Ghada M. 459
Alsaadi, Amjad 176
Alsadiq, Hussain 82
Alsaeedi, Lenah 145
AlSafran, Dannah 375
AlSafran, Sarah 375
Alshareef, Joud 145
Al-Shareef, Sarah 145
Alshehri, Shujon A. 274
Alsirhani, Shorouq 82
Alsubhi, Arwa 145
Alsudais, Mohammed 413
Altarteer, Samar 247, 403
Al-Thawadi, Dalal M. 274
Althobuti, Shahad Mashal 56
Alwajeeh, Tami 357
Alzahrani, Maram Abdullah 190
Alzamil, Waleed S. 12
Andrews, Samuel Samiái 452

D. Berkouk et al. (Eds.): CCTS 2024, PTT, pp. 487–489, 2026.
https://10.1007/978-981-95-7607-4

Aqaqah, Fatimah Saeed 190
Askool, Sanaa S. 168
Atique, Mohammad 206
Ayed, Tahar Lazhar 435

B
Bahri, Afnan 435
Bakry, Saad 341
Balfaqih, Fatima 247
Basodan, Zelal 366
Begum, Shahina 125

C
Cetindamar, Dilek 304
Chekfehand, Hussein 21

D
Dakhil, Omar 3
Dhaou, Imed Ben 157, 206, 215, 468

E
El Khateeb, Samah 29
Elsaid, Saad Mohamed 256
Elshafei, Rana 29
El-Taj, Homam 397, 443
Ezzat, Aya 29

F
Fallatah, Arwa 91
Farhana, Mosarrat 125
Farouk, Eman 29
Fathi, Heba 264
Fawzy, Mostafa 375
Figliola, Angelo 66

G
Gao, Jian Gang 197
Ghaleb, Maram 384
Gomaa, Mohammed M. 384
Guo, Xing 215

H
Hanaysha, Jalal Rajeh 424
Hanbali, Faten Sameer 477
Hassan, Aisha 82
Hassan, Maha M. 168

I
Isoaho, Jouni 215

J
Jewani, Varkha K. 206

K
Kaafil, Shifana Fatima 91
Khalid, Komal 238
Khalifa, Saida 333
Khan, Amani 349

M
M.E.A. Saleem, Alaa 75
Mahmutovic, Adnan 134
Makki, Wed 21
Masood, Amira 403
Melibari, Basnt 145
Mir, Arshad Ahmed 197
Mohamed, Khaled 66

N
Nada, Elshaimaa 46, 107
Najjar, Firas 443
Nasseraldin, Ahmed Khaled 435
Nasseraldin, Ahmed 397

P
Pileggi, Salvatore F. 304

R
Rashed, Haitham 82
Rashid, Zahid 295

S
Saci, Karima 324, 333
Salem, Amr 29
Sayari, Sonia 477
Shah, Syed Bilal Hussain 197
Shaheen, Rozina 315
Shaikh, Fatema Sabeen 231
Sharma, Bhisham 157
Soliman, Anne A. 29

T
Tanwar, Vishesh 157
Tenhunen, Hannu 215
Thafar, Maha A. 56
Thawani, Pragati V. 206

W
Walker IV, Nathaniel B. 349

Wanakuta, Henry 349

Y
Youldash, Mustafa 190

Z
Zahid, Umara 295
Zheng, Shun 197
Zou, Zhuo 215

GPSR Compliance
The European Union's (EU) General Product Safety Regulation (GPSR) is a set of rules that requires consumer products to be safe and our obligations to ensure this.

If you have any concerns about our products, you can contact us on

ProductSafety@springernature.com

In case Publisher is established outside the EU, the EU authorized representative is:

Springer Nature Customer Service Center GmbH
Europaplatz 3
69115 Heidelberg, Germany

www.ingramcontent.com/pod-product-compliance
Ingram Content Group UK Ltd.
Pitfield, Milton Keynes, MK11 3LW, UK
UKHW020952220926
475831UK00001B/33

9789819576067